ition

Principles of Refrigeration

C. Thomas Olivo

 Delmar Publishers Inc. ®

NOTICE TO THE READER

Delmar Staff
Senior Executive Editor: David C. Gordon
Production Coordinator: Helen Yackel
Design Coordinator: Susan Mathews

For information, address Delmar Publishers Inc.
2 Computer Drive West, Box 15-015
Albany, New York 12212

Cover photo credits, clockwise from top right:
Carrier Corporation
CTL-Cryogenics
Tecumseh Products Company
The Trane Company
Back cover photo:
courtesy of T.I.F. Instruments Inc.

Printed in the United States of America
Published simultaneously in Canada
by Nelson Canada,
A division of The Thomson Corporation

10 9 8 7 6 5 4 3 2 1

Library of Congress Cataloging-in-Publication Data

Olivo, C. Thomas.
  Principles of refrigeration / C. Thomas Olivo. — 3rd ed.
    p.   cm.
  Rev. ed. of: Principles of refrigeration / R. Warren Marsh, C.
Thomas Olivo. 2nd ed. 1979.
    ISBN 0-8273-3557-1
    1. Refrigeration and refrigerating machinery.   I. Marsh, R.
Warren. Principles of refrigeration.   II. Title.
TP492.O55   1990
621.56—dc20

DEDICATION

This third edition of *Principles of Refrigeration* is dedicated to my wife, Hilda. She, again, contributed her expertise to the editing of the manuscript.

Equally important, appreciation is expressed for lifetime support, continuing encouragement, and understanding. This helpfulness was evidenced at each successive stage from research, to organization, to content writing, to editing, and to final production . . . from the original publication to this present edition.

CONTENTS

Dedication iii
Preface xi
About the Author xii
Acknowledgements xiii
Introduction xvii
General Safety Considerations xx

SECTION: 1 REFRIGERATION

UNIT 1: INTRODUCTION TO REFRIGERATION 1

Brief Historical Sketch of "Man-Made" Cold 1   Early Experiments with Food Preservation 2
Experimenters of Ice-Making Machines 2   Modern Uses of Refrigeration 4   Industrial
Applications of Refrigeration 5   Summary 7   Assignment: Unit 1 Review and Self-Test 7

UNIT 2: REFRIGERATION SYSTEMS, CYCLES AND CLASSIFICATION 9

Refrigeration by Absorption and Evaporation 9   Compression System Using Capillary Tube
Control for Refrigerant 14   Compression Systems Using Thermostatically Controlled
Expansion Valve (TEV) 14   Mechanical Refrigeration Utilizes Temperatures Increased by
Pressure 15   Applying Early Principles of Pressure to a Refrigeration System 17
Refrigeration Cycles in Commercial and Industrial Units 18   Classification of Refrigerating
Systems 19   Refrigeration and Reverse Cycle Refrigeration Systems 20   Summary 28
Assignment: Unit 2 Review and Self-Test 20

SECTION 2: REFRIGERATION TOOLS AND MATERIALS

UNIT 3: REFRIGERATION TOOLS AND TEST EQUIPMENT 24

Refrigeration Hand Tools 24   Categories of Tools, Equipment and Test Instruments 24
Types, Features and Applications of Wrenches 25   Pliers and Strippers 33   Hammers 33
Chisels and Punches 35   Twist Drills and Bits 36   Hand Hacksaws and Sawing 38   Files
and Filing 39   Thread-Cutting Tools 39   Tables of Tap Drill Sizes 41   Standard 60° Thread
Form 41   Thread Designation on Drawings 41   Features of Tap Sets 41   Pipe Taps and
Thread Forms 43   Sheet Metal Cutting 43   Screwdrivers: Types and Functions 43   Power
Tools 45   Linear Measuring Tools 45   Precision Linear Measurements 46   Refrigeration
Gauges and Test Instruments 48   Solid-State Electronic Thermistor Vacuum Gauge 49
Other Air-Conditioning Instruments, Tools and Accessories 51   Summary 51   Assignment:
Unit 3 Review and Self-Test 52

UNIT 4: REFRIGERATION MATERIALS AND FITTINGS 55

Tubing for Refrigeration Systems 55   Refrigeration Fittings 60   Threaded Fittings 60
Purpose and Methods of Flaring 61   Principles of Soldering 63   Soldering and Brazing
Alloys 64   Brazing Techniques 65   Swaging 67   Summary 68   Assignment: Unit 4 Review
and Self-Test 69

SECTION 3: HEAT TRANSFER AND TEMPERATURE

UNIT 5: THE NATURE AND EFFECT OF HEAT ENERGY IN REFRIGERATION 72

Refrigeration Depends on Energy 72   Relationship of Matter to Refrigeration 72   The
Structure of Matter 73   Arrangement and Movement of Molecules 73   British Thermal Unit

Contents

of Measurement 75 Application of Heat and Heat Energy in Refrigeration 75 Comparison of Temperature and Heat 75 Relating Temperature to Refrigeration 76 Superheating and Subcooling 76 The Mysteries and Effects of Latent Heat 76 Latent Heat of Vaporization 77 Heat of Condensation 77 The Effect of Heat Energy on Pressure and Volume 78 Pressure Used for Refrigerant Controls 80 Heat Energy Causes Expansion and Changes in Volume 80 Summary 80 Assignment: Unit 5 Review and Self-Test 81

UNIT 6: BASIC METHODS OF HEAT TRANSFER **83**

Part A: Concepts and Principles of Heat Flow and Heat Transfer 83 Heat Transfer by Conduction 85 Heat Transfer by Convection 90 Forced Convection Condensers 92 Water-Cooled Condensers 93 Water Chillers 94 Cooling Coil Design 96 Heat Transfer by Radiation 97 Summary (Part A) 99 Assignment: Unit 6 (Part A) Review and Self-Test 100

Part B Insulation 103 Conditions Within Insulation 104 Moisture in Insulation 104 Moisture Vapor Seals 105 Insulation Against Conduction 107 Insulation Against Convection 110 Reflective Insulation Against Radiant Heat Energy 111 Summary 113 Assignment: Unit 6 (Part B) Review and Self-Test 114

UNIT 7: SENSIBLE HEAT IN A REFRIGERANT **117**

Sensible Heat of a Solid 117 Sensible Heat of a Liquid 117 Sensible Heat of Vapor 118 The British Thermal Unit (English System of Heat Measurement) 118 The Calorie (Metric System of Heat Measurement) 120 Measuring the Amount of Heat Energy 118 Refrigerant Tables 123 Summary 125 Assignment: Unit 7 Review and Self-Test 126

UNIT 8: LATENT HEAT AND PRESSURE IN REFRIGERATION **129**

Relationship of Latent Heat and Change of State 129 Total Heat—Enthalpy 131 Boiling Point and Condensing Temperatures Related to Refrigerants 133 Principles of Operation of Pressure Gauges 136 Pressure Gauges in Refrigerating Systems 139 High Pressure and Compound Gauges 139 Saturated Refrigerants 142 Net Refrigerating Effect 142 Rate at which a Refrigerant is Circulated 145 Comparison of Ice-Making and Refrigerating Capacity 146 The Mollier Diagram 147 Summary 147 Assignment: Unit 8 Review and Self-Test 148

SECTION 4: THE HEART OF THE REFRIGERATION SYSTEM

UNIT 9: THE COMPRESSOR IN MECHANICAL REFRIGERATION SYSTEMS **153**

Types of Compressors: Functions and Operating Parts 154 Reciprocating Compressors 154 Rotary Compressors 163 Centrifugal Compressors 166 Hermetic Compressor Troubleshooting (Service) Chart 168 Compression Ratio 170 Methods of Controlling Compressor Capacity 170 Lubrication of Compressors 173 Summary 174 Assignment: Unit 9 Review and Self-Test 175

UNIT 10: EVAPORATORS **178**

Types of Evaporators 178 Manifolding 181 Design and Operation of Direct-Expansion Evaporator Coil 181 Direct-Expansion Water Chiller and Secondary Refrigerants 181 Flooded Evaporators 183 Humidity and Its Relation to Evaporators 185 Importance of Defrosting Evaporators 186 Basic Methods of Defrosting Evaporators 188 Summary 196 Assignment: Unit 10 Review and Self-Test 197

UNIT 11: CONDENSERS **200**

Air-Cooled Condensers 200 Remote Air-Cooled Condensers 201 Water-Cooled
Condensers 204 Evaporative Condensers 208 Noncondensable Gases 209 Methods of
Cooling Water for Condensers 210 Spray Ponds 210 Summary 215 Assignment: Unit 11
Review and Self-Test 215

UNIT 12 METERING DEVICES FOR THE CONTROL OF REFRIGERANTS **219**

Types and Functions of Refrigerant Control Devices 219 Hand-Operated Expansion
Valves 220 The Low-Side Float System 220 The High-Side Float System 220 Automatic
Expansion Valves 222 Thermostatic Expansion Valves (Constant Superheat) 224 Capillary
Tube Controls 228 Summary 229 Assignment: Unit 12 Review and Self-Test 230

UNIT 13: REFRIGERANTS AND DRIERS **234**

Fundamentals of Chemistry Applied to Common Refrigerants 234 Qualities of
Refrigerants 236 Refrigerants for Domestic and Commercial Systems 237 Refrigerants
and Moisture 240 Properties of Desiccants 242 Drier Size 244 Drier Location 244
Driers with Moisture Indicators 244 Summary 246 Assignment: Unit 13 Review and
Self-Test 246

**UNIT 14: PRINCIPLES OF CHARGING AND TESTING REFRIGERATION
       SYSTEMS** **250**

The Function of Valves in Charging Refrigeration Systems 250 Suction, Oil, and Discharge
Gauges 253 Service Gauge and Testing Manifolds 257 Principles Relating to the
Evacuation of Refrigeration Systems 258 The Transfer of Refrigerants and Their Addition
to Systems 259 Refrigerant Measurement Using an Electronic Sight Glass Indicator 261
Solid-State, Programmable Air-Conditioning Charging Station 263 Tests for Refrigerant
Leaks 265 Electronic, Solid-State Halogen Leak Detector 266 Troubleshooting
Refrigeration Systems (Service) Chart 267 Summary 270 Assignment: Unit 14 Review and
Self-Test 271

SECTION 5: COMPONENTS OF REFRIGERATION SYSTEMS

UNIT 15: BASIC ELECTRICITY FOR REFRIGERATION SYSTEMS **276**

The Nature of Electricity 276 Producing Static Electricity 278 Conductors, Insulators and
Semiconductors 280 Potential Difference and Electromotive Force 281 Identification of
Alternating and Direct Current 283 Transmitting and Distributing Electricity 283 Circuit
Protection 285 Circuit Requirements for Refrigeration Systems 288 Summary 292
Assignment: Unit 15 Review and Self-Test 293

UNIT 16: ELECTRICAL COMPONENTS OF REFRIGERATION SYSTEMS **297**

Behavior of Electrical Devices in Refigeration Systems 297 Resistors 298 Solenoids and
Relays 298 Principles Affecting Inductive Devices 299 Electric Power 302
Summary 307 Assignment: Unit 16 Review and Self-Test 308

UNIT 17: ELECTRIC MOTORS FOR REFRIGERATION SYSTEMS **311**

Direct Current 311 AC Distribution Circuits 312 Measuring Power in AC Circuits 313
Four Wire System of Distribution 313 AC Potential Differences 315 AC Induction

Contents

Motors 315   Principles of Generating a Rotating Magnetic Field 316   Motors in Hermetic Units 322   Solid-State Starting Relays 324   Overload Protectors 326   Thermal Plug-on Motor Protector 327   TRIAC: Bidirectional Current Flow Control 327   AC Motor Controls 328   Solid-State Photoelectric Digital Tachometer 333   Summary 334   Assignment: Unit 17 Review and Self-Test 335

UNIT 18:   BASIC REFRIGERATION CONTROLS **339**

Classification of Controls 339   Humidity Controllers 340   Operating Controls 341   Secondary or Actuating Controls 343   Limiting or Safety Controls 346   Summary 352   Assignment: Unit 18 Review and Self-Test 353

UNIT 19:   TEMPERATURE INDICATORS, MEASUREMENT AND CONTROLS **355**

Temperature Measuring Systems 355   Absolute Temperature Scales 359   Temperature Indicators 361   Other Temperature Indicators 363   Electronic Solid-State Thermistor Sensor 366   Digital, Battery- (Electro-) Operated Test Thermometers 366   Solid-State Digital Thermometer/Pyrometer 367   Temperature Responsive Controls and Recorders 368   Motor Control Thermostats 368   Summary 373   Assignment: Unit 19 Review and Self-Test 374

UNIT 20:   SUPPLEMENTARY REFRIGERATION CONTROLS **377**

Modulating Controls 377   Water Regulating Valves 379   Refrigeration Timers and Controls 380   Pneumatic Control Systems 380   Multiple Unit Installations 382   Evaporator Regulator Valves (Snap-Action Type) 385   Thermostatic Suction-Pressure Valve 385   Solenoid Valves in Multiple Evaporator Installations 386   Multiple Condensing Units 388   Summary 390   Assignment: Unit 20 Review and Self-Test 391

SECTION 6:   ALL-WEATHER AIR-CONDITIONING SYSTEMS

UNIT 21:   HEAT PUMPS **394**

Compression Cycle Refrigeration Machines 394   Design Features of Heat Pumps 395   Principles of Operation of a Heat Pump 395   Indoor and Outdoor Heat Pump Coils 396   Heat Pump Metering Devices 396   Liquid Line Filter-Driers 398   Air-to-Air Heat Pumps 399   Ground- or Water-Coil Heat Pumps 399   Energy Sources for Ground-Source Heat Pumps 401   Water-to-Water Heat Pumps 402   Air-Conditioning Options 402   Thermostatic Controls 403   Summary 403   Assignment: Unit 21 Review and Self-Test 404

UNIT 22:   RESIDENTIAL AND COMMERCIAL AIR-CONDITIONING SYSTEMS **407**

Factors Affecting System Design 407   Duct Sizing and Location 410   Air-Conditioning Control Terms 410   Single-Package and Split-Package Air-Conditioning Units 410   Split-Package Heating/Cooling Units 410   Windows and through-the-Wall Mounted Air-Conditioning Cooling Units 412   Central Air-Conditioning Systems 414   Evaporative Condenser Systems 415   Cooling Towers and Water-Cooled Condensers 415   Chiller/Heater Units: Gas Fired 416   Cooling Cycle of a Chiller/Heater Unit 419   Heating (Hot Water) Cycle of a Chiller/Heater Unit 420   Construction of Gas Chiller/Heater Air Conditioners 421   Built-Up Winter-Summer Air-Conditioning Systems 421   Mechanical, Electrostatic, and Electronic Filters 423   Summary 424   Assignment: Unit 22 Review and Self-Test 425

UNIT 23:   NONMECHANICAL REFRIGERATION SYSTEMS **428**

Absorption Systems 428   Large Commercial Absorption Refrigeration Systems 430
Domestic Absorption Refrigerating Systems 431   Comparison of Nonmechanical and
Mechanical Systems 436   Steam-Jet Refrigeration Systems 436   Magnetic System of
Producing Low Temperatures 441   Summary 443   Assignment: Unit 23 Review and
Self-Test 444

UNIT 24:   NONMECHANICAL TRANSPORT REFRIGERATION **447**

Water-Ice Nonmechanical Transport Refrigeration System 447   Dry Ice (Carbon Dioxide)
Refrigerant 450   Eutectic Solution Refrigeration 452   Expendable Refrigerant
Refrigeration (Liquid Nitrogen System) 456   Combinations of Systems: Liquid Nitrogen,
Holdover Plate, Blower Coil Systems 460   Summary 462   Assignment: Unit 24 Review and
Self-Test 463

UNIT 25:   MECHANICAL TRANSPORT REFRIGERATION **466**

Principles of Mechanical Refrigeration 466   Self-Contained Refrigeration Systems 467
Capacity of Mechanical Transport Refrigeration Units 469   Large Capacity Nosemount
Units 470   Cooling Cycle of a Nosemount Unit 475   Electromechanical Systems 476   Other
Applications of Self-Contained Units 476   Compressor Driven by the Vehicle Engine 477
Compressor Driven by Power Takeoff 478   Combination Mechanical and Nonmechanical
Transport Refrigeration Systems 478   Summary 479   Assignment: Unit 25 Review and
Self-Test 480

UNIT 26:   AUTOMOTIVE AIR-CONDITIONING **482**

Human Considerations in Designing Air-Conditioning Systems 482   Applications of
Refrigeration to Automotive Air-Conditioning 483   Operation of the Automobile
Air-Conditioning System 484   Types of Automotive Air-Conditioning Systems and
Components 485   Air Distribution 486   Mechanisms Controlling Compressor Functions 489
Basic Compressor Designs and Operating Characteristics 490   Design/Operating Features of
Swash (Wobble) Plate Compressors 492   Condensers 494   Receiver-Driers, Refrigerant
Lines and Evaporators 494   Automatic Controls of Temperature and Climate 497
Application of Air-Conditioning to Vehicles Other Than Passenger Cars 497   Automobile
Air-Conditioning Troubleshooting (Service) Chart 498   Summary 502   Assignment: Unit 26
Review and Self-Test 503

UNIT 27:   REFRIGERATED DISTRIBUTION OF ELECTRICITY **507**

Cryogenics 507   Superconductors 508   Refrigerated Distribution of Electricity 509
Cryogenic Envelope 512   Superinsulation 512   Critical Temperature 514   Liquefaction of
Air and Nitrogen 515   Liquefaction of Helium for Superconducting Cryogenic Cables 517
Multi-Stage, High-Vacuum Helium Refrigeration Systems 518   Summary 519   Assignment:
Unit 27 Review and Self-Test 520

APPENDIX: HANDBOOK TABLES **523**

GLOSSARY OF TECHNICAL TERMS **533**

INDEX **547**

BASIC REFRIGERATION CYCLE DIAGRAM .... **Inside Front Cover**

HANDBOOK TABLES

TABLE		PAGE
1	Physical Characteristics of Eight Common Refrigerants	523
2	Properties of Saturated Water (R-718)	524
3	Properties of Saturated Ammonia (R-717)	524
4	Properties of Saturated R-12	525
5	Properties of Saturated R-22	525
6	Ventilation Requirements for Compressors	526
7	Vapor Pressures of Common Refrigerants	527
8	Thermal Transfer Characteristics	528
9	Wire Sizes and Electrical Characteristics of Copper and Aluminum	529
10	Temperature Conversion to Degrees Fahrenheit (°F) or Degrees Celsius (°C)	530
11	Copper Conductor Sizes Based on Voltage Drop:	
	(A) 115-Volt Branch Circuit	532
	(B) 230-Volt Branch Circuit	532

PREFACE

Refrigeration and air-conditioning contribute to the health, welfare, comfort, safety, and productivity of individuals and nations. Refrigeration is essential to the transporting, processing, storing, and preserving of products ranging from foods and other perishables to medicines and health supplies.

Within industry, greater capabilities to produce and control temperature, humidity, and other environmental conditions need to be researched, as does the development and manufacturing of new materials, techniques, and products. For example, the ability to produce extremely low temperatures in the treatment of metals to produce stability in grain structure and control metal hardness is important in manufacturing products to finer dimensional and geometric tolerances.

Using cryogenic temperatures approaching absolute zero and new high-temperature superconductor ceramics operating at liquid nitrogen temperature add to the effectiveness of transmitting and distributing electrical power. Other experimentation, research, product design, and manufacturing are all interlocked with new developments in refrigeration and air-conditioning.

* * * * * *

This book presents foundational principles and functional information (about components, units, tools, instruments, systems, and other products and processes) that are applicable to residential, medium commercial, and large industrial refrigeration and air-conditioning installations and servicing requirements.

IMPORTANT FEATURES OF THIS EDITION

- While the text maintains its emphasis on updated techniques, equipment, and applications in refrigeration, greater depth is provided to cover air-conditioning components, operating characteristics, and systems.
- Entire content updating to reflect current materials, products, processes, and systems is based on extensive occupational studies complemented by careful reviews by expert practitioners.
- Most of the line art, industrial photos, and other illustrations (that appeared in the second edition) are changed. The new and revised artwork reflects the latest products/systems designs, new materials, and state of current technology.
- Changes in refrigerants and characteristics of new products are reflected throughout the book and in the refrigerant tables.
- Design features; principles of operation; and applications of many new solid-state electronic, programmable, digital readout measurement instruments, control devices, and other equipment and systems are added throughout the text. These additions reflect changing on-the-job emphases from conventional mechanical-, electrical-, and pneumatic-actuated devices and instruments.
- A new section (6), *All-Weather Air-Conditioning Systems*, includes two additional units. Unit 21 covers *Heat Pumps;* Unit 22, *Residential and Commercial Air-Conditioning Systems.*
- Throughout the book there is equal emphasis on the application of principles, products, and systems among residential, commercial, and industrial uses of refrigeration and air-conditioning.
- Three new "troubleshooting charts" are added to provide new opportunities to identify, diagnose, and take corrective action, simulating on-the-job practices.
- The unit on *Automotive Air-Conditioning* incorporates design features, principles of operation, and applications of components within all-weather automotive systems.
- Unit 27, *Refrigerated Distribution of Electricity*, displays the potential of evolving helium gas, closed-cycle, extremely low-temperature environment refrigeration systems.

Preface

- Latest, full-color *Basic Refrigeration Cycle* diagram is cross-referenced throughout the text contents.
- Additional "review and self-test" items permit further enrichment of the comprehensive testing materials within the unit assignments.
- The contents are written in a refreshing manner using the terminology of the workplace. New "terms" are defined and applied in the units. They also appear in the *Glossary of Technical Terms*.
- A *Handbook Reference Table* on "Temperature Conversion to Degrees Fahrenheit or Degrees Celsius" is now included.

•   •   •   •   •   •

Principles of Refrigeration is the end product of the occupational and professional experiences of the author, complemented by similar input from many individuals, organizations, and product manufacturers. Collectively, these persons and groups developed standards, safety regulations, codes, and handbook tables; prepared technical/trade journals, data sheets, and significant product/systems/processes bulletins; and supplied other valuable resource materials.

These significant contributions are recognized in the *Acknowledgments*. A deviation is made in this Preface by recognizing R. Warren Marsh and Mrs. Vivian Marsh. The late R. Warren Marsh, Professor Emeritus, Erie County Technical Institute, New York, is cited at this time. Mr. Marsh's studies of the instructional and training needs of the workplace resulted in establishing the scope and sequence of the original book. He served a lifetime in the refrigeration and air-conditioning industry through effective teaching, instructional supervision, curriculum planning, and administration at secondary, post-secondary, and college/community levels, and through his work in industrial training and contributions to professional papers. These experiences established the foundations in the writing of the *Principles of Refrigeration* book. A sincere "thanks" is extended to Mrs. Vivian Marsh for her supportive assistance over the years from the first through the second editions.

C. THOMAS OLIVO

Albany, New York

ABOUT THE AUTHOR

Dr. C. Thomas Olivo ("Dr. Tom") ranks among the nation's foremost technical writers, curriculum development and instructional materials experts, administrative leaders, and occupational competency testing authorities.

Entering teaching from work within industry as an apprentice, journeyman and master craftsperson, Dr. Olivo served in successively responsible leadership positions. As Bureau Chief, he headed the New York State Education Department's Bureau of Vocational Curriculum Planning and Instructional Materials Development and Bureau of Industrial Teacher Education, and provided unusual leadership as State Director of Industrial-Technical Education.

He served with distinction as teacher, Director of a forerunner State Technical Institute, and as graduate professor, Temple University. Dr. Olivo provided exemplary services in conceiving, organizing, administering, and establishing the National Occupational Competency Testing Institute Consortium of States as a respresentative body concerned with student/trainee, teacher, industrial worker, and military occupational specialty testing materials and processes.

Dr. Olivo is recognized at local, state, national, and international levels in human resource development. His many publications are widely used throughout the United States, Canada, and the world.

ACKNOWLEDGEMENTS

Principles of Refrigeration incorporates changing knowledge and technology about new materials, new equipment, new processes, and new systems to meet the ever-increasing and more exacting demands of people and the business marketplace.

Each succeeding edition requires the input of different combinations of experts. Some practitioners serve as technical reviewers; others as occupational consultants who relate to design, manufacturing, marketing, installation, and servicing; and many more who deal with training. As leaders, each keeps abreast of evolving developments and makes a contribution that is important to keeping a publication technically accurate and up to date.

When the original *Principles of Refrigeration* was developed, acknowlegement for technical assistance was made to Amos H. Keirn, Senior Laboratory Engineer, Fedder Corporation; Robert L. Jameson, Manager, Minneapolis Honeywell Regulator Company; Frank Notaro, Supervisor, Linde Division, Union Carbide Corporation; and Steve Pasco, Pasco Refrigeration Company. The important work of technical editing by Vincent P. Lang, Carrier Corporation, was recognized. Credit was given to Louis J. Siy for original artwork and to the Delmar Publishers editorial staff for preparing the manuscript for publication.

Joe Edwards, Head, Air-Conditioning, Heating and Refrigeration Department, Tri-County Technical College, South Carolina, was credited in the second edition for significant contributions in updating the publication. Stephanie B. Olivo provided indexing and editorial services.

In this third edition, two individuals are specially cited for their precise work in carefully editing the contents of the final manuscript for state-of-the-technology relevance and accuracy. Recognition is made of their exceptional contributions in the area of solid-state electronic controls, devices, and instruments, and new and evolving developments in air-conditioning and refrigeration systems, equipment, processes, and products. The two men are Dr. Ralph D. O'Brien, Chairman and Professor of Industrial Technology, and William H. Gibson, Assistant Professor and Consultant, Refrigeration, Heating, and Air-Conditioning, both of Northern Kentucky University, Highland Heights, Kentucky.

Special recognition is made of John E. Burns, Heating, Air-Conditioning and Refrigeration Engineer, Gennesee Refrigeration Company, Albany, New York, for meticulously assessing this edition for content up-to-dateness, relevance, and technical accuracy; for providing excellent resource materials; and for excellent technical assistance. Recognition is made of Harold Schecter, former Chairman, Department of Air-Conditioning and Refrigeration, Mohawk Valley Community College, Utica, New York, for work on the revised outline and other content suggestions.

Initially, Marjorie Bruce and Todd Dreyer of Delmar Publishers provided editorial leadership in securing content and organization reviews. Credit is given to Cameron Anderson for final editorial services. The information provided and the personal services of these three editors were invaluable in producing this edition. Appreciation is also expressed to the following reviewers.

Initially, Marjorie Bruce and Todd Dreyer of Delmar Publishers provided editorial leadership in securing content and organization reviews. Credit is given to Cameron Anderson for final editorial services. The information provided and the personal services of these editors were invaluable in producing this edition. Appreciation is also expressed to the following reviewers.

The author would like to thank the following individuals for helpful suggestions in reviewing the manuscript:

Herb Molke
Bergen County Board
 of Vocational Education
Hackensack, NJ

Julius Sojas
Goodwin Technical School
New Britain, CT

Acknowledgements

Robert Farley III
Jarrettsville, MD

Prof. Henry Puzio
Lincoln Technical Institute
Union, NJ

Prof. Arthur Porter
St. Philips College
San Antonio, TX

Marvin Maziarz
Niagara County Community College
Sanborn, NY

RECOGNITION OF BUSINESS, INDUSTRY AND PROFESSIONAL ASSOCIATIONS

A special note of appreciation is expressed to each one of the business, industry, and professional associations that contributed to this book. The overwhelming, enthusiastic support was a very gratifying experience. The quality and quantity of the state-of-the-technology materials and the wealth of the art copy that was provided is testimony to the vigor and health of the industry. Equally significant, the contributions reflect these associations' concerns in developing highly qualified craftpersons, service technicians, and others.

A *courtesy line* appears throughout each unit and the book crediting the source of artwork. In addition, through the recognition of the following companies, the author expresses a special "thank you" to the individuals within each organization who so generously contributed his or her time, talents, and material resources.

Airserco Manufacturing Company, Inc.
  Dayton, Ohio
Alco Controls Division,
  Emerson Electric Company
  St. Louis, Missouri
Allen-Bradley, Rockwell International
  Company, Industrial Control Group
  Milwaukee, Wisconsin
American Standard Company
  Piscataway, New Jersey
Amprobe Instrument Division,
  Core Industries, Inc.
  Lynbrook, New York
ASHRAE, American Society of Heating,
  Refrigeration, and Air-Conditioning
  Engineers
  Atlanta, Georgia
Bacharach, Inc.
  Pittsburgh, Pennsylvania
Binks Manufacturing Company
  Franklin Park, Illinois
Bristol Babcock, Inc.
  Watertown, Connecticut
Cadillac Motor Car Division,
  General Motors Corporation
  Detroit, Michigan

Carrier Corporation:
  Advanced Technology Unit,
  Carrier Transicold Division,
  William Buynum Training Center
  Syracuse, New York
Comfort-Aire/Century,
  Heat Controller Inc.
  Jackson, Michigan
Control Products Division,
  Johnson Controls, Inc.
  Milwaukee, Wisconsin
Cooper Instrument Corporation
  Middlefield, Connecticut
Copeland Refrigeration Corporation
  Subsidiary of Emerson Electric Company
  Sidney, Ohio
Corken Steel Products Company
  Covington, Kentucky
Cryogenics: IPS Science and Technology
  Press
  Surrey, England
CTI-Cryogenics
  Waltham, Massachusetts
Dole Refrigerating Company
  Lewisburg, Tennessee

Dometic Corporation
 La Grange, Indiana
Dunham-Bush, Inc.
 Harrisonburg, Virginia
Electric World,
 McGraw-Hill Publishers, Inc.
 New York, New York
Ford Motor Company
 Customer Service Division
 Dearborn, Michigan
General Electric Company:
 Cryogenics Branch
 Schenectady, New York
Motor Manufacturing Department
 Ft. Wayne, Indiana
Heinemann Electric Company
 Lawrenceville, New Jersey
Greenfield Tap and Die Division,
 Greenfield Industries, Inc.
 Augusta, Georgia
Halstead Industries, Inc.
 Greensboro, North Carolina
Harrison Radiator Division,
 General Motors Corporation
 Lockport, New York
Honeywell Residential Division,
 Honeywell, Inc.
 Golden Valley, Minnesota
Imperial Eastman Corporation
 Chicago, Illinois
Johnson Controls, Inc.
 Milwaukee, Wisconsin
Kerotest Manufacturing Company
 Pittsburgh, Pennsylvania
Kock Processes Systems
 Westboro, Massachusetts
Kramer Trenton Company
 Trenton, New Jersey
Marsh Instrument Company
 Skokie, Illinois
Mason Supply Company
 Cincinnati, Ohio
Mc Quay/Snyder General Corporation,
 Refrigeration Products Group
 Scottsboro, Alabama

Mueller Brass Company
 Port Huron, Michigan
Parker Hannifin Corporation
 Wheeling, Illinois
Peerless of America
 Chicago, Illinois
Ranco North America
 Plain City, Ohio
Refrigeration Service
 Engineers Society
 Des Plaines, Illinois
Robinair Division, STX Corporation
 Montpelier, Ohio
Russells Technical Products, Inc.
 Holland, Michigan
Schaefer Brush Manufacturing Company
 Waukesha, Wisconsin
Sedco, Inc.
 Adrian, Michigan
Seeled Units Parts Company Inc.
 Allenwood, New Jersey
Simpson Electric Company
 Elgin, Illinois
S-K Hand Tool Company
 Chicago, Illinois
Snap-on Tools Corporation
 Kenosha, Wisconsin
Sporlan Valve Company
 St. Louis, Missouri
Standard Refrigeration Company
 Melrose Park, Illinois
Superior Valve Company
 Washington, Pennsylvania
Taylor Wharton Company
 Indianapolis, Indiana
Temprite Company
 West Chicago, Illinois
Tecumseh Products Company
 Tecumseh, Michigan
Tenny Engineering, Inc.
 Union, New Jersey
Texas Instruments Inc.,
 Motor Controls Department
 Attleboro, Massachusetts
The Cooper Thermometer Company
 Middlefield, Connecticut

Acknowledgements

The L.S. Starrett Company
  Athol, Massachusetts
Thermo King Corporation
  Minneapolis, Minnesota
Thermometer Corporation of America
  Fletcher, North Carolina
The Trane Company
  La Crosse, Wisconsin
TIF Instruments Inc.
  Miami, Florida
Trailmobile, Inc.
  Pullman Company
  Chicago, Illinois

Tranter Inc.,
  Platecoil Division
  Augusta, Georgia
Union Carbide Corporation,
  Linde Division
  Danbury, Connecticut
Virginia KMP Corporation
  Dallas, Texas
Wingaersheek Division
  Victor Equipment Company
  Danvers, Massachusetts
York International Corporation
  York, Pennsylvania

INTRODUCTION

Principles of Refrigeration serves three major needs.

✓ The content provides a broad, functional background of the principles and practices required of service technicians, craftspersons, planners, designers, and other related industry personnel. The technology and processes information is fundamental to all types of residential, commercial, and industrial systems.

✓ The instructional units are designed to provide *terminal* preparation for those individuals who move directly from training to on-the-job activities. These workers need to apply the "Why-to-Do" technology and "How-to-Do" basic skills developed through learning experiences to everyday specific applications in the field. The term *terminal* means a juncture from training to occupational employment. Once employed, it is necessary for the worker to participate continuously in occupational extension courses or upgrading training in order to keep abreast of changing technology and to prepare for higher levels of skills and job advancement.

✓ The instructional units are *preparatory* for those who plan to move into advanced phases of refrigeration and air-conditioning work, including design, engineering, or any scientific endeavor, or teacher training, where the outcomes are related to new materials, new methods, or new products.

ORGANIZATION OF *PRINCIPLES OF REFRIGERATION*

This book is organized into six sections. Each section represents a major block, division, or area of instruction. The six sections include:

- 1 Basics of Refrigeration
- 2 Refrigeration Tools and Materials
- 3 Heat Transfer and Temperature
- 4 The Heart of the Refrigeration System
- 5 Electrical Components of Refrigeration and Air-Conditioning Systems
- 6 All-Weather Air-Conditioning Systems

Each section contains a number of *Instructional Units* within which principles, related technical information, and applications are included for components, tools, processes, and systems. The instructional units are arranged in a logical order of dependence of one principle, law, or set of conditions upon another. The content within a unit proceeds in an organized and helpful teaching/learning pattern.

Basic Principles and Related Technical Information

Each unit includes a number of parts. Each part contains basic principles, technical information, and related skills that are specific to the area being studied. The *technical terms* (that are representative of those used daily in industry to communicate trade information about tools, instruments, products, processes and systems) are described and illustrated in each unit. Further, major technical terms are included for ready reference in the *Glossary*.

An important *Summary* at the end of each unit emphasizes significant items and serves as a teaching/learning review.

Introduction

Line Drawings, Industrial Photos, and Artwork

Line drawings of complete installations, cutaway sections of mechanisms and devices, and photographs of parts, components, and systems were specially selected to complement the text matter. Within this context, all artwork was planned to improve instructional effectiveness by simplifying a description or by showing parts and components in relation to the complete functioning of a system.

Planned Safety Program

Safety Precautions are included to direct attention to safe working conditions and practices in order to avoid personal injury and damage to tools, instruments, equipment, systems, and other property. *General Safety* conditions are first introduced. These are followed by specific safety precautions. Safety considerations provided by manufacturers, trade and industry, state and local codes, and other professional association publications need to be studied and followed.

Unit Assignments

The *Unit Assignments* are designed to provide a comprehensive review of the contents of each unit in a way that simulates on-the-job conditions. Each unit assignment contains a number of types of objective test items. This variation in testing methods provides motivation and stimulation. Test items may be selected from each unit assignment to meet specific testing needs.

Tables in the assignments permit the recording of calculated data and other technical information. Values obtained may be used for comparison purposes when studying the effects of changed quantities, materials, or other conditions. The tables also suggest an orderly pattern of recording data that, if followed, simplifies the gathering and interpreting of information, ensures accuracy, and conserves time.

Troubleshooting Charts

The use of *Troubleshooting Charts* is a functional and widespread trade practice. Examples of troubleshooting charts are included for hermetic compressors, refrigeration systems, and automotive air-conditioning systems. Each chart (in a simple and easy way) provides information in relation to (1) major problem areas, (2) nature or cause of each problem and (3) corrective action or remedy.

Handbook Tables

The *Appendix* provides additional learning experiences. Eleven specially selected *Handbook Tables* appear in the Appendix. Text references are made to specific information in the tables in order to solve typical on-the-job problems.

Glossary of Technical Terms

Technical Terms contained throughout the text (and referred to in the field) are included in the Appendix. The *Glossary*, provides a ready reference to terms used daily to communicate information about tools, instruments, processes, products, and systems of refrigeration and air-conditioning.

Index

The *Index* serves the important function of accessibility to the contents. Important items are listed alphabetically, so information that is frequently needed may be quickly identified through cross-referencing.

The Basic Refrigeration Cycle Diagram

This diagram provides the backbone of all instruction in refrigeration and air-conditioning. The four-color diagram is conveniently located on the inside front cover of the book. The *Basic Refrigeration Cycle Diagram* is referred to in many of the instructional units throughout the text.

SUGGESTED APPLICATIONS

Variations exist within training programs; instructional materials must meet exacting occupational demands. The organization of this book permits flexibility to adapt the contents to meet many different types of training programs and to a number of levels on which instruction may be given. Over the years, the previous editions of this book have been used successfully to meet the needs in the following programs.

- As a textbook in institutional programs where the learner must develop a broad understanding of refrigeration and air-conditioning terms, principles, devices, components, and systems. Equally important is the application of this technology to the acquisition of hands-on skills required on the job.
- As a basic textbook in vocational–industrial–technical education programs for use either in organized class, group, or individual study.
- As a practical refrigeration and air-conditioning textbook and resource manual used within the industry for apprentice training and additional in-plant training courses for craftspersons, service technicians, development, sales, and other persons.
- In military occupational specialty training courses.
- As a textbook or source book for adult programs, occupational extension, or supplementary training where a sound, practical working knowledge of refrigeration and air-conditioning is foundational to work in related occupations.
- As a textbook in teacher-education programs and as a resource guide in preparing courses of instruction and teaching plans and in developing tests for assessing learning experiences.

INSTRUCTOR'S GUIDE

An *Instructor's Guide* for the text contains solutions to all of the test items and problems. Suggested answers are provided wherever there may be variations in the answers given by students/trainees. The *Instructor's Guide* is intended to conserve valuable teaching time and to provide a uniform basis for the solutions to the problem material.

GENERAL SAFETY CONSIDERATIONS

The refrigeration and air-conditioning industry involves a great number of possible safety hazards to individuals and equipment and damage to property.

- Hazards result from working with machinery in motion and complex automated electrical/electronic transmission and controls systems. The dangers are further compounded by refrigeration processes that require the use of different gases and liquids under changing pressures and varying temperatures.
- Refrigerants are toxic and suffocating to differing degrees, and all present high-pressure and explosion hazards. Also, the products that are being refrigerated are sometimes explosive, toxic, or corrosive and may be hazardous to handle.
- Safety rules and practices must be observed in the planning, installing, servicing, and maintenance of refrigeration and air-conditioning components and systems. These services require the correct use of tools, instruments, and gauges; observation of safe work procedures; and continuous adherence to safety requirements.
- In general, there are five classes of *personal accidents*. These relate to: (1) improper and unsafe working practices in dealing with mechanical devices and conditions; (2) electrical components and conditions; (3) temperature controls and conditions; (4) pressure controls and conditions; and (5) toxicity conditions.
- While the industry provides considerable literature on safe practices, laws are passed, and codes are established to protect workers, equipment, and property. The key to safety remains with the individual worker and the exercise of vigilance in constantly checking occupational hazards and in working safely.
- Although guards and other mechanical and electrical devices are designed to improve the factor of safety, some accidents may be avoided by studying and knowing *in advance* what dangers are involved and then proceeding to work cautiously, following safe practices.

SPECIFIC SAFETY PRECAUTIONS

A great number of *Safety Precautions* are identified throughout the text. Supplemented by other safety resource materials from industry and professional sources, the aim is to emphasize the importance of following safe practices and to develop *safe habits* and *safe workplace attitudes* by all who serve the refrigeration and air-conditioning industry.

SECTION 1: REFRIGERATION

UNIT 1:
INTRODUCTION TO REFRIGERATION

Refrigeration contributes to the raising of living standards for the peoples of all lands. The advances made in refrigeration in recent years are the result of a team approach in which technicians, craftspersons, engineers, scientists, and others, pool their skills and knowledge.

The foundations on which new substances and materials are built are provided by science. This knowledge is applied to the refrigeration field by those who design, manufacture, install and maintain refrigeration equipment. It is then made useful through subsequent planned research, development and practical application.

The application of the refrigeration principle is limitless. The most common use, and one that is readily recognized, is the preservation of food. Almost all products in the home, on the farm, in business, in industry or in the laboratories are in some way affected by refrigeration. Thus, refrigeration has become an essential commodity in modern living.

This first unit describes some of the romance of refrigeration through the ages and in several parts of the world. Early experiments with ice-making and refrigeration machines are identified. Several applications of modern refrigeration systems are also included to show the scope and widespread importance of this basic principle.

BRIEF HISTORIAL SKETCH OF "MAN-MADE" COLD

The story of ice dates back as far as recorded history. While the Stone Age caveman knew what ice was, there was no thought about using it to preserve food. Thousands of years later the Chinese learned that ice improved the taste of drinks. So they cut ice in winter, packed it with straw and chaff, and sold it during the summer.

The early Egyptians found that water could be cooled by placing it in porous jars on rooftops at sundown, figure 1–1. The night breezes evaporated the moisture which seeped through the jars, making the water inside the jars cooler. The Greeks and Romans had snow brought down from mountaintops to cone-shaped pits which were lined with straw and branches and then covered with a thatched roof.

As civilization advanced, people learned how to cool beverages and foods for enjoyment. This knowledge increased the use of ice and snow.

FIG. 1-1   Cooling water by evaporating moisture   FIG. 1-2   Transporting ice by Clipper ship

EARLY EXPERIMENTS WITH FOOD PRESERVATION

Some of the earliest recorded experiments with food preservation date back to 1626 when Francis Bacon attempted to preserve a chicken by stuffing it with snow. In 1683, Anton van Leeuwenhoek opened up a whole new scientific world. This Dutchman invented a microscope and discovered that a clear crystal of water contains millions of living organisms. Today, these are known as microbes.

Scientists studied these microbes and found that rapid multiplication took place in warm, moist conditions such as provided in food materials. This multiplication of microbes was soon recognized as the major cause of food spoilage. By contrast, the same types of microbes in temperatures of 50°F or less did not multiply at all.

Through these scientific studies, it became apparent that fresh foods could be safely preserved in temperatures of 50° or less. It was now possible to preserve food by drying, smoking, spicing, salting or cooling.

Since little was known about how to create temperatures low enough to freeze water into ice, ice was transported from its source by Clipper ships to the principal cities of the world, figure 1–2.

EXPERIMENTERS OF ICE-MAKING MACHINES

One of the first patents (1834) for a practical ice-making machine was granted to Jacob Perkins, an American engineer living in London. These machines were used successfully in meat-packing plants. Within the next fifty years ice makers were produced in the United States, France and Germany, figure 1–3. In this period about 3,000 patents on refrigeration systems had been applied for in the United States.

While progress was made in producing ice by artificial means, nearly everyone favored natural ice, believing that artificial ice was unhealthful. Eventually, this superstition was overcome because: (1) artificial ice was produced from purer water than that usually found in lakes and ponds; (2) it could be made as needed; and (3) it did not need to be stored for long periods of time.

Even after the superstitions concerning artificial ice were overcome, it was not until the warm winter of 1890 when a shortage of natural ice occurred that the demand for artificial ice became significant. This event helped to start the mechanical ice-making industry. Thus, by the end of the 19th century, ice and refrigeration were becoming commonplace in the American home.

Another factor that contributed greatly to the further development of dependable refrigeration equipment was the availability of inexpensive electric power and the development of small electric motors. These were important mechanical milestones. Simultaneous with these developments, scientists constantly searched for simple truths about cause and effect upon which all of refrigeration depends.

Modern air conditioning was introduced to the world in 1902 by Willis Carrier in response to an industry problem of a Brooklyn, New York, printer who was having difficulty getting printing inks to register accurately on high-humidity, summer days. Carrier built a machine to blow cool air through the plant.

About 1910, functional, mechanical domestic refrigeration appeared. By 1918, Kelvinator produced the first automatic refrigerator for the American consumer. That year, 65 machines were sold. Today, over 10 million units are produced each year. Beginning with the 1920s, domestic refrigeration started to become one of the United States' important industries.

Fast food freezing processes and equipment were developed around 1923 and led to the beginning of the modern frozen foods industry. As for refrigerators, an automatic domestic *absorption unit* was incorporated in the Electrolux refrigerator of 1927. The *monitor top, sealed, hermetic system* was introduced and manufactured by the General Electric Company in 1928. By the 1940s, practically all refrigerator units were of the hermetic type.

The need for refrigerating large food storage systems, for comfort cooling of large buildings and auditoriums, and for the production of low temperatures for many industrial manufacturing processes required the design and manufacture of large commercial units. Also, automobile air conditioning, which had started slowly in the 1930s, began to move at an accelerated growth rate in the 1940s.

FIG. 1-3   Early artificial ice-making machine

Heat pumps were acknowledged in the late 1950s as an efficient year-round system for heating and cooling. Today, heat pumps are installed in over one third of all newly built homes and in increasing numbers in commercial buildings. Water heating is integrated by some manufacturers into space heating and cooling heat pumps, resulting in greater efficiency and economy.

The units which follow in this book deal with: (1) the scientific laws governing refrigeration; (2) the fundamental theories about the construction and operation of refrigeration equipment and the major units (components) of which they are built; and (3) the procedures used for the maintenance of equipment systems.

MODERN USES OF REFRIGERATION

Refrigeration may be defined very simply as the process of removing heat under controlled conditions. Cold is merely a relative term referring to the absence of heat. Thus, to produce "cold," heat must be removed. This may be accomplished by using ice or by mechanical means.

While refrigeration has been discussed in terms of food preservation, this is only one of many principal applications. It is also used to provide comfort conditions and to produce raw materials and finished products. A series of different refrigeration units such as the helical rotary water chiller illustrated in figure 1–4, is used in industrial/commercial installations. This particular model is manufactured in sizes ranging from

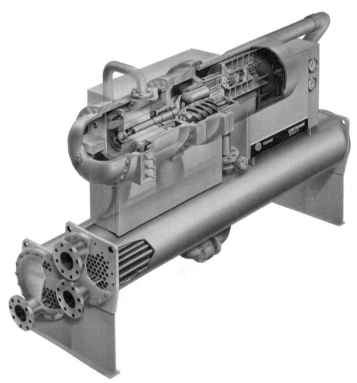

FIG. 1-4 **Modern helical rotary water chiller refrigeration unit** *(Courtesy of The Trane Company)*

FIG. 1-5 Environmental temperature and humidity climate-controlled walk-in room *(Courtesy of Tenney Engineering, Inc.)*

100 to 300 tons. The capacities of still other centrifugal water chillers are from 100 to 1750 tons.

Air-conditioning Applications

Air-conditioning units provide safety and comfort in nearly all forms of transportation including automobile, bus, truck, ship, submarine, aircraft, and aerospace. Air conditioning provides conditions of comfort in the home, office, business, hotel, apartment building, hospital and industry.

Air-conditioning equipment is capable of performing the following functions:

- Filtering out soot, dirt, and dust
- Humidifying and dehumidifying the air (adding or removing moisture)
- Heating and cooling the air
- Circulating the air within a prescribed space
- Removing undesirable inside air and taking in fresh outside air

Air conditioning thus makes it possible to do something about "the weather."

Proper air conditioning during and after operations is important in communicable disease wards of hospitals. The rapid removal of infected air helps speed recovery. Refrigeration of blood plasma makes blood banks possible, to meet emergency needs.

Industrial Applications of Refrigeration

Refrigeration is needed to produce correct climatic conditions and for environmental testing, temperature cycling, and humidity cycling (figure 1–5).

Other manufacturing processes require refrigeration: for example, cooling cutting oils helps in machining operations by lowering the temperature of the workpiece to prevent overheating. Quenching baths for heat treating operations may also be controlled through refrigeration processes.

In the microbiology, medical, research, and pharmaceutical fields, refrigerating units are used to store, process and test chemical and biological materials.

Refrigeration, as a quick cooling process, speeds production, cuts moisture losses in foods and reduces mold. The large frozen food industries, and others engaged in the preparation, marketing and purchasing of foods, all depend on refrigeration.

Important studies of the exact nature of electron movement are now being undertaken through a process which demands that the material being studied be subjected to the lowest possible temperature, a temperature at which electron movement slows down to the point where it may be observed.

Steels that must be aged to retain shape and dimensional accuracy are now refrigerated under new and rapid deep-freezing treatments. Similarly, aluminum is kept from aging too rapidly.

In these and other industrial applications, refrigeration units capable of reducing temperatures to –150°F and below are used in metalworking plants, tool shops and metallurgical laboratories for heat-treating and hardening operations. An example of a movable industrial refrigeration unit is displayed in figure 1–6.

FIG. 1-6 Movable industrial refrigeration/heating unit. Range: –100°F to +350°F *(Courtesy of Russells Technical Products, Inc.)*

As scientists, technicians and craftspersons experiment at still lower and colder temperatures approaching –273°C (–460°F), the new science of *cryogenics* (refrigerants) will reveal materials in a state that is not solid, liquid or gas.

SUMMARY

- The early history of refrigeration reveals that this principle was used primarily for cooling liquids and foods.
- The invention of the microscope opened up the scientific study of food preservation.
- One of the earliest methods of refrigeration used throughout the world was packing with natural ice.
- The first practical ice-making machines were produced during the early 19th century. However, the use of artificial ice was not accepted until the end of the century.
- The availability of electricity and the development of compact and dependable motors contributed to the widespread use of refrigerating equipment.
- Mechanical refrigeration is the process of transferring heat from one place to another.
- Some typical refrigeration applications include: air-conditioning units for transportation, comfort and safety in work; industrial processes; food processing; laboratory produced materials, and the like.
- The challenges to the technician, craftsperson, engineer and scientist for identifying new methods, processes and applications of refrigeration are limitless.
- There is a growing need for craftspersons with know-how to install, maintain and service refrigeration equipment and for others skilled in design, development, research, business, and marketing.

ASSIGNMENT: UNIT 1 REVIEW AND SELF-TEST

A. INTRODUCTION TO AND HISTORICAL SKETCH OF "MAN-MADE" COLD

Circle the condition, value or date that correctly completes statements 1 through 5.

1. The Stone Age caveman knew what ice was and (a) used it for cooling liquids, (b) used it for cooling and food preservation, (c) used it for food preservation only, or (d) did not use it at all.
2. Early Egyptians cooled water by (a) evaporating moisture that seeped through jars, (b) covering water jars with straw, (c) putting water jars in holes dug in the sand, or (d) using snow.
3. Scientists found that food spoilage occurred more rapidly at temperatures above (a) 15°F, (b) 50°F, (c) –20°F, or (d) –159°F.
4. The first automatic refrigerator was produced for the American consumer in (a) 1918, (b) 1929, (c) 1936, or (d) 1941.
5. Modern refrigeration units (a) have limited usage, (b) have unlimited usage, (c) can be used only for food preservation, or (d) can be used only for comfort cooling.

B. EARLY EXPERIMENTS WITH FOOD PRESERVATION

Provide a word or phrase to complete statements 1 to 6.

1. Natural ice had to be transported from cold to warm climates because man had not yet learned _____ .
2. Early scientists found that minute organisms multiply rapidly in _____ and _____ surroundings.
3. Organisms do not grow or spread at a temperature of _____ or lower.
4. One of the first patents for an ice-making machine that was used successfully was granted in the first half of the _____ century.
5. Superstitions about using artificial ice were overcome when people were convinced: (a) _____ and (b) _____ .
6. The growth of mechanical refrigeration was assured by two factors: (a) _____ and (b) _____ .

C. MODERN USES OF REFRIGERATION

1. State briefly what the refrigeration process means.
2. List three functions of air-conditioning units.
3. Name two commercial refrigerating and air-conditioning applications other than those given in the unit.
4. Name two food preservation applications.
5. Cite two industrial refrigerating and air-conditioning applications.
6. State briefly how refrigeration affects the new science of cryogenics.

D. SAFETY PRECAUTIONS

1. Cite two classes of accidents encountered by workers in the refrigeration industries.

Circle the condition that correctly completes the following two statements.

2. Most accidents which occur while working on refrigeration systems are due to (a) mechanical defects, (b) the technician not working safely, (c) electrical problems, or (d) industry hazards.
3. Most refrigerants are to some degree (a) toxic and suffocating, (b) nonhazardous, (c) noncorrosive, or (d) nonexplosive.

UNIT 2:
REFRIGERATION SYSTEMS, CYCLES AND CLASSIFICATION

Refrigeration systems consist of a series of main units (components), each of which performs a special function or job. The engineering, design, and planning that go into the manufacture of this equipment are very elaborate and costly. However, regardless of how complex the refrigeration equipment or the operations, each of the separate units as well as the entire system depends on simple scientific principles.

The work of early scientists to establish some of these principles is briefly described in this unit. Later, these principles are brought together in mechanical refrigeration systems. Such systems are then classified according to use. Finally, a comparison is made between refrigeration and reverse refrigeration.

REFRIGERATION BY ABSORPTION AND EVAPORATION

Refrigeration has been defined as the process of transferring heat from one place to another. The space from which heat is removed is said to be *cooled*, or *refrigerated*, or *air-conditioned*. Ordinarily, when a hot object is placed near or touches a cold object, heat flows from hot to cold. In studying refrigeration, however, it is helpful to take the point of view that a cold object removes heat from the hot object. Thus, the hot object becomes cooled.

For example, an individual experiences a feeling of being cooled when swimming in water that is cooler than body temperature. As this person stands up with part of the body above the water, the individual feels chilled even more. This is especially true under certain conditions when the air is dry and a wind is blowing.

This increased cooling results from the evaporation of water on the skin. *Evaporation* is the process whereby a liquid changes to a vapor. This change takes place because the liquid has absorbed heat. In the case of the swimmer, the heat absorbed by the evaporating water comes from the body. Hence, a cooling of the body takes place.

Rubbing alcohol vaporizes faster and has a more pronounced cooling effect on the skin than water. For this reason, alcohol is used for cooling the fevered body of a sick person. Ether has an even greater cooling effect. Fanning increases the effectiveness of the cooling in each instance.

One of the early refrigerants was ammonia. The term *refrigerant* refers to the fluid used in the refrigerating system to produce cold by removing heat. The well-known household ammonia is a very weak solution containing a small amount of ammonia in a large quantity of water.

When pure ammonia is placed in an open container, it evaporates so rapidly that it boils. The ammonia picks up heat much faster than alcohol or ether and in the process, changes from a liquid to a gas. When pure ammonia boils at atmospheric pressure, it cools to 28 degrees below zero on the Fahrenheit temperature scale (–28°F). The area from which the boiling ammonia removes heat is thus cooled or refrigerated.

Section 1: Refrigeration

One of the simplest ways in which this cooling action may be employed is illustrated in figure 2–1. The refrigerating unit consists of an insulated refrigerator box (A), an ammonia tank with the liquid under pressure (B), and a cooling unit (C). The cooling unit absorbs heat as the ammonia liquid is changed to a gas at atmospheric pressure. Such a system is not practical because of the cost of the refrigerant and the volume of harmful ammonia gas that would have to be piped safely away. What is needed, instead, is a method by which the refrigerant may be saved and restored to its liquid condition. In the liquid state the refrigerant is again ready to be changed to a gas and thus be capable of absorbing more heat.

There are two primary methods by which a refrigerant gas is reclaimed so that it can be used again in the refrigeration cycle. The first of these is based upon a discovery made over a century ago by Michael Faraday.

Faraday's Early Experiments with Absorption and Evaporation

The scientists of Faraday's time knew that many gases could be changed to a liquid state (liquefied) under certain conditions of temperature and pressure. With their limited equipment and techniques, they were not able to liquefy some of the then-known gases. One of the most stubborn of these gases was ammonia.

Faraday knew from his experience in the laboratory that silver chloride had a special capacity for absorbing ammonia gas. He exposed some silver chloride powder to

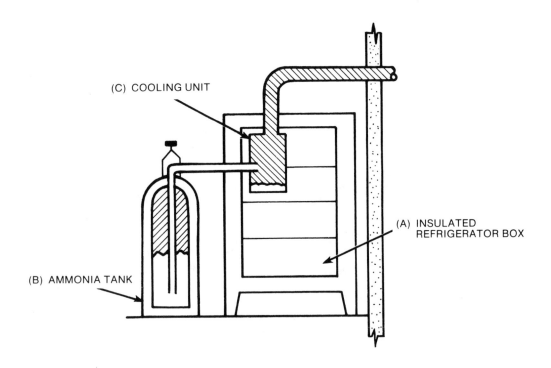

FIG. 2-1 Elementary refrigeration system

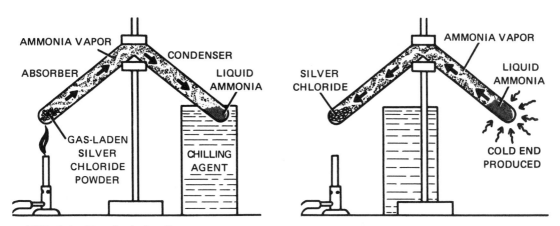

FIG. 2-2   Faraday's distilling apparatus   FIG. 2-3   Vapors absorbing and carrying away heat

ammonia gas until it had absorbed all it could hold. This gas-laden powder was then sealed in a test tube.

The test tube was formed into an inverted *V* and became Faraday's distilling apparatus, figure 2–2. When heat was applied under the powder, it drove the ammonia fumes out of the silver chloride powder.

The fumes were cooled by immersing the other end of the tube in a container of cold water. The water served as a heat-removal agent for the crude distillery.

As the hot ammonia vapors (fumes) entered the chilled end of the test tube, drops of liquid ammonia formed. The process was continued until Faraday had collected a sufficient quantity of liquid ammonia, figure 2–3.

The heat and cooling agent were then removed and Faraday began studying the characteristics of this newly-produced substance. It was the first time that ammonia had been changed from a gas to a liquid.

A strange thing happened. Almost immediately the liquid ammonia started to bubble and boil and to change back into a vapor. Heat was absorbed and carried away in the process. Again, the vapor was absorbed by the silver chloride powder. When Faraday touched the liquid ammonia end of the test tube, he found it was intensely cold. But, even more surprising was the fact that the coldness had been produced by the liquid as it boiled, even without any visible source of heat!

Each time the process was continued, Faraday noted the same change. What was significant was the fact that icy temperatures could be created in the laboratory any number of times without losing the ingredients in the test tube.

As crude as this experiment may seem, even today these same operating principles are used for refrigeration. While one end of a refrigerating unit may be heated, the other end may be cooled in a container of water to change the refrigerant to a liquid. The device, if then attached to an insulated box, may serve as a refrigerator by allowing the liquid to boil. The drawback, of course, is the fact that the operation is only intermittent and inefficient.

11

SAFETY PRECAUTION

This experiment should only be attempted in a laboratory by a person who is familiar with chemicals and heat resistant glass.

Simple Absorption System Refrigerator

Absorption system refrigerators have been designed to run continuously, automatically and economically. A modern household refrigerator which operates on the absorption principle is illustrated in figure 2–4.

Quite a number of refinements have been made. Water is now used to absorb the vapors of ammonia or other gaseous fumes instead of silver chloride crystals as were used originally.

The schematic drawing in figure 2–5 illustrates the basic cycle of a household refrigerator operating on the absorption principle. The basic operating cycle shows that the

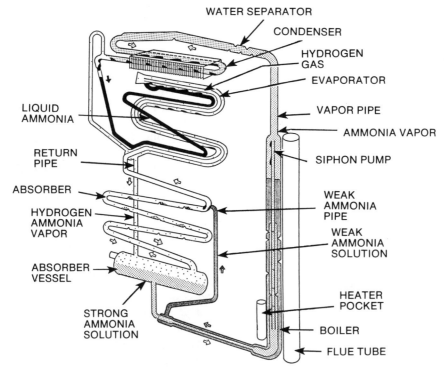

FIG. 2-4   Components of a modern absorption system refrigerator *(Courtesy of Dometic Corporation)*

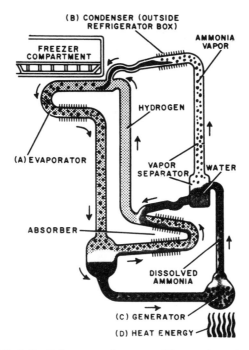

FIG. 2-5 Schematic diagram of basic absorption cycle

system picks up heat in the evaporator (A) and carries it outside the insulated refrigerator box to the condenser (B). Here, heat is removed by the room air passing over the condenser fins. Although a small unit is shown, larger refrigeration units are available to meet any need. Many units are currently manufactured with two evaporators, one for a separate freezing compartment.

The cooling effect is produced in the evaporator by the boiling of the refrigerant. This boiling refrigerant absorbs heat just the same as occurred in Faraday's experiments. Note that the hydrogen is circulated over the liquid ammonia to speed up the boiling and to carry away the heat at a faster rate.

The refrigerant is changed to a liquid by removing heat in and around the condenser. Another common way of expressing that heat is removed is to say that the refrigerant gives up heat.

Obviously, in this system heat is moving (being carried) from a low temperature to a higher one. This is contrary to the natural flow of heat which is from a higher to a lower temperature. Thus, it is necessary to add energy to the system. The energy in an absorption refrigerator is supplied in the form of heat energy resulting from the burning of a suitable fuel or from an electric heating element.

Thus, modern know-how applied to the simple fundamentals discovered by Faraday has resulted in the effective operation of the *absorption system.* This is a system in which the transfer of heat is accomplished without the aid of any moving parts.

Absorption systems are widely used in recreational vehicles, mobile refrigeration vehicles and systems, and in still other systems where alternative energy sources are required.

SAFETY PRECAUTION

Absorption refrigeration units are sealed at the factory. No attempt should be made to service such units because of their contents.

COMPRESSION SYSTEM USING CAPILLARY TUBE CONTROL FOR REFRIGERANT

The capillary tube system is the most common and popular of compression-type systems. This system is used in household refrigerators, freezers, air conditioners, dehumidifiers and small commercial applications.

The design of the capillary tube provides for the maintaining of a pressure difference while the compressor is running. A low pressure is maintained in the evaporator, causing the refrigerant to boil and absorb heat. The low-pressure vapor moves through the suction line back to the compressor. Here it is compressed to a high pressure and discharged into the condenser. In the condenser the high pressure vapor is cooled and returns to a liquid. Liquid refrigerant flows from the condenser through the liquid line to the filter or filter drier. From there the high pressure liquid flows through the capillary tube metering device to the evaporator where the pressure drops and heat is absorbed.

When the required amount of heat has been removed, the *thermal sensor* shuts off the compressor. The refrigeration cycle stops and remains *off* until the thermal sensor requires more heat to be removed.

When the unit is off, the capillary tube allows the pressure to balance between the high and low sides. This method of a *balanced system* usually eliminates using a motor with a high starting torque.

COMPRESSION SYSTEMS USING THERMOSTATICALLY CONTROLLED EXPANSION VALVE (TEV)

As the compressor is running, liquid refrigerant flows from the condenser to the liquid receiver, through the liquid line to the filter drier, and to the thermostatic expansion valve.

The operation of the *thermostatic expansion valve* is controlled by both temperature at the TEV control bulb and the pressure in the evaporator. The temperature of the *TEV control bulb* must be higher than the evaporator refrigerant temperature before the valve will open. How much the needle will open depends on the temperature of the evaporator. The warmer the evaporator, the greater the needle opening, allowing for a rapid flow of refrigerant and more rapid cooling.

The refrigeration cycle is similar to that of the capillary tube system. The major differences are a liquid receiver and a high starting torque motor, which are required in the thermostatic expansion valve system. The TEV system is used for installations ranging from large commercial units to residential units. Newer automotive air-conditioning systems are using restrictor metering devices.

MECHANICAL REFRIGERATION UTILIZES
TEMPERATURES INCREASED BY PRESSURE

The second method by which the refrigerant may be reclaimed for its role of carrying heat is based on another discovery. Even before Faraday's findings, two other scientists, Black and Watt, set out to determine how much heat is absorbed when a substance changes from a liquid into a gas.

The simple experiments which they conducted consisted of noting the effect of heat on water. As the water became hotter and hotter, they noticed that the mercury in the thermometer kept rising until the water started to boil. Then, regardless of the increased amount of heat that was applied, the thermometer never rose above the 212°F mark. What Black and Watt learned was that no matter how hot the applied heat is, the temperature of water never goes beyond 212°F, figure 2-6. More importantly, they proved that as a liquid changes to a gas, it absorbs unusually large amounts of heat without getting any hotter.

Black called this kind of heat the *latent heat of vaporization* because it absorbed heat energy in large quantities without raising the temperature and seemed to disappear. This explains why a boiling refrigerant is so effective as a cooling agent.

SAFETY PRECAUTION

This heat which reappears when the vapor condenses produces serious burns because of the heat released. Great care should be used with boiling water, especially when it is confined in a closed system and under pressure.

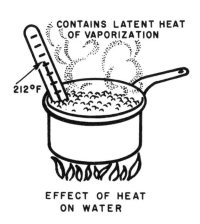

FIG. 2-6 Effect of heat on water

Heat of Vaporization Equals Heat of Condensation

The Black and Watt discoveries were followed by the work of other scientists. These men learned how to change a vapor back into a liquid by chilling it with a mixture of salt and ice. They were able to prove that the heat which was removed equalled, exactly, the amount of heat needed to vaporize the substance at the same pressure. So, the heat that apparently disappeared when a liquid boiled into a vapor reappeared when the vapor changed to a liquid.

Changing a Boiling Point by Applying Pressure

Still other scientists used pressure to convert a vapor back into a liquid. They knew that each substance condensed at the same temperature at which it boiled. Also, this boiling point formed a clear-cut dividing line in determining whether a substance would be a liquid or a vapor.

What they established was that under pressure this dividing line or boiling point could be changed. Because water boiled at 212°F these scientists expected steam to condense at the same temperature. When pressure was applied to steam, however, it condensed at a temperature higher than 212°F. The scientists learned from this fact that the greater the pressure, the higher the temperature at which a vapor condenses (changes from a vapor to a liquid), figure 2–7.

SAFETY PRECAUTION

This illustration is given to explain the relationship between pressure and boiling point. This kind of experimenting should not be attempted because of the dangerous conditions created.

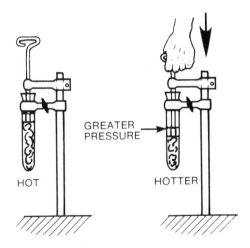

GREATER PRESSURE →

HOT HOTTER

FIG. 2-7 Vapors condense at higher temperatures under increased pressure

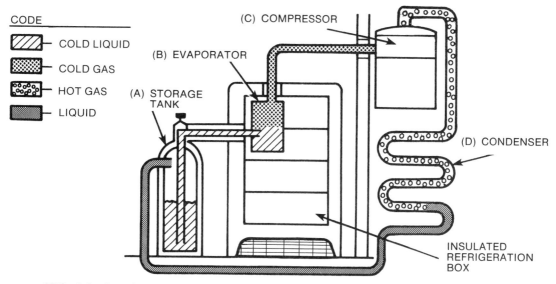

FIG. 2-8 **Simple mechanical refrigeration system** *(Courtesy of The Trane Company)*

Applying Early Principles of Pressure to a Refrigeration System

As elementary as the work of the early scientists seems, the principles which they established are the foundation for modern refrigeration, which depends on pressure to operate. For instance, ammonia boils at –28°F at normal atmospheric pressure. The vapors, even though they have absorbed large amounts of heat, must be made warmer than the temperature of the room air if heat is to flow out of them. Also the condensing point temperature must be higher than the room air, otherwise the vapors will not condense.

The temperature of a vapor can be increased by concentrating or compressing the vapor. This can be done through the use of pressure. Vapor, then, can be made hotter by applying pressure.

It is possible to build a mechanical refrigerator because (1) all liquids absorb a great amount of heat when they boil into a vapor, and (2) pressure may be used to make the vapor condense back into a liquid to be used over and over again.

A simple refrigeration system combining these principles is illustrated in figure 2–8. The liquid ammonia from the storage tank (A) is vaporized and absorbs considerable heat in the cooling unit (B). The compressor (C) takes the vaporized ammonia from the cooling unit and compresses it under a high pressure. The compressed ammonia gas is then forced to the condenser (D). Here it is first cooled to its condensing temperature, the latent heat is removed, and the ammonia now condenses, returning to the storage tank as a liquid.

What the mechanical refrigerator does is simply draw heat away from everything inside the cabinet. Again, heat is moved from a low to a high temperature against its natural tendency. The mechanical energy to do this is provided by the compressor which, in turn, receives its energy from an electric motor.

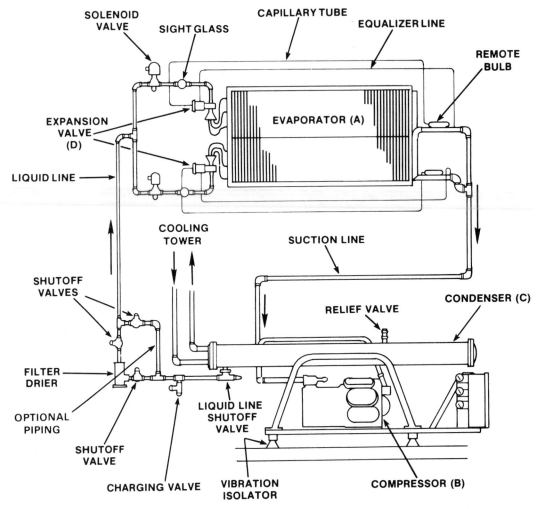

FIG. 2-9 Typical commercial mechanical refrigeration system *(Courtesy of The Trane Company)*

REFRIGERATION CYCLES IN COMMERCIAL AND INDUSTRIAL UNITS

A typical commercial mechanical refrigeration system is illustrated in figure 2–9. This system is made possible by the proper use of temperature, pressure and latent heat of vaporization. Instead of ammonia, other substances having certain advantages related to safety, storage, handling, and noncorrosiveness are widely used.

In the system shown, water is circulated through the condenser. The water removes heat from the hot, compressed refrigerant to condense it. Thus, the water carries away the heat that is picked up by the evaporator as the refrigerant boils. The refrigerant is then recirculated through the system again to carry out its function of picking up heat in the evaporator.

Note that at the top of the drawing an expansion (metering device) valve is placed in the line between the condenser and the evaporator. This expansion valve provides a restriction (reduced sized opening) so that there is a steady flow of refrigerant into the evaporator coil. The compressor maintains the difference of pressure required to change the state of the refrigerant, that is, from a liquid to a gas.

Figure 2–10 shows a refrigeration cycle in simplified terms. Since this is a basic vapor compression cycle, the following facts must be known:

▸ Heat is picked up by the boiling refrigerant regardless of the shape of the evaporator.
▸ Heat which is rejected at the condenser may be carried away by air, water, the evaporation of water, or other means.
▸ The function of an expansion valve or other metering device is to control the flow of liquid refrigerant to the evaporator.
▸ The heart of the system provides the energy for its operation whether it be by mechanical, heat, or other energy.

The different components (main parts) which are identified and described in detail in succeeding units are related to this basic refrigeration cycle.

CLASSIFICATION OF REFRIGERATING SYSTEMS

In general, refrigerating systems fall (according to use) into three groups: (1) high-temperature systems, (2) medium-temperature systems, and (3) low-temperature systems. The high-temperature refrigeration systems are used for air-conditioning equipment where temperatures between 25°F and 45°F are needed. Medium-temperature systems are used for food storage and other applications requiring temperatures between 25°F and 0°F. Low-temperature systems are for those applications requiring temperatures of 0°F or below.

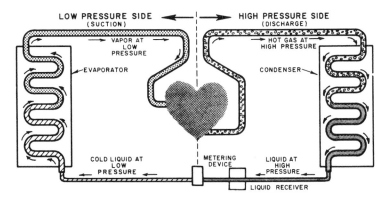

FIG. 2-10 Schematic diagram of refrigeration cycle

Refrigeration and Reverse Cycle Refrigeration Systems

A study of the basic refrigeration cycle reveals that heat is picked up at the evaporator and heat is given up at the condenser. The amount of heat rejected (given up or dissipated) is greater than that which is picked up at the evaporator. This heat that is given up makes it possible to use a refrigeration system for heating. When it is applied in this way, it is called *reverse cycle refrigeration*.

Even in this application each main part of the refrigeration system functions according to its position in the overall refrigeration cycle. The only difference is that the evaporator is located outside and picks up heat from the air or a water well. The heat is then rejected at the condenser which may be located inside in order to heat a definite area. Reverse refrigeration is gaining in popularity because it serves a two-fold purpose, providing either a cooling effect or a heating effect. Modern heat pumps use the principle of reverse cycle refrigeration.

SUMMARY

- Evaporation or vaporization is the process by which a liquid changes into a vapor as a result of absorbing heat.
- A refrigerant is a substance which is circulated in a refrigeration system to transfer heat.
- The condenser is that part of the refrigeration system in which the refrigerant condenses and, in so doing, gives off heat.
- The evaporator is that part of the refrigeration system in which the refrigerant boils and, in so doing, absorbs heat.
- An absorption refrigeration system is one in which the refrigerant, as it is absorbed in another liquid, maintains the pressure difference needed for successful operation of the system.
- The heat of vaporization is the heat required to change one pound of a liquid into its vapor or gaseous form at atmospheric pressure without changing temperature.
- Condensation is the process by which a vapor is changed into a liquid without changing temperature.
- The heat of condensation refers to the heat that is removed per pound of vapor to cause it to condense. It has the same numerical value as the heat of vaporization.
- Refrigeration systems are classified as high (from 45°F to 25°F), medium (from 25°F to 0°F) and low (0°F or lower).
- Reverse cycle refrigeration uses rejected heat to produce warmth.

ASSIGNMENT: UNIT 2 REVIEW AND SELF-TEST

A. REFRIGERATION PRINCIPLES, SYSTEMS, AND COMPONENTS

Circle the letter of the condition or process that correctly completes statements 1 through 10.

1. The boiling of refrigerant can be changed by (a) changing its pressure, (b) using a different heat source, (c) applying more refrigerant, or (d) changing the vaporization.

2. Temperature is used in mechanical refrigeration to (a) promote boiling, (b) increase pressure, (c) create the refrigeration effect, or (d) all of these.

3. Mechanical refrigeration (compression systems) use primarily two types of meter devices: (a) high and low side float, (b) AEV and high side float, (c) AEV and TEV, or (d) TEV and capillary tube.

4. Heat is absorbed in a compression system by the (a) compressor, (b) condenser, (c) liquid receiver, or (d) evaporator.

5. Heat is transferred from the refrigeration system to the surrounding air by the (a) evaporator, (b) metering device, (c) condenser, or (d) compressor.

6. The capillary tube is used (a) for change of state of refrigerant, (b) to carry vapor to the condenser, (c) to meter refrigerant to the evaporator, or (d) for none of these functions.

7. The compressor is sometimes called (a) the heart of the system, (b) the eliminator, (c) the metering device, or (d) the heat remover.

8. When the compression system shuts off, (a) the TEV system will balance, (b) all systems will balance, (c) the capillary system will balance, or (d) no compression system balances.

9. In a compression system the condenser changes the refrigerant from (a) low-pressure liquid to high-pressure vapor, (b) high-pressure liquid to high-pressure vapor, (c) low-pressure vapor to high-pressure vapor, or (d) high-pressure vapor to high-pressure liquid.

10. The compressor changes the low-pressure vapor to high-pressure vapor by (a) adding heat of compression, (b) removing heat from the evaporator, (c) forcing refrigerant to the condenser, or (d) all of these conditions.

Select the correct word or phrase to complete statements 11 through 17.

11. Refrigeration is the process of (adding) (transferring) heat to/from one place to another.

12. Evaporation takes place in refrigeration when there is a change in state from (gaseous to liquid) (gaseous to solid) (liquid to gaseous).

13. Of the following three, the liquid having the greatest cooling effect on an individual is (water) (ether) (rubbing alcohol).

14. At atmospheric pressure, liquid ammonia boils at (40°F) (0°F) (–28°F) (–60°F).

15. Faraday's distilling apparatus was used to test the effect of heating and cooling on a silver chloride powder laden with (ammonia) (freon) (hydrogen) gas.

16. A pressure difference is produced in an absorption type refrigerator by (absorbing) (heating) (cooling) the refrigerant.

17. In any refrigerating system heat is moved from a (low temperature to a higher one) (high temperature to a lower one) by adding energy.

B. REFRIGERATION UTILIZES TEMPERATURES INCREASED BY PRESSURE

Provide a word or phrase to complete statements 1 to 15.

1. Two principal types of refrigeration systems are (a) _____ and (b) _____ .

2. As more and more heat is applied to water at atmospheric pressure, its temperature _____ at 212°F.
3. A liquid, as it changes state to become a gas, _____ heat energy.
4. A vapor may be changed back into a liquid in a refrigerating system by _____ heat.
5. A substance at atmospheric pressure condenses at the _____ at which it boils.
6. Steam may be condensed at a temperature higher than 212°F when _____ is applied.
7. The greater the pressure, the higher _____ at which a vapor condenses.
8. In a simple mechanical refrigerator:
   a. The liquid _____ is vaporized.
   b. The refrigerant _____ heat in the cooling unit.
   c. A compressor compresses the refrigerant in the _____ state under a high pressure.
   d. In the _____ heat is removed from the compressed refrigerant. It then is returned to a storage tank in a _____ state.
9. The energy which moves heat from a low to a high temperature level is supplied by a _____ .
10. Large commercial refrigerating systems circulate _____ in the condenser to remove heat from the compressed refrigerant.
11. The restriction which is included in the line between the condenser and the evaporator is called a _____ .
12. The names of the parts of the mechanical refrigerator which perform the following are:
    a. _____ absorbs heat.
    b. _____ gives off heat.
    c. _____ maintains pressure difference.
13. To condense a vapor _____ heat.
14. To vaporize _____ heat.

C. REFRIGERATION CYCLES AND CLASSIFICATION OF REFRIGERATING SYSTEMS

1. List the three groups into which refrigerating systems fall.
2. Identify the temperature range for which each system may be used.
3. Describe briefly what takes place in the reverse refrigeration cycle.
4. Give two reasons why reverse refrigeration is a practical system.

D. SUMMARY OF REFRIGERATION PROCESSES AND SYSTEMS

1. Match each item in Column I with the correct description given in Column II.

Column I	Column II
a. Condensation	(1) A restriction which permits the compressor to maintain a pressure difference.
b. Heat of vaporization	
c. Evaporation	(2) Part of the system where the refrigerant boils and absorbs heat.
d. Evaporator	
e. Heat of condensation	(3) The heat required to change one pound of a liquid to a gas.
f. Condenser	
g. Metering device	(4) The heat which must be removed to change one pound of a vapor to a liquid at the same temperature.

(5) A refrigerating system producing temperature ranges from 45°F to below 0°F.

(6) Changing a vapor to a liquid at the same temperature.

(7) A change in state resulting when a liquid absorbs heat at any temperature.

(8) Part of the system where the refrigerant is condensed and discharges heat.

2. Secure a schematic diagram of a simple refrigerating system from a manufacturer or a distributor.
   a. Name six important parts of the system on the drawing (if they are not already lettered on the diagram).
   b. Explain in not more than five steps what takes place during a refrigeration cycle.

SECTION 2:
REFRIGERATION TOOLS AND MATERIALS

UNIT 3:
REFRIGERATION TOOLS AND TEST EQUIPMENT

The correct installation, maintenance, operation, and servicing of household, light commercial, and industrial refrigeration and air-conditioning products and systems depend on basic skills in the use of hand tools, test equipment, and materials.

Before starting any job, the refrigeration technician considers four major activities:

- Diagnosing the work assignment and establishing that all required components and materials meet specifications and are available, as needed.
- Selecting refrigeration tools, supplies, and test equipment that are appropriate to the work to be done.
- Following all personal occupational safety precautions and protecting tools and property against damage.
- Cleanliness standards while working on a system and keeping the work dry.

REFRIGERATION HAND TOOLS

Since personal injury, work failures, and damaged equipment result from improper tool usage, it is important to understand design features and limits within which hand tools may be used safely. Such tools as wrenches, screwdrivers, pliers, and cutting and forming tools are generally made of alloyed steels that are heat treated for added hardness and strength. However, surfaces and cutting edges on some of these tools deform, shatter, or dull unnecessarily when forces in excess of the tool capacity are applied.

Consideration must be given to the quality of the material, tool size, and shape appropriate to the needs of the job and the amount of force that may be safely applied.

CATEGORIES OF TOOLS, EQUIPMENT, AND TEST INSTRUMENTS

Tools and testing equipment may be classified under three broad categories:

- Basic hand tools, small power tools, and measuring tools. This category includes tools for securing parts and fasteners (wrenches); for bending, holding, and stripping (pliers and crimpers); hacksaws for cutting through metals, plastics, and other hard materials; files for metal removal; turning tools (screwdrivers); hammers for driving and forming surfaces; punches, drift pins, and chisels; twist drills for hole forming; thread-cutting tools, and sheet-metal cutting snips; and rules and micrometers for linear measurements.

- Basic equipment. These items are designed for charging, purging, and servicing air-conditioning systems.
- Testing and servicing instruments for use on different compressors, leak detectors, systems analyzers, vacuum pumps, and digital thermometers/pyrometers for measuring air, surface, and liquid temperatures in fahrenheit and celsius degrees.

Design features, principles, applications, and safety considerations for using hand tools are treated in this unit. Thermometers, high-pressure gauges, vacuum gauges, and compound gauges are introduced. Additional equipment and instruments are covered in later units in direct reference to specific units and components of heating, refrigeration, and air-conditioning systems.

TYPES, FEATURES, AND APPLICATIONS OF WRENCHES

The service technician uses many different types of wrenches. Some are common to a number of trade and technical occupations. Other types and sizes are specially adapted to heating, refrigeration, and air conditioning. In general, all wrenches have the following qualities and design features:

- Wrenches are made of alloy steels for toughness and resistance to becoming deformed under heavy loads.
- Wrenches are machined accurately for size and shape to receive standard and metric nuts, bolts, fittings, and other parts.
- The shape of the opening of a wrench fits over a maximum surface area of a fastener or part in order to provide the greatest gripping area, thus preventing rounding over or the wrench from slipping as force is applied.

Wrenches are one of the prime tools for refrigeration and air-conditioning work. Wrenches that retain an accurate shape and size when used properly are in the long run the least expensive and safest to use. Good quality wrenches are made of chrome-vanadium steels, are accurately machined to shape and size, and are heat treated. The outer surfaces may be plated or coated to resist rusting.

The service technician needs to know the special design features and how and when to use each of the following types of wrenches:

- Box, open end, and flare nut
- Socket (or socket set)
- Spline
- Ratcheting
- Torque
- Pipe
- Adjustable open end
- Allen (hexagon) set screw
- Service valve
- Spanner

Box Wrenches

Box wrenches are used in close quarters where it is impractical to use any other type. Box wrenches generally have two heads (double ended) that may be the same or two

FIG. 3-1   Fractional size, offset, double end, 12-point box wrench *(Courtesy of S-K Hand Tool Corporation)*

different sizes. The body may be flat or offset. Wrench sizes are stamped on each end. Figure 3-1 identifies a 12-point, double end, offset box wrench.

Bolt head dimensions (for which a wrench size must be selected) are given as the width across flats. On bolt sizes from 1/4″ to 7/16″, the wrench opening is 3/16″ larger than the bolt diameter. Bolt sizes in the 1/2″ to 3/4″ diameter range (in increments of 1/16″) require wrench openings that are 1/4″ larger than the bolt diameter.

Open-End Wrenches

Open-end wrenches, figure 3-2A, are designed so the end faces slide easily over the sides of screw, nut, bolt heads, and other parts that have parallel flat surfaces. For safe operation and to prevent damage to parts (due to wrench slippage), attention must be paid to ensure a correct fit between the wrench faces and part as illustrated.

Flare Nut Wrenches

This wrench derives its name from its use. Flare nut wrenches, figure 3-2B, are used to loosen or tighten flare nuts on fittings that are connected to tubing. The jaw opening permits the wrench to be slid past the tubing to fit over a flare nut.

The quick-opening/closing flare nut wrench design shown in figure 3-2C permits this wrench to slip easily over tubing and onto fittings, valves, and fasteners. The jaw is then adjusted quickly to fit accurately and to grip all the surfaces of a hexagonal flare nut. The 12-point contact ensures maximum contact and distribution of torque forces on the flat hexagonal surfaces and not on the corners. Another feature of this wrench is that it is adaptable for operations in limited quarters.

Ratchet Flare Nut Wrench

This wrench design is practical when there is limited space in which the handle may be moved, figure 3-3. Sockets are available to fit all standard sizes of nuts used in refrigeration work.

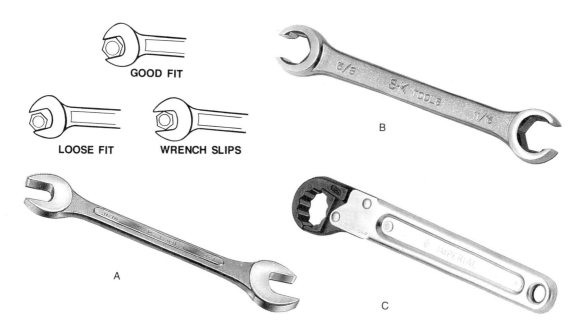

FIG. 3-2 Open-end and flare nut wrench designs. (A) Open-end wrench *(Courtesy of S-K Hand Tool Corporation).* **(B) Fractional size, double end, 6-point flare nut wrench** *(Courtesy of S-K Hand Tool Corporation).* **(C) Quick-adjusting flare nut wrench** *(Courtesy of Imperial-Eastman Corporation)*

Torque Wrenches

Torque wrenches, figure 3-4, are used whenever it is necessary to tighten nuts, bolts, fittings, and other parts to a particular degree of tightness. The torque wrench makes it possible to measure the force and helps prevent distortion and misalignment of parts.

The display model has a range of 40 lb. ft. to 250 lb. ft. in increments of 5 lb. ft. The force arm distance is 21″. This wrench can be angled up to 15° to provide clearance from obstructions. Comparable metric reading torque wrenches are graduated in kg. m. (kilograms per meter).

Torque wrenches are designed so that the force applied on the wrench handle is transmitted through the socket to a bolt or nut. The amount of torque may be read directly on a scale or measurement indicator on the handle, figure 3-4.

FIG. 3-3 Ratchet-type, flare nut wrench

FIG. 3-4 Torque wrench: fixed-ratchet head model (40 lb. ft. to 250 lb. ft. range)
(Courtesy of Snap-on Tools Corporation)

Torque is usually expressed in foot-pounds, inch-pounds, or as kilograms per meter in metric. The recommended *torque* (twisting force) that should be applied to tighten various parts and for different materials is found in service manuals.

The torque is computed by multiplying the pull exerted against the handle by the length of the handle (called the *force arm*), figure 3-5. The arm may be measured in inches, feet, or meters from the point at which the force is applied to the point at which it is transmitted.

On some designs, a light automatically flashes as the desired force is reached. In other models, a slip-clutch feature operates when a predetermined torque is reached and no greater torque may be applied.

Open-End Adjustable Wrenches

These wrenches (commonly referred to as Crescent wrenches) slide on a bolt or nut from the side and are used in close quarters where a socket wrench cannot be placed in position. The adjustable models, figure 3-6, permit setting the jaws to accommodate many different sizes using just the one wrench. It is important in using these wrenches to adjust the jaws to fit the work part as snugly as possible. Adjustable wrenches are designed in sets so that the larger sizes are heavier and stronger, consistent with the torque requirements.

Great care must be exercised in applying force to ensure that the movable jaw does not spring.

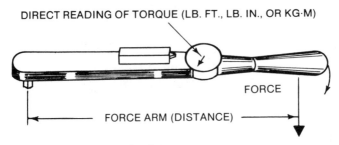

FIG. 3-5 Principle of torque measurement

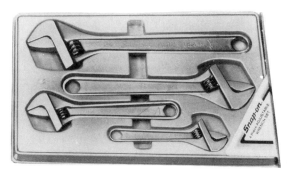

FIG. 3-6   Adjustable open-end wrench set *(Courtesy of Snap-on Tools Corporation)*

Socket Wrenches

There are four common forms of socket wrench heads: hexagon, six-point, twelve-point, and spline, figure 3-7. Sockets may be built solid, figure 3-8A, or *swivel* (B). The driving end may accommodate a *sliding T-handle* (C) or a *ratchet handle* (D). T-handles are made with 1/4", 3/8", and 1/2" square drives to correspond with similar openings in solid or swivel socket heads. The spline design in many sockets accommodates the same fastener head shapes as hexagon, 6-point, and 12-point sockets.

Ratchet Handles

Ratchet handles are used with many different types and sizes of socket wrenches, figure 3-9. A ratchet within the head makes it possible to apply force in one direction for any fractional part of a revolution. As the handle is returned to a starting position from which force may again be applied, an arm in the head slides over a toothed ratchet. This makes it possible to return the handle to a starting position without moving the socket.

The direction of ratcheting may also be changed in case a force must be applied to loosen a bolt or nut. Thus, the ratchet handle is convenient for either loosening or tightening operations.

Allen Set-Screw Wrenches

Hollow set-screw (Allen) wrenches are hexagonal in cross section and designed with two legs at a right angle. Allen set-screw wrenches are of short length, usually the width of a hub on a pulley, gear, or other equipment part. Generally, when fully tightened, the

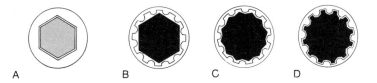

FIG. 3-7   Common forms of socket wrench heads: (A) hexagon, (B) 6 point, (C) 12 point, and (D) spline (also service 6- and 12-point fasteners)

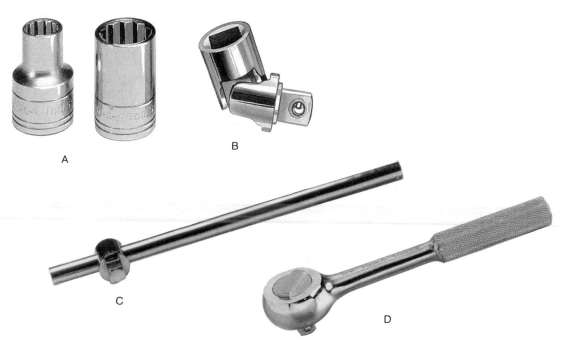

FIG. 3-8   Basic socket head and handle designs: (A) solid, (B) swivel (universal), (C) sliding T-handle, and (D) ratchet handle *(Courtesy of S-K Hand Tool Corporation)*

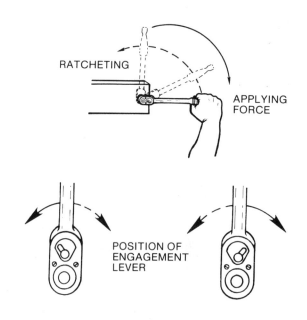

RATCHETING

APPLYING FORCE

POSITION OF
ENGAGEMENT
LEVER

FIG. 3-9   Action of ratchet handle

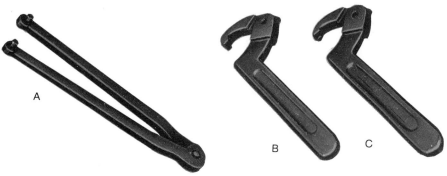

FIG. 3-10 **Adjustable spanner wrenches: (A) face pin, (B) hook, and (C) pin** *(Courtesy of Snap-on Tools Corporation)*

slightly crowned screw head is flush with or slightly below the surface of the threaded part as a safety precaution.

Allen set screws are used to secure parts and prevent them from turning. The wrenches are available singly or in sets covering a range of sizes. The size indicates the distance (measurement) across the flats of the hexagon screw opening.

Adjustable Spanner Wrenches

Adjustable spanner wrenches, figure 3-10, are designated as *hook spanner, face pin spanner,* and *pin spanner.* Hook and pin spanner wrenches are of single (adjustable) pin and roll pin construction. Both types are commercially available in sets that range from 3/4" to 6 1/4" capacity.

Face pin spanners, designed for use in servicing seals and packing nuts on hydraulic cylinders and other equipment, have a 2" to 4" capacity.

Adjustable hook spanners are used to service bearings, rings, and collars. The *lock pin* type is used to lock nuts, collars, and rings. The leg or pin fits into a hole or slot on the outer surface (periphery) of a round part.

Pipe Wrenches

Pipe wrenches are designed for gripping a pipe from the inside (internal pipe wrench as shown in figure 3-11A) or around the outside, figure 3-11B. *Internal pipe wrenches*

FIG. 3-11 **Pipe wrenches: (A) internal pipe wrench and (B) external pipe wrench** *(Courtesy of Snap-on Tools Corporation)*

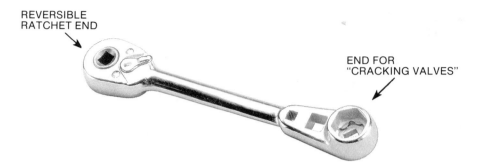

REVERSIBLE
RATCHET END

END FOR
"CRACKING VALVES"

FIG. 3-12   Refrigeration/air-conditioning service valve wrench *(Courtesy of Imperial-Eastman Corporation)*

are available to fit national pipe sizes from 3/8″ to 1″ diameter. *External pipe wrenches* have jaw capacities from 3/4″ to heavy-duty sizes up to 6″.

Refrigeration Service Valve Wrenches

A special wrench known as a *service valve wrench* is required to turn a valve stem. One end of the wrench, figure 3-12, is designed with a ratchet. The other *fixed end* has 3/16″, 1/4″, and 5/16″ square openings, 6- or 12-point socket openings, and accommodates the valve stem.

Valves are *cracked* with a slight opening. The opening allows the plunger (valve needle) to become unseated, permitting a very small quantity of refrigerant to flow. Using the fixed end, the service technician is able to quickly control the cracking and closing of a valve.

The ratchet end is used for rapidly opening and closing a valve stem. Packing gland-nut and valve-nut sockets are driven by the square drive. Reversible ratchets permit reversing the turning direction without removing the wrench from the valve stem.

Strap Wrenches

A high-tensile urethane coated *strap wrench*, figure 3-13, is used on nonmetallic plastic pipe fittings and other irregular-shaped parts.

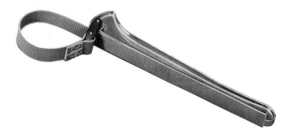

FIG. 3-13   Strap wrench *(Courtesy of Snap-on Tools Corporation)*

SAFETY PRECAUTIONS

- Select the design of wrench that conforms to the shape and size of the surface or part that is to be tightened or loosened. As force is applied, an incorrect shape or size may cause slippage and spoil the corners of the work part. Rounded corners make it difficult to properly apply the force needed to tighten or loosen the parts.
- Exert a *pulling force*. Pulling (as contrasted with pushing) helps to prevent injury if the bolt, nut, or workpiece loosens suddenly.
- Work within the torque or turning power of the wrench. Using an additional length of pipe or hammering to produce added force should be avoided. In addition to being a personal safety hazard, the wrench head and/or the part may become distorted or cracked.
- Use penetrating oil to soak the threads of hard-to-remove bolts and nuts. If conditions are safe, a torch may be used on corroded fittings and overly tight screw fasteners. A third technique is to tap the bolt or nut lightly to remove corrosion and loosen the parts.
- As force is applied, the teeth on pipe wrenches cut into the outside surface of the pipe. It is important that pipe wrenches be used on the pipes they are designed for and not on bolts and nuts where the flat surfaces may be damaged.
- Adjust the jaw openings to fit as closely as possible to the pipe size before applying force. This will permit good gripping and lessen any possibility of slippage.

PLIERS AND STRIPPERS

Pliers are used to grip, turn, bend, and cut parts, and for stripping insulation from wires. Pliers are formed with serrated jaws, which provide a good gripping surface. Some pliers are built solidly to provide a limited opening between the jaws. Adjustable joint models permit parallel jaw openings over a number of different settings. Pliers that are commonly used by the service technician are displayed in figure 3-14.

Cutting and Crimping Pliers

The multipurpose electrician's pliers are designed to handle wire-stripping, cutting, crimping, and splicing operations, figure 3-15. These pliers are used to crimp electrical insulated and noninsulated terminals. There are a number of openings that accommodate wire sizes from #10 to #22 and metric diameters from 1.3 mm to 8.0 mm.

HAMMERS

The service technician uses three basic head designs: *soft face, ball peen,* and *claw hammers,* figure 3-16. The *soft face hammer* is used to strike a solid blow without

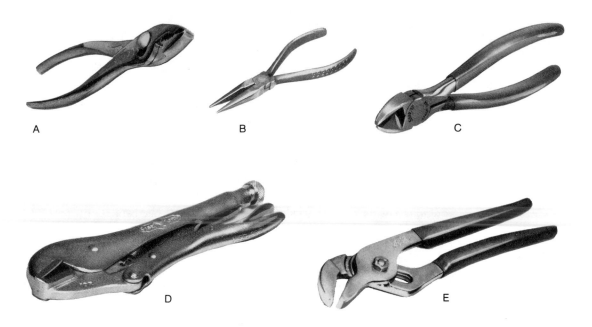

FIG. 3-14 **Basic types of pliers: (A) general purpose (combination) pliers, (B) needle nose pliers, (C) side (diagonal) cutting pliers, (D) vise grip, and (E) adjustable (slip-joint) pliers** *(Courtesy of Snap-on Tools Corporation)*

damage to the object. Soft face hammers may have plastic, bronze, or other soft metal faces that yield when an object is struck.

Hammer heads are flat shaped for soft face hammers, figure 3-16A, crown faced and ball shaped for ball peen hammers, figure 3-16B, for working with metals, and claw/flat faced, figure 3-16C, for use with building-construction materials. Hammer handles are made of hickory, wood, metal, urethane, fiberglass, and other plastics. Handles are contoured to provide a comfortable grip, ease in striking a blow, and safety in handling.

A sheet metal worker's hammer may be included as a fourth type of hammer. It is used to form edges on duct work and for other sheet metal processes. One end of the head is square and crowned (with the edges beveled). The other end consists of a long, tapered section with a rounded edge (at a right angle to the handle).

FIG. 3-15 **Stripping, cutting, and crimping pliers** *(Courtesy of Snap-on Tools Corporation)*

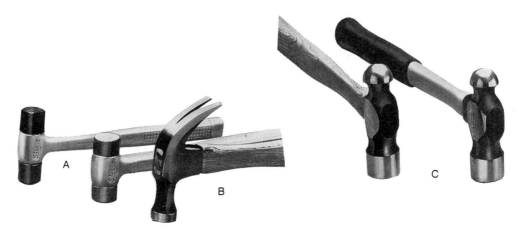

FIG. 3-16 Common types of hammers: **(A)** soft face, **(B)** claw, and **(C)** ball peen *(Courtesy of Snap-on Tools Corporation)*

SAFETY PRECAUTIONS

The following are some of the personal and work safety precautions that must be observed when using hammers:

- Use sanitary safety eye protection whenever there is any possibility of producing flying chip particles.
- Strike solidly and squarely to avoid glancing blows.
- Avoid striking the hardened face of a hammer against the hardened face of another tool or object.
- Discard a hammer that chips away or is cracked.
- Drive pins with the pin punch in direct line with the pin that is to be removed.
- Dulled chisel edges require grinding to restore them to their original shape and correct cutting angles.
- Remove (grind away) any mushrooming that may form around the head end of a chisel.

CHISELS AND PUNCHES

Chisels and *punches* are made of tough alloy steels that when hardened and tempered have excellent qualities to resist dulling or becoming deformed by heavy blows. The *cutting edges* are hardened to provide for good cutting action. The anvil or opposite end is drawn to a lesser degree of hardness to sustain the striking force of repeated hammer blows. The anvil end is usually ground to a slight crown. This design directs the hammer blows from the center of the tool to the cutting edge. A ground rim at an angle to the crown prevents the anvil (head) from mushrooming.

Chisels and punches range in size and shape of point or cutting edge depending upon the amount of material to be removed and the form of the required cuts. Sets of punches and chisels such as those shown in figure 3-17 provide for most service technician's needs.

TWIST DRILLS AND BITS

Twist drills and *bits* provide a functional, practical method for forming holes. Drills/bits are made from carbon, high-speed, and other alloy steels and are hardened and tempered. High-speed steel drills are preferred as all-round drills due to their ability to resist shock, excellent wear life, and ability to sustain good cutting edges at higher than normal drilling temperatures.

Twist drills, figure 3-18, are available in two common lengths called *jobber* and *mechanic's length*. The mechanic's length is preferred for its greater rigidity and strength. Twist drills have a *shank*. The straight shank is preferred in sizes up to 1/2", as these may be held and driven by a three-jaw chuck. The smaller sizes may be used to drill "by hand" using a hand drill; larger sizes generally require the use of a power drill.

There are three standard drill size systems: *fractional, number,* and *letter.* Common *fractional drill sizes* range from 1/16" diameter to 1/2" diameter in increments of 1/64". Larger sizes are available. *Number drill sizes* begin with #1 (0.228") and decrease in size to #80 (0.0135"). The most common range is from #1 through #60. *Letter drill sizes* increase in diameter from letter A (0.234", almost 1/4") to letter Z (0.413", nearly 7/16").

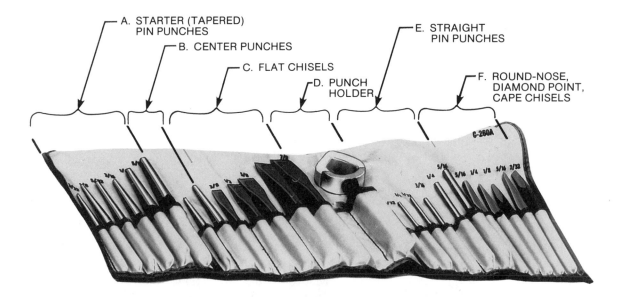

FIG. 3-17   Chisel and punch set (A) starter (tapered) pin punches, (B) center punches, (C) flat chisels, (D) punch holder, (E) straight pin punches, and (F) round-nose, diamond point, cape chisels (*Courtesy of Snap-on Tools Corporation*)

FIG. 3-18   Mechanic's length drill set *(Courtesy of Snap-on Tools Corporation)*

Number and letter drills are designed for drilling special sizes of holes as required for threading parts. Charts are provided to identify the size of the tap drill that will provide sufficient material to cut a thread to proper depth.

Cutting Edges and Point

Drills are ground at the point end to form *cutting edges* and a *chisel cutting point.* Figure 3-19 identifies these parts and cutting and rake angles to which the drill point is ground for cutting steel and cast iron. The drill point angle is 118°; the chisel point angle ranges from 120° to 135°; and the average clearance angle, from 8° to 12°. The clearance angle is increased for small diameter drills; for example, a 1/8″ diameter drill requires a 16° clearance angle. The drill flutes serve to form the lips for the cutting edges and point and to provide a channel through which chips flow out of and away from the work.

Drill sizes are stamped on the shank. Cutting speeds and cutting angles depend on the kind of material being drilled, hole diameter, and drilling method. A cutting fluid is used

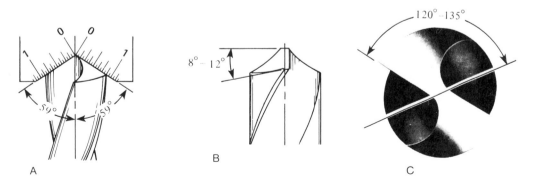

FIG. 3-19   Correctly ground cutting and clearance angles: (A) drill point angle, (B) cutting edge clearance angle, and (C) chisel point angle

37

with metal drilling to reduce cutting friction, aid in chip removal, and produce a finer machined hole. Manufacturers provide tables of recommended drill speeds and cutting fluids.

SAFETY PRECAUTIONS

- Use a three-prong ground socket with an electric drill. If one is not available, use a grounded adaptor. Electric drills require grounding for safety.
- Check the hole size before drilling too deeply into a workpiece. An oversized hole indicates that the drill lips are ground to different lengths.
- Incorrectly ground drills with too small a rake (clearance) angle produce friction and excessive heat from rubbing. If ground at too steep a rake angle, the cutting lips tend to dig (thread) into a workpiece.
- Safety eye protection is required against flying chips.

HAND HACKSAWS AND SAWING

When heavy gauge or thick metal sections are to be cut, a standard frame hacksaw and a correct pitch hacksaw blade are used. The blade length and the set of the saw teeth are selected according to the thickness to be cut, the kind of material in the part, and the nature of the cutting. In general, 10" to 12" length blades of fine pitch (32 and 24 teeth per inch) are used for cutting thin gauge metal parts. In principle, more than two teeth are in contact with the section being cut at all times. Coarser pitch saw blades (8, 10, 12, and 14 pitch) are designed for cutting through thick sections.

Some hacksaw blades are marked on the side for size. For example, a blade with the markings $12 \times \frac{1}{2} \times .025 - 24\,T$ means the blade is 12" long, 1/2" wide, 0.025" thick, and has 24 teeth per inch (pitch).

The *hacksaw frame* shown in figure 3-20A provides excellent blade tension control, accommodates 10" to 12" length blades, and may be positioned for normal straight (flush) cutting, 45° angle cutting, and horizontal cutting. A *low clearance hacksaw frame*, figure 3-20B, permits cutting in confined spaces where cuts must be taken through narrow, limited height openings.

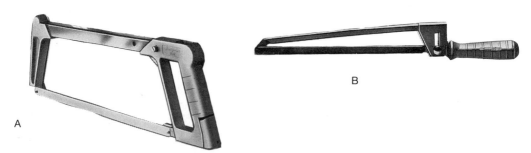

A   B

FIG. 3-20   Examples of hacksaw frames: (A) standard and (B) low clearance *(Courtesy of Snap-on Tools Corporation)*

FIG. 3-21   Four general-purpose files and file shapes *(Courtesy of Snap-on Tools Corporation)*

FILES AND FILING

Filing is often required to remove burrs from surfaces and ends of workpieces and to file away limited amounts of material so mating parts fit together. Basically, files are designated according to the shape of the cutting teeth, the cross-sectional form, and the length. The tang end is tapered to receive and to secure a comfortably fitting file handle. Fine, needle file sets, commonly called *Swiss files*, have round handles that are knurled for gripping.

The *bastard cut file* is a common purpose file for fast metal removal. 10″ and 12″ length files are popular. Finer single-cut files, known as *mill files*, are used for fine-finish filing and burring operations.

Flat, half-round, three-square (triangular), square, and *round files* are the most widely used shapes. Four of these file shapes are illustrated in figure 3-21.

THREAD-CUTTING TOOLS

Parts are secured on a great number of air conditioning parts by bolts and screws that fit accurately in threaded holes. Holes are threaded *(tapped)* by turning a hardened formed cutting tool, called a *tap*, into a particular size hole *(tap drill hole)*.

Outside threads are formed around the periphery (circumference) of a cylindrical surface. The cutting tool for producing external threads on the job is called a *threading die*, or just plain *die*. A *tap wrench* is used for holding and turning a tap by hand. Threading dies for external threads are mounted in and turned by a *die stock (die holder)*.

There are two common die stock and die holder designs. Each design permits the thread cutting edges of a die to be adjusted for minor variations in screw sizes at the workplace. In one model, figure 3-22A, the die is split on one side. Two adjusting screws in the die stock are used to apply force to slightly close the die. The third screw fits the split or groove to expand the die to the required size.

A second design, figure 3-22B, accommodates two adjustable die sections (halves). A *guide plate* under the die sections serves to correctly position the starting cutting teeth

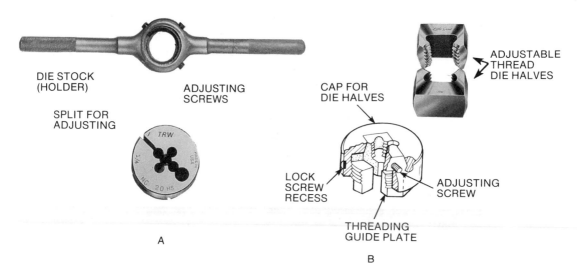

DIE STOCK
(HOLDER)

SPLIT FOR
ADJUSTING

ADJUSTING
SCREWS

CAP FOR
DIE HALVES

ADJUSTABLE
THREAD
DIE HALVES

LOCK
SCREW
RECESS

ADJUSTING
SCREW

THREADING
GUIDE PLATE

A

B

**FIG. 3-22   Common types of thread dies and split die stock: (A) split, adjustable die and three-section
round die stock and (B) dies and holding cap** *(Courtesy of Greenfield Tap and Die Division, Greenfield
Industries, Inc.)*

on the die over the workpiece in order to produce straight, accurately formed threads.
On coarse threads that require accurate cutting to depth and a good quality of threads,
the die sections are first set to cut shallow threads. The die is reversed and removed, and
then the sections are reset to match the required thread size. Like taps, the direction in
which the die is being turned to thread a part is reversed to "break the chips" every few
turns. A cutting oil is applied at the cutting edges.

National Standard Taper (NPT) Pipe Threads			
Nominal Pipe Size	**Threads per Inch**	**Outside Diameter**	**Tap Drill Size**
1/8	27	0.405	11/32
1/4	18	0.504	7/16
3/8	18	0.675	37/64
1/2	14	0.840	23/32
3/4	14	1.050	59/64
1	11 1/2	1.315	1 5/32
1 1/4	11 1/2	1.660	1 1/2
1 1/2	11 1/2	1.990	1 47/64
2	11 1/2	2.375	2 7/32

FIG. 3-23   (Partial) tap drill size table

Tap and Die Sets

Taps and dies are made of carbon steel, high-speed steel, and alloy steels. Taps and dies are designed for hand cutting or machine cutting of threads. While the straight flute tap is generally used on the job, spiral flutes and specially formed cutting faces are ground for use in other applications. Thread sizes and 60° included thread angle conform to American National and Metric Standards. While some thread sizes correspond and are interchangeable, care must be taken to distinguish between the two systems to ensure proper fit and use.

TABLES OF TAP DRILL SIZES

Tables of *tap drill sizes* for the required percent depth of thread are available. These tables give the correct fractional, number, or letter size drill to use for a particular thread size. Usually, threads that are cut to 75% of thread depth are satisfactory. When cut to this percent depth, a bolt or nut will fail from excess force before the threads fail. A portion of a tap drill size table is included in figure 3–23.

STANDARD 60° THREAD FORM

Each thread has a particular *thread form (profile)* and is designed for certain applications. As stated earlier, the 60° included angle thread form is widely used in heating, refrigeration, and air conditioning parts and equipment. Threads that are cut parallel to either the inside diameter or outside diameter are known as *straight (parallel) threads*. Screw threads that are cut on a cone-shaped surface are called *internal or external pipe threads* when they are included as part of a piping system or connection.

Thread sizes are standardized and fall into a number of *thread series*. The *coarse-thread series* and the *Unified series* are designated **NC** and **UNC**, respectively, according to American National Standards. Machine screws and nuts in the **NC** series are commonly used. When a greater number of threads per inch or threads of lesser depth are required for finer adjustment, thread sizes in the *American National fine* **(NF)** and *Unified fine* **(UNF)** series are selected. Similarly, *ISO Metric coarse* and *ISO Metric fine series threads* are used.

THREAD DESIGNATION ON DRAWINGS

Drawing notations, such as $NC \frac{1}{2} - 13$ identify the outside diameter and pitch in the American National System. Metric threads are specified in a different manner. A drawing dimensioned, for example, as $M8 - 1.25$ means that the outside thread diameter is $8\,mm$ and the tap drill size is $8mm - 1.25mm$ or $6.75\,mm$ diameter.

FEATURES OF TAP SETS

A *tap set* for a particular size, usually above 1/4″, consists of three taps: *taper, plug,* and *bottoming*, figure 3–24. A *taper tap* or *starting tap* is made so that the first five to seven teeth are ground taper. This taper permits the tap to fit into a tap drill hole, and as the tap is turned, each tooth cuts away a small amount of material. Less force is required to thread a hole, as the cutting action is distributed among the tapered tap threads.

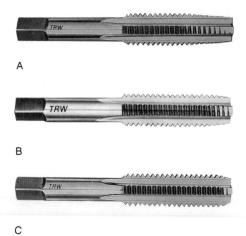

FIG. 3-24 **Set of taps: (A) taper tap, (B) plug tap, and (C) bottoming tap** *(Courtesy of Greenfield Tap and Die Division, Greenfield Industries, Inc.)*

The second tap *(plug tap)* is ground with the first few threads tapered. This tap is used to cut threads to full depth through a part or to a given depth when the last couple of threads need not be cut to full depth.

The third tap *(bottoming tap)* is designed with only the first one to one-and-one-half threads tapered. This tap is used in blind hole tapping to cut threads to full depth as far as is practical. Extreme care must be used in blind hole tapping to prevent jarring the tap at depth. Taps are hardened and are liable to fracture when subjected to sudden and sharp impact.

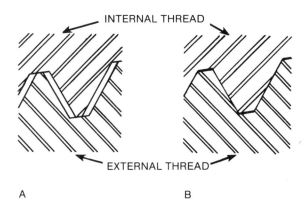

FIG. 3-25 **Principles of forming a Dryseal pressure-tight connection: (A) initial hand-tight contact and (B) pressure-sealed wrench tight**

PIPE TAPS AND THREAD FORMS

The three common 60° angle American National thread forms includes: (1) *standard taper pipe threads*, (2) *standard straight pipe threads*, and (3) *dryseal pipe threads*. Dryseal pipe threads are used when a *pressure-tight seal* is required. The seal is produced by cutting the threads, with the flat of the crest of one thread slightly smaller than the flat on the root of corresponding threads on a mating part. A pressure-tight seal is formed as the flats and roots are sealed by wrench-tightening the parts. This principle is illustrated in figure 3-25.

The taper on American National form taper pipe threads is 3/4" per foot. A gas-proof or liquid-proof connection is made by applying a pipe compound sealer or a thread seal tape a few threads in from the end of the male threads. This prevents the compound from entering the pipe. As the female threaded part is turned onto the male threads, the connection becomes leakproof.

SAFETY PRECAUTIONS

Personal and on-the-job safety practices that must be carefully observed follow:

- Break the chips after each two or three turns in order to clear them from the tap or die cutting edges and to reduce the cutting force necessary to thread a part.
- Continue to cut a thread until the end of the threaded part is flush with the top face of a cutting die. This indicates that there are an adequate number of threads cut to ensure a good threaded joint.
- Apply a cutting fluid liberally throughout the cutting process.
- Thread cutting produces sharp burred edges on the end of the workpiece and the thread crests. Avoid brushing over a thread or handling. Use a brush or other burr-removing tool to make the workpiece burr-free, if required.

SHEET METAL CUTTING

Metal cutting shears (tinner's snips) are used to cut straight, circular, arc, or other irregular patterns by hand. Shears are especially adapted to duct and other sheet metal work. The cutting blades are formed as *straight cut shears, right cut shears*, and *left cut shears*, figure 3-26A.

Aviation snips, figure 3-26B, have offset blades and are easily manipulated for cutting angles and other configurations. The sheared material flows beneath the worker's hand without curling. Aviation snips are designed to cut up to the equivalent of 20 gauge mild steel.

SCREWDRIVERS: TYPES AND FUNCTIONS

Screwdrivers are designed primarily to loosen or tighten screws and other types of fasteners. The kind of screwdriver to use depends on the nature of the fastening

FIG. 3-26   Common sheet metal cutting shears: (A) standard tinner's snips and (B) aviation snips (right-hand) offset blade *(Courtesy of Snap-on Tools Corporation)*

operation and the particular shape of the slot or indentation in the screw head.

A screwdriver should be made of high-quality steel with a hardened and tempered tip capable of withstanding maximum twisting forces and wear. The condition of the tip affects the appearance and condition of a screw head after force has been applied and also the fit of the tip in the screw head, figure 3-27.

A screwdriver is generally designed with a round blade or a square blade (shank). The square shank is useful on larger sizes, where it may be necessary to use a wrench on the shank to produce a greater torque. The blades are nickel/chrome plated or otherwise coated to resist rust and clean easily. Some screwdrivers have a permanent magnet anchored in the shank or have the point magnetized. This convenient feature prevents small screws from dropping off and helps to position screws for threading into a part.

Screwdriver handles are made of tough plastics, wood, and metal. Handles are made in square, triangular, and rounded, flat shapes, and are grooved in many cases for better gripping. *Screwdriver tips* are formed flat for cross-slotted screws or to specific shapes

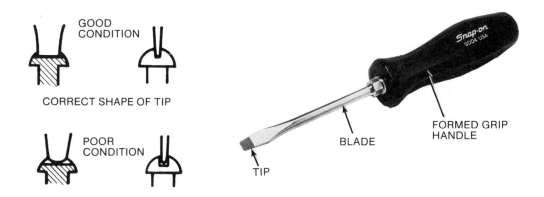

FIG. 3-27   Parts of a screwdriver and importance of correctly formed tip *(Courtesy of Snap-on Tools Corporation)*

FLAT   PHILLIPS   REED
AND
PRINCE

REMOVABLE BITS

FIG. 3-28 Basic forms of screw head slots and recesses

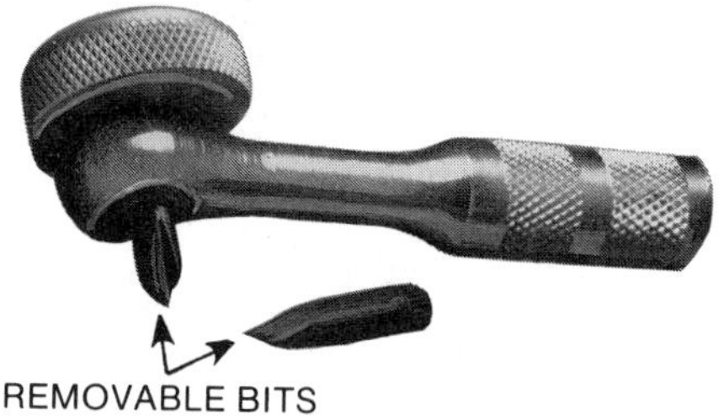

FIG. 3-29 Right-angle ratcheting screwdriver *(Courtesy of Snap-on Tools Corporation)*

to fit screws with Phillips, Reed and Prince, and other recessed indentations in the screw head, figure 3-28.

Offset Screwdrivers

Offset screwdrivers are used when there is an obstruction that prevents the use of a standard screwdriver. Offset screwdrivers may be solid and offset or of right-angle ratcheting design. In the solid offset type, the ends are bent to an angle of 90° to the body. The tip on one end is ground parallel to the body; the second end, at a right angle.

Ratcheting screwdrivers, figure 3-29, provide for the use of removable bits of different sizes. A bit is inserted in the ratchet handle, which turns (in the model illustrated) in one direction. For turning in the reverse direction, the removable bit is inserted in the opposite side of the handle bearing. Pressure (force) is exerted on the knurled knob to firmly seat the bit in the screw slot. Other types of bits are commercially available for use with power drivers.

POWER TOOLS

Air and power electric tools are popular for driving tools that require high operating speeds and/or torque. Power tools are adaptable to operations requiring the use of wrenches, ratchets for rapidly spinning fasteners on and off for assembly work, cutting and nibbling, hammering and chiseling, drilling, grinding, and other applications.

A pneumatic air drill is illustrated in figure 3-30A. An air power impact wrench, designed to accommodate many different accessories, is shown in figure 3-30B.

Most power tools have a variable speed trigger, a built-in adjustable torque regulator, a safety disengagement feature, and a forward and/or forward and reverse control.

LINEAR MEASURING TOOLS

Steel rules (9" and 12") and flexible steel measuring tapes are everyday measuring tools. Graduations of 1/32" on steel rules and 1/16" on steel tapes permit measurements to within tolerance on many jobs. With these graduations, the craftsperson is able to estimate measurements within 1/64" and 1/32", respectively.

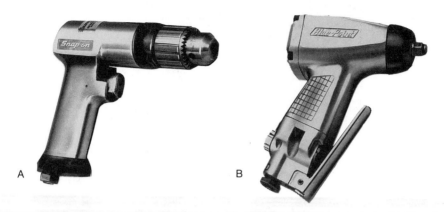

FIG. 3-30 **Air power drill and impact wrench: (A) air power drill and (B) air power impact wrench** *(Courtesy of Snap-on Tools Corporation)*

PRECISION LINEAR MEASUREMENTS

More accurate linear measurements to 0.001″ and 0.0001″ are made with standard *micrometers* and vernier calipers. Metric micrometers are calibrated to read to accuracies of 0.5 mm and 0.01 mm. A standard 0-1 micrometer with part names is illustrated in figure 3-31.

Precision Measurement with a Micrometer

The part to be measured is carefully calipered by positioning the anvil against the workpiece so that the spindle can be adjusted to fit over the part. The thimble is then carefully turned until the spindle barely grazes the part.

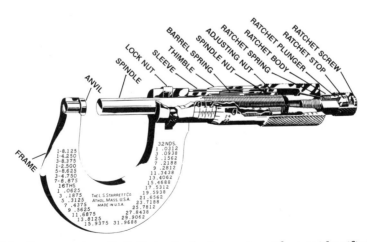

FIG. 3-31 **Cutaway view of a 0-1″ standard micrometer with parts identified** *(Courtesy of The L. S. Starrett Company)*

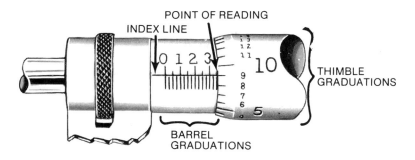

FIG. 3-32 Reading a measurement on a 0-1″ standard micrometer (0.359″ shown) *(Courtesy of The L. S. Starrett Company)*

The part size (dimension) is established by adding the value of the last visible gradua-tion on the barrel at the index line to the graduation reading on the thimble at the index line. Each vertical graduation on the barrel represents 0.025″; each thimble graduation, 0.001″. Each fourth graduation on the barrel is numbered. For example, 1 means 0.100″; 2, 0.200″, etc. In the micrometer reading illustrated in figure 3-32, the last numbered vertical graduation is 3 (representing 0.300″); *plus* 0.050″ for the additional two visible line graduations; *plus* the thimble graduation at the index line of 9 (for 0.009″). This particular reading is 0.359″ (0.300″ + 0.050″ + 0.009″).

Metric Micrometers and Readings

The same principles of reading a micrometer apply to metric micrometers, except for the values of the line graduations and the measurement unit (mm). The vertical gradua-tions on the barrel are in increments of 0.5 mm, and each horizontal line graduation on the thimble represents 0.01 mm. An example of a 5.87 mm metric measurement is shown in figure 3-33. The last visible numeral on the barrel is 5 (5 mm). There is one line

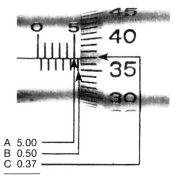

A 5.00
B 0.50
C 0.37

5.87 MM (READING TO NEAREST ONE-HUNDRETH)

FIG. 3-33 Reading a measurement on a 0-25 mm metric micrometer *(Courtesy of The L. S. Starrett Company)*

visible beyond the 5 graduation (= 0.5 mm). The 37 graduation on the thimble is at the index line (= 0.37 mm). Adding these values produces a measurement of 5.87 mm (5.0 mm + 0.5 mm + 0.37 mm).

When micrometer measurement readings are required for parts that are larger than the standard 0-1 or 0-25 mm micrometer capacity, another addition is made to the barrel and thimble reading. The size of the micrometer opening is added. For instance, if a part measures between four and five inches, a 4" to 5" range micrometer is used. In this case, the exact micrometer reading includes the opening size of 4.000" + the barrel reading + the thimble reading.

REFRIGERATION GAUGES AND TEST INSTRUMENTS

Instruments and Gauges

The technician must select and use instruments and gauges that are designed to establish pressure and temperature conditions inside the refrigerating unit. Four of the most common instruments used include: thermometers, pressure gauges, vacuum gauges, and compound gauges.

To maintain measurement accuracy:

- Use each instrument with care and within its capacity.
- Store each instrument after use in its proper case.
- Check and calibrate each instrument regularly.

Thermometers. The *thermometer* is used to check and to record the temperature of the refrigerator cabinet, evaporator, liquid and suction lines, and the condensing temperature unit. Many sizes, shapes, and types of thermometers are available to meet specific job requirements. The glass stem thermometer and the dial stem thermometer are two of the most common types.

The *glass stem thermometer* reading range is from -40°F to 120°F in increments of two degrees (2°). *Dial stem thermometers,* figure 3-34, are the favorite with refrigeration technicians, as they are not as easily broken as the glass stem type. The usual measurement range of dial stem thermometers is from -40°F to 120°F and 0°F to 220°F. Both the glass stem and dial stem thermometers come with a case and pocket clip for carrying on the job.

Gauges. Three basic types of gauges are used in service work: (1) *high pressure gauges,* (2) *vacuum gauges,* and (3) *compound gauges.* Pressure gauges help to establish what is happening inside the system. Most pressure gauges have a 2 1/2" dial and are connected to the refrigeration system with a pipe thread.

SAFETY PRECAUTION

Gauges may be used to measure pressures safely when the maximum pressure of the system falls within 75% of the maximum pressure capacity of the gauge.

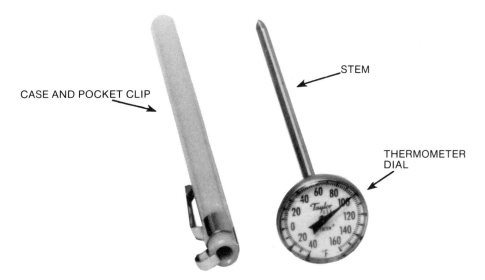

FIG. 3-34 Dial stem type temperature measuring thermometer *(Courtesy of Thermometer Corporation of America)*

Vacuum Gauges. The *vacuum gauge* measures lower-than-atmospheric pressure. Such gauges have one of four calibrations:

- *Inches of mercury* (Hg)
- *Pounds per square inch absolute* (psia)
- *Millimeters of mercury* (mmHg)
- *Microns* for extremely high vacuum ranges

The *mercury manometer* is sometimes used for normal refrigeration vacuum ranges; the *McLeod gauge* is used for very high vacuum range applications.

Compound Gauges. The *compound gauge* measures both pressure and vacuum. The calibrations generally range from 0″ to 30″ Hg and 0 psi to 200-500 psi. Also, some compound gauges have scales that are calibrated to measure temperature and pressure of different refrigerants, such as R-11, R-22, and R-502. The extra scales permit direct reading and eliminate the need to refer to pressure-temperature charts of different refrigerants.

Solid-State Electronic Thermistor Vacuum Gauge

A *thermistor* is a control device whose electrical resistance depends on temperature variations. When the temperature increases, a typical thermistor experiences an increased value of electrical resistance. As the temperature decreases, there is a decreased value of electrical resistance.

This phenomenon is used in the solid-state thermistor vacuum illustrated in figure 3-35. This particular gauge measures extremely low pressures (from normal atmospheric pressure down to 50 microns) beyond the capability of manifold gauges. When the

FIG. 3-35   Solid-state, electronic thermistor vacuum gauge *(Courtesy of Robinair Division, Sealed Power Corporation)*

thermistor is placed in a vacuum manifold, its temperature (and, correspondingly, its resistance) depends on the degree of heat transfer (by conduction) to the remaining air in the manifold.

Heat transfer away from the thermistor is at maximum at *one atmosphere of pressure*. Under this condition, both the temperature and resistance are low. Heat transfer is poor at very low pressures; temperature and resistance are high.

The change in resistance of the thermistor is measured by the instrument as it displays a corresponding range of pressure. The model thermistor vacuum gauge illustrated in figure 3-35 has ten individual light-emitting diodes (LEDs). The LEDs correspond to ten steps of measurement between 50 microns and 25,000 microns (1" Hg vacuum). The tolerances of this electronic equipment range from ± 10% of the measurement value at 50 microns (low end) to ± 20% at the high reading of 25,000 microns.

Thermistor Vacuum Gauge Operation

The gauge tube of the instrument is connected to the system being measured; the tube socket on the cable, to the mating plug on the gauge tube. When the slide switch is moved to the "ON" position and all the LEDs are illuminated, the gauge tube is above 25,000 microns. When *evacuating the system*, LEDs will go "OUT." The *vacuum level* of the system is indicated at the position next to the LED that goes "OUT."

In changing the system vacuum from 50 to 25,000 microns, the LEDs come "ON." The vacuum level is indicated by the LED next to the LED that is illuminated. A very dimly lighted LED (when all other LEDs are of normal brilliance) indicates that the vacuum level is approaching that specific LED value.

One advantage of this solid-state instrument over conventional meters with calibrated dials is that the LEDs have preset values. As such, vacuum levels are easier to read, simpler to take, and consistently accurate.

OTHER AIR-CONDITIONING INSTRUMENTS, TOOLS, AND ACCESSORIES

The hand tools and power tools and gauges that have been identified in this unit are complemented by still other tools and equipment that the air-conditioning technician uses. These apply to the detecting of leaks; the measurement of air, surface, and liquid temperatures; injecting compressor oil; removing and working with valve components; evacuating and charging valves; analyzing; installation work; maintenance; and servicing processes related to refrigeration, heating, and air-conditioning systems.

The three basic types of gauges and additional instruments and accessories are illustrated and described in detail in later units.

SUMMARY

- Refrigeration, heating, and air-conditioning workers are responsible for the proper use of tools and equipment in preventing damage to parts and units, avoiding personal injury, and protecting tool surfaces from becoming distorted or excessively worn.
- The selection of tools made of high-quality steels, of functional design, and correctly hardened, tempered, and formed is a wise investment.
- Common hand tools include wrenches of many types and sizes, pliers and crimpers, screwdrivers, chisels and punches, hacksaws, files, hammers, and sheet metal shears.
- Wrenches have machined solid jaws or adjustable jaws that grip surfaces on fasteners and other parts. Many flat, hexagon, 6-point, 12-point, and spline surfaces are of standard size. Others vary in size and require some type of adjustable wrench.
- Common designs of wrenches include (but are not limited to) box, flare nut, adjustable jaw, socket, spline, service valve, and others. Wrenches may be solid and straight, offset, or designed for ratcheting or providing uniform torque.
- Pipe wrenches and strap wrenches are general purpose wrenches for piping and irregularly shaped parts.
- Threaded refrigeration/air-conditioning pipe fittings conform to National Pipe Thread (NPT) standards for thread form, size, and pitch series.
- The most common thread form has a 60° included thread angle, flat crest and flat root, and falls into a standard pitch series.
- Pitches (threads per inch) are standardized for common screw thread fasteners, threaded holes, and commercially available parts.
- Thread notations on drawings and in specifications provide essential information about thread form, major diameter, and pitch series.
- Generally, threads for standard nuts, bolts, screws, and other common fasteners fall in the American National, Unified, or ISO Metric coarse and fine series of thread pitches.

- National Pipe Standards apply to straight, taper, and Dryseal pipe threads.
- Holes that are tapped on the job are tapped by hand. External threads are cut using dies that are mounted in holders.
- Tables of tap drill sizes provide size information for correct drill diameters to meet different size screw threads.
- Twist drills (and shorter insert drills) are ground so the cutting lips are equal in length, and the cutting angle, point angle, and clearance are correct.
- Fractional size twist drills up to 1/2" (in increments of 1/64") are most common. Number size (#0-#80) and letter size (A-Z) are designed primarily for tap drill sizes for threads under 1/2" in diameter.
- Twist drills are usually power driven. Cutting speeds should follow manufacturer's recommendations. High-speed steel drills are preferable to carbon tool steel drills.
- Cutting fluids for threading and drilling aid in the cutting process and in producing finer finish threads.
- Hand hacksaws provide workplace capability to make vertical, horizontal, and angle cuts in hard, tough materials.
- Screwdrivers may be straight, offset, or ratcheting. Common point forms include flat, Phillips, and Reed and Prince.
- Air- and electric-power tools are used on the job for operations ranging from high-speed drilling to lower-geared torque assembly processes.
- Linear measurements to accuracies of 1/64" may be obtained with a steel rule.
- Precision measurements to within 0.001" and 0.0001" are taken with the inch standard micrometer. Metric micrometers permit accuracies to within 0.5 mm and 0.01 mm.
- Glass stem and dial stem thermometers are used for temperature measurements.
- High-pressure gauges are used to establish what the conditions are within a system.
- Vacuum gauges are calibrated to measure pressures that are lower than atmospheric pressure.
- Compound gauges measure vacuums from 0" Hg to 30" Hg and pressures from 0 psi to 200 psi (and 500 psi).
- The solid-state electronic thermistor vacuum gauge reads pressures from normal atmospheric pressure (25,000 microns) down to 50 microns. Ten light-emitting diodes (LEDs) on the gauge, corresponding to ten steps of measurement within the 50 to 25,000 micron range, are designed to display the vacuum measurement.

ASSIGNMENT: UNIT 3 REVIEW AND SELF-TEST

REFRIGERATION TOOLS AND EQUIPMENT

Complete statements 1 through 10 by inserting the correct tool, condition, or required function.

1. Quality hand tools and cutting tools are constructed of _____ .
2. The cutting edges and bodies of hand tools are _____ to provide good cutting qualities and to resist being deformed through use.
3. The safe operation of cutting, forming, and torque-producing hand and power tools requires that each tool be used _____ .

4. Three design features of quality wrenches include (a) _____ ; (b) _____ ; and (c) _____ .

5. Box wrenches are designed so that a fastener may _____ in the specially shaped enclosed opening.

6. The jaw opening of a flare nut wrench permits the wrench _____ .

7. Ratchet flare nut wrenches are practical in applications where _____.

8. The function of a torque wrench is to _____ .

9. Excessive force applied on fasteners beyond the manufacturer's recommendations may produce (a) _____ ; and (b) _____ .

10. Adjustable wrenches are a practical tool to use when _____ .

11. Identify four common opening shapes for standard socket wrenches.

12. State briefly the function of a ratchet handle for a socket head.

13. (a) Identify the shape of Allen set-screw wrenches. (b) Give two examples of where an Allen set-screw wrench is used.

14. (a) State the function of a spanner wrench. (b) List three basic forms of spanner wrenches.

15. Identify the number of the function or design feature in Column II that correctly corresponds with each tool in Column I.

Column I	Column II
(a) Internal pipe wrench	(1) Teeth grip the outside surface of an object in order to turn it.
(b) Refrigeration service valve wrench	(2) Designed with fixed position settings to obtain a number of parallel jaw openings.
(c) Strap wrench	(3) Grip the inside surface to permit loosening or tightening.
(d) Adjustable joint pliers	(4) Fixed-size openings to permit wire stripping and cutting.
(e) Electrician's crimper	(5) Designed with a ratchet and a square and multi-point socket head.
	(6) Measure the amount of force required to tighten a fastener.
	(7) Flat jaw faces permit direct contact with fastener surfaces.
	(8) Tighten or loosen nonmetallic and/or irregularly shaped parts.

16. State three safety precautions to take when using hammers, chisels, and punches.

17. Give the correct angles for drilling steel and cast iron parts for the following drill point features: (a) cutting edges; (b) cutting point; and (c) clearance angle.

18. State what affect the following incorrectly ground angles have on the cutting action of a twist drill: (a) chisel point angle of 145°; (b) clearance angle of 15° to 20°; and (c) clearance angle of 2° to 4°.

19. List two functions that are served by the guide plate on a die stock.

20. State one advantage of using aviation snips over straight metal cutting shears in forming an irregularly shaped pattern in a sheet metal duct.

21. Name the tool that should be used for each of the following operations or conditions.
    (a) A cross-slotted screw head.
    (b) Screw head with only 3/4″ clearance above the head but with plenty of side clearance.
    (c) Tightening a setscrew with a recessed opening in its head.
    (d) Opening or closing a service valve.
    (e) Tightening an odd-sized nut.
    (f) Tightening a packing nut that is notched.
    (g) Tightening the head nuts on a compressor to a specified tightness (torque).
    (h) Removing the nut from a flare fitting on a system.
    (i) Cutting insulated electric wires.
    (j) Tightening a nut on a compressor mounting that is way down in between the compressor and condenser when there is no side clearance.

22. Identify the number of the function or design feature in Column II that correctly matches the test equipment in Column I.

Column I	Column II
(a) Dial stem thermometer	(1) Measures temperature and pressure from 0″ to 30″ Hg and 0 psi to 500 psi.
(b) High-pressure gauge	(2) Measures pressures up to 500 psi.
(c) Vacuum gauge	(3) Measures to accuracies of 0.001″ or 0.5 mm.
(d) Compound gauge	(4) Controls temperatures within a refrigeration system.
	(5) Calibrated to read in microns.
	(6) Adjusts pressures between 0 psi and 900 psi.
	(7) Measure temperatures within a $-40°F$ to $120°F$ range.

23. State two advantages of using a thermistor vacuum gauge with light-emitting diodes (LEDs) to measure a vacuum in contrast with the use of a standard dial-face pressure gauge.

UNIT 4:
REFRIGERATION MATERIALS AND FITTINGS

The refrigeration technician and craftsperson must learn correct work procedures, acquire basic skills, and master certain technical information. These are related to the proper use of tools, materials, instruments, and the major units within a refrigeration system. Having mastered the basic processes and acquired a broad technical understanding, the mechanic must then be able to follow written instructions intelligently. Such instructions are given in manuals furnished by manufacturers of refrigeration equipment and accessories and other technical material especially prepared for the refrigeration field.

The most commonly used form of a material used in the refrigeration system is tubing. Tubing provides a channel through which refrigerant liquid and vapor are transported between the major units of a refrigeration system. Tubing is used to make connections between many parts of the system. Tubes and gauges or major units are either joined together permanently by soldering or brazing, or temporarily by using *fittings*. Accordingly, this unit covers some of the properties of tubing and the principles of joining tubing with parts and accessories by soldering, brazing, or the use of fittings. The special tools and accessories and the regular cutting, forming, and fastening of tubing are illustrated and described.

TUBING FOR REFRIGERATION SYSTEMS

Properties of Refrigeration Tubing

A specially annealed copper tubing is often used in refrigeration work, except in absorption systems. The term *annealed* describes the degree of hardness. One peculiar property of copper is that when it is hammered, bent, or flexed, it becomes work-hardened and brittle. In this condition it cannot be formed or cut without breaking or cracking. This physical condition may be corrected by heating the copper to a dull red or a dark blue-purple color. When allowed to cool in air, the copper again becomes soft and pliable and may be bent or flared. This process of treating the copper with heat is called *annealing*.

The work-hardening or aging of copper may cause cracks to form when the tubing is flared, or it may buckle suddenly and tend to flatten out. Tubing is also *dehydrated* to free it of moisture. Manufacturers usually dehydrate copper tubing for refrigeration work and then seal the ends by either pinching or closing them with caps so moisture may not enter.

Copper oxidizes easily and turns from its natural brilliance of reddish-bronze to a dull greenish-brown color. This oxide is formed by the action of air or other chemicals on the

OD	Wall Thickness	Pounds Per Ft.
1/8	.030	.0347
3/16	.030	.0575
1/4	.030	.084
5/16	.032	.109
3/8	.032	.134
1/2	.032	.182
5/8	.035	.251
3/4	.035	.305

FIG. 4-1   Standard soft copper tubing sizes and weights   **FIG. 4-2   Soft copper tubing (coiled)**

copper. Before copper tubing may be mechanically bonded to another material, it is necessary that this oxide coating is removed. How this is done is treated later when soldering is discussed.

Deoxidized and dehydrated copper tubing is available in common sizes from 1/8″ to 3/8″ outside diameter (OD) and with wall thicknesses ranging from 30 to 55 thousandths of an inch. The standard dimensions and weights of copper tubing are given in figure 4-1. Soft copper tubing is furnished in 50- and 100-foot rolls, figure 4-2. Hard drawn copper is furnished in 20′ lengths.

In addition to soft-drawn copper, hard-drawn copper is used in a number of commercial refrigeration and air conditioning installations where rigidity is needed and tube sizes over 1/2″ are used. The hard-drawn copper and precharged tubing come in straight lengths rather than in rolls. The ends are either capped or plugged to keep the inside of the tubing clean and dry, figure 4-3.

Cutting the Refrigerant Tubing

Regardless of the manner in which tubing is connected to another part, it is important that the end be prepared properly. Two common methods of cutting tubing square and

FIG. 4-3   Hard-drawn tubing (capped)

FIG. 4-4   Hacksaw fixture (sawing vise) *(Courtesy of Imperial-Eastman Corporation)*

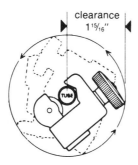

FIG. 4-5 Tube cutter designed for use in limited (tight) quarters *(Courtesy of Imperial-Eastman Corporation)*

by hand require the use of a hacksaw or a tube cutter, figures 4-4 and 4-5. The tube cutter is generally used for soft or hard copper tubing, aluminum, brass, stainless steel, and monel metal tubing.

During the cutting operation, a *burr* or fine feathered edge is raised on the inside of the tubing. Burrs may be removed with a fine-cut (mill) file, a scraping tool, or a reamer. The *reamer* is a cutting tool with a series of teeth or sharp cutting edges, figure 4-6. Some types are made to remove both internal and external burrs. The model illustrated has a capacity of 3/16" to 1 1/2" outside diameter.

While not a preferred process, care must be taken when hacksawing to use a fine-pitched wave-set form blade. Usually, a 32-pitch (32 teeth to the inch) blade is fine enough to cut easily. With the wave-set, there is sufficient clearance so the blade will not wedge or pinch in the cut.

The tubing should be positioned so that no chips fall into the part to be used. The remaining tubing should be pinched or capped immediately when it is not in use. This protects the inside against metal chips, moisture, and dirt.

Forming Refrigeration Tubing

There are a number of important steps to be taken and checks to be made when forming copper tubing:

✓ The cross-sectional area of the tubing at any bend must remain the same as the original tube size.

OUTSIDE REAMER BLADES

INSIDE REAMER BLADES

FIG. 4-6 Inside and outside reamer *(Courtesy of Imperial-Eastman Corporation)*

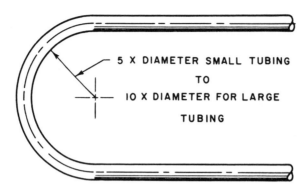

FIG. 4-7   Minimum safe bending radii

✓ The roundness of the tubing must be maintained. It should not flatten out or buckle as a result of a bending operation.
✓ The tubing must be formed so that no strain is placed on the fittings when the tubing is installed.
✓ Sudden stresses placed on copper tubing by trying to bend the metal quickly or by making too sharp a bend may cause the tubing to crack.

While it is not necessary to use special forming tools with the smaller sizes of copper tubing used on domestic refrigerators, a neater job may be done with them.

Tubes should be bent in as large a radius as possible. The minimum safe bending radius is usually five times the diameter of small tubing to ten times the diameter of large tubing, figure 4-7.

A number of special tools are available for bending operations on soft-copper and aluminum tubing. Either internal or external coil-spring bending tools are available in many sizes. The internal spring is designed to be used near the ends of tubes, figure 4-8. The external spring is best suited for the middle of tubes, figure 4-9.

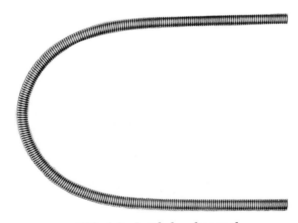

FIG. 4-8   Inside bending tool

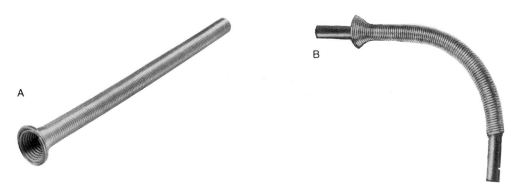

FIG. 4-9 **(A) Outside bend, spring-type tube bender and (B) forming an external bend** *(Courtesy of Imperial-Eastman Corporation)*

After a tube is formed with a coil-spring bending tool, the tool tends to bind, and removal from the tubing becomes difficult. Removal can be made easy by twisting the tool to expand the spring when used on the outside of the tubing, and twisting it to cause it to become smaller when used inside the tubing. In cases where the tubing is to be bent and flared at the same end, and an external spring bending tool is to be used, the tube should be formed first.

A lever-type tube bender (illustrated in figure 4-10) is used to form bends neatly and accurately without buckling the tubing. These tools will form bends up to 180° in one continuous operation. The forming wheel is calibrated to show the degree of bend attained. Manufacturers also furnish information for making bends to predetermined dimensions. Each of these tools is used with one size of tubing. Combination lever-type tube benders that are adapted to bending several sizes of tubing by changing the forming wheel and block are also available, figure 4-10.

FIG. 4-10 **Accurate tube bending tool with changeable forming wheel and block sizes** *(Courtesy of Imperial-Eastman Corporation)*

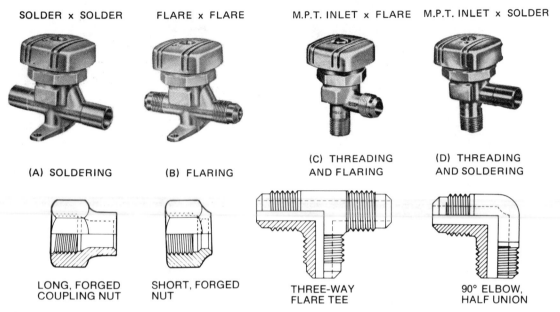

SOLDER x SOLDER   FLARE x FLARE   M.P.T. INLET x FLARE   M.P.T. INLET x SOLDER

(A) SOLDERING   (B) FLARING   (C) THREADING AND FLARING   (D) THREADING AND SOLDERING

LONG, FORGED COUPLING NUT   SHORT, FORGED NUT   THREE-WAY FLARE TEE   90° ELBOW, HALF UNION

FIG. 4-11 Basic design features of fittings used in refrigeration system applications *(Courtesy of Mueller Brass Company)*

REFRIGERATION FITTINGS

The application of common types of fittings is illustrated by the use of four valves, figure 4-11. Tubes are fitted to the line valve by soldered connections (A) or by flaring the ends of the tubing to fit over the flared connections (B).

Two different ways by which line valves may be connected are also described. A threaded fitting is provided on the bottom of (C), while a flared end is used for the other connection. At (D), a threaded and a soldered fitting are used. Different combinations are available. These photos show that there are two basic methods of forming or preparing the ends of tubing to connect them directly with or through the use of fittings. The two methods are flaring and soldering.

The lower threaded (pipe) parts shown on the two angle valves are called *male-pipe* (**MPT**) threads or fittings. They fit only into a matching internally threaded fitting called a *female pipe* (**FPT**) thread or fitting. Such fittings are provided on the compressor and service valves, and are called ports.

Tubing must be installed in such a way that there is minimum strain on the parts to which it is attached. This is why tubes are sometimes formed into loops. The loops tend to prevent copper tubing from hardening and breaking (crystallizing) when there is continuous vibration.

Threaded Fittings

Some refrigeration fittings have *national fine (NF)* threads that conform in shape and dimension to standards established for the industry for flared fittings. Other standard pipe threads are designated as National Pipe Threads. These threads have a taper of 3/4″

per foot. The taper provides a vapor-proof joint. NPT specifications are found in trade and engineering handbooks. The pipe threads and fine threads do not match.

Refrigeration fittings are usually made of brass and are cast or formed by drop forging. The threaded, flared, or soldered (as well as the internal moving parts) fittings are machined for accurate fitting and performance.

All fitting sizes are identified by the same size as the tubing size to which they are attached. For instance, a 3/8"-flaring tool is used to form a 3/8" tubing for a 3/8" flared fitting.

Purpose and Methods of Flaring

Leak-proof flares are important to the efficient operation of any refrigeration system. Flares, which fit against machined surfaces, are prepared by one of two common methods: They are either formed by a spinning action, or punched and blocked, figure 4-12. While there are many manufacturers of flaring tools, only one type is illustrated as being typical of the construction and operation of each, figure 4-13.

Single Thickness Flares. Flaring tools contain a series of blocks, usually ranging from 3/16" to 5/8", to accommodate tubing of these sizes. The tubing is inserted in the proper block and projects above the face of the block to whatever distance is recommended by the manufacturer. The better tools eliminate guesswork in positioning the tubing. Means are provided for positioning tubing that is formed to accurate flare diameters at

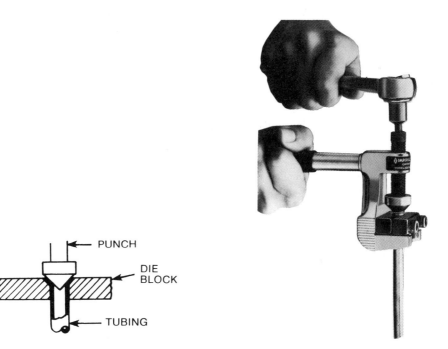

FIG. 4-12 Flaring tubing punch and block

FIG. 4-13 Flaring/swaging combination tool
(Courtesy of Imperial-Eastman Corporation)

the top of die blocks. The flare is formed as the cone is advanced into the tubing, forcing the tubing against the die block opening. A drop of lubricating oil should be placed between the cone and the tubing.

Precision flaring tools are also available for flaring tubing above the die blocks. This method maintains the original tubing wall thickness at the base of the flare, eliminating flare failure at this point. These tools are usually designed with a faceted flaring cone that tends to smooth out any surface imperfections on the flare seat by a burnishing action. The term *single thickness flare* means that the part of the tubing that forms the flare is the thickness of the tubing, figure 4-14.

The tool illustrated in figure 4-13 is part of a combination set that is designed for single flaring at 37° and 45° angles, double flaring at 45°, and swaging. Flaring cones, double-flaring adapters, swaging adapters, and individual diameter jaw sets (tube holders) permit single-tube flaring from 1/8" to 3/4" diameter, and double flaring and swaging on 3/16" to 3/4" diameter copper, brass, aluminum, and stainless steel tubing. This particular flaring/swaging tool is designed for hand-held or vise-mounted operation.

Double Thickness Flares. A *double thickness flare* indicates that the flare thickness is made up of two thicknesses of tubing. The double thickness flare is used on larger sizes of tubing and, especially, where an extra-strong joint is required to withstand vibration. The double thickness flare may be made by the punch and block method or by using a double flaring tool, figure 4-15. When using the flaring tool, figure 4-15B, the point of the flaring cone is brought down against an adapter that forces the tubing to become bell-shaped. The cone is backed off, the adapter is removed, and the flaring cone is rotated against the top edge of the tube until it forms the double shoulder flare.

Compression Fittings

There is still another common method of providing a leakproof connection. This method requires a specially machined nut. The cutaway section in figure 4-16 shows a capillary tube (A) that extends through the nut (B) into the connector fitting (C). As the nut is tightened, the nose section (D) is forced tightly against the connector fitting, squeezing the tip of the nose against the tubing. If such an assembly must be serviced at any time, the end of the tubing must be cut back and the special nut must be replaced to be certain that the connection is again leakproof.

LEFT: FACETED FLARING CONE ROLLS OUT FLARE ABOVE DIE BLOCK.

RIGHT: SMOOTH SUR-FACE DIES GRIP TUBING WITHOUT SCORING.

FIG. 4-14 Rolling a single thickness flare

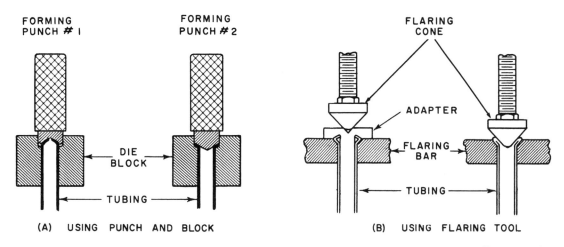

FIG. 4-15   Forming double thickness flares: (A) using punch and block and (B) using flaring tool

PRINCIPLES OF SOLDERING

Soldering refers to the bonding of molten solder (usually a low-melting temperature composition of two or more metals) to another material that is not molten. The solder flows into the grain structure of the metals being joined. Upon cooling, the solder solidifies and adheres to the surfaces, making a permanent joint. Brass and copper tubing, parts, and fittings that are not subjected to high temperatures may be soldered. A good soldered joint depends on:

√ Surfaces that are very clean, with no oil, grease, or oxides on them.
√ The use of a good flux that dissolves any oxides that may have formed on the surface.
√ Parts that are tight fitting at the places where they are to be joined.

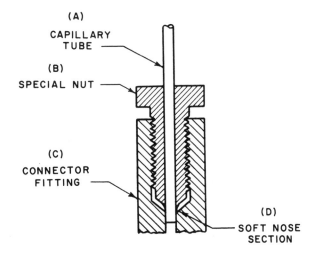

FIG. 4-16   Internal view of capillary tube connection

  ✓ A good quality of solder.
  ✓ An adequate supply or source of heat that can be directed to the surfaces being soldered.

Soldering and Brazing Alloys

Joints that are subject to moderate temperatures and pressures are generally soldered with a 50–50 tin-lead solder. A 95–5 solder is recommended for greater joint strength and higher pressure and temperature conditions. The 95–5 indicates a solder consisting of 95% tin and 5% antimony. These, and other solders, are referred to as *filler metals*. The process is called *soft soldering*. Soldering is performed at temperatures under 800°F.

Brazing is mainly used in large refrigeration and air conditioning systems where hard tubing, piping, and fittings require temperatures over 800°F due to the composition of the filler metal alloys. For example, a filler metal consisting of from 15% to 60% silver (or copper alloys containing phosphorous) is often used to braze copper tubing. These solders are called *hard solder* or *silver solder*.

Soldering and Brazing Equipment

A portable propane fuel tank with appropriate tip is commonly used by technicians for heating parts or surfaces that are to be soldered. The hand-held unit is easy to ignite, and it is easy to adjust the flame for the size and type of joint and to use to flow the solder into the joint.

An *air-acetylene (welding type)* unit is most often used to produce high temperatures for brazing processes. The unit consists of a torch and series of tips, regulator, hose, and a tank of acetylene gas. The tip sizes provide for small diameter tubing; large sizes are available for large diameter tubing and fittings and high temperature requirements. The high-velocity tip, figure 4-17, is preferred by technicians for general brazing needs.

Soldering Techniques

The metal in the parts to be soldered or brazed is called the *base metal*. The base metal tubing, piping, fitting, or other part is heated to the temperature at which the solder (filler metal) flows. At the melting point, the filler metal molecules are attracted to and flow into and along the base metal surfaces until the opening between the parts is filled

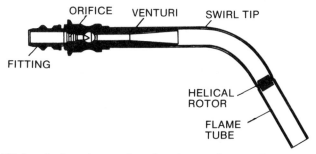

FIG. 4-17   High-velocity, air-acetylene brazing outfit tip *(Courtesy of Wingaersheek Division, Victor Equipment Company)*

with solder. The process of drawing the solder to fill the space between parts to be joined is another form of *capillary attraction.*

Soldered joints require:

- Clean mating surfaces.
- Parts that fit accurately to permit an adequate (not excessive) amount of solder to flow and fill the joint. Usually a clearance of 0.003" to 0.004" permits good capillary action.
- Use of the correct liquid, jelly, or paste flux. Fluxes are used to minimize oxidation and to float dirt and other foreign particles out of the joint assembly.
- Uniform application of heat across the tubing and fitting, avoiding overheating any area or pointing the flame into the parts.
- Use of the correct solder. The solder is touched to the joint so that it flows freely and is drawn into the area to be soldered. The solder is fed until the joint is filled.
- Wiping the joint clean.

Brazing Techniques

Brazing also depends on capillary attraction and procedures that are similar to those followed for soldering. However, the higher melting temperature filler materials require the heat of an air-acetylene torch. It is important that the tube to be brazed be heated first. Heat is then applied to the fitting. Then flame is continuously swept from the fitting to the tube.

Filler metal (applied to the joint) flows readily into the space between the mating parts when the melting point of the filler metal is reached. Oxidation at the joint is caused by the flux and the high temperature required for the brazing process. Where practical, the brazed joints are washed with soap and water.

Preparation of the Work for Soft Soldering and Brazing (Hard Soldering)

It is important that the metal parts to be soldered must fit accurately together and must be hot enough to melt the alloy. When the proper temperature is reached, and if the surfaces are clean and fluxed, the solder that is applied at the joint flows by capillary action between parts to be joined. Remember, soft solder should be heated by the parts and not by the direct heat of the torch.

When joining tubing and tubing fittings, there are some simple steps to follow. The tubing must be cut squarely and accurately to length. Any internal or external burrs must be removed. The outside of the tube must be cleaned at the place where it is to be soldered. This is done by brushing with a wire brush, figure 4-18A. The inside of fittings

---

SAFETY PRECAUTION

Chemicals in fluxes are very reactive and may cause skin irritations. For this reason, the flux should be applied by a brush and not with a finger. The flux should be used sparingly and kept away from the tube end. Also, carbon tetrachloride should not be used for cleaning parts as the vapor is very toxic.

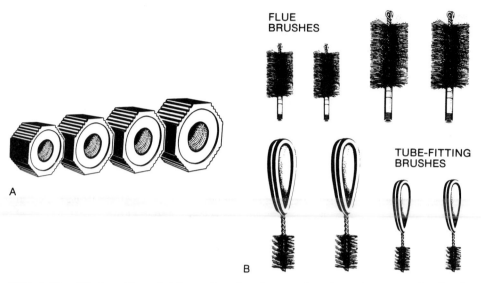

FLUE
BRUSHES

TUBE-FITTING
BRUSHES

A

B

FIG. 4-18 **(A) Outside and (B) inside tube-cleaning brushes** *(Courtesy of Schaefer Manufacturing Company)*

are cleaned with an internal wire brush, figure 4-18B. Any surface that has been cleaned should be covered as quickly as possible with a thin film of flux. A solder flux brush is convenient to use for this purpose.

The joint should be assembled so that the end of the tubing butts against the shoulder in the fitting. If the parts are turned slightly, this action causes the flux to spread over the surfaces.

Whenever possible, joints should be silver brazed by "looking down on the fitting." In this position, the flux and solder are kept from getting inside. A plastic sealing compound may be used to make a gastight joint between the tubing and the dry nitrogen, which is swept through the tubing to prevent oxide formation while soldering.

SAFETY PRECAUTION

The refrigerant container should not be connected to the high-pressure side of the system or any system of high pressure. Such a connection may cause excessive pressure. The excessive pressure may result in the violent bursting of the container, causing serious personal injury and property damage.

Design of Joints for Brazing

The parts to be brazed together must be well designed. In figure 4-19, a few examples are illustrated with line drawings to demonstrate what is considered good and poor design. Maximum strength depends on proper design and the accuracy with which the parts to be brazed fit together.

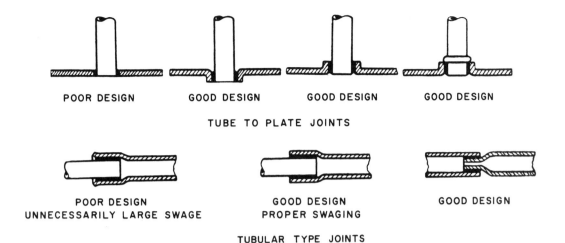

POOR DESIGN   GOOD DESIGN   GOOD DESIGN   GOOD DESIGN

TUBE TO PLATE JOINTS

POOR DESIGN
UNNECESSARILY LARGE SWAGE

GOOD DESIGN
PROPER SWAGING

GOOD DESIGN

TUBULAR TYPE JOINTS

FIG. 4-19 Joints designed for silver soldering or brazing

Swaging

Swaging is a means of shaping copper tubing so that two pieces may be joined without the use of a fitting. This operation is accomplished with a punch-type or screw-type swaging tool, figure 4-20. The tubing is clamped into the flaring bar, and the specially designed punch is hammered into the tubing, swaging or expanding the end so that it will fit over the end of another piece of tubing.

The screw-type tool produces the same swaged shape by pressure as the punch is forced into the tubing.

The fluxing of a brazed joint is important because any flux that enters the system cannot be removed easily. The flux is usually applied to the external surface of the part that is to slide into the joint so it does not get into the system. By turning the part, flux is spread on the inner surface of the outer tubing.

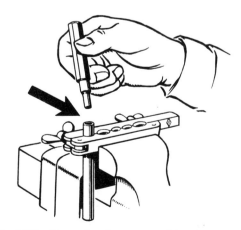

FIG. 4-20 Preparing to swage tubing

With practice, it is possible to determine the temperature of the parts being heated by the behavior of the flux. The flux turns a milky color at 212°F, bubbles at 600°F, and turns a clear liquid at 1100°F. Silver brazing alloy then melts at 1120°F and flows at 1145°F.

Because of the high temperatures required for silver brazing, the torch flame should be directed away from solenoids, shutoff valves, driers with replaceable cores, and other parts that may be damaged by overheating, figure 4-21A. Heat transfer may be prevented by placing water-soaked cloths over the fitting at places where heat is not needed or by removing the parts that must be protected, figure 4-21B and C.

All joints should be cleaned as quickly as possible to remove excess flux and to test the joint for leaks. Silver-brazed parts may be cooled gradually or quickly with water.

SUMMARY

- Copper tubing in the annealed (soft) form is used in domestic and small refrigeration systems. Hard-drawn copper tubing in larger sizes is used for commercial installations and air-conditioning systems. Tubing must be dehydrated as well as seamless.
- Copper tubing may become brittle from work-hardening caused by vibration, hammering, or aging. It may be annealed by heating to a dull red or a dark blue-purple color and cooling slowly.
- Tubing is generally cut squarely by hand using a tubing cutter. Internal and external burrs must be removed. Tubing may be formed into a desired shape with the aid of springs (external or internal) or a mechanical bender.
- Flaring tools are used to form the end of tubing. A flare may be formed by spinning or by forcing a punch into the tubing as it is held in a die block having the desired shape. A flare can be single or double thickness. Double thickness flares give added strength.
- Swaging tools are used for sizing, correcting out-of-roundness of the ends of tubes, and to form joints.

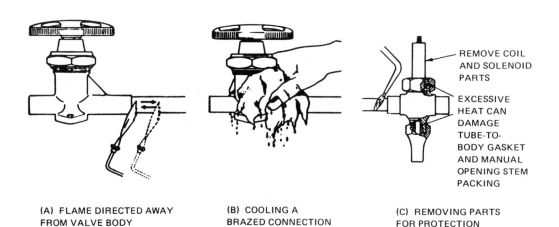

(A) FLAME DIRECTED AWAY FROM VALVE BODY

(B) COOLING A BRAZED CONNECTION

(C) REMOVING PARTS FOR PROTECTION

REMOVE COIL AND SOLENOID PARTS

EXCESSIVE HEAT CAN DAMAGE TUBE-TO-BODY GASKET AND MANUAL OPENING STEM PACKING

FIG. 4-21   Directing flame to protect units from heat

- Threaded refrigeration fittings conform to standards for National Pipe Threads (NPT).
- Soldering is a process of mechanically bonding together two pieces of material by the force of adhesion. This force is the result of attraction between the molecules of solder and the metal part being soldered.
  - ✓ The melting point of solder, which is the bonding agent, must be below that of the material to be soldered.
  - ✓ Solders are alloys (mixtures) of metals and are classed as either soft or hard.
  - ✓ Parts to be soldered must be free of oils and oxides, fit accurately, and be coated with the proper flux.
- Silver brazing is the process of joining parts using hard solders. The hard solders produce mechanically stronger joints than soft solders.

ASSIGNMENT: UNIT 4 REVIEW AND SELF-TEST

A. TUBING FOR REFRIGERATION SYSTEMS

Determine which statements (1 to 12) are true (T) and which are false (F).

1. Any good grade of copper tubing is satisfactory for refrigeration work.
2. Annealed means softened.
3. Bending copper tubing back and forth softens it.
4. Heating copper tubing until it turns a dark purple color softens it.
5. Refrigerated tubing must be dehydrated.
6. Refrigeration tubing comes only in rolls.
7. Once a roll of refrigeration tubing is cut, there is no further reason to cap the end.
8. Bending springs are available only for use on the outside of tubing.
9. Small tubing can be bent as sharply as needed.
10. A jig must be used with a hacksaw to cut tubing where the cut must be square.
11. After using a tubing cutting tool, there is no need to ream out the inside of the tubing.
12. Reamers are used and are available for inside and outside reaming.

Circle the letter of the material, tool, or valve that correctly completes statements 13 through 21.

13. The most commonly used type of material used in refrigeration is (a) plastic tubing, (b) copper tubing, (c) high-pressure flexible tubing, (d) low-pressure flexible tubing.
14. When annealed copper tubing is hammered, flexed, or bent, it becomes (a) soft and pliable, (b) kinked, (c) work-hardened and brittle, (d) none of these.
15. Special refrigeration tubing is required for servicing because it (a) has been dehydrated, (b) is easier to bend, (c) is easy to flare, (d) is easy to swage.
16. Soft copper tubing can be purchased in rolls of (a) 10 feet, (b) 30 feet, (c) 50 feet, (d) 100 to 250 feet.
17. If there is an unusual need to cut tubing with a hacksaw, the pitch of the teeth in the saw blade should be (a) fine: 30 or more TPI, (b) medium: 16 TPI, (c) coarse: 8 TPI, (d) doesn't matter.

18. One of the most important points to look for when bending tubing is (a) plenty of clearance between ends, (b) a big enough bend, (c) the tubing is to be flared at the same end as the flare, (d) the tubing is stress-hardened.
19. Two of the most common benders for small tubing are (a) gear and roller, (b) lever and bending spring, (c) hydraulic lever, (d) hydraulic spring.
20. The single thickness flare means that the part of the tubing forming the flare is equal to (a) the thickness of the tubing, (b) twice the thickness of the tubing, (c) 1 1/2 times the thickness of the tubing, (d) 3/4 the thickness of the tubing.
21. Swaging is used when (a) there is no need for flaring, (b) the joining of two pieces of copper without fittings is necessary, (c) recommended by the manufacturer, (d) it is easier.

B. TYPES OF REFRIGERATION FITTINGS

Add the correct term to complete statements 1 to 11.

1. Three most commonly used fittings in refrigeration work include (a) _____ , (b) _____ , (c) _____ .
2. Another type of fitting that is sometimes used for special applications is the _____ fitting.
3. The thread form and dimensions used on threaded fittings conform to the _____ .
4. A thread that is on the outside of a fitting is called a _____ and is abbreviated _____ .
5. An internal thread on a fitting is called a _____ and is abbreviated _____ .
6. There are two parts to a flaring tool: (a) _____ , (b) _____ .
7. A flare can be either _____ or _____ thickness.
8. The flare that provides greater strength against vibration is the _____ flare.
9. The tool used for making this greater strength flare requires a third part called the _____ .
10. The type of fitting in which a nut is squeezed against the tubing is the _____ fitting.
11. The tubing must be cut back and the nut replaced every time a _____ fitting is loosened.

C. PRINCIPLES OF SOFT SOLDERING AND SILVER BRAZING

Select the letter representing the word or valve that correctly completes statements 1 to 12.

1. Solder is classified as (a) only soft, (b) only hard, (c) both soft and hard.
2. Soldering is performed at temperatures (a) between 200°F and 250°F, (b) below 800°F, (c) between 850°F and 1,000°F.
3. Brazing metals consist of (a) 50% tin–50% lead, (b) 95% tin–5% antimony, (c) 15 to 60% silver and other alloying metals.

4. Air-acetylene units are designed for primary use in (a) soft-soldering, (b) brazing, (c) neither process.
5. A clearance is usually provided between mating parts that are to be soft soldered of (a) .003" to .004", (b) .010" to .015", (c) .016" to .032".
6. Flux is applied to a joint in order to (a) clean, (b) deoxidize, (c) protect the joint.
7. The flux that is used for silver brazing will also work satisfactorily for (a) 50–50, (b) 95–5, (c) neither solder.
8. Flux should be applied to (a) the outside, (b) the inside, (c) either the inside or outside, (d) neither the inside nor outside, part of the joint.
9. A swaging tool bears the closest resemblance to a (a) reamer, (b) flaring tool, (c) tubing cutter.
10. Soft solder should be melted by heat from (a) the flame, (b) the parts to be joined, (c) either the flame or the parts to be joined, (d) neither the flame nor the parts to be joined.
11. When soldering tubing onto a shutoff valve, the flame should be directed (a) toward, (b) away, from the valve.
12. If flux gets inside the refrigeration tubing, it will cause (a) considerable damage, (b) no damage, (c) an explosion.
13. Make a simple sketch showing how tubing should be properly (a) joined to a thin metal plate, (b) fitted inside tubing, (c) joined into a thick metal plate.

D. THE CRAFTSPERSON'S KIT OF REFRIGERATION TOOLS AND ACCESSORIES

Name the tool that should be used for each of the following operations or conditions.

1. Cutting 1/2" annealed copper tubing.
2. Preparing tubing for use with a flaring fitting.
3. Preparing tubing or joining tubing to another piece of tubing.
4. Provide heat to a fitting in preparation for soldering.
5. Aligning the hole in a mounting bracket with one in the frame so that a bolt or screw may be put in place.

SECTION 3:
HEAT TRANSFER AND TEMPERATURE

UNIT 5:
THE NATURE AND EFFECT OF HEAT ENERGY IN REFRIGERATION

Heat is a form of energy that can neither be created nor destroyed. Heat can only be produced by changing some other kind of energy into heat energy. Energy may be classified as solar (from the sun's rays), electrical, chemical, mechanical, heat, nuclear, and the like. The ultimate source of all energy is said to be the sun.

REFRIGERATION DEPENDS ON ENERGY

A refrigerating system depends upon different kinds of energy. A few simple examples may emphasize this point: (1) Electrical energy is changed (said to be *transformed*) to mechanical energy by means of the electric motor. This form of energy may be used to operate a mechanical (vapor compression) refrigerating system. (2) In an *absorption-refrigerating system*, the heat from a gas flame or electrical heating element supplies the energy to operate the system.

Since heat energy and the effect of heat within a refrigerating system must be understood by the technician, craftsperson, designer, and engineer, this unit deals with a simple explanation of matter and relates this to refrigeration processes. The effect of changes in the quantity of heat energy on temperature, pressure, and volume are also reviewed. These principles are later applied in a practical way to refrigerants as an answer to the question, "How can a refrigerant be conditioned to cause it to either absorb or give off heat?"

RELATIONSHIP OF MATTER TO REFRIGERATION

Heat is always said to be associated with matter. *Matter* is defined as anything that occupies space and has weight. Matter may be found in any one of three basic states: solid, liquid, or gaseous, figure 5-1. All matter, regardless of its state, is composed of small parts (particles) called *molecules*. What happens to the speed, freedom (or position), and number of these molecules determines (1) the state of the material, (2) its temperature, and (3) its effect upon other parts or mechanisms of which it may be a part.

72

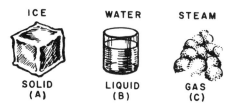

ICE   WATER   STEAM

SOLID   LIQUID   GAS
(A)   (B)   (C)

FIG. 5-1   Basic states of matter

THE STRUCTURE OF MATTER

Each molecule of matter is actually the smallest particle of a material that retains all the properties of the original material. For example, if a grain of salt was divided in two, and each subsequent particle was again divided (and the process was continued as finely as possible), the smallest stable particle having all the properties of salt would be a molecule of salt. The word *stable* means that a molecule is satisfied to remain as it is.

As fine as the molecule may seem to be from this description, each molecule is in itself made up of even smaller particles of matter. These particles are known as atoms. An *atom* is the smallest particle of matter having the properties of the material of which it is composed. By contrast, the atoms within a molecule are not always stable. Instead, atoms have a tendency to join up with atoms of other substances forming new and different molecules and substances, figure 5-2.

ARRANGEMENT AND MOVEMENT OF MOLECULES

The molecules in a given material are all alike. Different materials have different molecules. The characteristics and properties of different materials depend upon the

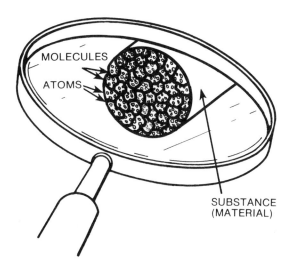

FIG. 5-2   The structure of matter

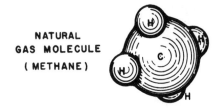

NATURAL
GAS MOLECULE
(METHANE)

FOUR HYDROGEN ATOMS COMBINED
WITH ONE CARBON ATOM

WATER MOLECULE

TWO HYDROGEN ATOMS COMBINED
WITH ONE OXYGEN ATOM

FIG. 5-3 Molecules

nature and arrangement of the molecules, figure 5-3. While millions upon millions of molecules form a material, the behavior of each molecule depends largely upon the material (substance) of which the molecule is composed.

Regardless of state, the molecules in a material are moving continuously. This movement or energy is called *kinetic* because it is an energy of motion. The addition of heat energy to a solid increases the kinetic energy of its molecules. In solids, this motion of the molecules is in the form of vibration where the particles never move far from a fixed position.

At a certain temperature, the addition of heat to a solid does not increase the motion or kinetic energy. However, the heat suddenly causes the solid to change to a liquid, figure 5-4. This is explained by saying that the addition of heat increases the freedom of the molecules (potential energy). There is greater freedom of motion of the molecules in liquids than solids. The molecules in gases have unlimited freedom to expand and fill their container.

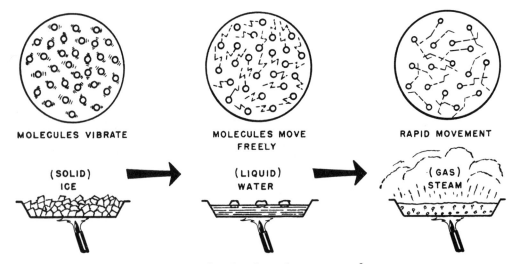

MOLECULES VIBRATE

MOLECULES MOVE
FREELY

RAPID MOVEMENT

(SOLID)
ICE

(LIQUID)
WATER

(GAS)
STEAM

FIG. 5-4 Freedom of molecules in basic states of matter

74

British Thermal Unit of Measurement

The *British Thermal Unit (Btu)* is a widely used standard of heat measurement in refrigeration and air conditioning. The Btu refers to the amount of heat that is contained in a substance and/or the level (quantity) of heat that is transferred from one body to another body. The standard of one British Thermal Unit (Btu) represents the amount of heat energy that is required to raise the temperature of one pound of water one degree Fahrenheit. For instance, 1 Btu of heat energy is absorbed by the water when one pound of water is raised from 100°F to 101°F. While the Btu is again defined with applications later in this book, this brief description provides an introduction to this unit of heat measurement.

APPLICATION OF HEAT AND HEAT ENERGY IN REFRIGERATION

Any one or combination of several changes may take place when heat, which is a form of energy, is added to a material or a substance:

- ▸ A rise in temperature may occur.
- ▸ The substance may melt, vaporize, or sublime.
- ▸ The substance may change size or color.
- ▸ The substance may be caused to exert a greater pressure.

When the heat energy is removed, the reverse of these effects takes place.

SENSIBLE HEAT

Sensible heat refers to that heat which produces a temperature rise as it is added to a material. The temperature rises because the molecules absorb heat energy and move faster when heated. As sensible heat is removed, the temperature drops. The molecules are said to move slower when there is a loss of heat energy.

COMPARISON OF TEMPERATURE AND HEAT

Temperature is the "hotness" of a material as compared to a fixed point in a temperature measuring system. What this means is that a temperature of 5° Fahrenheit (F) in a refrigerated space is five measuring units (degrees) above a fixed point of zero degrees in the same measuring system. (Temperature-measuring systems will be discussed in detail in another section.)

Using another example, when the specification plate on an electric motor indicates a temperature rise of 40 Celsius (C) degrees, the motor in normal operation may be expected to rise 40 Celsius degrees above the surrounding temperature. This temperature rise is produced by the addition of sensible heat developed from within the motor.

A clear distinction between heat and temperature may be made by using a simple illustration. Suppose a cup filled with warm water is removed from a container of warm water in which it has been immersed for some time. The temperature of the water in the cup and in the container is the same. In both cases, the molecules of water are moving at the same speed.

However, since the amount of water in the container is greater, it has a greater number of molecules and a greater quantity of heat. This example shows that the *quantity of heat* is related to the sum-total energy of the molecules. The average speed of the molecules is related to the *temperature*. This is an important distinction!

RELATING TEMPERATURE TO REFRIGERATION

When the compressor of a refrigerating unit is operating, the compressed refrigerant becomes heated and its temperature rises. Sensible heat, which results from the work done on the refrigerant to compress it (together with the heat of friction), is called *heat of compression*. It is this heat of compression that causes the temperature of the refrigerant to rise.

SUPERHEATING AND SUBCOOLING

In refrigerating systems, there is a so-called suction line that runs from the evaporator to the compressor. This line returns the heat-laden gases from the evaporator to the compressor. The line is so arranged that the refrigerant gas is warmed a few degrees as it picks up heat through the walls of the tubing and friction. Heat that is absorbed in this manner is known as superheat. *Superheat* refers to the heat contained in a gas beyond the amount required to maintain its boiling temperature. Since superheat causes a rise in temperature, it is sensible heat.

The absorption of heat in the suction line is deliberately planned to be sure that no liquid gets back to the compressor, since damage to the equipment may result if it does. In other words, *liquid floodback* and *liquid slugging* by the compressor are prevented by proper system design (metering device control).

The efficiency of some refrigerating systems is increased by precooling the liquid refrigerant before it goes through the restriction or metering device. This drop in temperature (*subcooling*) is accomplished by taking sensible heat away from the liquid through the walls of the tubing. Sometimes this is done by connecting the liquid line to the suction line. This plan works well because the returning refrigerant gas is considerably cooler than the entering liquid.

In brief review, sensible heat involves either a rise or fall in temperature. This temperature change may be felt or sensed by human touch or by the *sensing element* of an instrument. However, there is another kind of heat energy that seems to disappear. This is the same disappearing energy that Faraday discovered.

THE MYSTERIES AND EFFECTS OF LATENT HEAT

Latent heat has been spoken of as the kind of heat that results in a change of state without any change of temperature. The early scientists learned that the speed with which heat is added to a container produces no increase in temperature above the boiling point. The only increase is in the rate at which a liquid vaporizes and disappears. Since there is no change in temperature, and since the speed of molecules determines temperature, the average speed at which the molecules move is the same in either a liquid or gaseous state at the boiling point of refrigerants.

The question may be asked, "Then, what does heat accomplish?" The scientific explanation may be stated simply: The heat energy causes a change in the potential energy of the molecules as they change from one state to another. This fact is important because the heat energy in the refrigerant to be considered is the total energy. The total

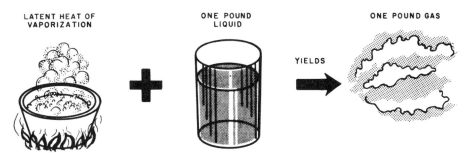

FIG. 5-5   Changing the state of a refrigerant

energy is the sum of the kinetic and the potential energies of all the molecules. These changes of state account for the ability of the refrigerant to transport heat energy from one part of the refrigerating system to another.

LATENT HEAT OF VAPORIZATION

This heat-carrying ability of the refrigerant is called the *refrigerating effect.* The refrigerating effect is found by determining the amount of heat that one pound of the refrigerant is capable of absorbing as it passes through the evaporator.

The refrigerant, as it boils in the evaporator and changes state from a liquid to a vapor, must absorb its *heat of vaporization.* Thus, the refrigerant vapor contains a great deal more heat energy than does the refrigerant in liquid form. This is true even though the temperature of both the vapor and the liquid is the same.

This amount of heat is the *latent heat of vaporization.* It differs in the amount per pound for each refrigerant, figure 5-5. The net refrigerating effect may be (and often is) considerably less than the heat of vaporization.

HEAT OF CONDENSATION

Another kind of latent heat that has only an indirect relation to refrigeration is *latent heat of fusion.* Because mechanical refrigeration systems followed refrigeration where natural ice was used, it is understood why the capacity of a system is measured in terms of how its cooling ability compares with that of ice.

The standard of measurement that is used is known as the *cooling effect* (ice-melting equivalent). The cooling effect refers to the effect produced by the melting of one ton of ice as it absorbs heat energy over a 24-hour period, figure 5-6. A machine with the capacity to produce this same cooling effect is classed as a *one-ton machine.* This does not mean that the machine can produce one ton of ice in a similar period. (It should be noted at this time that when British Thermal Units (Btu) of measurement are used, 1 ton of cooling capacity = 12,000 Btu.)

The ability to produce ice is referred to as *ice-making capacity,* figure 5-7. A refrigeration system may produce ice by removing latent heat to bring about a required change of state. However, there must first be the removal of a large amount of sensible heat in order to cool water down to the temperature at which it freezes (solidifies). The water freezes as a result of releasing the potential energy of the molecules. Thus, one pound of

EFFECT IN 24 HOURS

ONE TON OF ICE

FIG. 5-6   Cooling effect of one-ton machine

ice has less energy than one pound of water, even though they are both at the same temperature. This difference in energy is the *heat of fusion*.

Latent heat of fusion is the amount of heat that must be removed from a pound of material to cause it to solidify (change from a liquid state to a solid state without any change of temperature), figure 5-8.

In review, it may be said that the heat energy is a combination of the kinetic energy and the potential energy of molecules. Further, the molecules of any substance, such as a refrigerant, are in continuous motion.

THE EFFECT OF HEAT ENERGY ON PRESSURE AND VOLUME

Since there are millions of molecules of a refrigerant gas in even the smallest container, the molecules move or bump against each other and hit into the walls of the

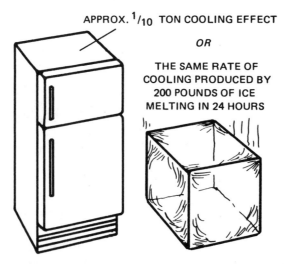

APPROX. $^1/_{10}$ TON COOLING EFFECT

OR

THE SAME RATE OF COOLING PRODUCED BY 200 POUNDS OF ICE MELTING IN 24 HOURS

FIG. 5-7   Cooling effect of a medium size household refrigerator.

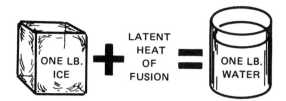

FIG. 5-8   Change from solid to liquid state

container. Any rise in the temperature of the gas is accompanied by an increase in the speed of the molecules. As the bombardment against the walls of a container increases, so the pressure increases, figure 5-9.

SAFETY PRECAUTION

The simplified illustration is used to explain relationships between heat and a refrigerant gas. Such an experiment should be performed in a laboratory with properly protected equipment and under careful supervision of experienced persons.

The cylinders or containers of a gas must be built strong enough to resist the greatest force or pressure that may be exerted against the walls. Such a pressure builds up according to the highest temperature to be met. This highest temperature must take into consideration that for some operations, where tanks of a liquid or gas must be

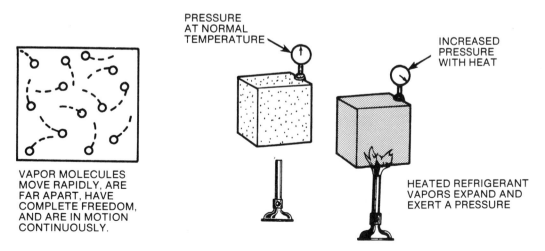

FIG. 5-9   Temperature rise affects movement and pressure

transported, there is direct exposure to the sun's rays. Two of the safety precautions that must be taken are:

▸ Cylinders should be protected against exposure to these rays.
▸ Care must be taken in disposing of used throwaway containers to be certain they are not thrown on trash fires.

PRESSURE USED FOR REFRIGERANT CONTROLS

The change of pressure that accompanies change of temperature may be put to good use in operating refrigerant controls. For instance, the *thermostatic expansion valve*, which is one of many metering devices, responds to pressure change. This change is caused by the change in superheat of the refrigerant gas.

HEAT ENERGY CAUSES EXPANSION AND CHANGES IN VOLUME

Change in temperature affects the size of solids and liquids. In general, as the temperature rises, a solid increases in size and is said to expand. This may be explained by the fact that as the speed of molecules increases, more space is needed for them to move. As the temperature goes down, the molecules slow down, and the solid contracts. This effect (among other applications) is used to operate thermostats for controlling refrigeration equipment.

The effect of temperature change is more pronounced in liquids than in solids. The force resulting from the expansion of a heated liquid is many times greater than that caused by the heating of a solid. This force is called *hydrostatic pressure*. If a refrigerant cylinder is filled at normal room temperature and then placed outdoors at a higher temperature, the cylinder may burst. This may happen if the increase in temperature is great and the internal pressure builds up beyond the strength of the container.

It is not practical to build cylinders strong enough to withstand such pressures since this cannot be done without unnecessary expense and other manufacturing difficulties. The simplest solution is to partially fill the cylinders so that a space is left at the top. This space permits the refrigerant to expand with greater safety.

Strict federal, state, and local government regulations must be observed for the safe filling, storing, transporting, and disposing of refrigerant cylinders. Periodic inspection and testing are important safety precautions.

SUMMARY

- A molecule is the smallest stable particle of a material having all of its physical properties.
- The arrangement and nature of molecules determines the characteristics and properties of each different material.
- Kinetic energy is an energy of motion; potential energy is stored up energy.
- Heat energy is the total energy of all the molecules in a given substance.
- Sensible heat is that heat which causes a change of temperature.
- Temperature is the relative hotness of a material with respect to a fixed reference point.

- The addition or subtraction of heat energy can cause a change of temperature, state, pressure, or volume of a refrigerant or other material.
- Change of temperature occurs in the superheating and the subcooling of refrigerants.
- Change of state of a refrigerant provides for the refrigerating effect.
- Change of pressure and volume (size) are used in the operation of refrigeration controls.
- Latent heat is heat energy needed to change the state of a substance without changing its temperature.
- The latent heat of vaporization is the amount of heat energy in a gas which is in addition to that found in the liquid at the same temperature.
- The latent heat of fusion is the amount of heat that must be added to a pound of material to change its state from a solid to a liquid or subtracted from a pound of the liquid to change it to a solid.
- The capacity of a refrigeration system is determined by comparing its ability to absorb heat to that of the heat absorbed by the melting of one ton of ice in a 24-hour period of time.
- Safety precautions and regulations must be followed in the filling, storing, transporting, and disposing of refrigerant containers and in the design and maintenance of refrigerating units.

ASSIGNMENT: UNIT 5 REVIEW AND SELF-TEST

A. RELATIONSHIP OF MATTER TO REFRIGERATION

1. Name four different kinds of energy.
2. Define matter.

Select the letter accompanying the words that best complete statements 3 to 7.

3. Matter exists in (a) three states, (b) two states, (c) one state.
4. A molecule is the (a) smallest, (b) largest, particle of a material having the physical characteristics of the material.
5. The molecules of a given material are all (a) different, (b) alike.
6. When heat is added to a body and causes only a change of temperature, the (a) potential energy, (b) kinetic energy, of the molecules is increased.
7. An increase in the movement of the molecules within a solid, liquid, or gas corresponds to (a) a drop in temperature, (b) no change in temperature, (c) a rise in temperature.

B. APPLICATIONS OF HEAT AND HEAT ENERGY IN REFRIGERATION

1. State three changes that take place as heat is added to a material.

Provide a word or phrase to complete statements 2 to 9.

2. Sensible heat, when added to a material, causes the temperature to
   _____ .

3. The heat of compression causes the temperature of a refrigerant to _____ .

4. The efficiency of refrigerating systems is improved by _____ the liquid refrigerant.

5. Latent heat results in a change in state without _____ temperature.

6. The total energy of a refrigerant is equal to the sum of the _____ and _____ energies of all its molecules.

7. When a liquid refrigerant changes state in the evaporator from a liquid to a gas, it must absorb the _____ _____ .

8. A two-ton refrigerating unit produces the same cooling effect as _____ as it absorbs heat energy over a _____ hour period.

9. The latent heat of fusion refers to the quantity of heat to be added to one pound of material to cause it to change from a _____ state.

C. THE EFFECT OF HEAT ENERGY ON PRESSURE AND VOLUME

Determine which statements (1 to 6) are true (T); which are false (F).

1. Heat energy is a combination of the kinetic and potential energies of a molecule.
2. Temperature rise is accompanied by a decrease in the speed of molecules.
3. When heated, a liquid exerts the same force as a solid.
4. Changes of pressure are used to operate control valves or systems.
5. Space is left at the top of cylinders for liquid refrigerants as a safety precaution against build-up of hydrostatic pressure by temperature changes.
6. Heat energy does not depend on the amount of material involved.

D. SUMMARY OF THE EFFECT OF HEAT ENERGY IN REFRIGERATION

Match each term or condition in Column I with the correct description or explanation given in Column II.

Column I	Column II
a. Molecule	(1) Energy needed to change the state of a substance without affecting its temperature.
b. Kinetic energy	
c. Sensible heat	
d. Latent heat	(2) A standard that compares the capacity of a machine to absorb heat energy over a 24-hour period with a similar effect of one ton of ice.
e. Latent heat of fusion	
f. Cooling effect	
	(3) Smallest stable particle of a material retaining all of its physical properties.
	(4) Causes a change of temperature.
	(5) Added amounts of heat energy in a gas.
	(6) Amount of heat that must be added to a pound of a material to change its state from a solid to a liquid.
	(7) Energy of motion.

UNIT 6:
BASIC METHODS
OF HEAT TRANSFER

The prime purpose of refrigeration is to produce desired temperatures within a specific area by transferring unwanted heat to a location where it is not objectionable. To understand how these processes are carried on, it is necessary to have a working knowledge of heat flow, how heat may be transferred, and how heat enters a refrigerated space.

There are three basic methods of heat transfer: conduction, convection, and radiation. Each of these methods is described and illustrated in this unit.

Next, the materials and operating mechanisms that serve as either protection against or part of control devices for heat transfer are reviewed in terms of each basic method. A few formulas and typical problems are included. These show some of the places where the service technician, designer, or manufacturer must be able to use simple mathematical calculations in the testing, servicing, design, or production of refrigerating equipment.

PART A: CONCEPTS AND PRINCIPLES OF HEAT FLOW AND HEAT TRANSFER

The natural tendency of heat is always to flow from a warm body to a cooler one. Heat is thus said to flow downhill. In this respect, heat flow may be compared to water flow. When water in two separate containers is at the same level, no water flows from one to the other. When the water level is either raised or lowered in one of the containers, the water flows from the higher level to the lower one, figure 6-1.

In a like manner, if two materials or substances are at the same temperature, no heat flows between them. However, if there is a temperature difference between them, heat flows from the warm body to the colder one. Figure 6-2A shows two liquids at the same temperature. If the temperature of one liquid is lowered to 32°F and forms ice, heat flows from the liquid at the 70°F temperature to the ice at 32°F, (B). The temperature of

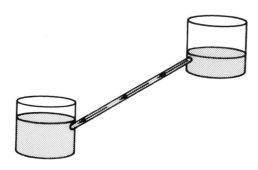

FIG. 6-1 Water flow

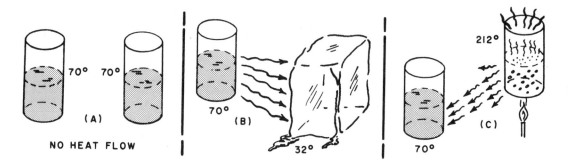

FIG. 6-2   Heat flow from warm to cold body

the liquid at (C) has been increased to 212°F. The heat now flows from this higher temperature of 212°F to 70°F.

The heat that must be transferred enters a refrigerated space in three main ways:

- The heat leaks through the walls surrounding the space to be cooled.
- The heat rushes into the refrigerated area as a door is opened.
- The heat is introduced into the refrigerated area in the material that is to be cooled, frozen, or stored.

Several steps are involved in the transfer of heat energy from the time it is picked up in the evaporator until it is given up (transferred or removed) at the condenser. The schematic drawing in figure 6-3 shows that in the transfer of heat, different temperatures are involved as the heat travels through different materials. In addition, the

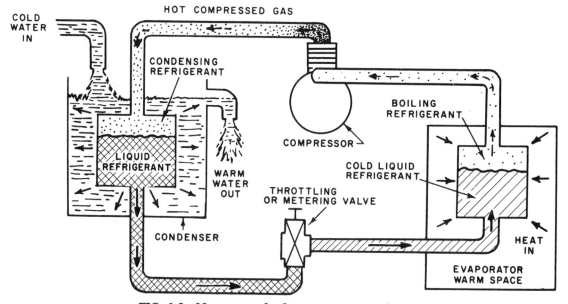

FIG. 6-3   Movement of refrigerant, water and air

refrigerant must be moved through the system, water or air must be circulated through a jacket surrounding the refrigerant condenser, and air or brine must be moved in the evaporator. All of these steps or processses depend on an understanidng of the three basic ways that heat moves or is transferred.

HEAT TRANSFER BY CONDUCTION

The flow of heat energy is always from hot to cold or from a high intensity to a low intensity. This flow continues until the speed of the molecules at the cold end increases with the absorption of heat and it reaches the same temperature. At the same temperature, the molecules in all parts have the same average activity. This method of transferring heat from molecule to molecule is known as *conduction*.

The conduction process may be illustrated with a heated steel part. As such a part is placed in cold water, the fast-moving molecules on the outer surface of the steel transmit the heat to the slower moving molecules of water and increase their speed. In turn, as the steel cools to the temperature of water, the speed of its molecules decreases.

While the outer surface of the steel is cooled, the center is still hot. The rapid motion of the molecules is transmitted to the outer surface by conduction. The collision of a fast-moving molecule at the center with a slow-moving one from the outer surface causes the one to lose speed while the other gains speed. This process continues until a uniform temperature is reached throughout the steel part.

Heat Transfer Through Conductors

While most metals are so-called good conductors of heat, different metals differ in their ability to conduct heat (conductivity), figure 6-4. In contrast to metals, most liquids, gases, and nonmetallic solids are partial *insulators*, with gases being among the best. Glass wool, rock wool, and loose asbestos packing are effective insulators because air is entrapped in the small cells of these materials. It acts as an insulator to increase the materials's effectiveness to prevent the transfer of heat by conduction. The order of heat conductivity and electrical conductivity for metals is the same.

Among the metals, silver is the best conductor, followed by copper and aluminum. Iron is low in comparison to copper. Copper is commonly found in refrigerating systems

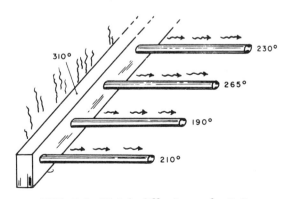

FIG. 6-4 Metals differ in conductivity

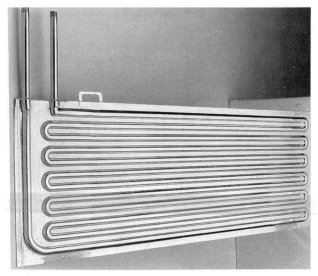

FIG. 6-5   Plate-type evaporator *(Courtesy of Tranter inc., Platecoil-Division)*

at places where the heat must flow through the walls of tubing. Iron must be used at times because of the refrigerant in the system.

Good conduction of heat is important in refrigeration systems. Conduction occurs through the walls of the evaporator, the condensor, and the tubing when the suction and liquid line are soldered together.

The photograph in figure 6-5 shows a plate-type evaporator. Heat is transferred by conduction from the refrigerated space to the boiling refrigerant inside the hollow ribs of the plate.

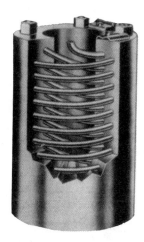 

FIG. 6-6   Internal construction details with refrig- FIG. 6-7   Dry-type cooling coil for liquids *(Cour-*
erant and water in separate coils   *tesy of Dunham-Bush, Inc.)*

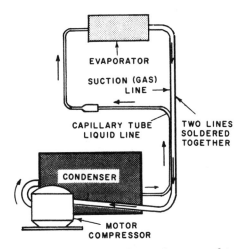

FIG. 6-8   Subcooling liquid and superheating refrigerant gas

Sometimes, simple coils of tubing are used for the same purpose. Figure 6-6 shows the internal design features of a dry-type cooling coil for liquids. The outside appearance (with other accessories) follows in figure 6-7. An aluminum casting surrounding the refrigerant coils and the liquid coils permits rapid heat transfer. In this type of unit a thermostatic expansion valve is used to control the refrigerant.

Another place where the heat of a liquid is transferred by conduction (molecular collision) is when the suction line is soldered to the liquid line. This is done to subcool the liquid refrigerant by the transfer of some of its heat energy to the cooler gaseous refrigerant in the suction line. Superheating of the suction gas is accomplished at the same time, figure 6-8.

The molecules of the liquid are at a higher temperature than those of the gas in the suction line. Therefore, the more rapid motion of the liquid molecules is transmitted by collision (conduction) through the walls of the tubing to the gas inside the suction line.

Another way to accomplish this same result is by using a *heat exchanger*, figure 6-9. The place of the heat exchanger in a refrigerating system is shown in the schematic drawing, figure 6-10.

Heat Conduction in the Operation of a Condenser

The conduction of heat also plays an important part in the operation of the refrigerant condenser. To repeat, the function of a condenser is to reject (transfer or remove) the

FIG. 6-9   Heat exchanger (*Courtesy of Superior Valve Company, Division of Amcast Industrial Corporation*)

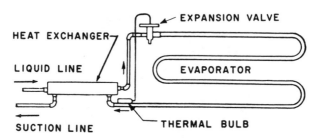

FIG. 6-10   Schematic drawing showing instatllation of heat exchanger

heat brought to it by the refrigerant gas. This gas is at a high temperature. The molecules of the gas are moving much more rapidly than the molecules of the material outside the tubing in which the gaseous refrigerant is contained.

Under these conditions, the heat flows from the inside out through the metal walls of the tubing by conduction. Heat is absorbed by the water surrounding the condenser tubing in which the refrigerant is contained. The energy of motion of the molecules, and their potential energy, is transferred through the tubing by conduction. Thus, condensation of the refrigerant is made possible by this removal of heat.

Figure 6-11 shows some of the internal construction of a typical water-cooled condenser. The heat travels by conduction through the walls of the tubes. The location of a shell and tube condenser in a compact refrigeration system is shown in figure 6-12.

Two other common forms of water-cooled condensers are illustrated in figure 6-13.

Area and Thickness Influence Heat Flow

The examples that have been used thus far operate on the principle of transferring heat by conduction. In each case, a difference of temperature has been cited as causing the flow of heat. A second factor that influences heat flow is the kind of material used in the construction of the parts. Then, there are the two other factors of area and thickness of material to consider. Drawings of heat exchangers show that there are many fins in the gas section and a *turbulator* in the liquid tube. The turbulator is a spiral-wound or spiral-shaped piece that is inside the liquid tube. It insures that warm liquid contacts the sides of the tube as much as possible. This improves the heat transfer.

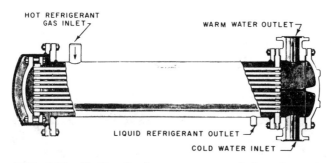

FIG. 6-11   Shell and tube type water-cooled condenser

FIG. 6-12 Water-cooled condensing unit *(Courtesy of Copeland Refrigeration Corp.)*

The fins serve another purpose. They increase the area of the metal in contact with the cold gas. This increases the effectiveness of transferring heat from the metal to the gas. The area of the fins provides a surface that is anywhere from 10 to 20 times as great as that in contact with the liquid. This increased surface area makes up for the difference in ability between the gas and the liquid to transfer heat from or to the metal.

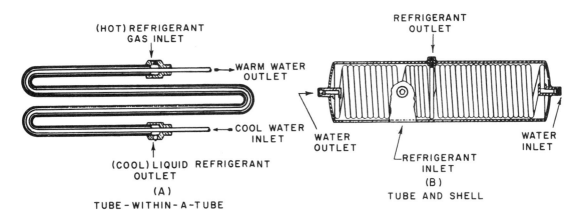

FIG. 6-13 Types of condensers

Section 3: Heat Transfer and Temperature

Calculating the Rate of Heat Transfer

The rate at which heat is transferred by conduction is calculated by using the simple formula:

$$H = \frac{(K) \times (A) \times (TD)}{(d)}$$

The letters denote the following:

H = The number of Btu that would be transferred by conduction in one hour.
K = The factor best used to describe the thermal characteristics (conductivity) of the material (metal) used.
A = The surface area affected by heat.
TD = The temperature differences between the inside and the outside surfaces of the metal.
d = The thickness of the metal.

What the formula says is that the heat transferred in one hour depends on the temperature difference. The greater this difference is, the faster the rate of heat transfer by conduction. Normally, a difference of from 8 to 15 degrees exists between the refrigerant and the outside material. The formula also indicates that the greater the area, the faster the transfer of heat. This explains why fins are used on tubing.

Heat is also transferred more rapidly in a material such as copper or aluminum than in iron. This means that the value of K in the formula varies with the material. A few values of K for different materials are given in table form. The materials, such as corkboard, rock wool, styrofoam, and fiberglass, are called *insulators* because of their low conductivity.

The K value indicates the Btu per square foot of area for each Fahrenheit degree of temperature difference per hour for each inch of thickness of the material. Heat conductivity values may be found in tables that are marked Btu/ft.^{2} (F°/hr.) (in. thickness), figure 6-14.

The denominator of the formula stands for thickness. This means, when numerals and values are substituted for letters in the formula, that three inches of a material transmit heat one-third as rapidly as one inch of the same material. An understanding of the conditions that affect heat transfer is valuable in diagnosing operating troubles and coil performance.

The *R-factor* is best used to describe the thermal characteristics of *insulators*. The R-factor = $\frac{1}{\text{K-factor}}$. For example, referring to figure 6-14, the R-factor of the Styrofoam = $\frac{1}{0.22}$, or 4.55 per inch.

HEAT TRANSFER BY CONVECTION

Convection is the second of the three methods to be described for heat transfer. *Convection* refers to the transferring of heat by the circulation of heated portions of a fluid. As a gas or a liquid gets heat from a hot surface, the heat expands the fluid, causing it to become less dense. The cooler and denser surrounding fluids settle and push the

Material	Conductivity (K Factor)
silver	2834
copper	2724
aluminum	1530
iron	331
steel	314
concrete	6.5
window glass	7.1
water	4.2
wood	1.00*
corkboard	0.27
rock wool	0.30
fiber glass	0.30

*varies according to kind of wood

FIG. 6-14   Heat conductiivty of materials Btu/ft.² (F°/hr. (inch of thickness)

less dense portion upward. This method of transmitting heat by upward currents *(convection currents)* caused by heat is called convection.

One of the best examples of convection is the cooling of a refrigerator, figure 6-15. The heat is transferred upward by convection, explaining why the warmest air is usually

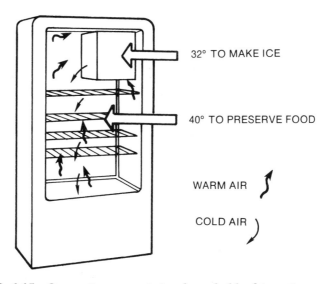

FIG. 6-15   Convection currents in a household refrigeration

found near the top. This warm air near the top comes in contact with the cold evaporator. As it cools, it becomes denser and starts to settle.

Convection currents are thus set in motion. As the cool air settles, it picks up heat from the warm food and the walls of the refrigerator through which the heat has leaked. This heat is carried to the evaporator by the convection currents.

Sometimes a baffle is placed to direct the convection currents in a large refrigerated space, figure 6-16. The drawing shows such a baffle installed in a walk-in cooler. A drip pan and drain is provided at the bottom of the baffle to take care of the condensation and/or the moisture from defrosting. The circulation of air in an overhead-coil refrigerator cabinet and the difference in temperatures are indicated at different places in the cabinet.

Many refrigerators depend on natural convection currents to carry away the heat rejected, given up, removed, or transferred at the condenser, figure 6-17. It is easy to understand that anything which interferes with convection currents should be eliminated. Such interference may be produced by dust, papers, or rags.

Forced Convection Condensers

The transfer of heat by convection currents is greatly speeded up by using a fan or pump to circulate the medium that carries the heat. In the refrigeration cycle, the refrigerant is circulated from the evaporator to the condenser. The refrigerant picks up heat in the evaporator and carries it to the condenser, where the heat is rejected. The refrigerant is then pumped to the evaporator, where it picks up more heat, and the convection circuit is completed. The water flowing over the condenser coil to carry away the rejected heat also constitutes a convection current.

The phantom view in figure 6-18 shows how in actual practice a fan draws cold air under the freezer and up over the condensing unit. The warm air is discharged through

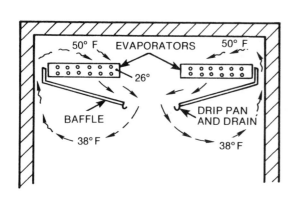

FIG. 6-16 Baffles direct convection currents in large refrigerated cabinet

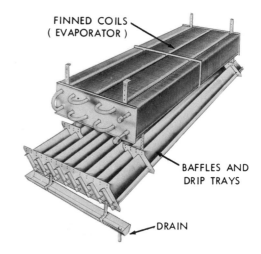

FIG. 6-17 Natural convection coils (*Courtesy of McQuay, Inc.*)

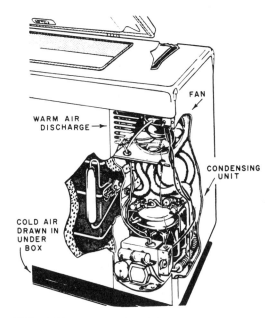

10 LBS. REFRIGERANT
85.7 Btu/LB.

27 LBS.
WATER
80°F

27 LBS.
WATER
65°F

10 LBS. REFRIGERANT
45.2 Btu/LB.

HEAT GIVEN UP   =   HEAT GAINED BY
BY REFRIGERANT   COOLING MEDIUM

FIG. 6-18   Fan used with forced convection condenser

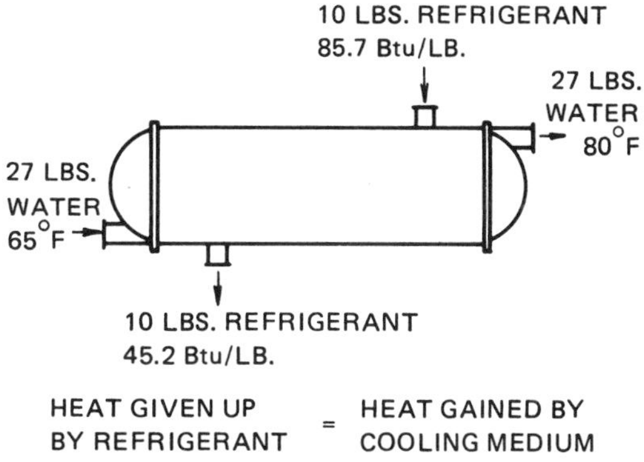

FIG. 6-19   Balance of heat in a water-cooled condenser

the louvers in the cabinet. The action of the fan in drawing air over the condenser provides for the more rapid removal of heat by forced convection currents.

This same action may also be used on large commercial air-cooled units. The air on the entering and discharge side can be measured with a thermometer. The discharge temperature is higher than that of the entering air. This indicates that the rapid air flow (convection currents) has picked up heat. This heat is carried from the hot refrigerant gas through the metal tubes by conduction. It is then removed from the tubes by convection currents of air.

Water-Cooled Condensers

Forced convection currents of water and refrigerant may be circulated through water-cooled condensers. The currents carry away the heat transferred from the hot refrigerant gas through the walls of the condenser to the cold water. The actual quantities of water used by a condenser are given in figure 6-19. One of the basic principles of condenser operation is that the heat given up by the refrigerant equals the heat gained by the cooling medium. Note that 10 pounds of refrigerant with 85.7 Btu of heat enters the condenser and, after producing a temperature change in the 27 pounds of water from 65°F to 80°F, leaves the condenser with only 45.2 Btu/lb.

Taking these values and working out mathematically how much heat is gained or lost, the results show that the answers are equal, figure 6-20. Each pound of refrigerant passing through the condenser loses 40.5 Btu (or 405 Btu for the 10 pounds).

At the same time, the 27 pounds of water pass through the condenser, where its temperature is raised 15°. The heat gained is equal to the product of the weight of water,

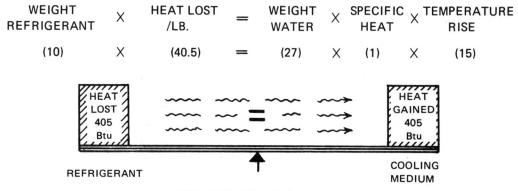

WEIGHT REFRIGERANT	X	HEAT LOST /LB.	=	WEIGHT WATER	X	SPECIFIC HEAT	X	TEMPERATURE RISE
(10)	X	(40.5)	=	(27)	X	(1)	X	(15)

FIG. 6-20   Heat balance

its specific heat (1), and the rise in temperature (15), or 405 Btu. This is the same value as the heat lost by the refrigerant. Although there actually is some heat lost by radiation, the quantity is very small and is usually not considered in such calculations.

This concept of heat balance, where the heat gained equals the heat lost, applies to all condensers including those that are air cooled and evaporative cooled.

Water Chillers

The water chiller also operates on principles of forced convection currents. The water to be chilled is circulated through tubes around which a liquid refrigerant passes. The moving refrigerant carries away by convection the heat that it receives from the water through the walls of the tubing. The example in figure 6-21 shows that the heat gain of the refrigerant is from 32.84 Btu/lb. to 72.68 Btu/lb. In this same period, 9.6 gallons of water is cooled from 57°F to 52°F.

The simple computations in determining the heat balance through this water chiller are shown mathematically in figure 6-22. Note that the weight of the water is found by multiplying the number of gallons of water by the weight of one gallon, which is 8.33 pounds. Again, in this example, the heat gained by the refrigerant equals the heat given up by the medium being cooled.

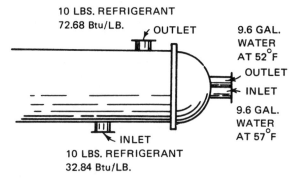

FIG. 6-21   Heat balance in a water chiller

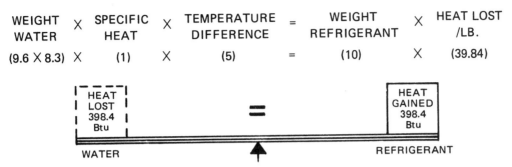

WEIGHT WATER	X	SPECIFIC HEAT	X	TEMPERATURE DIFFERENCE	=	WEIGHT REFRIGERANT	X	HEAT LOST /LB.
(9.6 X 8.3)	X	(1)	X	(5)	=	(10)	X	(39.84)

FIG. 6-22   Heat balance through flooded evaporator

Basic Types of Evaporators

The two basic types of evaporators in common use today are (1) flooded, and (2) direct expansion. Another classification refers to the evaporator surface; for example, bare pipe, finned-tube, plate, or adaptations of these, figure 6-23.

Forced Convection Cooling Coils

Evaporators frequently depend upon forced convection currents to bring heat to them for absorption. A *cooling coil* with motor-driven fans to provide forced circulation is illustrated in figure 6-24. Such a unit is effective in a refrigerator where the doors are frequently opened, allowing warm air to enter.

Ventilation of Compressors

The transfer of heat by convection currents is also important to the cooling of compressors of refrigeration units. A water-cooled unit may be placed in a room with a limited natural circulation and a small volume. However, an air-cooled unit must be

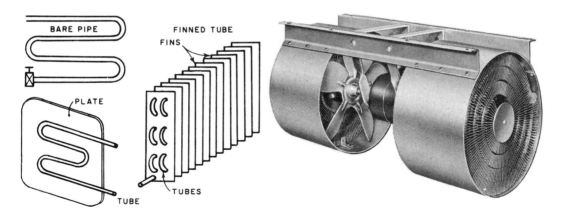

FIG. 6-23   Types of evaporator surfaces

FIG. 6-24   Circular forced convection cooling coil
(Courtesy of Peerless of America, Inc.)

95

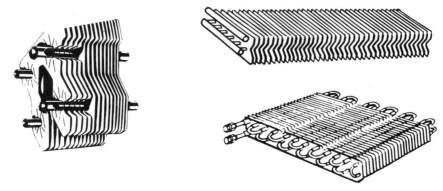

FIG. 6-25 Examples of convection cooling coils

located in a room with a high enough volume to prevent overheating the motor. Tables are found in handbooks that deal with compressor ventilation requirements for various sizes of condensing units. Such information is included in the Appendix in Table 6.

Cooling Coil Design

Attention is directed to the fact that in all devices using air convection currents, fins are attached to the tubing, figure 6-25. The purpose of fins to provide adequate surface areas in contact with moving air so as to rapidly transfer heat. The heat still needs to travel by conduction through the metal in the tubing and in the fins. It is important, therefore, that there be a solid and permanent mechanical contact between the fins and the tubing.

Many types of coils are being made using these combinations: copper tubing and aluminum fins; copper tubing and copper fins; and aluminum tubing for ammonia. The fins are bonded to the tubing either by mechanical drive fits or mechanical devices or by chemical processes such as dipping the whole unit in a tinning bath after assembly, figure 6-26.

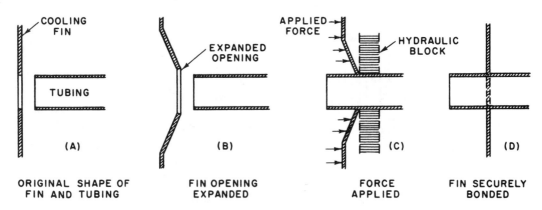

FIG. 6-26 Bonding cooling fins mechanically to tubing

Some manufacturers use a blow or hydraulic force to expand the tube against the fin. One method of bonding tubing to off-center fins is described in figure 6-27.

The fins are spaced to vary the capacity of the coil and to compensate for the depth of the coil. The fin spacings vary from 1/2" to 1 1/2". Greater fin spacing is required on deep coils in order to minimize air restriction. Some coils are designed and built of one piece of tubing, while other manufacturers make all bends separately. The bends are later joined to the straight lengths by silver brazing or other brazing processes.

Internal devices are sometimes included within the tubing to swirl the refrigerant and thus improve heat transfer.

Calculating the Transfer of Heat by Convection

Transfer of heat by moving fluids (air or water) may be easily calculated. In each case, the carrier of heat goes through a change of temperature, which is classed as sensible heat. Such heat may be found by the formula:

(H)	=	(M)	×	(c)	×	(TD)
Heat in Btu		Mass in lbs. of the heat transfer medium		Specific heat		Temperature change of the medium in the process

Air circulation is usually measured in cubic feet per minute (cfm). Therefore, the weight of a cubic foot of air at a given temperature must be known. When this weight is multiplied by the cfm, the result indicates the pounds of air circulated per minute. This value may be substituted for M in the formula.

Water flow is found in much the same manner. The water flow in gallons per minute (gpm) is multiplied by the weight of each gallon (8.33 lb. for water). The product, which may be substituted for M, gives the number of pounds per minute (ppm).

As a review, one example is given to show how the formula is used.

EXAMPLE

Water is circulated through a water-cooled condenser at the rate of 10 gallons/minute (10 gpm). The temperature as it enters the condenser is 65°F. As it leaves it is 75°F. Calculate the heat removed by the condenser per minute.

ANSWER:

$$\text{Heat} = (M) \times (c) \times (TD)$$
$$= (10 \times 8.33) \times (1) \times (10)$$
$$= 833 \text{ Btu/min.}$$

HEAT TRANSFER BY RADIATION

A third method of transferring heat is known as *radiation*. Unlike both conduction and convection (which require solids or fluids to transmit energy), the energy waves called

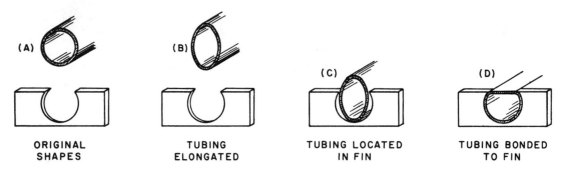

| (A) ORIGINAL SHAPES | (B) TUBING ELONGATED | (C) TUBING LOCATED IN FIN | (D) TUBING BONDED TO FIN |

FIG. 6-27 Bonding off-center fins mechanically to tubing

radiation move freely through space. The ordinary incandescent lamp provides an example of radiation. Within the lamp, all of the air is removed (evacuated) and there is no connection between the filament and the outer glass shell. The heat energy cannot be transmitted by conduction or convection, but it is transmitted by waves.

Reflecting, Transmitting, or Absorbing Heat Rays

Three things may happen to energy waves. They may:

1. Be reflected as they strike a body
2. Pass through the material
3. Be absorbed

Energy absorption and reflection depend on certain laws. A material usually has good absorption properties when it is dark and rough, figure 6-28. Surface area and temperature also determine the rate at which a body may lose heat. The greater the surface area and temperature, the greater the cooling rate. Color and finish influence the ability to reflect. The smoother a surface is finished and the brighter and lighter its color, the better the conditions for reflecting heat rays.

To summarize, radiant heat has many unusual properties. As it strikes a mirrored surface, it may be reflected. By contrast, if it hits an opaque object that casts a shadow, it

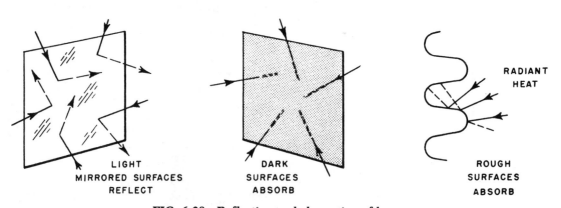

| LIGHT MIRRORED SURFACES REFLECT | DARK SURFACES ABSORB | RADIANT HEAT ROUGH SURFACES ABSORB |

FIG. 6-28 Reflection and absorption of heat rays

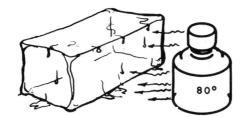

FIG. 6-29   Cold surface absorbs radiant heat

may be stopped. Then, too, as radiant heat passes through materials that transmit light, it does not heat such materials to any extent. In other words, it is only when a material interferes with the free passage of heat waves that it is heated.

A cold surface absorbs (takes) radiant heat from a warmer body, making it cooler, figure 6–29. A shiny white enameled surface reflects radiant heat energy. The transfer of heat energy by radiation plays only a small part in refrigeration and never enters into calculations of heat transfer. However, the tendency of materials and surfaces to reflect heat waves is used to good advantage in preventing heat transfer.

SUMMARY

PART A: CONCEPTS AND PRINCIPLES OF HEAT FLOW AND HEAT TRANSFER

- Heat transfer refers to the movement of heat energy from one place to another.
  - A temperature difference (TD) is necessary before heat energy flows or moves.
  - The natural tendency of heat flow is from a high to a low temperature level.
  - The compressor (or boiler) provides the energy necessary to transfer heat from a low to a high level.
- Conduction is a method by which heat is transferred by actual collision of the molecules. Conduction occurs best in solids through the metal walls of:
  - Plate evaporators; bare pipe coils; liquid and suction lines; the tubing of water- or air-cooled condensers.
  - The formula used to find the rate of heat transfer by conduction is:

$$H = \frac{(K) \times (A) \times (TD)}{(d)}$$

- Convection is a method of transferring heat by the actual movement of heated molecules (convection currents).
  - Natural convection currents of air carry heat:
    - ✓ From the source (food or walls) to the evaporator so that it can be removed.
    - ✓ Away from condensers.
  - Forced convection currents are produced by the circulation of:
    - ✓ Refrigerant in the system.
    - ✓ Air through a finned condenser.
    - ✓ Air through finned cooling coils.

✓ Air to operating units.
✓ Water through water-cooled condensers.
✓ Water and refrigerant through water chillers.
- The transfer of heat by moving fluids may be calculated by the formula:

$$H = (M) \times (Sp.\ Heat) \times (TD)$$

- Radiation is a third method of transferring heat, using energy waves that move freely through space. These energy waves may be reflected, penetrate the material, or be absorbed.

ASSIGNMENT: UNIT 6 REVIEW AND SELF-TEST

PART A: CONCEPTS AND PRINCIPLES OF HEAT FLOW AND HEAT TRANSFER

A. HEAT FLOW AND HEAT TRANSFER

Provide the terms or conditions that best complete statements 1 to 5.

1. The natural tendency of heat is to flow from a _____ to a _____ .
2. There is no heat flow unless there is a _____ between two materials or substances.
3. The liquids in two containers are at 205°F and 85°F. Heat flows from the liquid at _____ to the liquid at _____ .
4. Two ways in which heat enters a refrigerant are: (a) _____ and (b) _____ .
5. The term that describes the transfer of heat by molecular collision is _____ .
6. List three basic methods of transferring heat.
7. Name the method by which heat flows through (a) a solid, (b) a liquid, (c) a gas.

B. HEAT TRANSFER BY CONDUCTION

1. Explain briefly what takes place in heat transfer by conduction.

Select the letter representing the term that best completes statements 2 to 7.

2. An aluminum surface is cooled to the same temperature as a water bath. At this temperature, the molecules of both materials have (a) different speed, (b) the same average speed, (c) high and low intensity.
3. Which of the following provides the best insulation medium? (a) metals, (b) liquids, (c) gases, (d) nonmetallic solids.
4. The best conductor in the following group is (a) iron, (b) lead, (c) copper, (d) aluminum, (e) wood.
5. In a plate-type evaporator, heat is transferred from a refrigerated space to the boiling refrigerant by (a) radiation, (b) convection, (c) conduction.
6. The unit in the refrigeration system in which heat is rejected is the (a) condenser, (b) evaporator, (c) compressor.

7. Heat flows through the metal walls of tubing by (a) radiation, (b) convection, (c) conduction.
8. List three metallic conductors in their order of conductivity.
9. Explain briefly what occurs when the molecules of a refrigerant gas move more rapidly than the liquid surrounding the condenser tubing.
10. Make a simple schematic sketch of a water-cooled condenser. Label the refrigerant and water inlets and outlets.
11. List three factors affecting heat flow.
12. Calculate the rate at which heat may be transferred by a conduction for each set of valves given in the table. Use the formula: H = (K) × (A) × (TD) ÷ (d).
    NOTE: Determine K factor from the table. Round off the answer to nearest whole number.

Material	Conductivity (K)	Exposed Surface Area (A in. Sq. Ft.)	Temp. Range (F) TD	Thickness of Material (in.)	Rate of Heat Transfer Btu/hr.)
A Copper		10	8	1/4	
		20			
B Aluminum		10	8	1/4	
		20			
C Rock wool		36	12	1	
D Wood		36	12	1	

C. HEAT TRANSFER BY CONVECTION

1. State whether heat flows by convection, conduction, or radiation through: (a) copper tubing, (b) water, (c) Refrigerant 12.
2. What must be present for heat to flow by convection?
3. Make a simple line drawing of a large refrigerated cabinet with overhead coil. Label all the important parts, and show the convection currents and different temperatures.
4. Match each function in Column II with the correct term in Column I.

Column I

a. Forced convection condenser
b. Water-cooled condenser
c. Water chiller
d. Forced convection cooling coils

Column II

(1) A unit in which convection currents of liquid carry away the heat transferred from a refrigerant gas.
(2) A unit placed in a room with limited space and natural circulation.
(3) A unit provided with a fan that draws air in over the condenser to provide for more rapid removal of heat.

(4) A unit in which the moving refrigerant carries away by convection the heat it receives from the liquid to be cooled.

(5) A unit in which a fan circulates heated air to the evaporator for absorption.

5. Make a simple line drawing of a water-cooled condenser. Label the main parts and values. Use values of 100 Btu, 20 Btu, and 5 lb. for the refrigerant; 100 lb., 80°, and 76° for the cooling medium. Prove mathematically the concept of heat balance.

6. Identify the function of fins.

7. Name the method by which heat travels through tubing and fins.

8. Name two combinations of fins and tubes used on dry coils.

9. State how a deep coil may be designed to cut down air restriction.

10. What design feature is included to improve the heat transfer to the boiling refrigerant?

11. Calculate how much heat is transferred by convection, according to the data contained in the table.

    NOTE: Use the formula: H = (M) × (c) × (TD); the weight of water at 8.33 lb./gal.; the weight of air at the temperatures indicated as 0.077 lb./cu. ft.; (M) for air = weight at temperature × cfm; for water, weight/gal. × gpm.

Heat Transfer Medium		M	Specific Heat (c)	Temp. Change of Medium (TD) (F)	Transfer of Heat by Convection (H) (Btu/min.)	Rate of Circulation
A	Water		1	15		20 gpm
			1	30		15 gpm
B	Air		0.24	6		510 cfm
			0.24	8		850 cfm

D. HEAT TRANSFER BY RADIATION

1. Describe two things that may happen to heat energy waves.

2. State by what method heat may flow through a vacuum.

3. State the effect on a refrigeration system of each of the following conditions:

   a. Smoothness of surface.

   b. Rough, dark-colored materials that interfere with the free passage of heat waves.

   c. Shining water enameled surface.

   d. A cold surface.

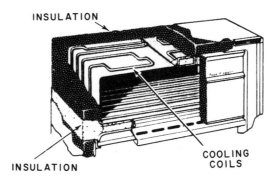

FIG. 6-30   Insulation around food freezer

PART B: INSULATION

Insulation is the name given to any material that helps prevent the transfer of heat by any one or combination of the three different methods: conduction, convection, or radiation. If an ideal insulation existed, it would be comparatively easy to refrigerate a space and keep it refrigerated. However, there is no perfect insulation material. Insulation materials (figures 6-30 and 6-31) merely slow down the transfer of heat so that the refrigeration system moves the heat faster than it leaks in. Enclosures are made of the best insulation material available at a reasonable cost.

The effect of temperature and moisture on insulating materials, the function of moisture vapor seals, and insulation for ultra-low temperatures are treated in this section. This technical knowledge then becomes background information to be applied

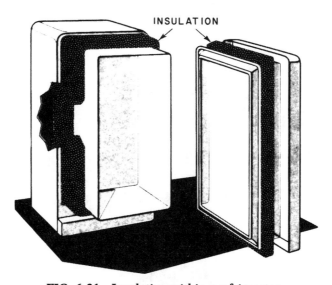

FIG. 6-31   Insulation within a refrigerator

against conduction, convection, and radiation. A few formulas are given, and a series of steps are developed to show how the rate of heat flow and the total heat leakage may be determined in a refrigeration system.

CONDITIONS WITHIN INSULATION

Insulation is not a positive barrier to heat flow but serves to retard the flow of heat. As such, a temperature on the warm side of a sheet of insulation does not suddenly become a low temperature on the cold side. The rate of heat flow is reduced gradually. The temperatures within the insulation drop gradually from the temperature of the hot outer surface to that of the cool inner surface.

Figure 6-32 shows the gradual change of temperatures within an insulating material. These temperatures change with any changes in a warm outer temperature or a cool inner temperature. If a room temperature drops from 80°F to 60°F while the refrigerator temperature stays at 40°F, the temperatures within the insulation move toward the warm side. On the other hand, if the refrigerator temperature drops and the room temperature remains the same, the temperatures inside the insulation move toward the cool side. Heat always leaks toward the cooler area.

Moisture in Insulation

Water is a good conductor of heat. For example, water conducts heat about twelve times as fast as cork. Thus, if water gets into an insulation, the insulating value of the material is greatly reduced.

It is necessary, therefore, that insulation be absolutely dry when first installed and be sealed perfectly so that it stays dry. The cell walls of all insulations absorb slight amounts of moisture. The possible exceptions include some of the cellular foam insulations made of glass, rubber, plastic, and the like. Even the cell walls of cork (a good insulator) absorb a slight amount of water, which lowers its insulating value.

Water vapor may condense between the fibers of fibrous-insulating materials, and moisture will be absorbed by the cell walls. Insulations of this type, regardless of

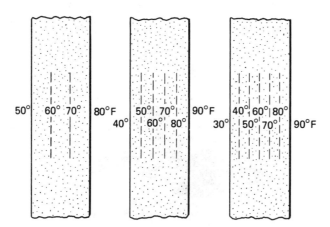

FIG. 6-32 Gradual temperature changes within an insulating material

whether they are made of animal, vegetable, or mineral fibers, are very susceptible to moisture. Such installations must be sealed perfectly whether they are used in loose or in batt form.

Water that gets into insulating material as a liquid is a rare and accidental happening. This sometimes occurs when washing the inside of the refrigerated space, or through spillage. It may also occur from rain beating against the outside wall of an insulated space.

The major moisture problem, however, is water vapor contained in the air, which is the source of most of the water than enters insulation. Strictly speaking, air is a mixture of air and water. The air and water vapor exist independently at the same temperature, but each with its own separate pressure. The pressure of the air alone is very large when compared to the pressure of the water vapor. The combination of air and water vapor pressure produces atmospheric pressure.

Moisture Vapor Seals

Actually, the vapor pressure outside a refrigerator pushes the moisture right through the insulation. This action may be prevented by a tight barrier placed outside the insulation. This barrier is called a *vapor seal.* Vapor seals are made of materials that moisture cannot penetrate. These can be metal foil, a layer of asphalt, paper soaked in asphalt, or some similiar moisture-vapor proof material. Metal foil is considered one of the best seals.

A vapor seal must be sealed tightly at the edges where it laps over the seals of the sides, top, and bottom of the refrigerator. The moisture seal must be continuous. Even a pinhole defeats the purpose of the vapor seal, so there must be no breaks of any kind.

The vapor seal must be on the warm- or high-vapor pressure side of the insulation, since this is the side through which water vapor enters. In the summer, the warm side of a refrigerator, cold storage room, or an air-conditioned house is the outside of the insulation. In these applications, therefore, the vapor seal should be on the outside. By contrast, a refrigerator operating in very cold weather (when the medium surrounding the refrigerator is colder than the refrigerated space) has it warm side on the inside of the insulated wall. The vapor seal in this case should be on the inside wall.

Styrofoam and ureafoam do not require vapor barriers. These materials maintain R-factors of 4–5/in. and 6/in. in most environments.

Dew Point Temperature

The temperatures inside the insulation drop gradually from the warm side to the cold side, figure 6-33. If there is no vapor seal on the warm side (or if it is broken or punctured), air or vapor enters the insulation and flows toward the cold side until it reaches its dew point temperature. At this point, the vapor in the air condenses into water. The water is then absorbed by the insulation.

The location of the dew point temperature inside the insulation shifts with temperature difference. Where the insulation is too thin or is wet, or for any other reason does not adequately insulate the wall, the dew point location may actually be outside the insulation. The warm side may be so cool that it is below the dew point. Under such a condition, the warm outside air condenses on the outside of the cabinet and gathers as *sweat.*

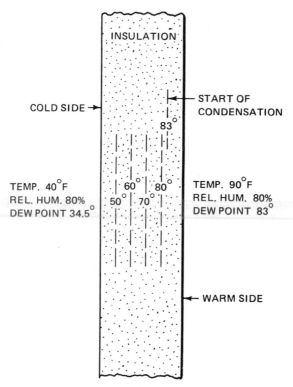

FIG. 6-33 Effect of vapor pressures within insulation

When sweating occurs, the insulation is either not thick enough for the temperature difference or it has deteriorated (usually because it has become wet). Sweaty spots appearing on the outside shell of the cabinet indicate:

- That the insulation was not packed properly into the wall originally, leaving open places with insufficient or no insulation.
- That the vapor seal is not tight, allowing water to form.
- That in constructing the frame, metal braces or other conductors extend from the cold inside liner to the warm outside shell.

A refrigerated cabinet must be well enough insulated:

- To keep the heat load that must be removed to a minimum.
- To always keep the dew point location inside the insulation and the moisture vapor seal, thus preventing the outside from sweating.

The dew point temperature determines the thickness and insulating value of the insulated wall to prevent sweating of the cabinet.

The difference in vapor pressures between the conditions outside and inside the refrigerator can push a surprisingly large quantity of vapor through the insulation. For instance, if the vapor seal of an eight- or ten-foot freezer is badly torn so that it has practically no sealing effect, one quart of water per month may be condensed in the insulation. This much water ruins the insulating value of the cabinet, causes it to sweat in damp weather and makes the machine run continuously as it is unable to keep the inside temperature at the required level.

In addition to decreasing the insulating value, water may cause actual damage to the insulating material. Some types of materials deteriorate if wet, while others sag and leave open spaces. The water in the insulation of ice cream cabinets, home freezers, other refrigerators, or rooms held below 32°F freezes into ice. The ice grows in size and eventually bulges the cabinet walls or causes damage within the unit.

The need for a perfectly tight vapor seal can hardly be overemphasized. The colder the inside of the cabinet, and the greater the temperature and vapor pressure differences between the outside and inside, the greater the need for a seal that is tight originally and that stays tight indefinitely. Also, it is necessary to seal carefully any opening where tubing enters the cabinet or insulating material.

Insulation for Ultra-Low Temperatures

Ultra-low temperature cabinets are designed for use at temperatures down to −150°F or lower. Such units are being used more and more by industry as research reveals processes that require these temperatures. It is evident that such cabinets require better insulation and construction than do the 0°F cabinets. Insulation thicknesses up to 12 inches are used, and extra care is taken with vapor-proofing to prevent the entrance of moisture.

When a refrigerator is to maintain a low temperature for a long time, it is usually desirable to use an insulation that has a high thermal capacity (mass-type insulation), such as corkboard. Any interruption to refrigeration will not be serious, because the insulation absorbs the heat into itself. On the other hand, when rapid fluctuations in temperature are required, an insulation of low-thermal capacity (storage ability) is used. The reflective type of insulation is the answer to this problem.

INSULATION AGAINST CONDUCTION

Some materials are far better insulators than others. The insulating value of a material is rated by its *conductivity* or *K* factor. The K factor represents that part of a Btu which in one hour will pass through one square foot of a material, one inch thick, if the temperature difference is 1°F.

For most commonly used insulations, this conductivity is between .20 and .30. This is to say that for a material such as cork or rock wool, about 1/4 of a Btu per hour will be transmitted through one square foot of insulation that is one inch thick, if one side is one degree warmer than the other. The lower the conductivity (K), the better the insulation.

The thickness of an insulating material, measured as the distance from the hot to the cold parts, affects the rate of heat flow. The closer the hot and cold parts are together, the greater the rate of heat flow. If these parts are one inch apart, the rate of flow will be twice as great as when they are two inches apart; three times as great as when they are three inches apart, and the like.

Putting the conductivity of a material and its thickness together, the amount of heat transmitted by one square foot of insulation in one hour may be found by multiplying the temperature difference (TD) in F degrees by K and then dividing by the thickness (d) of the insulation in inches. This is expressed in the following formula for rate of heat flow:

$$\frac{\text{Rate of Heat Flow}}{(\text{Btu/ft.}^2/\text{hr.})} = \frac{(K) \times (TD)}{(d)}$$

Amount of Heat Reaching a Refrigerator

The total amount of heat passing through the walls of a refrigerator in one hour may be found by multiplying the rate of heat flow per square foot per hour by the total outside area of the refrigerator in square feet. The result is the total heat leakage of the refrigerator in Btu/hour. This value, when multiplied by 24, gives the total Btu/day. The total amount of heat that passes through the walls per day is called the *total heat leakage per day*.

SAFETY PRECAUTION

Walk-in coolers and other refrigerated enclosures should be equipped with safety hardware so that the doors may be opened from the inside.

The formula assumes that the only insulating value the wall has is due to the insulation in the wall and that the insulation is all one kind. This may not always be true, as is shown in figure 6-34. Sometimes the wall is a built-up one with wood facings on both the inside and the outside. The wooden facing boards have insulating value that is in addition to the insulation itself. For economy, some walls are designed using two or more kinds of insulation.

The K values of the several materials used in constructing an insulated wall are different. Therefore, in order to get the conductivity of a built-up wall, each part must be computed separately. The term *conductance* (C) is used in place of conductivity to denote the Btu that pass through a material of a specified thickness rather than for a thickness of one inch. The conductance (C) is determined by dividing the conductivity (K) by the actual thickness. Thermal transfer characteristics of common building materials are given in Table 8 of the Appendix.

Total Heat Leakage

When computing total conductance for a given wall, the mathematical steps may be simplified if a *coefficient of transmission* is first determined. The total coefficient of transmission is the Btu/ft.² of wall for a one-degree temperature difference and is identified as *U*. This coefficient is found by dividing the conductance of each insulating material into 1. The results are then added together. This sum is then divided into 1. The result is the heat transmission coefficient (U). These statements may be reduced to a simple formula:

Total Heat Leakage = (U) × (A) × (TD) × (t) Time in Hours

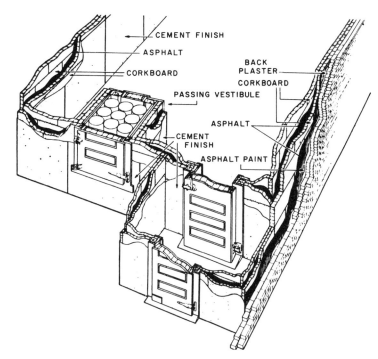

FIG. 6-34   Insulating materials in a refrigerator wall

EXAMPLE

A wall is composed of two inches of corkboard, two inches of foamglass, one-half inch plaster inside, and one inch of soft pine sheathing outside. The outside temperature is 100°F; inside, 40°F. The outside surface area is 500 square feet. Determine the total heat leakage value for a day.

PROCEDURE:

Step 1:	Locate in a table the K factor (conductivity) for each material.	Corkboard (.27) Foamglass (.40) Plaster (12.00) Soft Pine (.88)	
Step 2:	Determine the conductance for each material.	2" corkboard (.27 ÷ 2	= .135)
		2" foamglass (.40 ÷ 2	= .20)
		1/2" plaster (12.0 ÷ 1/2	= 24.0)
		1" soft pine (.88 ÷ 1	= .88)

Step 3:	Compute the resistance for each material by dividing 1 by the conductance value.	Corkboard 1 ÷ .135 Foamglass 1 ÷ .20 Plaster 1 ÷ 24 Soft Pine 1 ÷ .88	= 7.4 = 5 = .0417 = 1.136
Step 4:	Determine the total coefficient of transmission (U) for the entire wall by adding the resistance values of the four materials and dividing the sum into 1.	Sum 1 ÷ 13.58	= 13.58 = 0.074
Step 5:	Compute total heat leakage. The total heat leakage value is in Btu.	H = (U) × (A) × (TD) × (t) = 0.074 × 500 × 60 × 24 = 53,280 Btu *Answer*	

The more conventional approach is to use the formula:

$$\text{Total Heat Leakage} = \frac{(A) \times (TD) \times (t)}{R_T}$$

$R_{Total} \ (R_T) = R_1 + R_2 + R_3 \ ..... \ R_N$, and skip step 3 of the table.

$$\text{R-factor units} = \frac{(°F) \times (ft^2) \times (hr)}{(Btu) \times (in.)}$$

The preceding example gave the total heat leakage and did not include load due to warm food, motors, lights, opening of doors and other sources of heat.

INSULATION AGAINST CONVECTION

Materials (except for reflective types) that have the ability to retard the flow of heat because they are composed of tiny totally enclosed air cells are called "mass-type" insulation. The cells of such materials are either in a solid mass, such as the cellular insulations of cork and foam, or in fibers or threads, such as in hair felt, mineral wool, and fibreglass, figure 6-35.

Air that is not in motion (dead air) is a very good insulator and has a theoretical K factor of .18 Btu/hr./°F/sq. ft. of area, one-inch thick. The more air space is separated into small spaces, the more the convection currents are broken up, and the greater the insulating value, figure 6-36.

Convection currents in the tiny air cells of insulating materials are very small, so these materials act as good insulators of heat. The difficulty with air is that it cannot be kept motionless. Once convection currents and radiation of heat occur, the air space then makes a poor insulator.

110

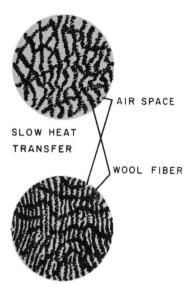

AIR SPACE

SLOW HEAT
TRANSFER

WOOL FIBER

RAPID HEAT TRANSFER

FIG. 6-35   Insulation within cabinet and door   FIG. 6-36   Effect of increased air space between fibers of insulation

In addition to good insulation, there is the need to design refrigeration compartments (such as those required in truck bodies for transport refrigeration) to avoid any metallic path through which heat can travel. The inner and outer shells of a refrigerated reefer, figure 6-37, illustrate how a urethane insulation foamed in-place adheres to the structure members. This type of construction combines excellent insulation with structural strength and rigidity of the reefer body.

The interior lining is rivetless. Polyurethane insulation is poured in-place and bonds to each preassembled wall and roof. The flow of insulation foam is controlled to produce an almost voidless insulated foam structure. This feature eliminates the need for metal fasteners on the interior walls.

When the panels are assembled into a reefer body, the interior joints are foamed in-place. Fibreglass spacers are also used to prevent heat transfer from the exterior to the interior cargo area. The combination of polyurethane insulation and fibreglass provide improved thermal characteristics for refrigerated vans. The R-factor for polyurethane foam is 6.6/in. (K = 0.15).

REFLECTIVE INSULATION AGAINST RADIANT HEAT ENERGY

The white shiny enameled finish on the outside of a refrigerator box has been mentioned as a technique for turning back radiant energy waves. Reflection prevents much of the heat energy from entering the refrigerated space.

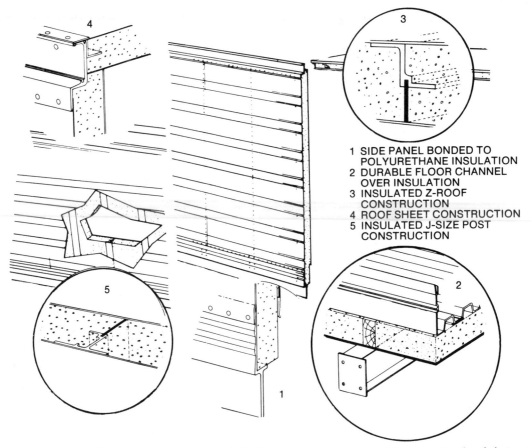

1 SIDE PANEL BONDED TO
  POLYURETHANE INSULATION
2 DURABLE FLOOR CHANNEL
  OVER INSULATION
3 INSULATED Z-ROOF
  CONSTRUCTION
4 ROOF SHEET CONSTRUCTION
5 INSULATED J-SIZE POST
  CONSTRUCTION

FIG. 6-37 Design of inner and outer reefer shell eliminates heat transfer *(Courtesy Trailmobile Inc.)*

The ability of a material to reflect radiant heat is called its *reflectivity;* to absorb heat, its *absorptivity.* The more heat that a material absorbs, the less it reflects, and vice versa. A material that has high reflectivity must have low absorptivity.

This principle is made use of in insulating houses, refrigerators, trucks, and the like by sheathing them in materials called *reflective insulation.* Usually, thin sheets of bright metals such as aluminum, lead-coated steel, and stainless steels that reflect the heat are used as reflective insulation.

From these examples, it is obvious that only sheets of materials having high reflectivity are of much use as reflective insulation. Several sheets of aluminum, lead-coated steel, or any other reflective material spaced from 3/8″ to 3/4″ apart may be built into the wall to be insulated. The number of sheets used depends upon the difference in temperature between the inside and outside of the house or refrigerator.

For temperature differences of about 50° (such as for a refrigerator wall), two or three sheets of reflective insulation may be used. Four or five sheets are used for 100°

temperature differences, such as for freezer installations. Roughly speaking, one sheet of reflective insulation is almost the equivalent of one-inch layer of corkboard. A sheet of aluminum foil reflects 86% to 90% of the radiant heat that strikes its surface.

Aluminum foil also makes a good moisture vapor seal when the edges are carefully lapped and cemented to form an air-tight seal.

It is highly desirable to be able to see the products stored inside many types of commercial refrigerators, test cabinets, and display cases. Several sheets of glass, usually two or three for temperature differences of about 40°, are used in such installations. Part of the insulating value of these multiple glass windows is due to:

▶ The reflective value of the glass surfaces (if clean).
▶ The air spaces between the panes of glass.
▶ The low conductivity of glass itself.

SUMMARY

PART B: INSULATION

- Insulation refers to any material that effectively slows down the transfer of heat in temperature.
- Moisture in an insulating material decreases its ability to insulate. Thus, all moisture must be excluded for insulators.
- Vapor seals that are continuous and tight provide a barrier from the vapor pressure outside a refrigerator, which tends to push the moisture through the insulation.
    ✓ An effective vapor seal must be used on the warm side of the insulation.
    ✓ Layers of asphalt or asphalt-coated paper are used as a vapor seal although metallic foils are superior.
- Dew point is that temperature at which the air (space) becomes saturated. When the air is cooled to the dew point, water vapor can condense into liquid form (provided its latent heat is removed.)
    ✓ The dew point temperature determines the insulating valve of a wall and the thickness of insulation required.
- The K factor, or conductivity, represents that part of a Btu which will in one hour pass through one square foot of a one-inch thick material for each degree of temperature difference.
- The rate of heat flow in Btu per square foot per hour may be found by the formula:

$$\text{Rate of Heat Flow} = \frac{(K) \times (TD)}{(d)}$$

- The total heat leakage = (TD) × (U) × (A) × Time in hours.
- Mass-type insulation, such as foam glass or corkboard, is effective in retarding heat transfer by conduction. Its effectiveness is due to the trapped air spaces.
- Reflective insulation is effective in retarding heat transfer by radiation. Aluminum, lead-coated steel, and other reflective materials are used.

ASSIGNMENT: UNIT 6 REVIEW AND SELF-TEST

PART B: INSULATION

A. CONDITIONS WITHIN INSULATION

Determine which statements (1 to 10) are true (T); which are false (F).

1. Insulation materials speed up the transfer of heat in a refrigeration system.
2. The rate of heat flow through insulation is reduced gradually.
3. The temperature throughout an insulating material is the same.
4. Temperatures within insulation remain the same regardless of outside temperature changes.
5. If a refrigerator temperature drops while a room temperature remains the same, the temperature plane inside the insulation moves to the warm side.
6. The insulating value of a material is reduced if the material absorbs water.
7. Cellular insulations such as glass, rubber, and plastic absorb large amounts of water.
8. Moisture in insulation is usually formed from water vapor that condenses from air.
9. The combination of water vapor and air pressures produce atmospheric pressure.
10. Cork is a better heat insulator than an equal thickness of wood.

Select the letter representing the term that best completes statements 11 to 21.

11. Moisture may be pushed through insulation by vapor pressure unless prevented by (a) a vapor seal, (b) the dew point, (c) a greater thickness of insulation.
12. Moisture enters the insulation on the (a) cold, (b) low, (c) high vapor, pressure side.
13. The warm side of a refrigerator in summer time is on the (a) outside, (b) inside, (c) in the center, of the insulation.
14. Moisture vapor continues as vapor in the insulation until it reaches the (a) vapor pressure, (b) dew point, (c) vapor seal, temperature.
15. A refrigerator cabinet must be insulated to keep the dew point location (a) outside, (b) inside, (c) on, the surface of the insulation and the moisture vapor seal.
16. The greater the temperature and vapor pressure difference between the outside and the inside of a cabinet, the greater the need for (a) a higher vapor pressure, (b) a higher dew point, (c) a perfectly tight vapor, seal.
17. Ultra-low temperature cabinets are used at temperatures of approximately (a) 0°F, (b) 32°F, (c) –150°F, (d) –480°F.
18. The insulation for refrigerating units that must maintain a low temperature for a long time is one that has (a) high vapor, (b) high thermal, (c) low thermal, capacity.
19. The effectiveness of heavy aluminum foil is (a) greater than, (b) less than, (c) the same as, that of asphalt impregnated paper when used as a vapor seal.
20. The vapor barrier side of the insulation should be installed facing (a) the warm wall, (b) the cold wall, (c) either wall.
21. The amount of heat transmitted in one hour through one ft.^{2} of a 3" thick material for a 1° temperature change is called its (a) C, (b) U, (c) K, factor.

B. INSULATION AGAINST CONDUCTION

1. Define conductivity or K factor.
2. State how the thickness of an insulating material is measured.
3. Determine the rate of heat flow for each set of conditions given in the table.

	Material	K Factor	Temperature Difference (TD)	Thickness of Insulation in In.	Hourly Rate of Heat Flow (Btu/ft.²)
A	Foam glass		27°	3	
B	Steel tubing		124°	1/8	
C	Window glass		27°	1/8	
D	Water		48°	6	
E	Corkboard		48°	6	
F	Polyurethane		48°	6	

	Insulating Material	Outside Area of Refrigeration (ft.²)
A	Foam glass	100
B	Steel tubing	10
C	Window glass	100
D	Water	50
E	Corkboard	50
F	Polyurethane foam	50

4. Compute the total heat leakage per day of a refrigerator using materials A through F in Problem 3 for the areas given in the table.
5. A refrigerator wall is built of 3" corkboard, 2" foamglass, 3/4" plaster, and 1" pine. The outside temperature is 120°F; inside, 35°F. The total outer surface area is 460 ft.². Compute the total heat leakage in 24 hours, correct to the nearest whole number (rounding-off the U factor correct to three decimal places). Show all computations and values obtained.

C. INSULATION AGAINST CONVECTION

Provide a word or phrase to complete statements 1 to 3.

1. According to the principles of mass insulation, heavy dense materials such as concrete, stone, and steel have _____ heat transfer values when com-

pared with light-weight porous materials.

2. Dead air has a _____ K factor than plaster.

3. When convection currents and radiation of heat takes place in an air space, it makes a _____ insulator when compared with tiny air cells of insulating materials.

D. REFLECTIVE INSULATION AGAINST RADIANT HEAT ENERGY

Provide the correct word or phrase to complete statements 1 to 7.

1. Reflection prevents _____ from entering a refrigerated area.

2. The more heat a material reflects, the less it _____ .

3. Two materials used as reflective insulation are _____ and _____ .

4. The spacing and the number of sheets of reflective insulation depends upon _____ .

5. A polished sheet of aluminum foil reflects about _____ of the radiant heat energy striking its surface.

6. Aluminum foil makes a good moisture vapor seal when the edges _____ to form an air-tight seal.

7. Dirt on a glass surface has a tendency to reduce its _____ ability.

8. State two reasons why multiple glass windows have insulating value.

UNIT 7:
SENSIBLE HEAT
IN A REFRIGERANT

The movement of molecules in a refrigerant is controlled by the heat energy that is either added or taken away. When the temperature of a solid, liquid or gas is raised or lowered, the energy that causes temperature changes is called *sensible heat*. This sensible heat is valuable to the designer, craftsperson and technician only if its effects are known and there is some way to measure it.

The more heat energy molecules receive, the faster they move. Such movement is accompanied by a change in temperature. The heat that is used to give the molecules movement and to raise the temperature is sensible heat because its effect can be determined by one of the senses—the sense of feeling.

This unit begins with a description of sensible heat as it is used in solids, liquids and gases. Then the English and metric systems of measuring sensible heat are described, and examples are used to show how the principles are applied. Explanations follow on what is meant by the specific heat of different refrigerants and other materials. This background technical information is presented in order to show how the physical properties of refrigerants relate to the amount of heat required to produce temperature changes. Finally, the value of Refrigeration Tables is stressed.

SENSIBLE HEAT OF A SOLID

As more and more sensible heat is added to a solid, the molecules move faster and faster within a limited space. This is so because of the solid state and the fact that there are strong forces holding the molecules together. The heat that causes changes in the temperature of a solid is known as the *sensible heat of a solid.*

SENSIBLE HEAT OF A LIQUID

The heated molecules of a liquid move about with greater freedom than those of a solid. The molecules hit into one another and also strike against the sides of the container. The speed of the molecules depends on the amount of heat energy that is given to or taken away from the molecules. This time, the heat energy that is added to or removed from a liquid is called the *sensible heat of a liquid.*

Subcooling the Liquid Refrigerant

As heat is removed from a liquid, its temperature drops and it becomes cooler. Since the change (drop) in temperature is produced by the removal of heat, sensible heat is removed. This same thing happens as heat is removed from the liquid refrigerant by the condenser and the liquid line heat exchanger as it approaches the metering device. Subcooling of the refrigerant also occurs in the receiver and liquid line when the ambient air temperature is lower than the liquid refrigerant temperature. This takes place between points T_8 and T_7 on the Basic Refrigeration Cycle Diagram.

The process of precooling the liquid refrigerant is referred to as *subcooling*. Subcooling adds to the efficient operation of the system. Such results may be produced by soldering the suction line (filled with relatively cold refrigerant vapor on its way back to the compressor) to the liquid line. The liquid is hotter than the vapor. Subcooling results when the vapor removes sensible heat from the liquid refrigerant.

Additional cooling occurs as the liquid refrigerant passes through the metering device. The temperature of the liquid at point T_7 on the Basic Refrigeration Cycle Diagram is approximately that of the room or a little lower. Additional cooling occurs as the liquid refrigerant passes through the metering device into the low pressure side of the system. At T_2, the refrigerant is at a temperature corresponding to the low side pressure in the evaporator, some degrees below T_7. This drop of temperature is caused by the removal of sensible heat.

SENSIBLE HEAT OF A VAPOR

Like a solid or a liquid, a vapor can be warmed by the addition of heat energy. The addition of heat energy to the molecules causes their speed (called *velocity*) to increase and the vapor to become warmer. The heat that is added to the vapor is still sensible heat. It is called either the *sensible heat of a vapor* or *superheat*.

Referring again to the Basic Refrigeration Cycle Diagram, superheating of the refrigerant vapor takes place in the suction line between points T_3 and T_6. The gas is superheated by the heat received from (1) the air around the line, or (2) a heat exchanger designed for the purpose, or (3) the liquid line.

MEASURING THE AMOUNT OF HEAT ENERGY

The amount of sensible heat to be applied to a solid, liquid or gas must be controlled if a refrigeration system is to operate efficiently. This means that the amount of heat energy must be measured. Heat energy is measured by the effect it produces on materials.

One effect of adding heat to a substance is to raise its temperature. Thus, it is natural to say that heat quantities should be defined in terms of how much a temperature rise they produce in some material. Water has been selected as one material that is widely used as a standard.

Quite naturally, the effect of adding heat to water is to increase its temperature, figure 7-1. How much heat it takes to raise the temperature of water may be measured easily. For instance, it takes a certain amount of heat to raise the temperature of one pound of water one degree, say from 50°F to 51°F. Although the amount of heat required to produce this change varies slightly for each degree rise of temperature, for all practical purposes it will be ignored in this unit.

The British Thermal Unit (English System of Heat Measurement)

A standard measurement may be adopted for measuring the amount of heat required to warm one pound of water one Fahrenheit degree. This is exactly what British scientists long ago decided upon as a standard unit for measuring heat. This standard (as stated earlier in unit 5) is known as the *British Thermal Unit* (thermal referring to

FIG. 7-1 Effect of adding heat

heat). The measurement is abbreviated *Btu*. Within the United States and many other countries using the English system of measurement, the Btu is the standard heat unit.

If one Btu is required to raise the temperature of one pound of water one Fahrenheit degree, it takes twice as much heat to raise the same quantity two Fahrenheit degrees, figure 7-2. So, it is possible to measure the amount of heat (Btu) required to warm any weight of water through any number of degrees of temperature. The Btu value is computed by multiplying the number of pounds of water by the number of Fahrenheit degrees of temperature through which the water is raised.

To raise the temperature of 2 pounds of water from 50°F to 150°F, as in figure 7-3, first determine the change in temperature. This equals 100 degrees of temperature rise (150 – 50 = 100). Since there are 2 pounds of water that undergo this temperature change, the difference in temperature (100) is multiplied by the weight (2).

$$Btu = (100) \times (2) = 200$$

The container illustrated in figure 7-4 holds four pounds of water, or twice as much as that in the container in figure 7-3. However, the temperature rise of 100°F is the same. The Btu value that will produce this effect is found by multiplying the temperature rise by the weight of water.

$$Btu = (100) \times (4) = 400$$

In summary, the heat energy added to water increases the movement of molecules and gives them greater speed of motion. This increased speed produces the effect that is called rise of temperature.

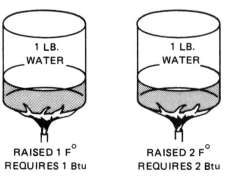

FIG. 7-2 Effect of temperature on Btu

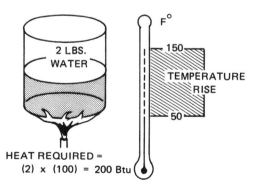

FIG. 7-3 Determining Btu required for a specific change

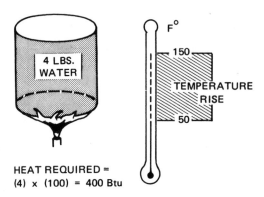

HEAT REQUIRED =
(4) x (100) = 400 Btu

FIG. 7-4   Btu required when weight is doubled

The Calorie (Metric System of Heat Measurement)

A second system is used in some countries to measure the quantity of heat. This system is referred to as the *metric* system. The unit used in heat measurement is the *calorie*. A calorie of heat indicates the amount of heat, which when added to one gram of water, raises its temperature one Celsius degree. Technical handbooks usually take this value as being between 15°C and 16°C because of the slight differences in value between different temperatures.

Since the gram is such a small quantity for comparison and practical use in refrigeration work, a larger unit known as the *large calorie, kilogram calorie* or *kilocalorie* is used. This is easy to understand because the units in the metric system are related to each other in multiples of ten. *Kilo* means one thousand, so one kilogram = 1,000 grams. The kilogram calorie represents the amount of heat required to raise the temperature of 1,000 grams of water one Celsius degree.

There are, then, two common units of heat measurement. The British thermal unit (Btu) is used with pounds and Fahrenheit degrees of temperature changes; the kilocalorie unit, with kilograms and Celsius degrees of temperature change.

Converting Units of Heat Measurement

There are times when a value in one system must be changed to its equivalent value in the other system. If two simple values are remembered, considerable time may be saved. 1 Btu = 0.252 kilocalories, and 1 kilocalorie = 3.968 Btu.

EXAMPLE 1:

▶ To change a heat value from English to the metric equivalent

1 Btu = 0.252 kilocalories. Therefore, multiply Btu by 0.252 to obtain kilocalories.

▶ Change:

20 Btu to kilocalories   20 × 0.252 = 5.04 kilocalories

EXAMPLE 2:

▶ To change a heat value from metric to the English equivalent

1 kilocalorie = 3.968 Btu. Therefore, multiply kilocalories by 3.968 to obtain Btu.

▶ Change:

10 kilocalories to Btu   10 × 3.968 = 39.68 Btu

Specific Heat

Experience proves that most materials require less heat per pound to raise their temperatures than does water. It requires about 1/11 as many Btu (0.0939 Btu) to raise the temperature of one pound of copper one degree Fahrenheit as compared with an equal amount of water. Similarly, it requires about 1/30 Btu (0.033 Btu) to raise the temperature of one pound of mercury one Fahrenheit degree.

When a comparison is made between copper and water and mercury and water, it can be seen that it takes only 0.093 and 0.033 times as much heat to change the temperature of these materials one degree F. This condition exists because materials vary in their ability to absorb and exchange heat. If equal weights of water, copper, liquid or gas refrigerant, or any other substances are heated through equal changes in temperature, each material will have absorbed a different amount of heat. Surprisingly, water requires the largest amount.

It takes 50 Btu to change the temperature of five pounds of water by ten degrees. However, for a similar weight of copper, it requires (50 × 0.0939), or 4.65 Btu. As a further contrast, it takes only (50 × 0.033), or 1.65 Btu, to raise the temperature of five pounds of mercury 10°F, figure 7-5.

Sample Specific Heat Values

The preceding examples indicate that water is taken as a standard unit of measurement. Water is considered to have a value of one (1.000) because it is being compared with itself. The values of other materials are in comparison to water. Copper, when

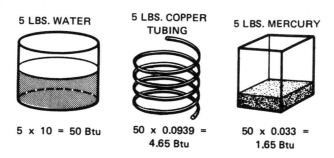

5 LBS. WATER   5 LBS. COPPER TUBING   5 LBS. MERCURY

5 x 10 = 50 Btu   50 x 0.0939 = 4.65 Btu   50 x 0.033 = 1.65 Btu

FIG. 7-5   Btu required to change temperature 10°F

compared with water, takes only about 1/11 units of heat to raise its temperature one Fahrenheit degree; mercury, 1/30. Since these values are relatively constant, a number of 0.0939 has been assigned to copper. This is known as its *specific heat*. The specific heat of mercury, as indicated before, is 0.0333. Figure 7-6 compares the specific heat of three substances to water.

Specific heat is the ratio of

- The amount of heat required to change by one degree the temperature of a unit mass of a substance to
- The amount of heat required to change the temperature of the same unit mass of water one degree.

The term ratio means *divided by*. Thus, specific heat is not expressed in any unit of measurement other than by a numerical value. This same specific heat value may be used in problems involving either the English or the metric systems.

Determining the Amount of Sensible Heat Required for a Given Change

The amount of heat required to produce changes of temperature in a substance depends upon three things.

▸ The amount of material as represented by its mass (m).
▸ The specific heat of the material (c).
▸ The size of the temperature change (represented by a delta symbol and the letter $t—\Delta t$)

The fourth value may be determined whenever three values are known. This may be done by using the formula:

$$
\begin{array}{ccccc}
\text{Heat} & & \text{mass of} & \text{specific} & \text{change in} \\
\text{(added or taken away)} = & \text{material} \times & \text{heat of material} \times & \text{temperature} \\
\text{(H)} & & \text{(m)} & \text{(c)} & (\Delta t)
\end{array}
$$

Note that when parentheses are used in this formula, they indicate that the values are to be multiplied. If m is in pounds, and Δt in F degrees, then H will be in Btu. On the other hand, if m is in kilograms, and Δt in C degrees, then H will be in kilogram calories.

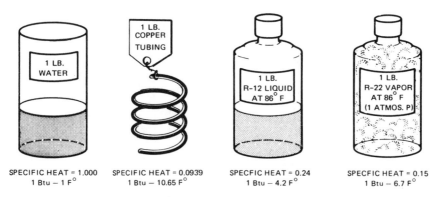

SPECIFIC HEAT = 1.000	SPECIFIC HEAT = 0.0939	SPECFIC HEAT = 0.24	SPECFIC HEAT = 0.15
1 Btu – 1 F°	1 Btu – 10.65 F°	1 Btu – 4.2 F°	1 Btu – 6.7 F°

FIG. 7-6   Comparison of effect of specific heat and Btu for different substances

EXAMPLE 3:

Determine the Btu required to raise the temperature of 100 lbs. of copper tubing in a condenser by 110 F°. The specific heat of copper is 0.0939. This means it takes that much of a Btu to raise the temperature of one pound of copper tubing one F degree. To raise the temperature of the 100 pounds through the 110 F degree change requires the application of more heat. Exactly how much more may be found by substituting values in the formula,

ANSWER:

$$H = (m) \ (c) \ (\Delta t) = (100) \ (0.0939) \ (110) = 1032.9 \ \text{Btu}$$

REFRIGERANT TABLES

Tables are available to show the specific heat (c) and other properties of many common materials. A table giving the physical characteristics of eight common refrigerants is included in the Appendix. Such a table, as related to R-11, appears in Figure 7-7. These data save considerable time because a known value may be easily substituted in a formula instead of computing it.

Suppose a refrigeration system that circulates five pounds of Refrigerant 11 (R-11) per minute must subcool the liquid 20 F° (from 86°F to 66°F) before it enters the metering device. The required Btu may be computed by following three simple steps.

STEP 1	Set down the formula	$H = (m) (c) (\Delta t)$
STEP 2	Find the value of c for the liquid (R-11)	$c = .24$
STEP 3	Substitute values in the formula	$H = (5) (.24) (20) = 24$
NOTE: The H value indicates that 24 Btu are needed to secure the required results.		

The refrigeration tables use a -40°F as a reference temperature for refrigerants regardless of whether it is so indicated. Years ago it was assumed for convenience that the refrigerant should contain no heat at this temperature (-40°F). Further, the tables were developed at a period when a temperature like -40°F was a low temperature for the operation of mechanical refrigeration units. Actually, the temperature that meets this condition according to scientific theory is absolute zero degrees (0°R, or -460°F).

The heat content of water and a selected number of saturated refrigerants is given in a series of handbook tables on the Properties of Refrigerants. These tables are included in the Appendix of this book. Part of one of the tables showing its content is illustrated in figure 7-8. It should be noted that the three columns on the right side contain information about the Heat Content in Btu per pound for the refrigerant identified in the table.

Using Refrigerant Table Data in Calculations

Refrigeration tables greatly simplify the solving of problems. For example, Appendix Table 5 on the Physical Properties of Refrigerant-22 shows that the heat content of liquid Refrigerant-22 with reference to -150°F is 34.9 Btu/lb. at 86°F. This same refrigerant has a heat content of 24.3 Btu/lb. at 50°F.

Selected Physical Data (Performance based on 5°F evaporator temperature and 86°F condenser temperature)	Refrigerant 11
1. Chemical Formula	CCl_3F
2. Molecular Weight	137.4
3. Boiling Pt. (°F) at 1 Atm. Pressure	74.9
4. Evaporator Pressure at 5°F (psig)	23.9*
5. Condensing Pressure at 86°F (psig)	3.49
6. Critical Temperature (°F)	388
7. Critical Pressure (psi absolute)	640
8. Compressor Discharge Temp. (°F)	111
9. Compression Ratio (86°F/5°F)	6.19
10. Specific Volume of Saturated Vapor at 5°F (cu. ft./lb.)	12.2
11. Latent Heat of Vaporization at 5°F (Btu/lb.)	83.5
12. Net Refrig. Effect of Liquid -86°F/5°F (Btu/lb.)	66.8
13. Specific Heat of Liquid at 86°F	0.21
14. Specific Heat of Vapor at Constant Pressure of 1 Atm. & 86°F	0.14
15. Coefficient of Performance	5.03
16. Horsepower/Ton Refrigeration	0.938
17. Refrigerant Circulated/Ton Refrig. (lb./min.)	2.99
18. Liquid Circulated/Ton Refrig. (cu. in./min.)	56.6
19. Compressor Displacement/Ton Refrig. (cfm)	36.5
20. Toxicity (Underwriter's Laboratories Group No.)	5
21. Flammability & Explosivity	None
22. Type of Suitable Compressor	Centrif.

*inches of mercury vacuum.

FIG. 7-7   Type of data found in refrigerant table

Temp.	Pressure		Volume Gas	Density Liquid	Enthalpy (Heat Content)		
					Liquid	Latent	Vapor
°F	psig	psia	cu.ft./lb.	lb./cu.ft.	Btu/lb. Above –40°F		

FIG. 7-8   Contents of table on properties of a refrigerant

Now the difference between these two values is 10.6 Btu/lb. (34.9 Btu/lb. minus 24.3 Btu/lb.). This is the amount of heat to be removed from each pound of refrigerant in subcooling it from 86°F to 50°F. If five pounds of refrigerant were being circulated per minute, the total amount of heat required would be 53.0 Btu/min. (5 × 10.6).

Often the term *enthalpy* is used instead of heat content. Enthalpy tables may be preferred to other methods of calculating data because they are more accurate. Enthalpy tables allow for changes in value of the specific heat for each temperature instead of using a fixed value.

The information contained in refrigeration tables is also presented in graphic form. Charts, known as *Mollier diagrams*, show the thermal properties of refrigerants at different temperatures.

SUMMARY

- Sensible heat is heat energy that causes a change of temperature of a solid, liquid or gas.
- Sensible heat changes the speed with which molecules move.
  - The addition of sensible heat causes a rise in temperature.
  - The removal of sensible heat causes a drop in temperature.
- Subcooling is the drop in temperature resulting from the removal of sensible heat from the liquid refrigerant.
- Superheating is the rise in temperature resulting from the addition of heat to the refrigerant vapor, either in the evaporator or the suction line.
- Heat quantities are measured in either English (Btu) or metric (calorie) units.
  - A Btu is the amount of heat energy required to raise the temperature of one pound of water through a change of one F degree.

    1 Btu = 0.252 kilogram calories (kilocalories)

  - A calorie is the amount of heat energy required to raise the temperature of one gram of water through a change of one C degree; the kilogram calorie, 1,000 grams.

    1 kilocalorie = 3.968 Btu

- Specific heat is the number that results from dividing the amount of heat (Btu or calories) required to change the temperature of a given weight of material by the amount of heat (Btu or calories) to cause the same change of temperature in the same weight of water.

- Heat added or taken away from a material equals the product of its mass, specific heat and change in temperature.

$$H = (m)\,(c)\,(\Delta t)$$

- Refrigeration tables use a -40°F reference temperature and allow for the variation in specific heat with temperature.
- Enthalpy is the name given to the total heat in the refrigerant at any temperature (with reference to -40°F).
- The enthalpy in a refrigerant is listed in refrigerant tables along with other properties at different temperatures.

ASSIGNMENT: UNIT 7 REVIEW AND SELF-TEST

A. SENSIBLE HEAT IN A REFRIGERANT

Determine which statements (1 to 10) are true (T); which are false (F).

1. The greater the amount of heat energy absorbed by molecules, the slower they move.
2. Heat energy that causes change in temperature is sensible heat.
3. The molecules of a liquid at 50°F are moving faster than those of a solid at 50°F.
4. The heated molecules of a liquid move about with more freedom than those of a solid at the same temperature.
5. Heat energy added to a liquid is sensible heat of a vapor.
6. Refrigerant tables use absolute zero as a reference temperature.
7. Precooling and subcooling of the liquid refrigerant refer to the same condition.
8. Subcooling occurs when the vapor removes sensible heat from the liquid refrigerant.
9. Superheat is the same as the sensible heat of a liquid.
10. Heat energy added to the molecules of a vapor causes their velocity to increase.

Match each heat term, Column I, with the correct description, Column II.

Column I	Column II
a. Sensible heat	(1) Heat required to raise 1 gram of water 1 F°.
b. Superheating	(2) A drop in temperature resulting from the removal of sensible heat from a refrigerant liquid.
c. Subcooling	(3) Heat required to raise 1000 grams of water 1 C°.
d. Btu	(4) Heat energy that causes a change in temperature.
e. Kilocalorie	(5) Heat required to raise 1 pound of water 1 C°.
	(6) A rise in temperature resulting from the addition of heat energy to a refrigerant vapor.
	(7) Heat required to raise 1 pound of water 1 F°.

B. MEASURING THE AMOUNT OF HEAT ENERGY

Provide the correct term or condition to complete statements 1 to 7.

1. Heat energy must be measured by the _____ it produces.
2. The name given to the amount of heat required to change the temperature of one pound of water from 59°F to 60°F is _____ .
3. The standard unit of heat measurement in the English system is the _____ .
4. Since the calorie is so small, the unit of heat measurement used for refrigeration work in the metric system is the _____ .
5. One Btu = _____ kilogram calories.
6. One kilogram calorie = _____ Btu.
7. The word used to describe the heat contained in a refrigerant is _____ .
8. State briefly how the sensible heat of water is computed.
9. Describe what a calorie means.
10. State briefly how sensible heat is computed for any material.

C. SPECIFIC HEAT

1. Tell how specific heat may be computed.
2. Identify what each letter or symbol represents in the specific heat formula.
3. Define enthalpy.
4. Refer to the Table of Physical Characteristics of Common Refrigerants in the Appendix. Then supply the information requested in the table for refrigerants A to D.

| | Refrigerant | Specific Heat at 86°F | | Liquid Circulated per Ton Refrigerant (cu.in./min.) |
		Liquid	Vapor at Constant Pressure of 1 Atmosphere	
A	R–12			
B	R–22			
C	NH_3 (R–717)			
D	H_2O (R–718)			

5. Indicate what unit of measure in the English system is used for each letter in the formula: $H = (m)(c)(\Delta t)$.
6. State the term used for the heat that is used to raise the temperature of a liquid.
7. What is the name for heat in Btu per pound per degree change in temperature?
8. What is the customary starting point for calculating the total heat of a refrigerant?
9. Give the numerical value of the specific heat of water.

10. What term is used to denote the amount of heat required to change the temperature of one pound of water one degree F at 59°F?
11. Determine the numerical value of the ratio when comparing the specific heat of iron in the metric and English systems.
12. Compute the values that are missing in the table.

	Substance	Mass	Temperature Change	Heat Transferred
A	Water	50 lbs.	70 to 120°F	
B	Water	50 g	50 to 51°C	
C	Water	540 g	60 to 84°C	
D	Water	25 lbs.	50 to 51°F in *two* hours	

13. Insert the missing values by either locating them in the refrigerant tables or by computation.

	Substance and/or Part	Mass	Temperature Change	Specific Heat	Heat Absorbed, Required or Abstracted
A	Brass radiator	120 lbs.	80 to 150°F	0.09	
B	Ice	65 lbs.	−20 to 32°F		1690
C	Ammonia (Liquid)		70 to 20°F	1.1	1100
D	R-12 (Liquid)	30 lbs.	90 to 34°F	0.24	
E	R-22 (Gas)		63 to 74°F	0.15	1.65
F	Copper Tubing	900 lbs.	120 to 80°F	0.094	

14. How much heat is absorbed by a 200-pound copper condenser when heated from 20°F to 90°F?
15. How much heat is absorbed by a 150-pound steel condenser (specific heat 0.12) in being heated from 80°F to 105°F?

UNIT 8:
LATENT HEAT AND PRESSURE IN REFRIGERATION

In review, it may be said that matter exists in any one of three states, depending on the energy in the molecules. This energy may be either *kinetic* (which is an energy of motion) or *potential* (an energy of position). The kinetic or potential energy of a material may be changed by adding heat.

The addition of heat can result in an increase in kinetic energy and temperature because the molecules move faster. As heat is removed, the opposite is true. It is sensible heat that causes changes in temperature.

Heat can be added to a material under certain conditions without changing its temperature. Heat thus absorbed causes the molecules to move at the same rate of speed but with greater freedom. This results in an increase in potential energy (stored up energy or energy of position). A great increase in potential energy results in a change in fluidity (freedom) which is a *change of state*. The heat that causes this change is called *latent heat*.

This unit deals with the effect of latent heat and pressure on the change of state of a refrigerant. New terms and conditions like pressures above or below atmospheric, saturated refrigerants, and the importance of vapor pressure in selecting a refrigerant are discussed. Examples are included of a number of typical problems with which the technician, craftsperson and engineer are concerned. These demonstrate how the principles of refrigeration, the terms, and the conditions are all related to the design and maintenance of refrigerating equipment.

RELATIONSHIP OF LATENT HEAT AND CHANGE OF STATE

There are two kinds of latent heat. The *latent heat of fusion* produces a change in state from a solid to a liquid. The *latent heat of vaporization* results in a change of state from a liquid to a gas. The principles relating to such changes and the conditions which each latent heat produces are covered in detail.

Latent Heat of Fusion

Water exists in the liquid state at ordinary temperatures. If heat is taken away from it, cooling occurs and the water remains in liquid state until it is cooled down to 32°F (or 0°C). As the process of taking heat away continues, the temperature of the water remains at 32°F but, gradually, the water changes into the solid state. This process is known as *solidification* or *freezing*.

The interesting fact is that water will not freeze when it is merely cooled down to 32°F, figure 8-1. An additional amount of heat must be removed before the potential (stored up) energy of the molecules is reduced to the point where there is a change to the solid state.

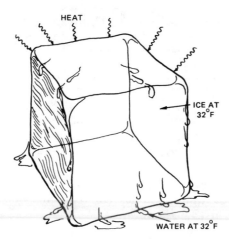

FIG. 8-1   Latent heat of fusion added to ice

It so happens that the amount of heat that must be added to change a solid to a liquid after it has been brought up to its melting temperature is the same as that required to cause solidification. For example, ice melts at the same temperature at which water freezes. The difference is determined by whether heat is added or taken away. It should be noted that different materials freeze at different temperatures.

Latent heat is measured in Btu. The latent heat of fusion is the amount of heat in Btu required to change a pound of a material from the solid to liquid state without changing its temperature. Through experiment it has been determined that it takes 144 Btu to change one pound of ice at 32°F to water at this same temperature. The 144 Btu is considered as a constant (unchanging) value for measuring latent heat of fusion.

Latent Heat of Vaporization

Boiling or *vaporization* are the terms used to express a change of state from liquid to vapor or gas. When a liquid changes into the vapor state, its molecules gain an even greater degree of freedom. As such, they contain a larger amount of potential energy, which is produced by the addition of heat energy.

Here, again, is this condition in which the addition of heat results in no further increase in temperature. What happens is that the additional heat causes the material to change from the liquid to the vapor state.

This important characteristic of all fluids may be easily demonstrated with water. Observation shows that a considerable amount of heat must be added to water at its boiling temperature before it changes state to steam at the same temperature. This heat, which is needed to cause the change of state without a change of temperature, is the *latent heat of vaporization.*

When a thermometer is placed in an open container like the one illustrated in figure 8-2 and heat is applied until the water boils, the thermometer reads 212°F. If the sensing element of a second thermometer is placed only in the path of the steam vapor, and steam is produced as the water changes state, the second thermometer will also read 212°F. As the boiling process continues and the water turns to vapor the temperature of both the boiling water and its vapor, remains at 212°F. So long as the pressure on the vapor does not change, the temperature of the steam and water remains the same.

When the source of heat is removed, the steam condenses rapidly on the sides of the container because the latent heat of vaporization is surrendered. The temperature of the remaining liquid stays the same until all the steam is condensed and no more latent heat is available. From that point on the temperature within the container falls as the liquid gives up heat.

Latent heat of vaporization is also measured in Btu. At normal atmospheric pressure (14.7 psia) the latent heat of vaporization for water is 970.4 Btu per pound. This is another constant that is used as a standard for comparison purposes. Tables showing corresponding latent heat values for several refrigerants at different temperatures are included in the Appendix, Tables 2, 3, 4 and 5.

Total Heat—Enthalpy

The terms *heat content, total heat* and *enthalpy* have the same meaning. The enthalpy of a refrigerant at a given temperature is the amount of heat that must be added to one pound of the refrigerant to bring it up to the required temperature as a liquid or vapor from the reference temperature of –40°F. Enthalpy or heat content is another characteristic that is listed in Tables 2, 3, 4 and 5 on Properties of Refrigerants. The total heat of the liquid is just sensible heat at that temperature. The total heat (enthalpy) of the vapor includes the sensible heat of the liquid plus the latent heat needed to form the vapor at that same temperature.

Two cases are cited in order to bring together technical information about latent heat and total heat.

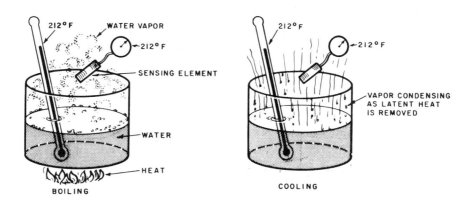

FIG. 8-2 Effect of latent heat

EXAMPLE 1:

Determine the latent heat required at 30°F to vaporize one pound of liquid Refrigerant 12.

PROCEDURE:

Refer to Table 4 on Properties of Refrigerants for the Btu per pound required to cause the Refrigerant 12 to vaporize at 30°F.

ANSWER:

This value is found to be 65.36 Btu/lb.

EXAMPLE 2:

Compute the heat that must be added to Refrigerant 12 to raise its temperature from 20°F to 70°F and then to cause it to vaporize.

PROCEDURE:

Step 1: Refer to Table 4 on Refrigerant 12 properties.

Step 2: Record the total heat of the vapor at 70°F   84.36 Btu/lb.

Step 3: Record the heat of liquid Refrigerant 12 at 20°F   12.86 Btu/lb.

Step 4: Subtract the two values to find the Btu of heat to be added per pound.

ANSWER:

84.36 – 12.86 =   71.50 Btu/lb.

The Effect of Pressure on Change of State

It is obvious that at the same pressure, the boiling point of a liquid and the condensing temperature of a vapor are identical. Evaporation or condensation depends on whether heat is being added or taken away. Refrigerants, as fluids, are affected by heat, temperature and pressure just the same as water. Although different refrigerants have varying boiling points and latent heats, each refrigerant reacts in the same manner.

Boiling Point and Condensing Temperatures Related to Refrigerants

One of the beginning units indicated that ether and ammonia were used as early refrigerants. Later, methyl chloride, carbon dioxide and sulfur dioxide came into use. The boiling points (temperatures) of these common refrigerants and a few in the refrigerant family are shown graphically in figure 8-3. The usefulness of a refrigerant in transporting heat depends primarily on its latent heat of vaporization.

A specific example may show how the operation of a refrigerating device depends on the boiling point of a refrigerant and its latent heat of vaporization. From the drawing, the boiling point of ammonia is found to be -28°F. In the tables, its latent heat of vaporization under normal atmospheric pressure at this temperature is given as 589.3 Btu/lb. This means that if heat is added to ammonia when its temperature is –28°F, it takes 589.3 Btu to change one pound of liquid ammonia at this temperature to a vapor. If only half this number (294.6) Btu are added to the ammonia at –28°F, only one-half pound of ammonia is changed from the liquid to the vapor state.

On the other hand, if at that same pressure, heat is taken away from ammonia which has been cooled to a temperature of –28°F, then some of the ammonia is changed from the gaseous into the liquid state. The quantity of ammonia that changes state depends on the amount of heat that is taken away. It follows that if 589.3 Btu were removed, one pound of the ammonia gas would be changed to the liquid state (liquefied). If 1178.6 Btu were removed, two pounds of the refrigerant would be liquefied.

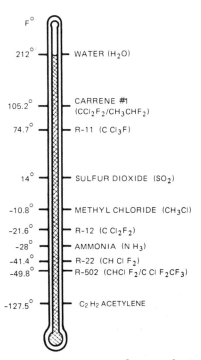

FIG. 8-3   Boiling temperature at normal atmospheric pressure

One important fact to remember is that these values apply only under certain specific conditions of pressure. Any variation of pressure affects the latent heat of vaporization. Latent heat of different materials is found in Tables 2, 3, 4, and 5 of the Appendix. They are also available in similar tables furnished by manufacturers, and in handbooks. Refrigeration systems depend on vaporization, a heat-absorbing process.

Normal Atmospheric Pressure

Constant reference has been made to normal atmospheric pressure and the possible effects of pressure on the temperature at which a refrigerant boils. Scientifically, *pressure* is a force per unit area. This means, in very simple terms, that if a force of one pound acts on an area of one square inch, the pressure is one pound per square inch. To simplify this measurement, it is written as (1 psi) because pressures are usually expressed in pounds (p) per square (s) inch (i).

A column of air, which is one square inch in area and extends from the earth to the upper limit of its atmosphere, has been calculated to exert a force of 14.7 pounds. This pressure is taken at sea level in order to establish a standard that may be easily reproduced. This pressure at sea level of 14.7 pounds is referred to as *normal atmospheric pressure*. At altitudes above sea level, the atmospheric pressure drops as illustrated in figure 8-4. At 1,000 feet above sea level the pressure is 14.0 psi; at 5,000 feet 12.6 psi, and so on. This fact must be considered when refrigeration pressure controls are set for operation at high altitudes. While the atmospheric pressure varies according to weather conditions, the 14.7 psi is used for all practical purposes in refrigeration work. Since these pressures are measured above absolute zero, they are expressed as *absolute pressure*.

Barometer Readings of Atmospheric Pressure

If a 30-inch-long narrow tube (which is closed at one end) filled with mercury is turned over into a dish of mercury, the mercury will remain in the tube. The column of mercury, if measured, would be found to be 29.92″ at sea level, figure 8-5. This is so, because the normal atmospheric pressure on the surface of mercury in the dish supports a column 29.92″ high.

This instrument is known as a *barometer*. Other materials may be used in barometers, but mercury is preferred because it provides a practical medium. A barometer measures pressure. Changes in pressure are reflected in changes in the height of the column of mercury which a given pressure may support.

SAFETY PRECAUTION

Safe practice requires that the glass tubing is thick-walled barometer tubing and the mercury should only be left in a closed container because it evaporates and the vapor is poisonous if inhaled.

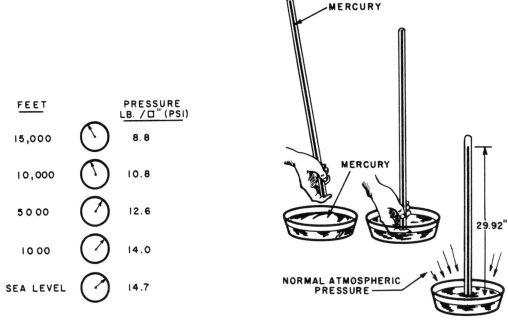

FIG. 8-4   Atmospheric pressure decreases with altitude

FIG. 8-5   Column of mercury supported by normal atmospheric pressure

Computing Units of Absolute Pressure

At 14.7 psi of absolute pressure (written 14.7 psia) the standard barometer reading is 29.92 inches of mercury. The chemical symbol for mercury is Hg. Therefore, the standard barometer reading may also be expressed as 29.92 in. Hg. In order to convert pressure in psia to in. Hg, divide the standard atmospheric pressure (14.7 psi) by the height of a column of mercury that it supports (29.92 in. Hg). The result of 0.491 psi for each inch of mercury represents a constant value.

For example, a barometer reading of 29.45 in. Hg can be converted to absolute pressure in psia by multiplying by the constant 0.491. The product (14.46) represents the pounds per square inch absolute (14.46 psia). By reverse process the pressure in psi can be converted to in. Hg by dividing the psi by 0.491.

Pressure above Atmospheric

Many refrigeration tables give values in absolute pressure. In order to use such tables, the pressure indicated on a gauge must be converted to absolute pressure, which includes the pressure of the atmosphere. This is done very easily by adding the gauge and atmospheric pressures. For instance, if an absolute pressure must be determined which corresponds to a gauge reading of 50 lb./sq.in. (psig), this value is added to (14.7). The answer (64.7) is psia.

Gauges that are used on industrial machinery like boilers and refrigerators read zero (0) when they are open to the atmospheric pressure. The reading on such gauges is called the *gauge pressure*. It is evident that when the gauge reads *0*, the pressure inside the gauge is the same as the atmosphere, or 14.7 psi. When such a gauge reads 10 psi, it indicates that the pressure is 10 psi above that of atmospheric pressure. The absolute pressure would be 24.7 psia (14.7 + 10).

Pressure below Atmospheric

Many refrigeration systems depend on pressures below atmospheric in the evaporator. Such systems are said to operate under a partial vacuum, which can also be measured in inches of mercury. The dials of instruments which measure the degree of vacuum are calibrated in inches of mercury and do not go beyond about 30″.

PRINCIPLES OF OPERATION OF PRESSURE GAUGES

Figure 8-6 shows three drawings in order to simplify the explanation about the use of inches of mercury (in. Hg) to designate pressures below atmospheric. Drawing (A) shows a closed container connected by a glass tube with a vessel of mercury. Since the vessel is open to the atmosphere, the gauge reads 0 psig. The level is the same both inside and outside the tube.

At (B) the vacuum pump reduces the pressure inside the container so that it becomes less than atmospheric. The mercury is pushed up the tube by the atmospheric pressure exerted on the mercury in the open vessel. When the level in the tube is 10″ higher than the level in the container, the gauge reads *10*. The pressure inside the container is said to be 10 inches of mercury vacuum (not pressure). Similarly, the pressure at (C) is reduced to the point where the gauge reads 20 inches of mercury (20 in. Hg vac.).

Reference Table 7 (which appears in the Appendix) indicates that the pressure in the evaporator of an R-11 system is 23.9 inches of mercury vacuum at a saturation (boiling)

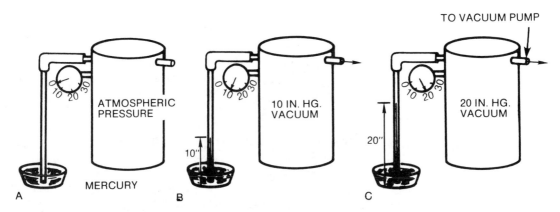

FIG. 8-6 Inches of mercury measure pressures below atmospheric

point of 5°F. This same table shows that the pressure in the evaporator of a system using water as a refrigerant is 29.7 inches of mercury vacuum at a saturation point of 40°F.

A pressure gauge attached to the suction-service valve of the compressor would indicate 23.9 and 29.7 in. Hg vac. for the R-11 and water, respectively. These readings are on the left side of the zero and indicate that the pressure in this part of the system is below atmospheric.

A second method of indicating amount of pressure is in terms of the *residual absolute pressure*. It must be remembered that mercury reaches a height of 29.92 inches in a barometer tube at sea level. For all practical purposes, there is a perfect vacuum above the mercury in the sealed tube. If the vacuum pump in drawing (C) continued to remove gas until it achieved a perfect vacuum, the mercury in the tube would reach 29.92" in height (assuming normal atmospheric pressure outside the container).

Actually, the mercury column stands only 20" high and lacks 9.92 in. Hg of being a perfect vacuum. Therefore, the 9.92 in. Hg of pressure inside the container is its residual absolute pressure. Thus, pressure is designated as 9.92 in. Hg absolute pressure or as 20 in. Hg vacuum. Both of these designations for pressures below atmospheric are important.

The vacuum reading is for pressures of refrigerant vapors at saturation and the residual absolute pressure is used to indicate the degree of vacuum in a system which is being pumped down in preparation for charging. It is now considered necessary to pump a system down to 50 *microns of mercury*, absolute pressure. A *micron* is a measurement of almost 1/25,000 of an inch (or 0.001 mm; 1/1,000,000 of a meter). The 50 microns represents the residual gas pressure (absolute) in the system. At this low pressure practically all of the water vapor has been removed. Incidentally, if a system cannot be pumped down to 50 microns it indicates a leak or the presence of water vapor or other contaminants.

The term *seizing-up* means that a movable part or unit becomes secured (temporarily fitted together tightly) and no motion is possible under normal operating conditions. A good refrigeration vacuum pump can "pull" a vacuum of five microns or less.

Condensing Pressures and Boiling Point

The boiling point of a liquid has been defined as that temperature at which the vapor pressure of the liquid is equal to the pressure on the liquid. What this law says is that all fluids are affected by pressure and temperature acting simultaneously. This may best be illustrated with water. Water is a liquid at all temperatures between 32°F and 212°F under normal atmospheric pressure. Water changes state and freezes into a solid when the temperature is reduced below 32°F. At normal atmospheric pressure the state of water is changed from liquid to vapor as it is heated at 212°F.

The boiling point is affected by pressure, figure 8-7. Water boils at a lower temperature than 212°F when the pressure on the water is decreased; and at a higher temperature, when the pressure is increased. To emphasize this point, water under a vacuum of 19 inches of mercury boils at 165°F. Thus, by decreasing the pressure, the boiling point is lowered. By contrast, at 10 pounds gauge pressure (10 psig) the temperature must be raised to 240°F before boiling begins.

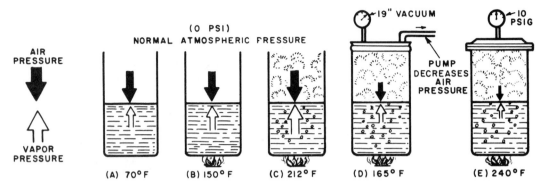

FIG. 8-7 Effect of pressure on boiling point

SAFETY PRECAUTION

The experiment was used to illustrate the principle involved. It should only be performed with proper equipment such as a thick-walled pressure cooker equipped with an adequate pressure release, etc.

While water has been used as an example, other refrigerants behave in the same manner. The tables of Properties of Refrigerants show that for Refrigerant 12 to boil at 86°F its vapor pressure has to be 93 psig. Conversely, if this refrigerant is to condense at 86°F it must be compressed to a pressure of 93 psig. If latent heat is removed at that temperature and pressure, the vapor condenses back into the liquid state. This is what happens in the condenser of the refrigerating system.

By comparison, a pressure of 154.5 psig is required for ammonia to condense at 86°F after which the removal of latent heat results in condensation. Sulfur dioxide (SO_2) requires only 58 psig, while carbon dioxide (CO_2) takes 1028.3 psig. Water requires a vacuum of 28.8 inches of mercury (or 1.12 in. Hg absolute) to condense at 86°F. Pressure values for other temperatures and refrigerants are found in the Appendix in Reference Tables, 2, 3, 4, and 5 on Properties of Refrigerants.

Pressure and Critical Temperature

The hotter a gas or a vapor, the more difficult it is to liquefy. This means that at higher temperatures it requires more pressure to liquefy a gas than at lower temperatures. Surprisingly, if a gas were at a high enough temperature, it would be impossible to liquefy it regardless of how much pressure is applied. The temperature at which this condition occurs is the *critical temperature* of the gas or vapor. Taking a few examples, the critical temperature of Refrigerant 12 is 234°F and ammonia, 271.4°F. This means that ammonia cannot be liquefied if its vapor temperature is over 271.4°F.

Importance of Vapor Pressure in Selecting a Refrigerant

Part of a refrigerant table for the older CO_2 refrigerant is reproduced in figure 8-8. The data shows that the pressure in the evaporator of such a system is 318 psig at the

Temp. F	* Vapor Pressure		
	R-22	Water	CO$_2$
5	28.3	29.87*	318
10	32.9	29.86*	346
20	43.3	29.82*	407
30	55.2	29.76*	476
40	69.0	29.71*	553

Note: Vapor Pressure in psig except where (*) indicates in. of Hg. Vac.

FIG. 8-8 Section of a refrigerant table

saturation (boiling) point of 5°F. The same table shows that for a system using water as a refrigerant, the pressure in the evaporator is 29.71" of mercury vacuum at a saturation point of 40°F.

A pressure gauge attached to the suction-service valve of the compressor (point P_1 on the Basic Refrigeration Cycle Diagram) would register 318 on the right side of zero for the CO$_2$ and 29.7 inches of mercury vacuum on the left side for the water vapor system. When the needle of the gauge is to the left of zero, it indicates that the pressure in this part of the system is below atmospheric. Any leaks in this part of the system would take air and moisture into the system, which is undesirable .

Thus, the vapor pressure of a refrigerant is an important factor in its selection for a system that is to be operated at a low temperature in the evaporator.

PRESSURE GAUGES IN REFRIGERATING SYSTEMS

High Pressure and Compound Gauges

What happens inside a refrigeration system is usually shown by pressure gauges. These gauges are so designed that they may be checked for accuracy and readjusted when necessary. Two of the most commonly used gauges are illustrated in figure 8-9. The high pressure gauge at (A) is graduated for readings of 0 to 300 psig. As the name indicates, this gauge is used on the high side of the system.

The compound gauge at (B) is used on the low side of the system. This gauge measures pressures above atmospheric and below atmospheric. The scale markings on compound gauges to the left of the zero graduation are usually in inches of mercury like: 1", 2" or 5" of Hg vacuum (below atmospheric pressure). The above-atmospheric graduations, located to the right of the zero graduation, are in terms of pounds per square inch (psi).

Location of Pressure Gauges in Refrigeration Systems

The compressor discharge pressure may be determined by connecting a high pressure gauge into the discharge service valve of the compressor. The suction pressure is

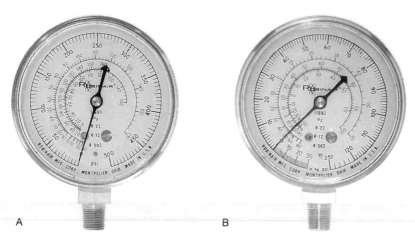

FIG. 8-9   Gauges for measuring refrigerant: (A) high-pressure gauge and (B) compound gauge *(Courtesy of Robinair Division, SPX Corporation)*

found by connecting the compound gauge into the suction-service valve of the compressor. (These connections are located at points P_2 and P_1, respectively, in the Basic Refrigeration Cycle Diagram.)

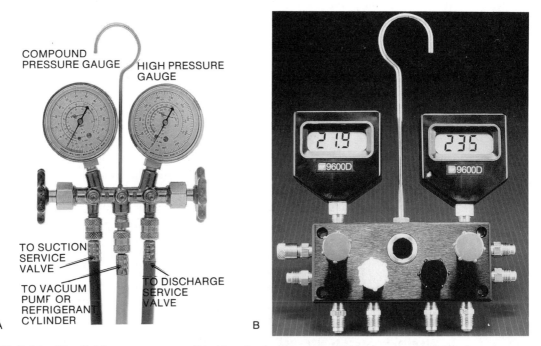

FIG. 8-10   Manifold gauge systems: (A) side wheel testing manifold gauge set and (B) digital manifold gauge with sight glass *(Courtesy of Robinair Division, SPX Corporation)*

FIG. 8-11   Portable recording gauge *(Courtesy of Bristol Babcock, Inc.)*

The maximum pressure gauge has a check valve built into it that stops the indicating hand on the dial at the place of maximum pressure. A push-button on the side of the gauge socket releases the pressure, permitting the pointer to return to zero.

Pressure gauges are frequently used with a manifold instead of being connected directly into the service valves. The manifold, in turn, is connected by flexible hosing to the two service valves. A third hose makes it possible to either evacuate or charge the system. A charging and testing gauge combination is illustrated in figure 8-10.

Digital Manifold Gauge with Sight Glass

The digital manifold gauge displayed in figure 8-10 at (B) features a large liquid crystal display (LCD) for easy-to-read information. The control valves and hoses are color coded for high side, low side, charge, and vacuum. The optical sight glass permits visual contact with the refrigerant.

The range of this model is from 0 to 29.9 in. Hg vacuum and 0 to 99.9 psi pressure on the low side; 0 to 999 psi on the high side (within a tolerance of ± 2%). A calibration screw provides for easy zeroing.

As an added safety feature, there is a high-pressure alarm warning for dangerous levels of 400 psi and over. A pressure/temperature conversion switch makes superheat calculations possible. The piston valve network in this gauge allows for 100% shutoff. The gauge operates on a 9-volt dc power supply (battery).

A more complete picture of the performance of refrigeration equipment may be obtained with a portable recording gauge like the one illustrated in figure 8-11. This instrument gives a continuous record of pressure, temperature, or a combination of pressure and temperature, and the time at which changes occur.

SATURATED REFRIGERANTS

A refrigerant vapor is said to be *saturated* when it is in contact with its liquid form or when it is at its boiling point for a specific pressure. This occurs in only two places in the system: (1) in the evaporator where the refrigerant is boiling as it absorbs latent heat, and (2) in the condenser where the refrigerant condenses as it rejects latent heat. The term *saturated* really refers to the space occupied by the refrigerant rather than the refrigerant itself.

Condition of Equilibrium

Saturation indicates that the space holds as much of the vapor as it can at a particular temperature. Suppose liquid ammonia is introduced into a cylinder, which has previously been evacuated, until the cylinder is about half full of the liquid. Under the reduced pressure, many molecules pass rapidly from the liquid to the vapor form, filling the space above the liquid. After conditions are stabilized, the molecules of the liquid and vapor move with the same average speed. Although the liquid is not visibly boiling, some molecules are escaping from the liquid and are joining vapor molecules.

At the same time, some vapor molecules are going back into the liquid form. When changes of state of the molecules of ammonia are taking place at the same rate, a condition of *equilibrium* exists, figure 8-12. The space above the liquid and the liquid are "saturated." The vapor is referred to as a *saturated vapor.*

The information given in refrigerant tables applies under these conditions of equilibrium. Thus, either boiling or condensation may take place at the saturation point. If latent heat is removed, some vapor condenses. Boiling takes place in the evaporator at saturation (equilibrium). Condensation occurs in the condenser at saturation as heat is given up.

NET REFRIGERATING EFFECT

The liquid refrigerant flowing through an evaporator is vaporized as it absorbs heat from around the evaporator. The quantity of heat that is absorbed by one pound of

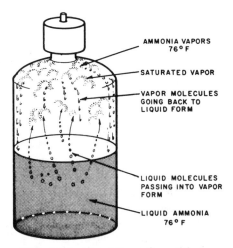

FIG. 8-12   Condition of equilibrium

refrigerant in this process is known as the *refrigerating effect.* Each pound of refrigerant flowing through the evaporator is able to absorb only the heat needed to vaporize it, provided no superheating takes place.

The net refrigerating effect depends upon (1) the temperature at which the liquid is vaporized in the evaporator and (2) the temperature of the liquid approaching the expansion valve. Thus, the refrigerating effect is the difference in Btu between the heat content of each pound of refrigerant vapor at a specified pressure and the heat content of the liquid refrigerant at a specified pressure and the heat content of the liquid refrigerant at its temperature as it approaches the expansion valve.

Another example may be used to show how this information is applied.

EXAMPLE 3:

Determine the refrigerating effect of one pound of Refrigerant 12 at a pressure of 37 psig in the evaporator. The temperature of the liquid at the expansion valve is 80°F.

PROCEDURES:

Step 1: Review the data in Table 4 on Properties for Refrigerant 12.

Step 2: Determine the temperature at which Refrigerant 12 vaporizes in the evaporator at 37 psig. This is found to be . . . .   40°F.

Step 3: Locate in the "heat content vapor" column for Refrigerant 12 the heat value of each pound of Refrigerant 12 vapor which leaves the evaporator at 40°F. This is . . . .   81.44 Btu.

Step 4: Locate in the "heat content of liquid" column the value at 80°F. This is the Refrigerant 12 temperature approaching the expansion valve. This is . . . .   26.37 Btu.

Step 5: Subtract the Btu valves.

ANSWER:

The difference in Btu value is the refrigerating effect per pound of Refrigerant 12. This is . . . .   55.07 Btu.

As refrigerating effect is being discussed it must be remembered that no substance can remain liquid at a temperature higher than the boiling temperature corresponding to its pressure. For example, if a closed container partially filled with 200°F water at atmospheric pressure is put under vacuum until the inside pressure drops to 1.0 psia, some of the water will flash into vapor. The flash vapor cools the remaining water until it reaches its boiling point. With a pressure of 1.0 psia (27.99 in. Hg vac.) the corresponding boiling point of water is 100°F.

These are the same conditions existing on the two sides of the expansion valve. If the pressure in the evaporator is 37 psig, then the highest temperature at which it is possible to have liquid Refrigerant 12 in the evaporator is 40°F. Assume that the liquid approaching the expansion valve is at 80°F. The refrigerant passing through the valve must be cooled from 80°F to 40°F. In surrendering heat, a small part of the liquid refrigerant vaporizes, absorbing heat. The sensible heat lost in this process of cooling the liquid is converted into latent heat of vaporization for the small quantity of refrigerant that flashes into vapor. However, the balance of the liquid vaporizes when there is sufficient latent heat. This takes place as the refrigerant flows through the cooling coil. Under these conditions, about 14% of each pound of Refrigerant 12 liquid which passes through the expansion valve becomes vapor while 86% remains liquid.

Thus, only .86 of each pound of Refrigerant 12 liquid in the system is valuable in absorbing heat in the evaporator. The other .14 pound is already vaporized and cannot absorb latent heat as long as it remains mixed with the liquid. It is for this reason that the refrigerating effect of a refrigerant is always less than its latent heat. In this example, the refrigerating effect may be computed by multiplying the latent heat of vaporization of the refrigerant by .86. Returning again to Refrigerant 12, its latent heat of vaporization at 40°F is 64.16 Btu/lb. The refrigerating effect equals: 64.16 × .86 = 55.2 Btu/lb., essentially the same value as obtained in Example 3.

Another example shows how data from tables are used to find both refrigerating effect and percent of liquid refrigerant vaporized.

EXAMPLE 4:

The temperature of liquid ammonia at the expansion valve is 80°F. In the evaporator, the temperature of the vapor is 30°F. Determine the refrigerating effect, and the percent of liquid vaporized while flowing through the expansion valve.

PROCEDURE:

Step 1: Secure a table of Properties for ammonia (Table 3 in Appendix).

Step 2: Check the "total heat of vapor" column for its Btu value at 30°F. This is .... 620.5 Btu/lb.

Step 3: Check the "heat of liquid" column for its Btu value at 80°F. This is .... 132.0 Btu/lb.

Step 4: Subtract these two values.

ANSWER:

The difference is the refrigerating effect of .... 488.5 Btu/lb.

Step 5: Determine the heat of liquid ammonia at 80°F. This equals .... 132.0 Btu/lb.

Step 6: Determine the heat of liquid ammonia at 30°F. This equals .... 75.7 Btu/lb.

Step 7: Subtract these values to get the heat lost by the liquid in cooling from 80°F to 30°F. This is .... 56.3 Btu/lb.

Step 8: Check the table for the latent heat of ammonia at 30°F. This is .... 544.8 Btu/lb.

ANSWER:

The percent of liquid ammonia vaporized = (56.3 ÷ 544.8) × 100, or 10.33% under these conditions.

RATE AT WHICH A REFRIGERANT IS CIRCULATED

A refrigeration system that removes heat at the rate of 12,000 Btu/hr. or 200 Btu/min. is said to have a capacity of one ton. A definite weight of refrigerant liquid must be vaporized in order to remove 200 Btu/min. This weight may be easily computed by dividing the 200 Btu/min. by the refrigerating effect of the refrigerant. This may be expressed as a formula: W = 200 ÷ RE, where (W) is the weight of the refrigerant to be circulated per minute, and (RE) is the refrigerating effect on the refrigerant in Btu/lb.

EXAMPLE 5:

The refrigerating effect of ammonia is 488.5 Btu/lb. Under this condition, what weight of ammonia refrigerant must be circulated per minute for a 15-ton plant?

PROCEDURE:

Step 1: Compute the weight of the refrigerant to be circulated per ton per minute.

$$W = 200 \div 488.5 = 0.409 \text{ lb./min./ton}$$

Step 2: Multiply the weight to be circulated in lb./min./ton by the 15-ton plant.

$$15 \times 0.409 = 6.14 \text{ lb./min.}$$

ANSWER:

The product of 6.14 lb./min. is the weight of ammonia refrigerant to be circulated to produce the refrigerating effect of a 15-ton plant.

COMPARISON OF ICE-MAKING AND REFRIGERATING CAPACITY

The ice-making capacity of a refrigerator was defined earlier in terms as the actual amount of ice that it can produce. Inasmuch as ice-making involves the reducing of the temperature of water to freezing and then the actual change of state, the ice-making capacity of a machine is not the same as its refrigerating capacity.

EXAMPLE 6:

Assume that a pound of water at 80°F is placed in a brine in a freezing tank which is at 15°F. Since water is being used, its specific heat is 1, the latent heat of fusion is 144 Btu/lb., and the specific heat of ice is 0.5. Another factor of say 15% is used to take care of heat losses in the system. To find the ice-making capacity, four steps must be taken.

PROCEDURE:

Step 1: Determine the Btu to be removed in cooling the water from 80°F to 32°F. Since one pound is used, this value equals:

$$1 \times 1 \times (80 - 32) = 48 \text{ Btu}$$

Step 2: Find the Btu's removed by solidifying water at the freezing point. This value for one pound is: $\qquad$ 144 Btu

Step 3: Determine Btu to be removed in cooling the ice to 15°F.

$$1 \times 0.5 \times (32 - 15) = 8.5 \text{ Btu}$$

Step 4: Multiply the total Btu per pound by 115% in order to take care of the assumed heat losses of 15%.

ANSWER:

$(48 + 144 + 8.5) \times 1.15 = 231$ Btu/lb.
or 462,000 Btu/ton

A refrigeration unit capable of removing 288,000 Btu in 24 hours is rated as having a capacity of one ton. However, such a unit is capable of freezing only 288,000/462,000 ton of ice, or 62.3%. In this instance the ice-making capacity of the unit in a one-hour period is said to be 62.3% of its refrigerating capacity.

THE MOLLIER DIAGRAM

A great deal of the information that is contained in tables on Properties of Refrigerants is also represented in graphic form. One such graph, which is known as a Mollier Diagram, figure 8-13, gives values from 1.0 psia to slightly above 600 psia for R-12. Note that by referring to such a chart it is possible to quickly determine psia, Btu per pound above –40°F, saturated liquid and saturated vapor temperatures. These diagrams are limited in accuracy because of the difficulty in interpreting exact values. However, such charts save considerable amounts of time when only fairly accurate values are required.

SUMMARY

- Latent heat is that heat energy which causes a change of state without any change of temperature.
  - Latent heat of fusion is the amount of heat to be added to (or subtracted from) one pound of a substance to cause it to melt (or solidify).
  - Latent heat of vaporization is the amount of heat to be added to (or subtracted from) one pound of the refrigerant to cause it to vaporize (or condense).

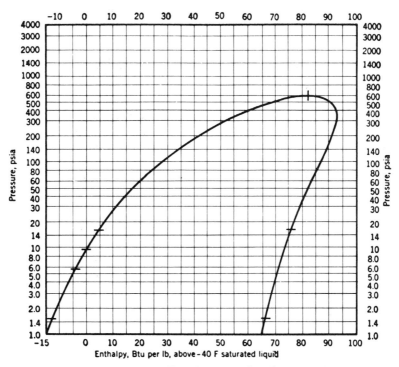

FIG. 8-13   Mollier diagram of Refrigerant 12

- Solidification and melting take place at the same temperature.
- Vaporization and condensation of a refrigerant take place at the same temperature provided the pressure is the same.
- The boiling point is that temperature at which the vapor pressure of the refrigerant is equal to the pressure on its surface.
- Saturation is a condition which exists in a refrigerant vapor when the space it occupies holds as much of the vapor as possible at that temperature.
  - Saturation occurs in the condenser and evaporator only when both liquid and vapor are in contact with each other.
- Pressure is the force on a unit of area exerted by refrigerant vapors. Gas pressure changes as the absolute temperatue changes.
  - Atmospheric pressure is the pressure exerted because air has weight. Under normal conditions this is 14.7 lb./sq. in. absolute.
  - Gauge pressure is the pressure above or below atmospheric pressure.
  - Absolute pressure is measured with reference to the point at which the refrigerant is considered to have no motion or pressure. Absolute pressures are measured in pounds per square inch absolute (psia), inches of mercury and microns.
  - Refrigerant pressures above atmospheric are measured in pounds per square inch gauge (psig).
  - Refrigerant pressures below atmospheric are measured in inches of mercury vacuum (in. Hg vac.).
- Equilibrium is the condition existing at saturation when the molecules of the refrigerant in liquid state are changing into the vapor state as rapidly as vapor molecules are changing into the liquid state.
- Enthalpy refers to the quantity of heat contained in a refrigerant at a given temperature with reference to –40°F. Values given in tables are the number of Btu/lb. of the refrigerant.
- Refrigerating capacity is the ability of a system to remove heat as compared with the cooling effect produced by the melting of ice. A one-ton refrigeration system is one that removes 12,000 Btu/hr. or 200 Btu/min.
- Refrigerating effect is the amount of heat transferred by one pound of refrigerant as it is circulated in the system.
- Ice-making capacity is the ability of a refrigerating system to make ice, starting with water at room temperature.
- Mollier diagram is a graphic method of representing the heat quantities contained in and the conditions of a refrigerant at different temperatures.

ASSIGNMENT: UNIT 8 REVIEW AND SELF-TEST

A. RELATIONSHIP OF LATENT HEAT AND CHANGE OF STATE

Provide a word or phrase to complete statements 1 to 8.

1. Heat, that when added to a material does not change its temperature, produces an increase in its _____ energy.
2. Heat which causes a change of state is called _____ .

3. The process of changing state from the liquid to the solid is known as _____ .

4. The amounts of heat to be added to or taken away from a material to change its state are _____ .

5. The number of Btu needed to change one pound of water at 32°F to ice at this temperature is _____ .

6. The process of changing state from a liquid to a vapor is called _____ .

7. The latent heat of vaporization for water at normal atmospheric pressure is _____ .

8. The reference temperature used in computing total heat or enthalpy of refrigerants other than water is _____ .

9. Match the condition or effect in Column II with the correct term in Column I.

Column I	Column II
a. Latent heat of vaporization	(1) The amount of heat to be added to one pound of a refrigerant in order to bring it up to the required temperature from the reference temperature (–40°F).
b. Boiling point	
c. Saturation	(2) The amount of heat which, when added to or subtracted from one pound of a refrigerant, causes it to become a vapor or liquid.
d. Equilibrium	
e. Enthalpy	(3) The temperature at which the pressure of the refrigerant equals the pressure on its surface.
	(4) A force on a unit area exerted by a refrigerant vapor.
	(5) The condition of a refrigerant vapor when the space it occupies holds as much vapor as possible.
	(6) A condition where there is an equal exchange of the molecules changing from one state to another, and vice versa.
	(7) The quantity of heat in a refrigerant at a given temperature with reference to –40°F.

10. Name the vessel in which the liquid is vaporized in a compression refrigeration system.

11. State the condition of the refrigerant in most of the evaporator.

12. What feeds liquid to the evaporator?

13. Name what removes vapor from the evaporator.

14. After the refrigerant becomes a vapor, what kind of heat does it absorb?

15. Give (a) the latent heat of vaporization and (b) the specific heat of water.

16. Refer in a handbook or the Appendix to a Table of Properties of Refrigerants. Then record the missing heat content (latent, liquid and vapor) for each of the three temperatures given for A, B and C in the table on page 150.

17. For Refrigerants A and B, determine from the data you recorded for Problem 16:
    a. Which refrigerant has the greatest latent heat at 30°F?

	Refrigerants	Required Temperature °F
A	Ammonia	30
		100
		-20
B	Saturated R-12	30
		100
		-20
C	Water at Saturation	30
		100
		-20

    b. How much more heat does one pound of this refrigerant vapor contain than one pound of the liquid at 100°F?

    c. Which of the two refrigerants contains the least heat at –20°F?

    d. How much heat does it take to vaporize one pound of this refrigerant at –20°F?

18. Determine the Total Heat from –40°F for both Vapor and Liquid and calculate the refrigerating effect for each refrigerant under the conditions listed.

	Refrigerant	Temperature (°F)		Total Heat from –40°F	
		Vapor	Liquid	Vapor	Liquid
A	Saturated R-12	30	80		
		-40	0		
B	Saturated R-22	30	80		
		-40	0		
C	Ammonia	30	80		
		-40	0		

19. Determine from the data calculated for Problem 18:

    a. Which of the three refrigerants has the greatest refrigerating effect?

    b. Under what conditions is the refrigerating effect greatest?

    c. The amount by which the refrigerating effect of the one exceeds the other of the following refrigerants:

        (1) R-22 than R-12 for 30° Vapor

        (2) Ammonia than R-22 for 30°F Vapor

        (3) Ammonia than R-12 for –40° Vapor

B. THE EFFECT OF PRESSURE ON CHANGE OF STATE

    1. Indicate the correct pressure or refrigerating condition which applies to each measurement or value given in the table on page 151.

Select the letter representing the condition which best completes statements 2 to 5.

    2. The vapor pressure of steam at 220°F is (a) greater than, (b) less than, (c) equal to, that at 260°F.

	Value or Measurement
A	14.7 lb./sq. in.
B	Above or below atmospheric pressure
C	Just gauge pressure in total answer
D	psia
E	12,000 Btu/hr. or 200 Btu/min.
F	Inches Mercury Vacuum (in. Hg vac.)

3. The vapor pressure of Refrigerant 12 at 5°F is (a) greater than, (B) less than, (c) equal to, that at 14°F.
4. The vapor pressure of a refrigerant is 93 psi in a condenser and 11.8 psi in the evaporator. Under these conditions, the temperature in the condenser is (a) greater than, (b) less than, (C) the same as, that in the evaporator.
5. The amount of heat to be removed from two pounds of steam at 212°F to condense it is (a) 288 Btu, (b) 144 Btu, (c) 970.4 Btu, (d) 1840.8 Btu.
6. Determine the gauge pressure (psig) corresponding to an absolute pressure of 24.7 psia.
7. What is absolute pressure (psig) corresponding to 15" of vacuum (in. Hg vac.)?
8. A gauge shows a pressure of 85 psig. Determine absolute pressure in psia.
9. A tank gauge records a vacuum of 12" of mercury. Give the absolute pressure in inches of mercury.
10. A pressure gauge indicates a vacuum of 8 inches of mercury. Find the psia.
11. A gauge reads 28.50" Hg vac. What is the temperature of the boiling water?
12. A pressure gauge on a boiler reads 0.05 inches of Hg absolute. Determine the temperature of the boiling water.

C. SATURATED REFRIGERANTS

1. Name the two places in the refrigeration system where the refrigerant vapor is saturated.
2. What condition exists with an ammonia vapor when the same number of molecules are changing to liquid as are changing from liquid to vapor?
3. What term denotes the vapor immediately after it has changed from a liquid?
4. When a liquid and its vapor are at saturation, and more heat energy is added to cause the liquid to change into vapor, what is that heat called?
5. Name the two classes of heat that go to make up the heat of saturated vapor.

D. NET REFRIGERATING EFFECT

Provide a word, phrase or value to complete statements 1 to 7.

1. The name given to the heat which is absorbed by one pound of R-12 when it evaporates in the evaporator is ——————— .

2. The refrigerant which has the greatest heat-carrying ability per pound is _____ .

3. The amount of heat to be removed from two pounds of steam at 40°F to condense it is _____ .

4. The amount of ice at 0°C that would be converted into water at 0°C by the addition of 28,000 Btu of heat is _____ .

5. The latent heat of Refrigerant 12 at 20° is 67.9 Btu/lb. The amount of heat that one pound of Refrigerant 12 liquid absorbs in changing to a vapor (also at 20°) is _____ .

6. The net refrigerating effect of Refrigerant 12 at –40° vapor and 0° liquid is _____ .

7. With the compressor running, the pressure in a fully active evaporator using Refrigerant 12 is 21 lb./sq. in. gauge. The temperature of the main part of the evaporator (to the nearest degree) is _____ .

8. Compute the heat content of the liquids and vapors given in the table for the temperatures and quantities given.

	Substance	Temp. (°F)	Quantity (lb.)	Heat Content
A	Liquid Ammonia	80	5	
B	Ammonia Vapor	80	5	
C	Water (liquid)	140	5	
D	Water Vapor	140	5	

9. List the procedure (briefly) and show the values of calculations when determining:
   a. The refrigerating effect, and
   b. The percent of liquid vaporized while flowing through the expansion valve.
      Data: (1) One pound of Refrigerant 12 is flowing through the expansion valve.
      (2) The pressure in the evaporator is 46.7 psig.
      (3) The Refrigerant 12 liquid enters the expansion valve at 86°F.

10. Develop the steps in your procedure to show how to compute:
   a. The Btu to be removed to freeze the water in one hour, and
   b. The equivalent refrigerating capacity of a unit, under the following conditions:
      Data: (1) Weight of water, 100 pounds   (4) Specific heat of ice, 0.5
      (2) Temperature, 90°F   (5) Heat losses, 15%
      (3) Brine in freezing tank, 10°F

SECTION 4:
THE HEART OF THE REFRIGERATION SYSTEM

UNIT 9:
THE COMPRESSOR IN MECHANICAL REFRIGERATION SYSTEMS

The compressor is referred to as the "heart" of mechanical refrigeration systems, figure 9-1. This comparison is made because the compressor pumps refrigerant through the system in the same manner as the heart pumps blood through the body. The function of the compressor may also be compared with the other major operating units (components) in the system.

Starting at the low side of the evaporator, the vapor, which is at a low temperature and pressure, flows through the suction line to the compressor. The compressor compresses this gas and by so doing raises its pressure and temperature. The hot, high-pressure gas then flows to the condenser where, as it gives up heat, the gas condenses to form a liquid.

The compressor also lowers the pressure in the evaporator. This causes the refrigerant to boil at the reduced pressure and temperature. The heat from the space to be cooled flows into the evaporator because of the resulting low temperature. This heat

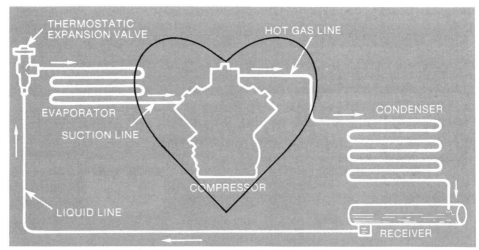

FIG. 9-1 The compresor in the refrigeration system

vaporizes the liquid refrigerant. The refrigerant vapor, which contains the absorbed heat from the evaporator, is pumped (pulled) back to the compressor. During this part of the cycle, the gas temperature is raised. This high temperature vapor is then discharged from the compressor. The heat from the hot gas flows into the water or air, which passes through or around the condenser. As a result, the refrigerant condenses to a liquid.

Stated briefly, the function of a compressor is to maintain a pressure difference between the high and low sides of the system. In this process conditions are created in which:

- The pressure and temperature of the refrigerant in the evaporator are lowered, allowing the refrigerant to boil and absorb heat from its surroundings.
- The pressure and temperature of the refrigerant in the condenser are raised allowing the refrigerant to give up heat at existing temperatures to whatever medium is used to absorb the heat.

To learn how these conditions are created, this unit discusses the following:

- Reciprocating, rotary and centrifugal compressors: the functions and the important operating parts of each mechanism.
- Compression ratios for compressors.
- The control of compressor capacity by varible speed motors and with such methods as hot-gas bypass, cylinder bypass and cylinder unloading.
- The need for adequate lubrication.

TYPES OF COMPRESSORS: FUNCTIONS AND OPERATING PARTS

The variety of refrigerants and the size, location and application of the systems, are some of the factors which create the need for many types of compressors. Since refrigerant properties differ, one compressor may be required to handle large volumes of vapor at small pressure drops, and another, small volumes of vapor at large pressure drops.

There are four main groups (classifications) of compressors: (1) reciprocating, (2) rotary, (3) centrifugal, and (4) the screw compressor. The action of the mechanical parts of the compressor determines its classification.

- In a reciprocating compressor, a piston travels back and forth (reciprocates) in a cylinder.
- In a rotary compressor an eccentric rotates within a cylinder.
- In a centrifugal compressor, a rotor (impeller) with many blades rotating in a housing draws in vapor and discharges it at high velocity by centrifugal force.

Reciprocating Compressors

Reciprocating compressors are usually a piston-cylinder type of *pump*. The main parts include a cylinder, piston, connecting rod, crankshaft, cylinder head, and valves.

These parts are labeled on the sketch in figure 9-2 which shows the operating cycle of a very simple compressor. On the down stroke of the piston, a low pressure area is created between the top of the piston, the cylinder head and the suction line of the evaporator. This causes the cool refrigerant vapor to rush into this low pressure area.

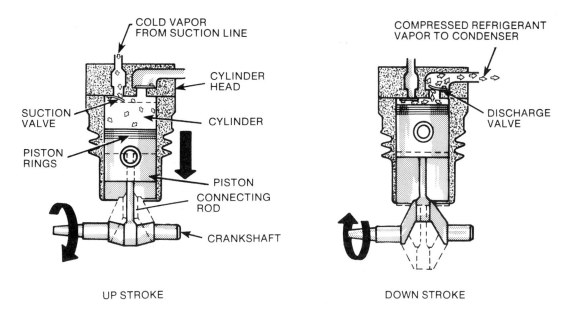

FIG. 9-2   Operating cycle of a reciprocating compressor

On the discharge (compression) stroke the piston, acting over a considerable surface area of the gas, compresses it and forces it at high pressure and increased temperature to move through a small valve opening to the condenser. The valves in the cylinder head are so designed that, depending on the part of the stroke, one is open while the other is closed. These valves control part of the refrigerant gas by directing it to either enter the hollow opening or discharge under pressure through the valve opening to the condenser.

Returning from the top of the stroke, the piston again draws in the cold refrigerant vapor and the cycle continues. Note that the connecting rod causes the piston to move up and down (reciprocate). The connecting rod is attached to a rotating crankshaft and serves to change rotary motion to straight line (rectilinear) motion.

The valve that controls the flow of refrigerant from the suction line into the cylinder head is known as the *suction valve;* the valve leading to the discharge line, the *discharge valve.* The rings on the piston prevent the gas from escaping between the piston and cylinder walls and improve the operating efficiency.

The housing for the compressor is referred to as a *crankcase.* It contains part of the bearing surfaces for the crankshaft and stores oil that is used to lubricate the crankshaft and the connecting rod.

There are many types of reciprocating compressors. One of the most common ways of classifying them is by the number of cylinders. While most refrigerating compressors are one (single) cylinder some models are two cylinders. The two cylinders run more smoothly and are more compact. Larger installations require compressors with three to ten cylinders and more.

The arrangement of the cylinders is still another method of classifying compressors. Some are *vertical* (figure 9-3); others are *horizontal, 45-degree inclined, V-type, W-type, radial,* and the like. Any one of these types may be single, double or any other number of cylinders, depending on size and the nature of the installation.

The way in which the rotary motion of the crankshaft is changed to the reciprocating motion of the piston is shown by the line drawing in figure 9-4. The wrist pin and the upper end of the connecting rod have an oscillating or reciprocating motion while the lower end of the connecting rod combines a reciprocating and rotary motion.

The eccentric is still another type of crankshaft. This crankshaft is a cast iron eccentric (an off-center disc) mounted on a steel shaft. It is comparatively inexpensive to manufacture, provides a larger wearing surface for the crankshaft, and runs smoother because it is better balanced.

On large installations the size of the lines requires the use of hard copper tubing. Vibration produced by the rotary, reciprocating and oscillating motion of the parts may cause excessive noises and may even break the soldered connections. Vibration absorbers of corrugated copper tubing with a braided bronze protecting cover may be installed in the liquid and suction lines to prevent vibration from traveling into these lines.

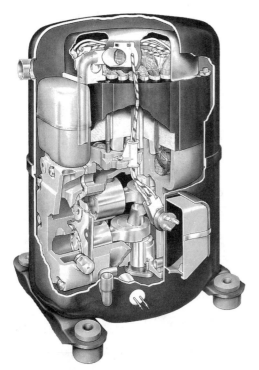

FIG. 9-3 Cutaway form of a two-cylinder reciprocating-type air conditioning and heat pump systems compressor *(Courtesy of Tecumseh Products Co.)*

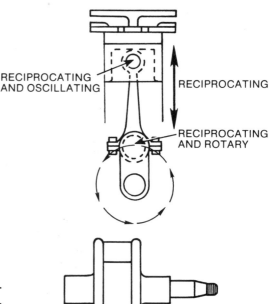

RECIPROCATING AND OSCILLATING

RECIPROCATING

RECIPROCATING AND ROTARY

FIG. 9-4 Motion in piston rod assembly

FIG. 9-5 Closeup of valve cage and ring plate valve *(Courtesy of The Trane Company)*

Compressor Valves

Nonflexing Ring Plate Type Valves.   Two common types of valves are used in compressors: the *nonflexing ring plate* and the *flexing disc*, figure 9-5. The position of the nonflexing ring plate suction and discharge valves in a compressor appear at in the cutaway section, figure 9-6.

The ring plate is a thin ring which is held closed over the circular discharge gas inlet in the top of the cylinder by springs. The suction valve is a ring plate that fits around the outside and just below the top of the cylinder. The valve is held closed by small springs.

The suction valve opens on the down stroke of the piston because the cylinder pressure is less than the vapor pressure in the suction line. The pressure in the crankcase and in the lower portion of the compressor is the same as the suction pressure at the inlet side.

On the upstroke of the piston the suction valve closes and the pressure within the cylinder causes the discharge valve to open. The high-pressure vapor then passes into the compressor cylinder head, figure 9-6, through the center holes in the valve cage as well as around the discharge valve cage. The metallic noise produced by the opening and closing of the discharge valve is cut down by a cushion of plastic material installed in the valve cage. The valve thus opens against plastic rather than metal. This also increases the valve life.

If a slug or liquid refrigerant were to enter a cylinder, the head might be blown off because of the small clearance space at the point where the piston reaches the top of its stroke. Instead of being fastened in position, the cylinder head in this illustration is held firmly in place by a strong spring. When a slug of noncondensable liquid enters the cylinder, the entire head lifts and passes the liquid to the discharge outlet. When there is

157

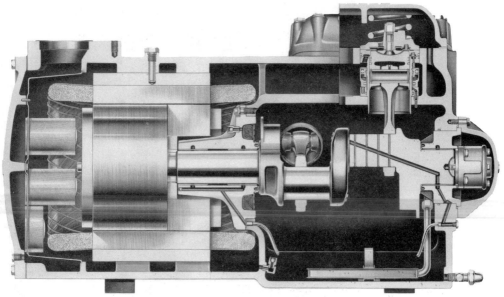

FIG. 9-6   Cutaway section of a 75-100 ton compressor *(Courtesy of The Trane Company)*

only compressible vapor in the cylinder, the pressure produced during ordinary operation is not great enough to lift the safety head.

The refrigerant vapor travels through the strainer or screen to the cylinder head where the suction valve is located. A metal section inside the cylinder head separates the suction and discharge valves. The screen housing has a small opening in the bottom which allows oil to be carried by the suction vapor back into the crankcase.

Flexing Disc or Reed Type Valves.   Small modern refrigeration compressors use high grade steel reed or disc valves, figure 9-7. These valves are quiet, simple, efficient, and long lasting. For these reasons they are especially adaptable to high speed compressors.

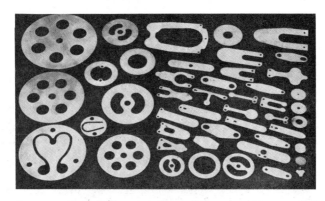

FIG. 9-7   Designs of small flexing valves *(Courtesy of Tecumseh Products Co.)*

Although some compressors have the suction valve in the piston head, the trend is toward locating both suction and discharge valves in the valve plate. A valve plate is placed between the body and the head. This design makes it possible to replace both valves with a minimum of difficulty.

The operation, seating and tightness of the valves are important. When the discharge valves leak, hot, high-pressure vapors seep back into the cylinder on the piston suction stroke. A suction valve leak will permit some of the hot vapors to go into the suction line on the compressor stroke. The hot gas leading by the valves and piston raises the temperature of the suction gas. Then, as the warmer suction gas is compressed, the discharge gas becomes hotter. On the next stroke, the higher temperature gas leaks back resulting in a still higher discharge temperature with a marked decrease in compressor capacity and efficiency.

Cooling Compressor Heads

During compression, the temperature of the refrigerant vapor rises. The temperature is controlled by cooling the upper part of the cylinder walls and the cylinder head to minimize the work required for compression and to keep the cylinder head from overheating. This cooling is done on ammonia compressors by jacketing a cylinder wall and head jacket through which water is circulated.

When fluorinated hydrocarbon is discharged from compressors, its temperature is much lower than that of ammonia. Consequently, water jackets are rarely provided on fluorinated hydrocarbon compressors. Instead, the cylinder walls and cylinder head are designed with fins to facilitate the transfer of heat to the surrounding air.

Safety Springs

There are times when liquid refrigerant or oil floods over into the suction line and compressor. This may be due to faulty adjustment of the expansion valve, a leaking float valve, or other similar difficulty. If the quantity of liquid is large and if it cannot get through the valve ports, serious damage may result.

The valve, the valve retainer, discharge valve coil springs, and shoulder screws are mounted (on old model compressors) on the valve plate, figure 9-8. The coil springs are strong enough to hold the valve retainer down during normal operation. Under these conditions, the discharge valve opens and allows gas to pass into the discharge chamber. The lift of the valve is limited by the valve retainer.

Whenever liquid or oil becomes trapped between the top of the piston and valve plate, a hydraulic pressure created in the cylinder forces the valve retainer to lift. This lift of the valve allows the liquid to discharge into the head where it either vaporizes or passes into the condenser. When the liquid has been cleared from the compressor, the valve retainer reseats. A liquid flooding condition often causes the breakage of valves, piston rods, crankshafts, and cylinder heads.

Shaft Seals

Open compressors are made with the crankshaft extending through the crankcase for direct motor, V-belt, gear, or chain drive. A crankshaft seal is required to prevent refrigerant leakage. Leakage may take place under both static and moving conditions at

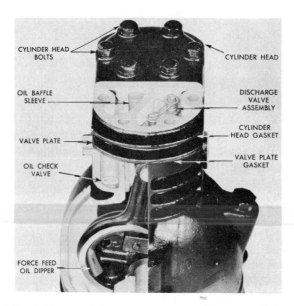

CYLINDER HEAD BOLTS

CYLINDER HEAD

OIL BAFFLE SLEEVE

DISCHARGE VALVE ASSEMBLY

CYLINDER HEAD GASKET

VALVE PLATE

OIL CHECK VALVE

VALVE PLATE GASKET

FORCE FEED OIL DIPPER

FIG. 9-8   Assembly of older model compressor parts *(Courtesy of Tecumseh Products Co.)*

the point at which the crankshaft passes through the housing. In practically all cases the crankcase is exposed to the circulating refrigerant vapor.

With horizontal double-acting compressors the piston rods slide back and forth through a stuffing box. This is usually sealed with graphite or a metallic or semimetallic packing. Most compressors that use a rotating shaft which projects out of the crankcase, utilize a bellows or syphon seal.

The crankshaft seal is made up of two gaskets and a flexible spring-expanded bellows which is attached to the crankshaft. A ring on the bellows rides against the seal face. This, in turn, forces against the seal face gasket and the seal shoulder of the shaft. With a spring pressure of 30 to 50 psi and with normal lubrication of the bearing surfaces, the seal reduces the refrigerant leakage to a minimum.

Service Valves

Service operations can be speeded up and simplified when conventional compressors are equipped with suction and discharge service valves. The discharge valve is fastened to the head and the suction valve to the body or head of the compressor.

The cross-sectional view in figure 9-9 shows a compressor service valve. The suction or discharge line connection is at (B); the connection to the compressor at (A). The third opening in the valve at (C) is for a gauge connection to permit charging or purging of the lines as well as to check pressure.

The valve is known as a *back-seating type*. This means that when the stem is turned all the way back, the gauge port is closed. The valve stem is usually back-seated so the line connection is open to the compressor. When the valve stem is in as far as it will go, the suction or discharge line is shut off while the gauge port is open to the compressor. At

160

midpoint the stem is so located that the compressor is open to the line connection, the gauge port, or both. Because of this design, the gauge connection must be plugged except when a gauge is being used. Double-seating valves permit the packing to be changed with little difficulty.

Hermetic Compressors

Extreme accuracy of finish and close dimensional tolerances are essential in the design and construction of hermetic compressors. Hermetic compressors are very quiet and efficient. There are two types that are in popular use: the reciprocating compressor and the rotary compressor.

Hermetic Reciprocating Compressors. Hermetic and open-type compressors are similar. The main difference is that the electric motor of the hermetic compressor is encased in the crankcase or within a sealed housing containing the compressor. With such a design, the compressor is driven directly by the motor, revolves at motor speed, and requires no shaft seal. Single cylinder models are available for small units while two-cylinder ones are used for larger units.

The three basic designs for hermetic compressors are illustrated in figures 9-10 (a semi-hermetic reciprocating compressor), 9-11 (a hermetic motor and compressor) and 9-12. The compressor body itself serves as the casing. The crankcase is extended to hold the motor. Since this type of unit is bolted together and can easily be taken apart and reassembled, it is known as a *semi-hermetic unit.*

The compressor and motor are enclosed in a steel casing (dome or hat). The stationary field (stator) of the motor may be pressed into half the dome. The compressor is secured to this stator. This unit is usually mounted on springs or rubber mounts, which dampen or absorb vibration.

The section in figure 9-11 shows a motor and compressor assembly mounted on springs inside the dome or casing. Such a casing is usually made of two pieces that are automatically welded together at the joint.

As can be imagined, one major problem with hermetic units is the cooling of the electric motor. In one design the stator is pressed into the dome to help cool the motor. This provides easy heat transfer from the windings to the case.

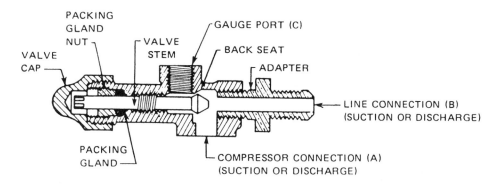

FIG. 9-9   Cross section of compressor service valve

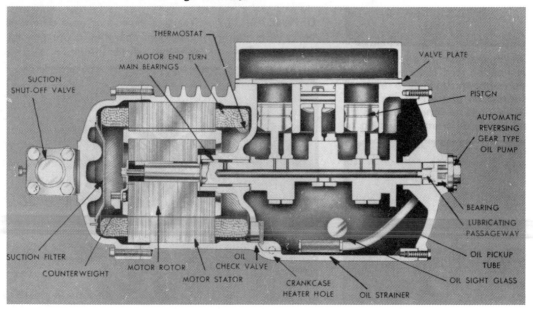

FIG. 9-10 Cutaway view of 6-cylinder, large capacity, semi-hermetic reciprocating compressor *(Courtesy of Dunham-Bush, Inc.)*

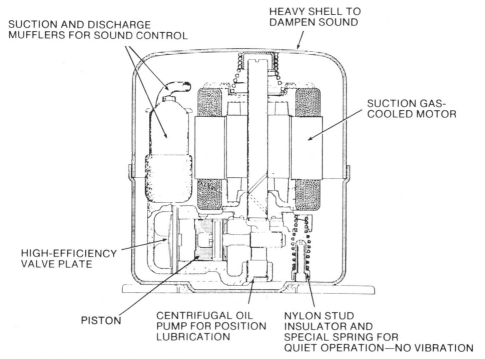

FIG. 9-11 Section of hermetic motor and compressor *(Courtesy of Copeland Corporation)*

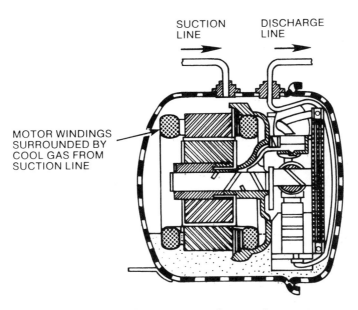

FIG. 9-12   Cooling motor windings with returning gas

A second design (illustrated in the section view, figure 9-12) provides a way of passing the returning vapor around the motor windings before the gas is compressed in the compressor. While the cool gas removes a great deal of heat, this design has the disadvantage of reducing compressor efficiency because it warms the returning gas.

Compressor displacement can be determined by the following formula:

$$\text{Displacement (in.}^3) = \frac{\text{Bore (in.}^2) \times 3.14}{4} \times \text{Stroke (in.)}$$

Mufflers.   Sound deadening devices are necessary on the smaller hermetic units for both the intake and exhaust openings. A device known as a muffler eliminates the gasping sound of the intake stroke and the exhaust gases. The mufflers are designed so that they permit the gases to increase in volume. This is accomplished through the use of baffle plates mounted in small cylinders in the muffler. Any increase in volume produces a decrease in velocity, thus reducing the pumping noises.

Rotary Compressors

Hermetically sealed rotary compressors are widely used for fractional-tonnage refrigeration and air-conditioning applications. These rotary compressors may be divided roughly into two types. In the first type one or more stationary blades are used for sealing the suction from the discharge gases. The second type uses *sealing blades,* which rotate with the shaft.

Rotary Compressor with Stationary Blade.   The operating parts and their functions are explained in figure 9-13. The moving parts in a rotary compressor include a steel ring, a cam (eccentric) and a sliding barrier.

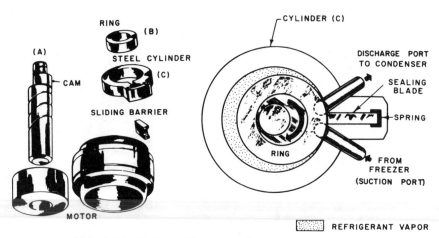

FIG. 9-13   Parts and function of a rotary compressor

The stationary parts consist of the motor, which drives the shaft and the steel cylinder. The ring is precision machined so it fits over and may be turned on the cam. The outside rim (periphery) of the ring also fits inside the cylinder. As the shaft cam rotates, it moves the ring so that one point on its circumference (periphery) is always in contact with the cylinder wall.

The shaded area on the drawing indicates that there is a crescent-shaped space between the ring and cylinder. As the ring turns and rolls on its rim against the inner wall of the cylinder, the space changes position. Within the cylinder head there is a suction port from the evaporator and a discharge port to the condenser. Between these two ports is a sliding barrier (sealing blade), which separates the two chambers thus formed. At the same time, the blade permits compressed refrigerant vapor at high pressure to be forced to the condenser on one side, and allows the low pressure vapor from the evaporator to enter the other side.

Three schematic drawings in figure 9-14 demonstrate how this is done. The start of the intake stroke (A) shows the sliding blade held against the ring. Refrigerant vapor is drawn into the lower pressure area between the ring and the cylinder wall. At the same

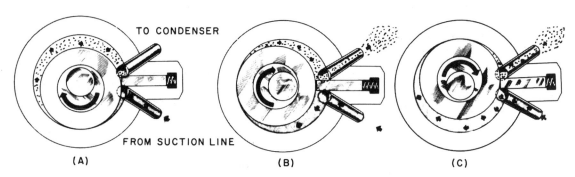

FIG. 9-14   Intake and compression phases in a rotary compressor

time, the refrigerant vapor ahead of the ring is being compressed into a constantly decreasing space.

As the ring turns toward mid-position (B) additional refrigerant vapor is drawn into the cylinder while the compressed gases are piped off of the condenser.

At the end of the compression phase (C), most of the compressed refrigerant vapor has passed through the discharge port to the condenser. A new charge of refrigerant is drawn into the compressor. This, in turn, is compressed and deflected against the sealing blade to the discharge port. In this manner the low pressure and temperature refrigerant is compressed gradually to a high pressure and temperature. Although a very simple single blade barrier was used as an example, the basic principles of operation of other types of rotary compressors are similar.

The second type of rotary compressor has a number of blades in the rotor, figure 9-15. These blades are forced against the wall by centrifugal action. The cylinder head is off center so there is, again, a crescent-shaped space between the rotor and the cylinder.

The low-pressure gas is drawn through the suction port (A). As the rotor (B) turns counterclockwise the gas is compressed because of the continually reduced space between the rotor and the cylinder (C). As the pressure increases, so does the temperature.

When the compressed gas reaches the discharge port (D) it passes into the high-pressure dome because there is no space ahead of the rotor into which the gas may go. At this point the clearance is about .0001". This design feature, plus the fact that there is also a fine film of oil lubricating the parts, makes it impossible during operating conditions for the refrigerant vapor to leak from the high to the low-pressure side. However, when the compressor is idle and there is no lubricating film between the rotor and cylinder, some vapor may leak out.

To control this, a check valve should be provided in the suction connection to the compressor. This valve prevents the hot, high pressure vapor from backing up into the evaporator. In this type of compressor it is also possible to use less expensive motors with low starting torque because the pressure within the compressor balances when it is not operating.

Importance of Lubrication in Rotary Compressors. Proper operation of rotary compressors depends on maintaining a continuous film of oil on the cylinder, roller and blade surfaces. This oil feeds into the cylinder through the main bearings. Sometimes a forced-feed lubrication system or separate oil pump are used. In still other cases an oil passage is included in the end plate at the point where the blade is moving out of the slot. When another passage is connected to the slots, a pumping action of the blades is created.

A good lubricant for compressors has these properties:

- It must be free of moisture, wax and foam.
- It must have the correct viscosity for the specific refrigerant used.
- It must be free of impurities which cause carbon to form around the exhaust valve.

Rotary compressors are quiet in operation, have limited vibration and may be used where a fairly high volume of refrigerant must be moved per ton of refrigeration. However, for the same size compressor, the centrifugal compressor delivers a greater volume than the rotary compressor.

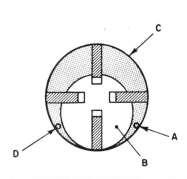

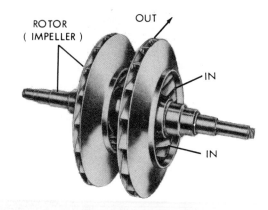

FIG. 9-15 Rotating blade type of rotary compressor

FIG. 9-16 Function of a centrifugal compressor *(Courtesy of Carrier Corp.)*

Centrifugal Compressors

A centrifugal refrigeration system depends upon centrifugal force to compress the refrigerant vapor. The rotor (impeller) of a centrifugal compressor draws in vapor near the shaft and discharges at a high velocity at the outside edge of the impeller, figure 9-16. The high velocity (inertia) is converted into pressure.

When the pressure drop is high, the compressor is built in stages. The discharge at one stage enters the suction inlet of the next until by the time the last stage is reached as much energy as possible has been used.

Centrifugal compressors, figures 9-17 and 9-18, are especially adapted for systems ranging as high as 5,000 tons and as low as 50 tons. They are also adaptable to temperature ranges between –130°F to 50°F. It should be noted that centrifugal compressors operate best with refrigerants having a high specific volume.

Centrifugal compressors are reasonably efficient even when operating with loads as low as 20% of normal. Because of their high operating speeds, centrifugal compressors may be connected directly to a steam turbine drive. Smaller sizes are usually driven by electric motors, some of which are equipped with standard gear driven speed increases. At 7,000 to 8,000 rpm some units develop capacities of 100 to 200 tons. Heavier units running at 3,500 to 4,000 rpm range from 1,000 to 2,000 tons in capacity.

Centrifugal Refrigeration Equipment. The position of the centrifugal compressor in a sealed refrigeration unit is shown in figure 9-18. These units are available in sizes varying from 50 to 5,000 tons. Note that the condenser, evaporator, compressor, and drive are all assembled as a single unit. Air and other noncondensable gases are automatically purged from the condenser. The cycle efficiency is increased by liquid intercooling. This system is designed primarily for air conditioning and water cooling.

Stated very simply, the operating cycle runs like this: The liquid refrigerant flows from the condenser to the evaporator. The float valve mechanism controls the liquid level in the evaporator.

The evaporated refrigerant then passes through the eliminator plates and past a vane control. This control regulates the flow of refrigerant vapor into the centrifugal com-

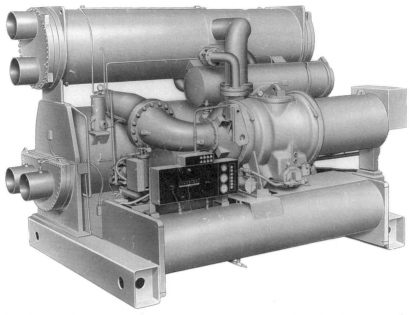

FIG. 9-17 **Heavy-duty commercial centrifugal (compressor) chiller** *(Courtesy of Carrier Corp.)*

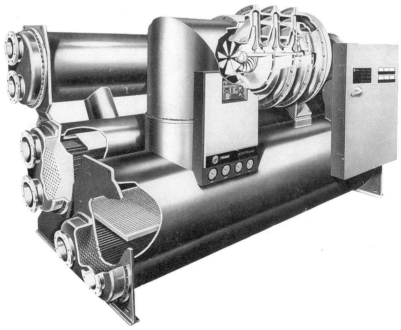

FIG. 9-18 **Internal design features of a hermetically sealed centrifugal refrigeration unit** *(Courtesy of The Trane Company)*

compressor (2 and 3). From this point the vapor is forced back to the condenser. The condenser (11 and 12) is water cooled. The high-pressure vapor that is forced back into the condenser is then piped directly to the evaporator (9). The refrigerants that are used mostly for these applications are Refrigerant 11 or Refrigerant 113.

HERMETIC COMPRESSOR
TROUBLESHOOTING (SERVICE) CHART

Troubleshooting charts are widely used throughout the refrigeration and air-conditioning industry. The charts provide a simple way of identifying common causes of many malfunctions and other problems and suggest how each problem can be corrected. The *hermetic compressor service chart* deals with seven basic compressor, capacitor and relay problems. Common causes of problems and remedies are provided.

TROUBLE/PROBLEM	COMMON CAUSE	REMEDY
1. COMPRESSOR WILL NOT RUN; NO SOUND	No power	Check fuses, wiring receptacle.
	Motor protector open	Reset; check current.
	Motor control contacts open	Check pressure settings on control.
	Open circuit in motor	Replace compressor.
2. COMPRESSOR WILL NOT START; INTERMITTENT HUMMING; (CYCLING-ON PROTECTOR)	Low voltage	Check voltage.
	Improper wiring	Check circuit according to wiring diagram.
	Open starting capacitor	Check capacitor; replace, if necessary.
	Relay not closing	Replace the relay.
	Open circuit in starting winding	Check the leads first. If there is no problem with the leads, replace the compressor.
	Compressor winding grounded	
	High head pressure	Check discharge shutoff and receiver valves to be sure they are open. Eliminate the high head pressure.
	Binding compressor	Check the oil level. Replace the compressor if the binding condition cannot be corrected.
	Weak capacitor	Replace the capacitor.
3. COMPRESSOR MOTOR WILL NOT GET ON "RUN" (STARTING) WINDING	Low voltage	Correct the low line voltage.
	Improper wiring	Check the wiring against the wiring diagram.
	Defective relay	Replace the relay.

TROUBLE/PROBLEM	COMMON CAUSE	REMEDY
	Running capacitor shorted	Check the running capacitor and replace, if necessary.
	Windings shorted	Check compressor; replace if defective.
	Starting capacitor weak	Replace capacitor.
4. COMPRESSOR RUNS BUT SHORT-CYCLES ON PROTECTOR	Low voltage	Check low voltage condition.
	Pressures too high	Check compressor for overheating and/or overcharge.
	Weak protector	Replace the protector.
	Running capacitor defective	Check capacitance. Replace capacitor, if necessary.
	Stator partially grounded or shorted	Check resistances and possible grounding. Replace stator, if defective.
	Unbalanced three-phase line	Check voltage of each phase. If unequal voltages, correct the unbalance condition.
	Discharge valve leaking or broken	Replace the valve plate.
5. STARTING CAPACITOR: BURNOUT	Short cycling	Cut down the number of starts to 20 or less per hour.
	Incorrect capacitor	Check manufacturer's specifications to identify correct capacitor rating and voltage requirements.
	Wrong capacitor voltage rating	Install capacitor with the required voltage rating.
	Shorting of capacitor terminals by water	Install capacitor so terminals remain dry.
	Sticking relay contacts	Replace the relay.
6. RUNNING CAPACITOR: BURNOUT	Excessive line voltage	Limit (reduce) the line voltage (not to exceed 10% above the motor rating).
	High line voltage and light load	Limit (reduce) the voltage if beyond the 10% excessive.
	Wrong capacitor; voltage rating too low	Check specifications. Install capacitor that meets required voltage rating.
	Shorting of capacitor terminals by water	Install capacitors so terminals will remain dry.
7. RELAYS: BURNOUT	Low line voltage	Increase the voltage (bring to not less than 10% under the compressor motor rating).

TROUBLE/PROBLEM	COMMON CAUSE	REMEDY
	Excessive line voltage	Reduce the voltage (keep within a maximum of 10% above the motor rating).
	Short cycling	Cut down on the number of starts per hour.
	Relay sticking or vibrating	Remount or replace the relay.
	Incorrect relay	Check manufacturer's specifications. Replace with relay as specified for motor compressor.

COMPRESSION RATIO

Compression ratio is the ratio of two pressures. The compression ratio number is found by dividing the discharge pressure (P_d) by the suction pressure (P_s), where both are given as absolute pressures.

Table 1 in the Appendix (Properties of Refrigerants) lists compression ratios for different refrigerants under normal conditions. The values given are for the temperatures designated for the standard ton, at 86°F condensing temperature and at 5°F evaporating temperature. A value of 4.06 for Refrigerant 22 means that the discharge pressure (absolute) is 4.0 times that of the suction pressure (absolute).

By referring to Table 5 in the Appendix, which lists the properties of Refrigerant 22, the pressure at 86°F is 172.87 psia and at 5°F, 42.888 psia. The compression ratio, according to definition, is 172.87 ÷ 42.888 (4 to 1).

Compression ratio is of prime importance only when it approaches a high limit. Such a high compression ratio denotes either high head pressure and/or a low back (suction) pressure. This results in loss of efficiency and excessive superheating of the discharge gas. Either condition may cause damage to the compressor. The compression must be accomplished in stages in those cases where there are excessively high compression ratios.

METHODS OF CONTROLLING COMPRESSOR CAPACITY

The three common methods of controlling compressor capacity are (1) by using a variable speed motor, (2) by bypassing the hot gas, and (3) by bypassing the cylinders (cylinder unloading). The capacity of a compressor must be controlled because refrigerant loads are seldom constant. Operating under partial loads and low back pressures creates a condition where the coil may freeze or damage may result.

Variable (Multi-) Speed Motors

The capacity of a compressor is proportional to the speed of the driving motor. When the suction pressure of the refrigeration system is high, the motor speed and compressor capacity must be increased. For this reason, electric motors with two or more speeds are used. The motor speed may be selected according to the wiring. The schematic drawing of the two-speed motor in figure 9-19, indicates that it can be operated at either 1200 or 1800 rpm, depending on the external wiring. Such a speed change may also be

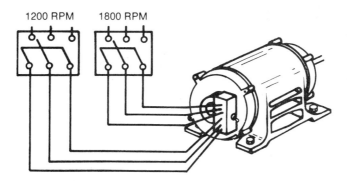

FIG. 9-19 Schematic drawing of a two-speed motor

made automatically. Variable-speed motors find limited application because the motors and their controls are expensive to use on large installations.

Hot Gas Bypass Method

A simple hot gas bypass compressor capacity control is shown graphically in figure 9-20. The name identifies the process used. The temperature or pressure of the refrigerant may be controlled by a solenoid stop valve (1) in the bypass line (2). When a capacity reduction is needed, the solenoid opens and permits some of the hot gas discharge by the compressor to be returned to the suction line (3).

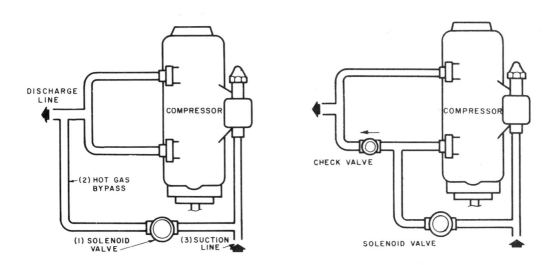

FIG. 9-20 Compressor capacity reduced by hot gas bypass method

FIG. 9-21 Compressor capacity controlled by cylinder bypass method

For full compressor capacity the solenoid stop valve closes so that the gas from both banks of cylinders passes through the discharge line. The solenoid is opened for half-capacity operation and closed for full-capacity operation.

When operating at reduced capacity for long periods of time, the cylinder heads become very hot; there are many lubricating problems and a great deal of noise. This method of controlling capacity is more practical where the reduction is of short duration and does not occur frequently.

Cylinder Bypass Method

For simplification, the drawing of the cylinder bypass system in figure 9-21 shows the parts on the outside of the compressor. Actually these can be positioned either internally or externally. The cylinder bypass method is activated by either temperature or pressure controls.

The solenoid valve opens whenever the controller requires a capacity reduction. By this action, the discharge gases from one bank of cylinders goes directly to the suction line. The check valve does not permit any gas at high pressure to reach the isolated bank. Then, since the lines are large, no high pressure is created in the bypassed cylinders. As a result the bypassed cylinders have a suction pressure above and below the valve plate so the cylinders do no work. The horsepower required in this method decreases in proportion to capacity reduction.

Cylinder Unloading Method

A fourth method of satisfactorily controlling compressor capacity is known as *cylinder unloading*. By this method the suction valves of some cylinders are held open, preventing compression. In an open position the piston draws gas from the suction manifold on the down stroke. On the up (return) stroke the piston returns the gas to the suction line without compressing it.

The mechanism used by one manufacturer is illustrated in figure 9-22. In so-called single-step unloader systems, one-half of the cylinders are unloaded. The horsepower required in the cylinder unloading method is decreased in proportion to the capacity reduction.

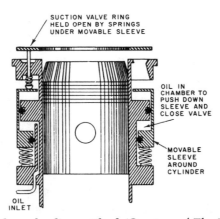

FIG. 9-22 Cylinder unloading method *(Courtesy of The Trane Company)*

LUBRICATION OF COMPRESSORS

A number of conditions must be considered in the lubrication of compressors. For instance, the oil must remain fluid at low temperatures. This is necessary in systems where the refrigerant and oil are miscible and where some of the oil in circulating with the refrigerant works its way into the evaporator. Unless the oil remains fluid, the low temperatures cause it to *congeal* (come together and lose its viscosity). This can cause a low oil level in the compressor and decrease evaporator efficiency.

Another requirement of the oil used in the compressor is that it must be free from moisture. Small quantities of moisture form a highly corrosive acid when in contact with SO_2. With methyl chloride and the R-refrigerants, moisture in contact with the refrigerant produces sludge and other undesirable deposits. Any large accumulation of moisture may also freeze in the expansion valve.

Many of the R-refrigerants are readily miscible with lubricating oil. For this reason a high viscosity oil must be used to compensate for its thinning out. However, since this is not true for all refrigerants, it is important to follow the manufacturer's recommendation on the proper oil to use for the specific conditions required.

The oil-miscible refrigerants permit greater flexibility in locating the condenser and evaportor units in relation to each other. Further, properly sized suction lines may be bent to conform to the mechanism and positioned as needed. The oil-miscible refrigerant also carries oil to lubricate parts of the compressor, the expansion and solenoid valves and other controls.

Viscosity of Oil

The term *viscosity* refers to the resistance of oil to flow and is designated by a numeral. This viscosity numeral is determined by the number of seconds it takes for a specific quantity of oil at a given temperatue to flow through the opening of a measuring device called a *viscosimeter* (viscosity meter). If a given quantity of oil at 100°F takes 60 seconds to flow through the opening, it is said to have a viscosity of 60 at 100°F. A heavier oil has a higher number; a lighter oil, a lower number.

Oil Separators

An *oil separator*, figure 9-23, is a device used to separate oil from refrigerant gas, returning the oil to the compressor and allowing the refrigerant to continue on its circuit through the refrigerating system. It depends for its operation on a reduction of gas velocity in the superheated state and is, therefore, located in the discharge line between the compressor and the condenser.

As the oil-laden refrigerant gas enters the oil separator its velocity is reduced. Since the oil particles have attained a greater inertia and are less inclined to change their direction of flow, the oil adheres to impingement screens allowing the gas to continue on its circuit through the refrigerating system.

The oil reservoir is that area in the base of the oil separator where oil is accumulated prior to its return to the compressor. When enough oil has accumulated to raise the float, the valve opens and since the pressure in the oil separator is greater than in the compressor crankcase, a positive oil return is accomplished.

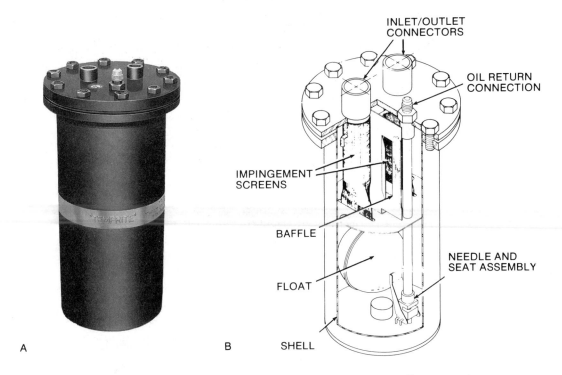

FIG. 9-23 (A) External and (B) internal features of an oil separator (flange type) *(Courtesy of Temprite Company)*

Some refrigerating systems use low-side oil separators. These are built into the crankcase at the point where the suction line enters the compressor. Although one type of oil separator and a simple installation are illustrated, there are many different types and sizes depending on the requirements of the refrigerating system.

SUMMARY

- The compressor takes a refrigerant vapor at a low temperature and pressure and raises it to higher temperature and pressure.
- Three basic types of compressors include: reciprocating, rotary and centrifugal.
  - The main parts of a reciprocating compressor include: a cylinder, piston, connecting rod, crankshaft, cylinder head and valves.
  - The main parts of a rotary compressor are: a cylinder, an eccentric that rotates in the cylinder around an off-center axis, and one or more sliding barriers.
  - The main parts of the centrifugal compressor are: the buckets on the rim of a wheel and in the casing around the wheel. The refrigerant is compressed by its own energy (inertia) as it is thrown off from the rapidly rotating buckets. Although the centrifugal compressor rotates at high speed, it is practically vibrationless.

- The largest single source of power for the compressor is the electric motor. Compressors may also be driven by steam turbine or internal combustion engine.
- Compressor crankshaft seals prevent air from entering the compressor, and oil and refrigerant from escaping.
- Discharge valves may be provided with a built-in safety feature to allow for oil pumping without damage to the compressor.
- Compression ratio is the ratio of two pressures: the absolute discharge pressure divided by the absolute suction pressure.
- Four common methods of controlling compressor capacity include variable (multi-) speed motors, cylinder bypass, hot gas bypass and cylinder unloading.
- Compressors may be lubricated with a splash system, pressure system, or a combination of both. Oils used in the system must have high viscosity and be free of moisture.
- Oil separators are installed to facilitate the return of oil to the crankcase of the compressor.

ASSIGNMENT: UNIT 9 REVIEW AND SELF-TEST

A. TYPES OF COMPRESSORS: THEIR FUNCTIONS AND OPERATING PARTS

Determine which statements (1 to 10) are true (T); which are false (F).

1. A reed type of valve is operated by a coil spring.
2. A shaft seal is required for an open-type compressor.
3. The speed of a compressor does not affect its capacity.
4. The compression ratio is found by dividing the reading of the discharge pressure gauge by that of the suction pressure gauge.
5. The slugging of a refrigerator is impossible if the vapor is superheated when it returns to the compressor.
6. The valves in a reciprocating compressor are operated by a cam (off-center) shaft.
7. A muffler is used on a hermetic unit to cut down the noise caused by slugging.
8. A hermetic unit may be either a reciprocating or rotary type of compressor.
9. The oil for lubricating a hermetic compressor collects in the crankcase during the off part of the cycle.
10. Centrifugal compressors are commonly used for small systems.

Add the correct term to complete statements 11 to 20.

11. The function of the compressor is to receive refrigerant vapor at a _____ temperature and pressure and change it to a _____ level.
12. The refrigerating unit that has the motor and compressor both enclosed in the same housing is the _____ type.
13. The compressor which has pistons and connecting rods is of the _____ type.
14. The compressor which compresses the refrigerant by its own inertia is the _____ type.

15. The two types of valves most commonly used in refrigeration compressors are the _____ type and the _____ type.

16. Damage to a compressor by slugging may be prevented by using_____ on the discharge valve or cylinder head.

17. During the off part of the cycle, the pressure on both sides of the_____ compressor is likely to equalize.

18. Leakage is most likely to occur in the _____ type of compressor through the _____ .

19. Compressors are cooled by equipping the upper part of the cylinder and the head with _____ or _____ .

20. Hermetic motors are cooled by the _____ .

21. Indicate by number the type of compressor which meets each design or operating condition given.
    (1) Reciprocating   (3) Rotary with rotating blades
    (2) Rotary with stationary blade   (4) Hermetic   (5) Centrifugal
       a. Sealing blade in the cylinder wall.
       b. Safety springs over the discharge valve.
       c. Balance weights on the crankshaft.
       d. Motor surrounded by refrigerant.
       e. No valves or sealing blades.
       f. Sealing blades located in the rotor.
       g. Vibration absorbers installed in the discharge and suction lines.
       h. Compressor runs at extremely high speeds.
       i. The pressure in the suction line equals that in the discharge line during the off part of the cycle.
       j. Compressor has pistons and cylinders.

B. HERMETIC COMPRESSOR TROUBLESHOOTING

Troubleshoot the ten hermetic compressor problems given in the table by stating the corrective action to take in each instance.

PROBLEM AREA	CAUSE OF ACTION	CORRECTIVE ACTION
Compressor not starting; no hum	Open motor protector	
Compressor not starting; cycling-in problem	Improper wiring	
Compressor motor not on starting winding	Weak starting capacitor	
Compressor runs; short cycles on the protector	Unbalanced three-phase line	
	Broken discharge valve.	
Starting capacitor: burnout	Short cycling	
Running capacitor: burnout	Excessive line voltage	
	Light load; high line voltage	
Relays: Burnout	Low line voltage	
	Short cycling	

C. COMPRESSION RATIO

1. Refer to Table 1 (Physical Characteristics of Eight Common Refrigerants) in the .Appendix or a handbook. Find the compression ratios at standard ton conditions for these refrigerants:
   a. Ammonia   b. Water vapor   c. R-503
2. Calculate the required values in the following problems using the data given:
   a. Type refrigerant: R-22; Readings (psig) Discharge gauge: 150, Suction gauge: 15. Determine the compression ratio of the compressor.
   b. Type refrigerant: NH_3; Reading (psig) Discharge gauge: 100, Suction gauge: 5. Determine compressor ratio of the compressor.
   c. Reading of Discharge gauge: 125 psig. Compression ratio of system: 3. Determine the reading at the suction gauge.

D. METHODS OF CONTROLLING COMPRESSOR CAPACITY

Add the correct refrigeration terms to complete statements 1 to 9.

1. Name the four methods by which the capacity of a compressor may be varied.
2. There is no apparent change in the horsepower requirement when the _____ system is used.
3. Name the two systems that use a solenoid in the refrigerant line _____ and _____ .
4. The system that requires a check valve in addition to the solenoid is the _____ system.
5. The system that uses a hydraulic method to hold the valves open is the _____ system.
6. The systems where there is a decrease in horsepower resulting from variations in load are: (a) _____ , (b) _____ , and (c) _____ .
7. The system that causes excessively high head pressures and temperatures is the _____ system.
8. The pressure in the suction and discharge lines equalizes in either the _____ or _____ systems.
9. The system that should be used when a lower capacity occurs at intermittent intervals and for a short duration of time is the _____ .

E. LUBRICATION OF COMPRESSORS

1. Describe briefly how reciprocating, rotary, hermetic, and centrifugal compressors are lubricated.
2. How is the viscosity number of a lubricating oil determined?
3. Explain the operation and construction of the high-side oil separator.
4. List two advantages of oil being miscible with refrigerants.
5. Name two disadvantages of oil being miscible with refrigerants.

UNIT 10:
EVAPORATORS

The evaporator is the device in the low-pressure side of a refrigeration system, figure 10-1, through which the unwanted heat flows. The evaporator absorbs the heat into the system in order that it may be moved or transferred to the condenser. The evaporator is known as a *cooling coil, blower coil* or *chilling unit*. Regardless of name, its function is to absorb heat from the surrounding air or liquid and, by means of a refrigerant, to move this heat out of the refrigerated area.

The major points covered in this unit include: basic evaporator theory, the classification of evaporators according to use and function, and the effect of humidity on the operation of evaporators. Considerable attention is given to methods of defrosting evaporators manually, semi-automatically and automatically.

TYPES OF EVAPORATORS

The two basic types of evaporators are the dry or direct-expansion and the flooded types. The dry or direct-expansion type is formed by a continuous tube, figure 10-2. The refrigerant from the metering device feeds into one end of the tube and the suction line connects to the other or outlet end. There is no provision in the direct-expansion evaporator to recirculate the liquid or gas within the evaporator, or to separate the liquid and the gas. The feed for such an evaporator may be either at the top or the bottom.

By contrast, the flooded evaporator, figure 10-3, recirculates the refrigerant by using a *surge chamber*. The surge chamber is a drum or container into which liquid enters

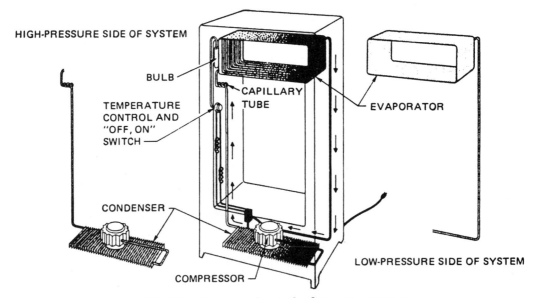

FIG. 10-1 Basic mechanical refrigeration system

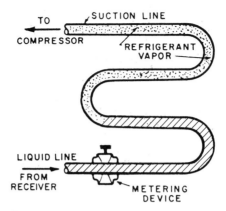

FIG. 10-2 Direct expansion "coil"

from the metering device. The liquid flows from the surge chamber to the evaporator where it boils and then returns to the surge chamber. There the liquid and the gas in the refrigerant are separated. The refrigerant vapor is drawn off into the suction line as illustrated, while the liquid is recirculated in the chamber.

The advantage of the flooded-type evaporator is that the whole surface of the evaporator coil is in contact with the liquid refrigerant under all load conditions. This is possible because the liquid level is controlled and the liquid which is not evaporated is recirculated.

Direct-Expansion Evaporators

The coil is one of the most widely used devices for cooling air. As warm air flows around the tubes in a coil, the air surrenders its heat to the boiling refrigerant inside the tubes. The flow of this liquid refrigerant is usually controlled by one of a number of

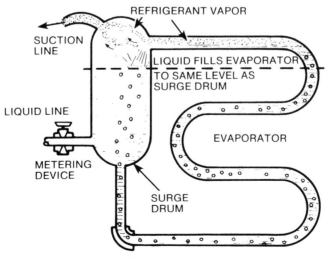

FIG. 10-3 Flooded "coil"

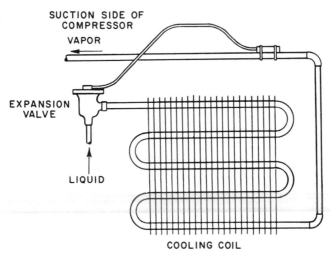

FIG. 10-4   Direct-expansion coil

types of valves: the automatic expansion valve, the float valve for flooded valves or the thermostatic expansion valve (TEV), capillary tube or AEV for dry expansion coils.

The schematic drawing shows a direct-expansion coil in figure 10-4. The tubing and fins or tubing and plate form the evaporator. The inside of the evaporator is always filled with a mixture of the refrigerant in liquid and vapor form. At the inlet side of the coil the refrigerant is mostly in liquid form (85% liquid, 15% vapor). As the liquid refrigerant moves through the evaporator to the outlet side, more and more of it is vaporized. At the outlet, the refrigerant is all vapor.

The important parts of a direct-expansion cooling coil are shown on the accompanying photo, figure 10-5. To repeat, the refrigerant absorbs heat as it evaporates (boils)

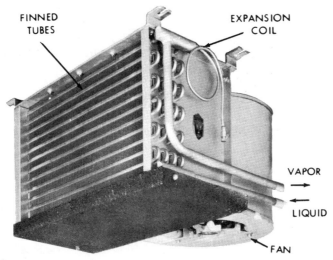

FIG. 10-5   Direct-expansion cooling coil *(Courtesy of Peerless of America, Inc.)*

inside the tubes. The tubes take heat from the surrounding air. This air may be blown through the coil by forced convection currents.

Manifolding

The flow of a refrigerant through the evaporator coil is divided into one or more *circuits*. Circuits means that the flow is channeled through separate rows of tubes rather than through one lone tube. This process of circulating the refrigerant in the direct-expansion or dry evaporators is referred to as *manifolding* and is of great importance.

Manifolding makes it possible to maintain the correct velocities and the desired pressure drop of the refrigerant within definite limits. Velocity is important because the refrigerant must have a sufficient force to *scrub* the walls of the tubes. Scrubbing the walls of the tubes prevents the film of liquid refrigerant and oil from adhering to the inside surfaces. The velocity must be such that it insures a continuous movement of the oil in the refrigerant. At the same time, the velocity should not be so low as to cause excessive pressure drop since this creates an added load for the compressor.

Pressure drop is merely the pressure difference required to maintain the desired refrigerant flow. It is recommended that the total pressure drop across a direct-expansion evaporator coil is between one and five pounds per square inch, approximately. When the pressure drops below one pound, the velocity of the refrigerant is too low for satisfactory heat transfer and for the movement of the coil. Pressure drop above five pounds is not economical. The pressure drop causes excessive superheat and prevents utilization of the coil area.

Design and Operation of a Direct-Expansion Evaporator Coil

The most familiar form of evaporator is the direct-expansion cooling coil, sometimes referred to as DX coils. These coils may be used in air-conditioning units, or as part of a central refrigeration system. The capacity of a DX coil depends on:

- The quality of air which is blown through the coil.
- The temperature of this air.
- The temperature of the boiling refrigerant.

Increasing the quantity of air or the temperature difference between the air and the refrigerant increases the coil capacity. Under these conditions, the coil is able to evaporate a maximum quantity of refrigerant in a given time. An increase in rate of evaporation causes the suction pressure to be raised. This is due to the fact that additional gas is formed while the compressor runs at a constant speed.

On the other hand, if the quantity of air or the temperature difference is decreased, the rate at which the refrigerant is evaporated decreases. This results in a lower suction pressure. The suction pressure is lowered because the compressor can pump the vapor away faster than it forms, until equilibrium is reached at the lower pressure.

Direct-Expansion Water Chiller and Secondary Refrigerants

Another type of evaporator is the shell-and-tube water cooler. In this evaporator water surrounds the outside of the tubes. Refrigerant liquid inside the tube becomes a vapor as it absorbs heat from the water. The cutaway drawing in figure 10-6 shows the main parts and internal construction of a shell-and-tube water cooler.

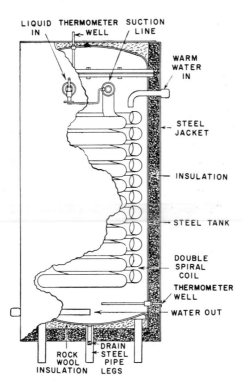

FIG. 10-6   Direct-expansion shell-and-tube water cooler

Large refrigeration and air conditioning installations require the use of evaporators at a considerable distance from each other. The piping problems connected with refrigerant flow and oil return are costly unless secondary refrigerants are used. In such an installation, water, which is the secondary refrigerant, is cooled by the evaporator. This cool water is then circulated through the cooling coils in the different parts of the large system. This method is reasonable in cost because water is inexpensive and easy to handle. Brine solutions are used instead of water for operation at below freezing temperatures.

While only two types of direct-expansion evaporators have been described, there are many others like bare pipe and plate evaporators, natural and forced convection-current evaporators, and cooling coils.

In summarizing, it may be said that in the direct-expansion evaporator the liquid refrigerant is fed to the evaporator only as it is needed. The liquid boils in the evaporator as it absorbs heat from the surrounding space. The refrigerant remains liquid only as it enters the evaporator, boiling away rapidly as it progresses along the evaporator tubes. It is for this reason that direct-expansion evaporators are sometimes called *dry-expansion evaporators.* By contrast, another type of evaporator operates with its coils full of liquid at all times. These are referred to as *flooded evaporators.*

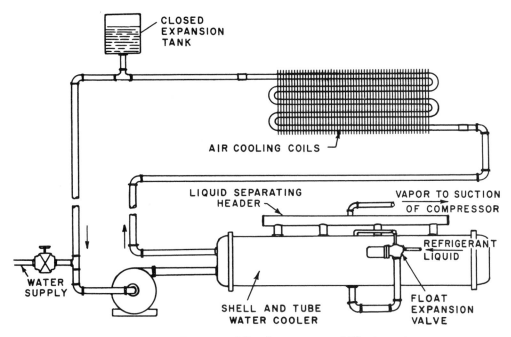

FIG. 10-7 Position of flood-type water chiller in system

Flooded Evaporators

The difference between the direct-expansion and the flooded evaporators is best illustrated by comparing water chillers for each type. The dry-expansion cooler carries the water between the shell and the tubes. Baffles are included to increase the velocity and the heat transfer. The refrigerant which passes through the tubes of the water chiller is metered at the expansion valve so that only the required amount is admitted. The liquid refrigerant starts to boil as it enters the tubes and continues to boil as it moves through the tube circuits. By the time the refrigerant reaches the suction end, all of the liquid is converted to a vapor.

By comparison, the water in a flooded chiller is passed through the tubes. The liquid refrigerant is carried between the shell and the tubes. A float valve is used to maintain the liquid level to replace any liquid that boils away. The diagram in figure 10-7 shows the function and relationship of the flooded-type of water chiller to the rest of a refrigerating system.

The more detailed drawing in figure 10-8 illustrates an older model flooded-coil evaporator with a float valve to maintain the liquid refrigerant at a constant level. The liquid level is such that it almost covers the entire evaporator.

It is for this reason that such an evaporator is known as the flooded type. As fast as the refrigerant evaporates, the float regulates the liquid until it reaches a specific level.

The flooded evaporator is efficient because the liquid refrigerant is in contact with the entire inside surface of the tubes. It is through this contact with the relatively warm

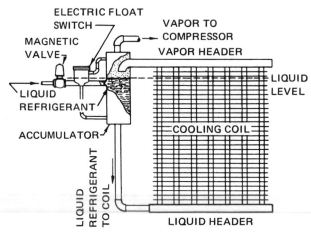

FIG. 10-8   Flooded coil evaporator with liquid level float

walls of the tubes that vaporization of the liquid takes place. It is apparent that a flooded-evaporator system requires more refrigerant than a dry system. The flow of liquid refrigerants to flooded evaporators is handled through float controls. Among these are the *low-side float*, and the *high-side float.*

Low-Side and High-Side Float Coils

The low-side float evaporator in figure 10-9 identifies an old system that is still used in commercial refrigeration applications. It is operated by a float which controls the flow of refrigerant to the coil. Notice that the float is contained within the evaporator and requires extra space. This low-side float is named for the low-pressure side of the valve where it operates.

On a high-side float evaporator, the connection to the liquid line is at the bottom of the coil. Introducing the liquid at this point provides a better agitation. The float valve for

FIG. 10-9   Old system of low-side float type of evaporator *(Courtesy of former Norge Refrigeration Company)*

184

the coil is located outside the coil itself. The enlarged areas at the top and sides are used to collect the vapor which rises through the liquid.

Low-side and high-side float coils are not as widely used as they were at one period of time.

HUMIDITY AND ITS RELATION TO EVAPORATORS

Two properties of air that need to be reviewed are relative humidity and dew point temperature. Relative humidity is a ratio of the amount of water in air space to the amount of water that the air space can hold at a given temperature. Humidity is important in refrigeration because fresh foods and dairy products require high humidity conditions to reduce moisture loss. Goods sealed in vapor-proof containers are not affected by moisture conditions.

A few basic principles are cited to show the effect of humidity on refrigeration. As warm air cools, its capacity to hold moisture decreases and its relative humidity increases. When the air temperature reaches the point where the relative humidity is 100% the air is *saturated* because it cannot hold any more moisture. This temperature is called the *dew point* temperature.

Air that is cooled below its dew point temperature gives up the moisture which it cannot hold and deposits it on the nearest colder surface. The colder surface in a refrigerated area is the evaporator. Therefore, the moisture deposits on the surface of the evaporator.

The evaporator serves the twofold purpose of (1) cooling the air temperature below the surrounding atmosphere and (2) reducing the moisture content of the air by removing moisture. The moisture that collects is called *condensation*. This condensation collects as water on the evaporator if its temperature is above 32°F and below the dew point. When the temperature of the evaporator falls below 32°F, the condensation freezes and forms a coating of frost, figure 10-10. The temperature difference between the surrounding air and the evaporator determines the rate at which the moisture is extracted and condenses.

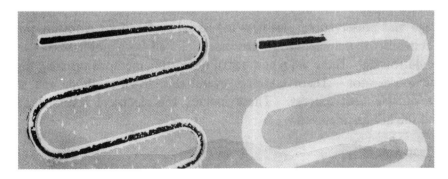

ABOVE 32°F   FROST AND ICE BELOW 32°F

FIG. 10-10 Moisture removal

When the condensation turns to ice and builds up on the evaporator surface, it acts as an insulation and retards heat flow. Therefore, when an evaporator must operate below 32°F, it must be *defrosted* periodically to prevent loss in capacity and efficiency of operation.

IMPORTANCE OF DEFROSTING EVAPORATORS

It is evident that anything which slows down the ability of an evaporator to transfer heat energy affects the capacity and efficiency of the refrigerating system. In review, it may be said that the ability of an evaporator to absorb heat depends on:

- Its size, surface condition and material.
- The method and amount of air circulated over it.
- The temperature difference between the evaporator and the surrounding air.
- The velocity turbulence.
- The kind of refrigerant in the evaporator.
- The quantity of oil in the refrigerant.

An evaporator with a clean surface which is without scale or film absorbs heat easily and rapidly. Any coating such as rust, frost, ice, or corrosion acts as an insulator and reduces the capacity of the system. Under such conditions the suction pressure drops to match the reduced quantities of heat absorbed by the evaporator. With decreased efficiency, the cost of refrigeration increases.

Ideal evaporator conditions require:

- The largest possible surface area and one that may be kept fully refrigerated.
- Rapid air circulation around clean and frost-free evaporation unit surfaces.
- A low temperature difference of 8° to 10° between the refrigerant and air.
- A low quantity of water vapor removal leaving a high humidity to preserve food appearance, moisture content and weight.

The erection of many frozen food plants, low temperature locker storage rooms, the widespread manufacture and use of low-temperature refrigerators, display cases for food storage and preservation, plus new industrial applications requiring precise temperature controls for manufacturing processes . . . all have emphasized the need for automatic defrosting. There are a number of defrosting methods utilizing heat sources within the refrigerator. There are also several shut-down methods and still others which use outside heat sources. These include the electrical resistance, the water defrost and the hot-gas methods.

Frosting Evaporators

Evaporators fall within three broad groups according to operating conditions: (1) frosting, (2) nonfrosting and (3) defrosting.

The frosting coil was one of the first types produced for use in refrigerated cabinets of all kinds. Since early cabinets had evaporators with limited areas, they had to be operated at very low temperatures to produce the required refrigerating effect. As such, they never became warm enough to permit the frost and ice to melt.

The frosting-type evaporator always operates at temperatures below 32°F. This means that the coil frosts continually when in use and the machine must be shut down at regular intervals to remove the frost. This condition occurs because the system runs at extremely low temperatures in order to keep the refrigeration fixture cool. As the frost grows in thickness, the coil efficiency decreases until the ice and frost are removed.

Nonfrosting Evaporators

Nonfrosting coils, figure 10-11, use only the thermostatic expansion valve type of refrigerant control. The area of these coils is so great that at a temperature between 21° and 32°F the evaporator is capable of cooling a refrigerator box down to approximately 37°F. For this reason, such coils are labelled as nonfrosting. The refrigerant inside the coils operates at temperatures between 20°F to 22°F. While there may be an occasional slight coat of frost on the coils with the compressor working, it disappears when the compressor stops.

The big advantage of a nonfrosting evaporator is its operation at a temperature close to freezing. At this temperature it does not draw the moisture out of the air rapidly. This permits a relative humidity in the cabinet of from 75 to 85%. At this condition, foods are kept fresh and do not lose weight.

Since the nonfrosting coils have greater area, they are bulkier than the frosting types. They also require baffles to direct the air flow in one direction past the coils. Coil efficiency is increased as baffles speed the air cooling process.

Defrosting Evaporators

A defrosting evaporator is one in which frost accumulates on the cooling coils when the compressor operates, and melts after the compressor shuts off.

The evaporator defrosting-type coil (with a greatly increased surface area) remains at a 20°F to 22°F temperature while the compressor is running. At these temperatures frost forms on the coil surface. When the compressor stops, the coil warms up to a temperature slightly above 32°F. The ice melts at this temperature before the compressor starts again.

While this method works in some cases, it does not in others where the moisture may not have sufficient time to escape before the next compressor cycle lowers the tempera-

CONDENSATE
TRAY

FIG. 10-11 Nonfrosting evaporator with condensate tray *(Courtesy Kramer-Trenton Company)*

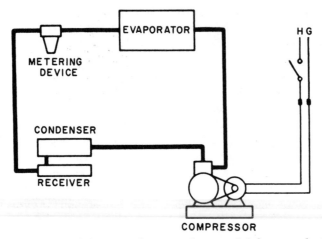

FIG. 10-12   Schematic drawing of manual defrost method

ture of the evaporator to the 20°F to 22°F range. When this happens, ice accumulates at the bottom of the fins to block the free circulation of air around the coils, thus reducing the effectiveness of the evaporator.

Defrosting-type evaporators provide uniform and efficient coil operation in transferring heat. The disadvantage lies in the fact that a larger unit is needed and that temperature must be sacrificed between the air in the refrigerated cabinet and the evaporator.

BASIC METHODS OF DEFROSTING EVAPORATORS

Manual Defrost Method

The earliest and simplest way of defrosting is the manual shutdown method, figure 10-12. Either the compressor is stopped or the refrigerant to one evaporator (in large

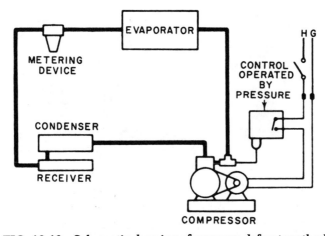

FIG. 10-13   Schematic drawing of pressure defrost method

installations) is closed off until the frost or ice is melted. At that point the compressor is started or the refrigerant valve is reopened. This method is still employed on large equipment using skilled operators.

If defrosting is neglected, the process becomes expensive, especially when food removal is necessary or the ice must be mechanically scraped.

Pressure Defrost Method

The *pressure cycle* or *pressure defrost* method is used on finned gravity or blower-type evaporators, figure 10-13. These are used where temperatures between 35°F and 60°F are needed for preserving fresh foods. The high humidities prevent loss of weight and preserve appearance in foods, flowers and other marketable products.

The temperature of the evaporator indirectly controls air temperature of the refrigerator in the pressure defrost method. The evaporator temperature, in turn, is controlled by the saturation pressure of the refrigerant. The saturation pressure corresponds to the required temperature of the evaporator. The control is set to permit the coil to warm up to 32°F and defrost before it starts again.

The pressure defrost method is not practical for refrigeration temperatures that must be held to temperatures below 35°F because of the difficulty of getting a complete defrost between cycles. Another disadvantage of this method is the difficulty of controlling air temperatures in refrigerators in unheated areas. If the temperature of the air surrounding the condensing unit is lower than the temperature in the evaporator, the refrigerant condenses into the compressor crankcase. This action prevents the temperature in the evaporator from rising high enough to close the pressure control.

Temperature Defrost Method

The temperature defrost, figure 10-14, and the pressure defrost methods are similar in that for every period that the compressor runs there is a defrost cycle. A remote bulb-type thermostat is attached to the evaporator. Its location on the evaporator must be selected with great care and depends on the temperature requirements of the system. The remote-bulb thermostat controls the cut-in and cut-out points of the

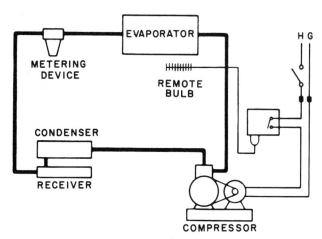

FIG. 10-14 Schematic drawing of temperature defrost method

compressor which are 34°F and 22°F respectively. The temperature of the evaporator, rather than pressure, controls such a refrigeration system.

The temperature defrost method makes it possible to maintain temperatures regardless of load variation. This method is preferred to the pressure defrost method for cold locations because the operation of the compressor is controlled directly by temperature. One disadvantage is that the temperature defrost method is not as dependable in assuring a complete defrost especially in hot weather or other high load conditions. This disadvantage may be overcome by an occasional manual defrost.

Time Shut-Down Defrost Method

In some systems where a required temperature range within the refrigerated area is between about 25°F and 35°F, it is possible to shut down the compressor during a low-service load period. Figure 10-15 shows that in this method, the shut-down period may be controlled by a clock during those hours when the doors of a refrigerator remain closed and nothing is added to the cabinet. Before the compressor cuts back in, the air temperature rises above 32°F in order to defrost the evaporator. The time switch in this case stops and starts the compressor according to a set schedule. This schedule is established on the basis of trials to determine the time for a complete defrost.

Supplementary Heat Defrost Methods

It is obvious that when the air temperature in a refrigerator must be kept below 32°F, it is not practical to shut down the system. Since the evaporator must get its heat from the air in the refrigerator, the temperature must rise not less than three degrees above 32°F for a long enough time for the evaporator to defrost. If this is not possible, additional heat must be supplied to that of the air in the refrigerator. When additional heat is provided, the method is known as supplementary heat defrost.

There are four common sources of supplementary heat:

- Hot discharge gases from a compressor
- Water
- Electrical resistance
- Outside air

Hot-Gas Defrost Method

The function of the condenser is to remove heat from the hot gas on the high-pressure side of the system. In this process, the vapor is changed back into a liquid. If instead, the vapor is passed through a bypass line to the evaporator at a point between the evaporator and the expansion valve, the heat from the gas could be fed directly into the evaporator. This is exactly what takes place in *hot-gas defrosting*, figure 10-16.

As the hot gas, which is bypassed, gives off heat to the cold evaporator the frost melts and at the same time a large quantity of the hot gas condenses into a liquid. This liquid moves out of the evaporator into the suction line from which it picks up heat. It again becomes a gas by the time it reaches the compressor. If this does not take place, damage to the compressor may result due to a liquid instead of a gas return at the beginning of the normal refrigerating cycle.

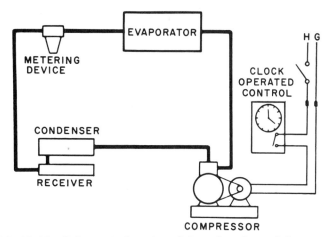

FIG. 10-15 Schematic drawing of time shut-down defrost method

At the start of the hot-gas defrost cycle, the hot gas contains the following:

- The heat from the refrigerator.
- The heat from the suction line.
- The heat of compression.

As the bypassed gas gives off some of its heat to the evaporator, this heat is lost in melting the frost to water. Although there is some heat picked up in the suction line and from the heat of compression, as defrosting continues, the temperature of the hot gas goes down. Since it cannot build up, the defrosting process continues to take more and more time.

The refrigerant that condenses in the cooling coil should be evaporated during the defrost cycle. To do this, heat may be applied to the hot-gas bypass line on its way to the evaporator or to the suction line beyond the evaporator.

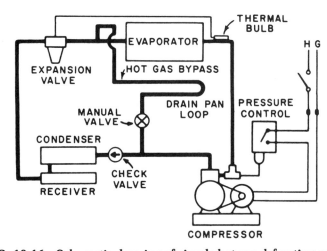

FIG. 10-16 Schematic drawing of simple hot-gas defrosting system

Vaporizing is accomplished by applying heat from some type of heating element. This may be electrical, hot air or hot water. Figure 10-17 shows that a blower evaporator is sometimes installed to serve the suction line. The air that is forced over this re-evaporator assures that only gas returns to the compressor. The blower operates only during the defrost cycle.

The drawing also shows a regulating control valve in the suction line. This controls the pressure of the gas on the low-pressure side of the system to operate the compressor. Unless only gas returns to the compressor, slugs of liquid refrigerant may form. If these slugs once enter the compressor, damage may result to the valves and other parts of the system.

Hot-Gas Thermobank Method

As the name suggests, the *thermobank* is a bank for storing heat, figure 10-18. The heat from the compressor discharge is stored in a small tank of liquid during the normal refrigerating cycle. During the defrost cycle, this heat goes to the suction line instead of directly to the hot-gas bypass line. This insures that the liquid from the evaporator is vaporized early in the defrost cycle. As such, liquid slugging of the compressor is prevented. It also helps to maintain the defrost heat level.

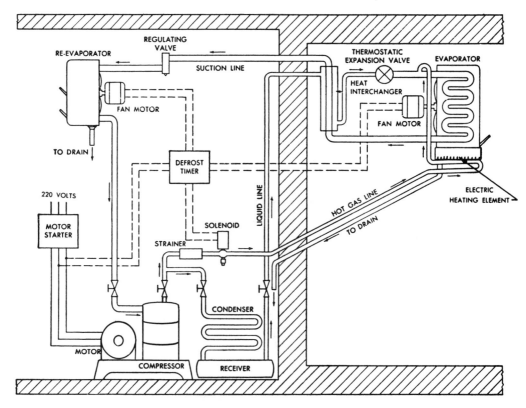

FIG. 10-17   Simple automatic hot-gas defrosting system with re-evaporator

FIG. 10-18 Major components of a thermobank system installation *(Courtesy of Kramer-Trenton Company)*

The *thermobank hot-gas defrost refrigeration system,* figure 10-18, is efficient during both wintertime and summertime operation. The major components of this particular model are shown schematically in figure 10-19.

One of the requirements for operation of a thermobank system is the subcooling of the liquid refrigerant, particularly for winter operation. Liquid is fed at a low temperature to the expansion valve by using a *thermolator* in the system. The thermolator

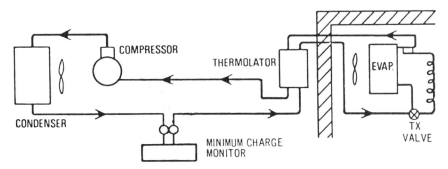

FIG. 10-19 Schematic showing heat transfer components in a thermobank hot-gas defrost refrigeration system *(Courtesy of Kramer-Trenton Company)*

193

provides protection against liquid return to the compressor and overheating of the suction line and discharge refrigerant during summer operation.

Note in figure 10-19 that the thermolator is located at the point where the refrigerant lines pass through the refrigerator wall. With this design, there is practically no refrigeration loss or unnecessary heating of the suction line.

During the defrost cycle, discharge hot gas is moved through the thermolator liquid line. The suction stream from the evaporator passes through the thermolator. Coil steaming is minimized by the reduction of the superheat of the hot gas due to heat transfer in the thermolator.

Evaporator Design Features

Industrial and commercial refrigeration industry needs are met by using a high-efficiency evaporator, figure 10-20. Different coil designs provide for low, medium and high temperature level system needs. Propeller or blower wheel types of fans are common for the fan section. These can be direct drive or belt driven.

Reverse-Cycle Defrost Method

An evaporator may also be defrosted by reversing the flow of the refrigerant in the system. The cooling coil becomes the condenser and the condenser becomes the cooling coil. Thus, the functions of the condenser and the evaporator are reversed or transposed. This reversing from the normal refrigerating cycle to the heating, hot-gas defrost cycle and back again is accomplished by installing a four-way valve, two check valves, an expansion valve, a time control and a low-pressure control valve.

In the defrost position, the hot gas from the compressor travels to the cooling coil. As it moves, the hot gas condenses in and heats the cooling coil and bypasses the thermostatic expansion valve. The refrigerant evaporates in the condenser and from there returns to the compressor as a gas. Freezing temperatures are prevented in the condenser by the automatic expansion valve. This valve is adjustable to maintain a maximum safe low-side pressure.

FIG. 10-20 Model of a draw-through air evaporator *(Courtesy of Kramer-Trenton Company)*

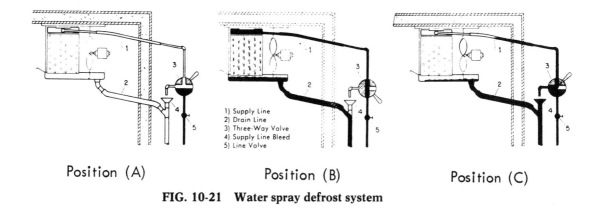

Position (A)   Position (B)   Position (C)

1) Supply Line
2) Drain Line
3) Three-Way Valve
4) Supply Line Bleed
5) Line Valve

FIG. 10-21   Water spray defrost system

Water Defrost Method

In a water-defrost system such as the one illustrated in figure 10-21, position (A) shows the system while the low-side of the refrigerator is pumped down. This action leaves the evaporator free of liquid refrigerant before defrosting begins. Then, both the compressor and evaporator fans are stopped and, if there are louvers, these are closed so that the evaporator coil is enclosed.

The three-way valve (position B) is then opened. Water is supplied to a spray head above the evaporator. The ice and frost are melted as the water washes over the evaporator. The water is caught in a drain pan and drawn out through a drain line. At position (C) the three-way valve is turned to a third position which permits the water to empty out of the supply line and other drain lines during the defrosting cycle.

Brine-Spray Defrost Method

While brine-spray units are seldom used, defrosting is done on some large installations by spraying a brine over the cooling coil. A method is provided in these systems for recovering the brine and recirculating it with a pump. It is also necessary to use eliminator plates which protect the refrigerated space and air from the brine spray.

Concentrators are also used because the water from the melted ice and frost dilutes the brine. The concentrators evaporate the excess water, thus keeping the freezing point of the brine below the temperature of the evaporator.

Electric Defrost Method

The last of the methods to be discussed is one that is increasing in importance, especially where fully-automatic defrosting is desirable. Heat for defrosting is needed at the cooling coils, the drain lines and the drain pan. Electrical resistance elements for heating are attached directly to or are built into the evaporator or within the refrigerant passages.

Special evaporators are required for installing this type of system. In operation, the low-side is first pumped down, the compressor and blower fans are stopped, and baffleclosed. Next, the resistance heaters are turned on to melt the ice from the evaporator and to heat the drain pan and line to prevent refreezing. After the defrost cycle when

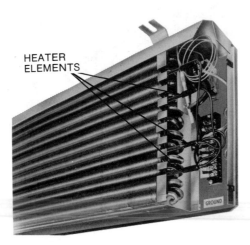

FIG. 10-22   Unit cooler with heater elements for defrosting *(Courtesy of Dunham-Bush, Inc.)*

the ice and frost are removed, a thermostat control returns the refrigerating system to normal operation.

Defrost Cycle.   During the defrost cycle of the model illustrated in figure 10-22, the compressor is turned off. The electric heaters are started for defrost. As the coil temperature rises to the preset temperature, the defrost is terminated. The compressor then starts. As the coils are cooled to below freezing, the fans start for the normal cooling cycle. This cycling method prevents *water carryover.*

The advantage of the electric-heater defrost method is that it has a great deal of flexibility. All types of timers, automatic control devices and safety limit switches, which open if the evaporator becomes too hot, may be combined to meet very specific job requirements.

SUMMARY

- Evaporators are classed as direct-expansion or flooded types according to operation.
- A direct-expansion evaporator is one that contains only enough liquid to continue boiling as heat is absorbed by it.
- A flooded-evaporator is one that is full of a liquid refrigerant at all times. Additional liquid is permitted to enter only to replace that which boils away.
- Refrigerant distributors are used to distribute the refrigerant simultaneously to several tubes in parallel. This avoids the need for excessive pressure (drop) to force the refrigerant through a single tube of an equivalent volume.
- Secondary-refrigerant is the name given to chilled liquid-like water which is circulated to distant units where the air is to be cooled in individual rooms. The secondary refrigerant is cooled in a chiller which is part of a compact, centrally located refrigeration system.

- The dew point is the temperature at which air space is 100% saturated by the water vapor it contains. When the air space is cooled to dew point temperature, the water vapor condenses.
- Moisture forms on an evaporator operating below the dew point temperature of the surrounding air. If the moisture freezes, it becomes frost. This frozen condensation reduces the heat transfer by conduction through the walls of the evaporator.
- Evaporators may be classed as frosting, nonfrosting and defrosting, according to operation conditions.
- Basic methods of defrosting evaporators include:
  - Manual defrost
  - Temperature defrost
  - Pressure defrost
  - Time shut-down defrost
- Supplementary heat defrost methods which depend on an additional supply of heat to that of the air in the refrigerator include:
  - Hot-gas defrost
  - Water defrost
  - Hot-gas thermobank
  - Brine-spray defrost
  - Hot-gas reverse cycle
  - Electric defrost

ASSIGNMENT: UNIT 10 REVIEW AND SELF-TEST

A. TYPES OF EVAPORATORS

Determine which statements (1 to 10) are true (T); which are false (F).

1. The evaporator is sometimes called the high-side of the system.
2. The refrigerant is inside the tubes of a direct-expansion chiller.
3. The evaporator is a device to reject heat from the refrigerating system.
4. The temperature of medium being cooled must be below that of the evaporator.
5. The two types of evaporators are the flooded and the direct-expansion.
6. The heat picked up in the evaporator is equal to that rejected in the condenser.
7. The selection of the proper evaporator depends primarily upon the application.
8. Water and brine are usually secondary refrigerants.
9. In flooded evaporators the liquid level almost covers the evaporator.
10. Primary surface evaporators operating below 32°F must be purged frequently.

Provide a word or phrase to complete statements 11 to 20.

11. The main function of the evaporator in the system is _____ .
12. The pressure drop is greater in a _____ circuit coil than in a split-circuit coil.
13. Another name that is sometimes used for an evaporator is _____ .
14. The heat absorbed by the evaporator is _____ that rejected by the condenser.

15. As the refrigerant absorbs heat in the evaporator the following change takes place: _____ .
16. The device that controls the level of the refrigerant in a flooded evaporator is called _____ .
17. The materials most frequently used as secondary refrigerants are _____ .
18. The device used with the evaporator to keep the pressure drop at a minimum is the _____ .
19. The device outside the evaporator which controls the refrigerant level is the _____ .
20. The control device located inside an evaporator is the _____ .

B. HUMIDITY AND ITS RELATION TO EVAPORATORS

1. What term identifies the temperature at which the water vapor in the air condenses out?
2. What term describes the condition of the air space when vapor condenses out?
3. What is the relative humidity level when water vapor condenses?
4. Give the temperature at which frost forms on the coil.
5. How does frost on the evaporator affect the rate at which heat is transferred to the refrigerant?

Indicate by letter the type of coil that matches each condition. Use (a) for frosting coil, (b) nonfrosting, and (c) defrosting.

6. Operates continuously below 32°F.
7. Operates close to 32°F and has a large area.
8. Frost melts when the compressor stops.
9. Very little, if any, frost forms on the coil.
10. Ice is likely to form on the lower part of the coil.

C. BASIC METHODS OF DEFROSTING EVAPORATORS

Indicate by letter the type of defrost system in Column II which meets the conditions given in Column I.

Column I	Column II
1. Controlled by a time-clock mechanism.	a. Manual
2. Control set to start a compressor at about 34°F	b. Pressure c. Temperature
3. Must provide for the evaporation of excess water.	d. Time shut-down e. Hot gas
4. Built into the refrigerant passages.	f. Water
5. Requires heating of the drain pan.	g. Brine
6. Condenser operates as an evaporator and vice versa.	h. Reverse refrigeration i. Electric

7. Controlled by suction pressure.
8. Requires mechanical removal of frost if neglected.
9. Uses a bypass around condenser and metering device.
10. Uses a thermobank
11. When is the best time for time shut-down defrost?
12. Describe briefly the difference between temperature and pressure defrost.
13. Name four sources of supplementary heat for use with the hot-gas defrost system.
14. What operating condition requires heat for the drain pan during defrost?
15. Name (a) the mechanical device used for reverse refrigeration defrost and (b) give its function.
16. State briefly two steps that are taken with a system before using water or brine spray.
17. What takes place in the evaporator during a hot-gas defrost?
18. What danger is there to the forming of liquid refrigerant slugs?
19. Which is the most flexible method of defrosting?
20. Name one disadvantage of manual defrost.

UNIT 11: CONDENSERS

The condenser is a device that transfers heat from the refrigeration system to a medium which can absorb and move it to a final disposal point. The condenser is the door through which the unwanted heat flows out of the refrigeration system. It is in the condenser that super-heated, high-pressure refrigerant vapor is cooled to its boiling (condensing) point by rejecting sensible heat. The additional rejection of latent heat causes the vapor to condense into the liquid state.

There are three types of condensers, figure 11-1. The name of each type is determined by the condensing medium. An *air-cooled condenser* uses air as the condensing medium; a *water-cooled condenser*, water; and the *evaporative condenser*, both air and water. The design and operation of the major components of each type of condenser is covered in this unit. The term *component* is used by designers, service technicians, engineers and others, to identify a major part, device or unit within a system.

The function of two broad groups of cooling towers is also treated in detail. This is done because of the importance of water as a condensing medium.

AIR-COOLED CONDENSERS

Base-Mounted Air-Cooled Condensers

A *condensing unit* consists of a compressor, a condenser and receiver. The two important factors governing the performance of an air-cooled condenser are (1) the square feet of cooling surface, and (2) the cubic feet per minute of air available for cooling. Air-cooled condensing units are designed for small system applications as well as for systems requiring in excess of 100 tons capacity.

Air-cooled condensers are generally made of copper tubes and aluminum fins and, when the condensing unit is mounted on a base, are called *base-mounted*. The fins may be bonded to the tubes by the old solder-dipping method, the modern oven type of

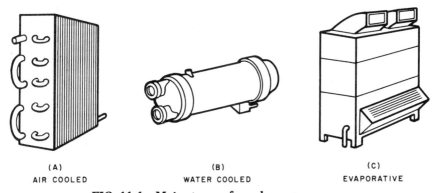

| (A) | (B) | (C) |
| AIR COOLED | WATER COOLED | EVAPORATIVE |

FIG. 11-1   Major types of condensers

FIG. 11-2 Air-cooled condensing unit with semihermetic compressor *(Courtesy of Dunham-Bush, Inc.)*

copper brazing, or by securing them mechanically to make the necessary thermal bond. The close fin spacing requires frequent cleaning to prevent dust and dirt from clogging the surface and reducing the condensing capacity.

Steel condensers are usually protected against corrosion by paint-dipping and, where possible, by mounting them indoors. Shelters should be provided if an outdoor location is necessary.

A common base-mounted, air-cooled condensing unit with a semihermetic compressor is illustrated in figure 11-2. These units are designed to meet commercial or high-temperature requirements for refrigeration and air-conditioning systems. R-12, R-22 or R-502 refrigerants are used in the units. Also, the units can be roof mounted in a ventilated penthouse, platform mounted or two-tier mounted to conserve floor space. This unit must be located away from a wall to permit the free flow of air into the condenser.

A condensing unit with a hermetic compressor is illustrated in figure 11-3. This unit requires a separate and slow-speed motor to drive the condenser fan. The slower speed helps to lower the noise level. Because air-cooled condenser units are located close to the floor, they pick up lint, dust, grease, dirt, and foreign matter. Regular cleaning can be done by brushing, vacuum cleaning or any other method. Whenever pressure is used to blow out the dirt, persons working in the area as well as surrounding equipment should be protected.

Remote Air-Cooled Condensers

The *remote* air-cooled condenser is very popular. Its performance and the added advantages of placing this type of unit in different locations account for its popularity. When a condenser may be located in another place away from the compressor, it becomes possible to select the best location for each. This factor permits placing the condenser outdoors. In this location there is an ample supply of cooling air available at the lowest possible temperature.

FAN AND CONDENSER

HERMATIC
COMPRESSOR

LIQUID RECEIVER

FIG. 11-3 Air-cooled condensing unit *(Courtesy of Copeland Corp.)*

An air-cooled condenser, figure 11-3, requires approximately 1,000 cfm of air per ton. With this large quantity of air needed, slow-speed belt-driven fans of the wide-blade propeller type must be used to keep the noise level as low as practical. Propeller-fan type condensers may also be mounted indoors provided the ducts are short and the air velocities are low. Propeller fans are efficient when operating against low static pressures. Since the remote condenser has its own fan and motor, any required cfm of air may be supplied without taking power from a compressor or motor.

Remote-type condensers, figure 11-4, may be raised off the floor or located outdoors on brackets for cleaner operation than base-mounted condensers. The capacity of a remote air-cooled condenser system may be built up by combining units of various sizes.

The unit cooler displayed in figure 11-4 is part of a series with capacities ranging from 5,200 Btu/hr. to 48,000 Btu/hr. The cabinet design permits easy suspended installation and disassembly. Draw-through fans permit uniform air distribution at extremely low noise levels. The cooling coils (housed within the shallow depth cabinet) are copper tubes with mechanically bonded aluminum fins. This design provides efficient heat transfer.

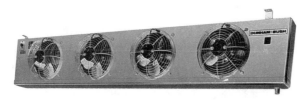

FIG. 11-4 Battery of remote air-cooled condensers *(Courtesy of Dunham-Bush, Inc.)*

Remote air-cooled condenser unit coolers are used commercially in supermarkets, food processing and other product storage room installations.

Refrigerant Piping for Remote Air-Cooled Condensers

The main piping for a typical remote air-cooled condenser is shown in the schematic drawing in figure 11-5.

Note that the condenser is mounted at a higher level than the compressor. It is good practice to provide an oil loop at the compressor discharge. If this is not practical, a check valve may be used instead. The oil loop or the check valve collects the oil and liquid which drain back from the discharge line on the off-cycle. Thus, the compressor is protected from starting against a liquid column. The loop should be six inches high for every ten feet of vertical discharge line. Note that the purge line is located at the highest position in the discharge line. The *winterizing* valve overcomes the low discharge pressure caused by cold winter temperatures on the refrigerant.

Discharge Line Pulsations

Remote condensers are subject to discharge line *pulsations* (beating). This is especially true when high-speed compressors are used. The long column of gas in the discharge pipe acts like a spring under pressure. If the timing of the compressor strokes is the same as the natural timing of the discharge line column, a pulsating cycle builds up to produce noises and vibrations which are objectionable. A muffler installed in the discharge line absorbs the compressor pulsations and eliminates virbration.

When a remote air-cooled condenser is installed indoors it is possible to use the rejected heat for room heating in the wintertime. The installation as illustrated in figure

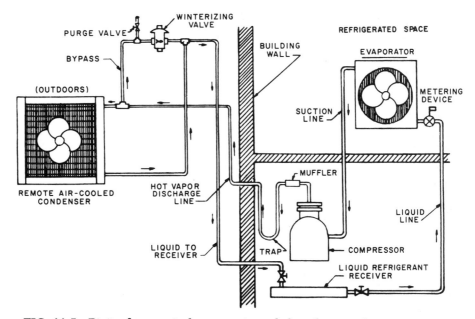

FIG. 11-5   Piping for a typical remote air-cooled outdoor condenser installation

11-6 shows a simple sheet metal duct construction with a hinged damper. The position of the damper controls the air flow so that it either supplies heat indoors or exhausts it outdoors.

Air-cooled condensers that are located outdoors should face the prevailing winds to help the movement of air through the fan. A wind scoop should be provided to deflect the wind downward and prevent overloading the fan motor. This is necessary whenever there is a possibility that strong winds may blow into the discharge and against the fan.

WATER-COOLED CONDENSERS

Water-cooled condensers are used with compressors of one horsepower and larger. Water-cooled condensers are most economical where there is an adequate supply of clean, inexpensive water of minimum corrosiveness and there is an efficient economical way of getting rid of the discharged, heated water.

The several types of water-cooled condensers commonly used include:

- Shell-and-coil
- Tube-in-tube
- Shell-and-tube

Shell-and-Coil Water-Cooled Condensers

Shell-and-coil water-cooled condensers are widely used in the smaller sizes up to a 10-ton capacity, figure 11-7. They have the advantage of being compact in size and serve the dual function of condenser and receiver. The cutaway photo shows a conventional design of a horizontal shell-and-coil water-cooled condenser. Such a condenser may be furnished in a vertical design as well.

As the name implies, a water coil is wound inside a shell. The refrigerant gas within the shell condenses on the outside of the water coils. The liquid refrigerant collects in the bottom of the shell where it is then removed through a bottom outlet or a dip-tube.

The water coils are so designed that the water may be drained off conveniently whenever the unit is not in operating condition. This prevents damage due to freeze-up.

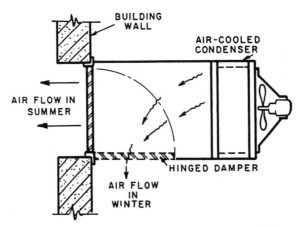

FIG. 11-6 Utilizing rejected heat from a remote air-cooled condenser

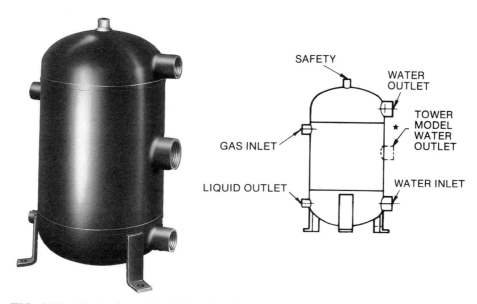

FIG. 11-7 Vertical, sealed shell-and-coil water-cooled condensers *(Courtesy of Standard Refrigeration Company)*

Many condensers also have fins on the tubes. These fins transfer heat better when the tubes are in a horizontal position and the fins are arranged vertically. This design permits the refrigerant to drain easily from between the fins. Horizontal condensers usually have the horizontal tubes in the coils arranged in a trombone shape. Vertical condensers have spiral coil tubes that are free draining and have fins arranged vertically.

Tube-in-Tube Water-Cooled Condensers

Tube-in-Tube condensers are made with one or more water tubes inside a refrigerant tube, figure 11-8. The coils of the simpler types of condensers are also wound in the

FIG. 11-8 Horizontal, cleanable, tube-in-tube water-cooled condenser *(Courtesy of Standard Refrigeration Company)*

shape of a trombone. The cooling water enters the center tube at the bottom of the condenser. The hot gases enter at the top and pass downward between the two tubes. The incoming water condenses and subcools the liquid refrigerant which is piped away at the bottom. Heat transfer efficiency is high because of the advantages of counterflow between the water and the refrigerant.

Condensers with trombone-shaped coils are usually limited to the smaller size due to problems of bending large tubes.

Tube-in-tube condensers are also built with straight tubes in a wide range of sizes, figure 11-9. Removable *header* plates at the ends of the rows of tubes provide a way to reach inside the tubes for mechanical cleaning. Individual banks of coils may be combined to give capacities up to approximately 25 hp.

Tube-in-tube condensers require a separate receiver for storing the condensed liquid refrigerant which drains freely from the condenser. The size of the liquid line between the condenser and receiver should be larger than the liquid line leaving the receiver.

The larger line between the condenser and receiver permits:

- Free drainage of liquid from the condenser to the receiver.
- Any gas in the receiver to rise to the condenser and be condensed back to a liquid state.

This sometimes happens when the receiver is located in the direct rays of the sun or some other hot place.

Shell-and-Tube Water-Cooled Condensers

Shell-and-tube condensers, figure 11-10, are built in a variety of sizes. Removable heads make it possible to clean the tubes mechanically. The cover plates over the water boxes on shells above 12″ diameter are bolted so that the head may be removed and the tubes cleaned without disturbing the water connections. Tubes are often silver-soldered into the tube sheets to make permanent joints on the smaller sizes. The tubes on large-sized condensers are usually rolled into thick tube sheets.

Shell-and-tube condensers are finned tubing for efficiency and economy. Some manufacturers provide for tubing replacement by designing the tubing with the ends flared out larger than the fin diameter.

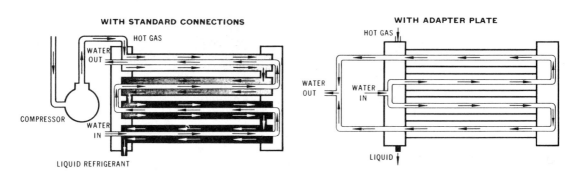

FIG. 11-9   Straight tube-in-tube condensers *(Courtesy of Standard Refrigeration Company)*

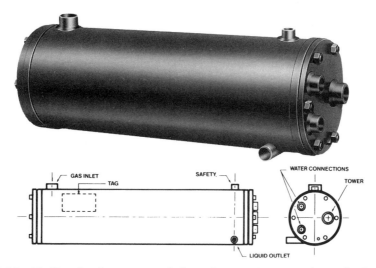

FIG. 11-10   Shell-and-tube water cooled condenser *(Courtesy of Standard Refrigeration Company)*

Cleaning the Tubes

Condenser tubes become coated with alkali and other foreign materials in water. These coatings reduce the heat conductivity through the tubing walls and cut down the efficiency of the unit. When condensers are cleaned regularly, mild chemical cleaning solutions can be used.

However (depending on water conditions), if the condensers are cleaned less than once a year, the tubes might need to be manually cleaned with abrasive materials. Care must be taken in cleaning tubes by scraping, using cleaning brushes, or with abrasives. In such processes, some metal is removed and, consequently, tubes might be weakened, figure 11-11.

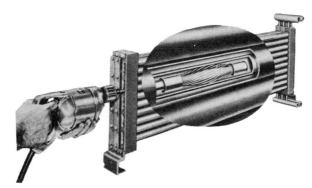

FIG. 11-11   Cleaning tubes of water-cooled condenser with special tool *(Courtesy of Standard Refrigeration Company)*

EVAPORATIVE CONDENSERS

Another method of removing rejected heat from a condenser combines the effectiveness of forced-circulation convection currents with the ability of a vaporizing liquid to absorb heat. The condenser which operates in this manner is classified as an *evaporative* type.

The evaporative condenser was developed to overcome the problems arising from the use of many water-cooled condensers in small air-conditioning systems. The water supply and drainage facilities of some communities became overloaded. In other localities the high cost of water prevented its extensive use for air-conditioning installations. The evaporative condenser which functions as a condenser and cooling tower was designed to overcome these objections.

The schematic drawing in figure 11-12 shows how a typical unit functions. Air is drawn through an opening near the bottom (A). It flows upward across the refrigerant condensing coils (B) through the water sprays (C) and liquid eliminators (D) and into the fan (E). The water vapor and air is discharged at the top outlet (F).

The refrigerant condensing coil is usually of the extended surface or finned type. The refrigerant enters at the top of the coils (G) and flows across them and downward. The condensed liquid then discharges into the receiver (H). When the receiver is located in the water tank it subcools the liquid refrigerant to an even lower temperature.

The hot condenser is cooled by both the air and the evaporation of water from the spray nozzles. As the water evaporates, each pound removes 970 Btu of the heat which the condenser has rejected. This type of condenser uses only from five to ten percent as much water as a water-cooled condenser of similar capacity. The effectiveness of the evaporative condenser in removing latent heat produces great savings of water in locations where water conservation is essential. Forced convection currents of air and

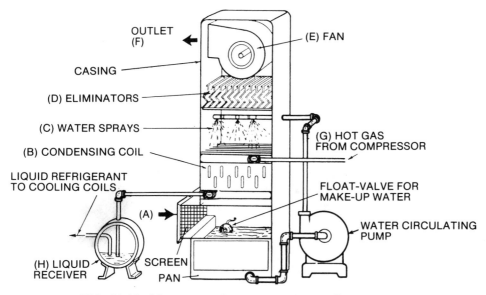

FIG. 11-12 Major units in an evaporative condenser

water vapor carry away the heat in the evaporative condenser. Water is admitted to the reservoir to make up for the quantity that evaporates. Water treatment is important in the prevention of scaling.

Pressure Drop

Well-designed condensers require only a small pressure drop to push the refrigerant through. This pressure drop is almost zero in water-cooled shell-and-tube condensers where the refrigerant condenses in the shell and in evaporative condensers. Air-cooled and tube-in-tube water-cooled condensers require a larger drop. However, pressure drop seldom exceeds 10 psi. Where the pressure drop is greater than 10 psi, it may be caused by a restriction or the use of a condenser which is too small for the load.

NONCONDENSABLE GASES

Noncondensable gases, like air, chlorine, oil, and water vapors, and combinations of oxygen, hydrogen and nitrogen, collect within the refrigerating system. Since these gases do not condense with the refrigerant, they usually collect in the condenser or receiver. Noncondensable gases raise the condensing pressure above that required for the condensing temperature. This is so because it has been proven scientifically that if several gases exist in the same closed container, the total pressure is the sum of the individual pressure that each gas would exert if it were alone in the container.

This higher pressure (due to noncondensable gases) increases the work of the compressor, reduces the refrigerating effect, and raises the power consumption. At higher discharge temperatures, oxidation of oil is likely to occur, especially at the compressor discharge valves.

The sources of noncondensable gases from outside the system are:

- Improper evacuation of a new system before charging.
- Drawing air into the crankcase of the compressor during servicing.
- Through leaks in the system when operating under a vacuum.

As a result of these conditions, noncondensable gases are produced within the system through:

- The decomposition of the lubricant.
- The breakdown of the refrigerant under high temperatures.
- By chemical reactions.

The presence of noncondensable gases may be checked in many systems. For an air-cooled system a thermometer is clipped to the condenser outlet. The compressor is then turned off until the thermometer reads the same as the surrounding air temperature. Noncondensables are indicated by a higher head pressure than the pressure corresponding to the temperature of the air within the room.

For a water-cooled system, a thermometer may be clipped to the liquid line leaving the condenser. The water valve is opened until the reading on the thermometer and the water inlet and water outlet temperatures are the same. Again, the presense of noncondensables is indicated by a higher head pressure than the pressure corresponding to the water temperature given in refrigerant tables. The excessive head pressures due to noncondensables in the system are almost twice as great when the compressor is running as when it is shut down.

Purging Valves

When done properly, purging may eliminate a great deal of the trouble caused by noncondensables. The position of the purging valves should be determined by the system. The purge valves on a condenser-receiver should be on the shell. For long shells with a gas inlet in the center, a purge connection should be located at both ends. A tubular type condenser with separate receiver should be designed so the receiver may be purged. On the remote condenser of either the air-cooled or evaporative type, the purge valves may be located either on the discharge line before the entrance to the condenser or at the highest point in the discharge line.

Care must be taken to shut down the system for a sufficient period of time to allow the gases to collect before opening the purging valves. Water-cooled condensers must not be purged too rapidly because the water in the tubes may freeze. While automatic purging methods are available, they generally are used on large tonnage systems.

METHODS OF COOLING WATER FOR CONDENSERS

As has been mentioned before, the cost of water in some communities is too high to permit wasting it after it has been used for cooling. In other localities the water available for condensing purposes is warm. Warm water requires the use of relatively large quantities to keep the condenser pressure down to a moderate point. The cost of water for the condenser of a refrigerating plant under either of these conditions may be so high that the same water must be used many times. This can be done economically by cooling the water after it leaves the condenser.

Spray Ponds

One of the oldest and simplest methods of cooling water is the *spray pond* technique. In this method, water is sprayed into a stream of outdoor air. The temperature of the water approaches the wet bulb temperature of the air, but never reaches it. Whenever warm water is brought into contact with air, a very small quantity of water evaporates into the air stream. The latent heat needed to evaporate this small quantity of water is provided by the large body of water which is cooled. It is this contact with the relatively warm water that produces a rise in the wet bulb temperature of the air. All the heat which is transferred from the condensing refrigerant vapor to the water is transferred to the outdoor atmosphere during this process.

Although spray ponds give satisfactory performance, they do require large areas. A spray pond consists of a large number of nozzles located from three to six feet above a collecting pan. The nozzles are positioned to spray the water up into the air and radially for some distance. Since the water is broken up into fine mist, a large surface area of water is brought into contact with air. As the moisture particles cool, they fall to the floor of the pond as drops of water and then flow to the suction line of the condenser pump.

Spray ponds must be located where the air can circulate freely through the water spray. For this reason, they are located on the roofs of buildings. Louver fences are installed around spray ponds in those locations where entrained moisture may be

210

carried away by wind. The low cost of installing and maintaining spray ponds makes their use desirable where space is available.

Cooling Towers

Cooling towers have been used for many years as a practical device for cooling water for condensers. These towers are also erected on the roofs of buildings or in such locations where air can circulate freely through them.

Cooling towers are divided into two broad classes:

- The natural draft or wind tower where the circulation of air through the tower depends upon natural wind movements.
- The forced or induced draft tower through which air is forced or drawn by fans.

Natural Draft Cooling Towers

The *natural draft* cooling tower cools water by moving air at low velocities through the tower. It is for this reason that the roof of a building is an excellent location. Weather bureau reports show that there are few places where the natural wind velocity does not reach at least five miles per hour.

Natural draft towers like the one illustrated in figure 11-13 are built of heavy wood or steel framework. They vary in size from four to twelve feet and are built up to thirty or more feet high, according to the required capacity. Where fire ordinances do not require metal construction, the towers are usually made of cypress or redwood timbers. The towers have a number of wooden decks that can be spaced a few feet apart to allow access to them. There are open spaces between the boards which form the decks in order that water may drip or run freely from deck to deck.

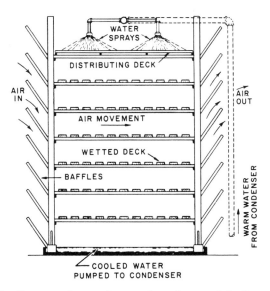

FIG. 11-13   Construction and operation of natural draft cooling tower

The warm water discharged from the condensers is first distributed over the top deck. The decks present a large wetted area to air which moves crosswise through the tower, and retard the fall of water.

As the water falls it takes time to travel between the decks. Also, in spreading out over the various decks a large surface area of water is exposed. This speeds up the cooling process.

The cooled water then collects in a water-collecting tank at the bottom. From this location it is piped to the suction connection of the circulating pump as shown in the piping sketch, figure 11-14.

The pump, in turn, forces the cooled water through the condenser. The water picks up heat in the condenser and is piped to the nozzles which spray it over the distributing deck of the tower, and the circuit is completed.

The sides of natural draft cooling towers are open to permit breezes from all directions to blow across all of the wetted decks. The air blowing through the open spaces carries a fine spray of water which it picks up from the falling water. As in the case of spray ponds, louvers are installed around the sides of the tower to prevent the wind from blowing the moisture spray and depositing it upon adjacent surfaces where it may be objectionable.

The louvers usually consist of sloping baffle boards which run around the edge of every deck. Since the baffle boards are slanted and overlap, the air which leaves each

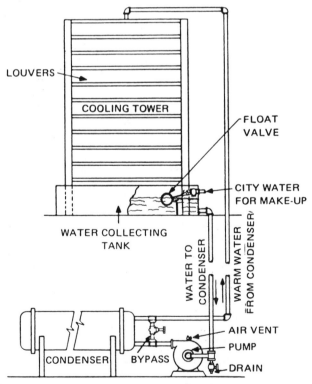

FIG. 11-14 Piping for natural draft cooling tower

surface is deflected upward. Thus, most of the entrained moisture is deposited on the baffle boards from which it runs off. In this process, the loss of water due to evaporation may amount to less than five percent of the total water circulated.

Small natural draft cooling towers of still another design are commonly used. Such towers consist of a louver enclosure around all four sides. The spray nozzles at the top break up the water from the condenser into a fine spray. Since there are no wooden decks in this type, the inside of the tower is filled completely with fine spray. The louvers permit the air to circulate through the fine mist. The warm water spray is cooled by evaportion as it falls in the cooling tower. The cooled water collects at the bottom in a water-collecting tank from which it is circulated through the condenser.

The natural draft cooling tower has been used more than any other type because it is inexpensive to operate and to maintain. One disadvantage is the large area needed for its installation. Where space or an open location facing the prevailing winds is not available, another cooling tower known as the *forced draft* type is used.

Forced Draft Cooling Towers

The forced draft cooling tower gets its name from the fact that air is forced mechanically through the tower, figure 11-15. Small towers may be located indoors and may be connected to a source of outdoor air. Larger towers (because of their height) are generally located on a building roof. The large capacity rooftop cooling tower, like the one illustrated in figure 11-15, has a top discharging plenum and is enclosed in a sound-absorbing wall.

The larger forced draft towers are constructed with wooden timbers or boards spaced at regular intervals to break the fall of the water and to spread it out over a large surface. The sides of the forced air tower are totally enclosed by a wooden or metal shell. Fans are located in the side or at the bottom as illustrated. The cooling air from the condenser is blown through the tower, leaving it at the top.

FIG. 11-15 A forced draft cooling tower (*Courtesy of Binks Manufacturing Company*)

The warm water from the condenser is sprayed on the distributing deck from which it falls from one deck to another. At the bottom it collects in a tank. A pump takes the cooled water from the tank and pumps it through the condenser. As the water is again heated in the condenser it is pumped back to the distributing deck at the top of the tower and the cycle continues.

Some small forced towers are built without decks. The large surface area of water is produced as a fine spray mist by nozzles located at the top of the tower. The heated water is cooled as air continuously circulates around the moisture particles as they fall through the tower. In the process, some of the water evaporates.

On inside installations cooling towers require two connections to the outside. The first is for drawing cool outdoor air into the tower; the second, for discharging warm air. Occasionally, when air is to be exhausted outdoors from an air-conditioned space, the exhaust is first drawn through the cooling tower. This provides air at a lower wet bulb temperature than is available from outdoors.

Forced draft towers are more expensive to operate than natural draft towers because of the cost of moving air through them. Increased operating cost is offset by three advantages:

- Greater flexibility of installation because the location does not depend on wind velocity.
- Less area is required for a tower of a given capacity.
- The possibility of cooling water to a slightly lower temperature than with a natural draft tower.

Water Treatment

The inside of the cooling tower or condenser may become coated with and attacked by undesirable foreign materials from both the air and the water. These may be removed by water softening or by metal-cleaning materials. When the interior surface of a cooling tower or condenser becomes coated with lime, fly ash, and the like, it becomes difficult and costly to remove the coating. In addition, the possibility of corrosion attacking the metal covered by this coating is increased.

A systematic water treatment program prevents the formation of scale and algae and other undesirable materials. In the absence of such a program the water should be drained regularly each week or every two weeks and replaced with clean fresh water. Cooling tower and condenser manufacturers suggest that the float valves be so adjusted as to permit a small quantity of water to flow constantly. This tends to keep the water fresh and clean. This practice, however, does not eliminate the need to treat the water and to flush the tank at regular intervals.

The entire cooling tower or condenser must be inspected for deterioration and rust spots. In addition there is also the problem of winterizing the equipment in certain geographic locations.

SUMMARY

- The condenser is the component that removes the heat from the refrigeration system. The refrigerant is both cooled and condensed in the condenser. The refrigerant vapor is at a high temperature and pressure level in the condenser.

- Air-cooled condensers depend on air drawn through tubes and fins for a good distribution of air to cool the refrigerant.
- The remote air-cooled condenser makes it possible to position the condenser and compressor in the best location.
- Water-cooled condensers are economical where there is an adequate supply of clean, inexpensive water. Common designs include the shell-and-coil, tube-in-tube, and shell-and-tube. Each tube is cleaned chemically or mechanically depending on construction.
- The evaporative condenser combines the principles of forced-circulation convection currents with the ability of a vaporizing liquid to absorb heat. In other words, the cooling effect caused by the evaporation of water takes place directly on the condenser surface.
- A receiver is a container for storing liquid refrigerant. It may be a separate component or it may be included as part of a condenser.
- Pressure drop refers to the pressure difference needed to push the refrigerant through a component like the condenser, evaporator or line.
- Noncondensable gases sometimes leak into a refrigerating system. They hamper the system by exerting part of the total pressure which should only be exerted by the refrigerant.
- Common methods of cooling water for water-cooled condensers include:
  ✓ Spray pond
  ✓ Natural and forced draft cooling towers
- Water treatment is required in the maintenance and servicing of cooling towers and evaporative condensers.

ASSIGNMENT: UNIT 11 REVIEW AND SELF-TEST

Determine which statements (1 to 10) are true (T); which are false (F).

1. The refrigerant absorbs heat in the condenser part of the system.
2. The refrigerant boils in the condenser.
3. Sensible heat is removed from the refrigerant by the condenser.
4. The refrigerant is superheated as it enters the condenser.
5. The refrigerant is subcooled as it leaves the condenser.
6. The refrigerant is a liquid as it leaves the condenser.
7. The condenser is usually located inside the refrigerated space.
8. The refrigerant goes directly from the condenser to the evaporator without going through any other part of the system except the tubing.
9. The refrigerant liquid in the condenser is at the same temperature as the vapor directly above it.
10. The heat rejected by the condenser is just equal to the amount of heat picked up in the evaporator.

A. AIR-COOLED CONDENSERS

Select the letter of the word or phrase which best completes statements 1 to 10.

1. A condenser is a device for removing (a) heat, (b) noncondensable gases, (c) water vapor from the refrigeration system.

2. The condensing medium for the condenser is (a) water only, (b) air only, (c) either air or water.
3. In order to reduce cleaning to a minimum, the fins should be spaced (a) close together, (b) far apart, (c) without concern for distance between fins.
4. The natural draft air-cooled condenser would normally be (a) too small, (b) too large, (c) just right for use in a large system.
5. A forced draft condenser uses (a) a propeller fan only, (b) a centrifugal fan only, (c) either a propeller or centrifugal fan, (d) neither a propeller nor a centrifugal fan.
6. Within the condenser the refrigerant (a) boils, (b) condenses, (c) boils or condenses.
7. The condenser removes from the refrigerant (a) sensible heat only, (b) latent heat only, (c) both latent and sensible heat, (d) neither latent nor sensible heat.
8. The heat rejected by the condenser is (a) equal to both the heat of compression and the heat picked up by the evaporator, (b) the heat of compression only, (c) the heat picked up by the evaporator only.
9. The greatest quantity of the heat that is removed by the condenser is (a) sensible heat, (b) latent heat, (c) neither sensible nor latent heat.
10. The type of condenser that makes use of latent heat of water to absorb heat from the refrigerant is (a) air cooled, (b) water cooled, (c) evaporative.

Provide the term or word needed to correctly complete statements 11 to 18.

11. Air-cooled condensers may be classified as _____ and _____ .
12. Pulsations of the vapor are likely to be set up if the discharge line is _____ .
13. The pulsations may be dampened out by installing a _____ in the discharge line of the compressor.
14. The major parts of the condensing unit are _____ .
15. When the remote condenser is mounted higher than the compressor, a trap or oil loop is provided in the discharge line in order that the compressor will not have to start against _____ .
16. Instead of an oil loop a _____ may be used.
17. One advantage of installing a remote air-cooled condenser inside is that it proca-pacity in a water-cooled condenser.

18. Since the remote condenser has its own fan motor, it can be placed where any required _____ may be supplied.

B. WATER-COOLED CONDENSERS

1. Name the type of condenser to use for systems larger than those handled by air-cooled condensers.
2. When condenser tubes are cleaned frequently, how is this operation done?
3. How may neglected or severely coated tubes be cleaned?
4. Name three types of water-cooled condensers.
5. Which of the three types of water-cooled condensers is the most difficult to clean?

6. What type of water-cooled condenser acts as a receiver and a condenser?
7. What precaution must be taken in designing condensers with fins added to the tubes?
8. Give one reason why the size of the liquid line between the receiver and condenser should be larger than the liquid line leaving the receiver.
9. By what method may tubes be joined to tube sheets?
10. Name one design feature which may be included to increase tubing efficiency and capacity in a water-cooled condenser.

C. EVAPORATIVE CONDENSERS

Provide a word or phrase to complete statements 1 to 6.

1. The condenser that uses both forced air circulation and water is known as the _____ type.
2. The kind of heat that water gains as it contributes to the operation of the condenser is _____ .
3. The approximate value of this heat for each pound of water is _____ .
4. The evaporative condenser is most effective under which condition of relative humidity?
5. The main advantage of an evaporative condenser over a water-cooled condenser is that _____ .
6. The pressure drop is _____ in air-cooled condensers than in water-cooled sheet-and-tube condensers.
7. Study the schematic drawing of the evaporative condenser. Then use arrowheads and letter the drawing to identify ten major units or parts of this condenser.

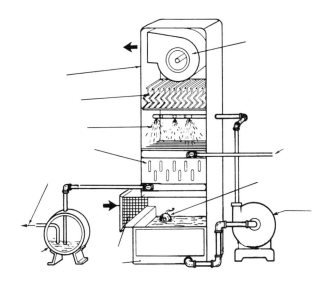

D. NONCONDENSABLE GASES

1. Name three gases that may collect within a refrigerating system.
2. List two sources of noncondensable gases from outside the system.
3. Give two reasons why noncondensable gases are produced within a system.
4. Name the process and a device which may be installed in the system for removing noncondensable gases.

E. METHODS OF COOLING WATER FOR CONDENSERS

1. Describe briefly in three steps how a spray pond works.
2. Give two advantages of forced draft towers over natural draft towers.
3. State two conditions that are created by water in cooling towers or condensers.
4. Name the parts, materials or processes indicated on the drawing of a natural draft cooling tower. After each name, describe the function of the part or the process that goes on.

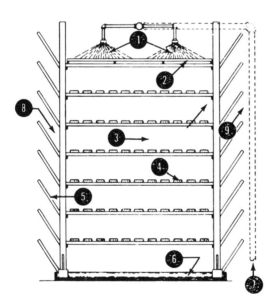

UNIT 12:
METERING DEVICES FOR THE CONTROL OF REFRIGERANTS

A high-pressure liquid refrigerant must be reduced to a low-pressure liquid refrigerant in the correct quantities to operate a system at maximum efficiency and without overloading the compressor. In the early stages of refrigeration, high- and low-pressure conditions and rate of flow were controlled manually by adjusting a needle valve. With improvements and experience in the design, operation and servicing of refrigeration systems, a number of refrigerant control devices, which were economical, efficient and automatic, were developed.

TYPES AND FUNCTIONS OF REFRIGERANT CONTROL DEVICES

The main types of refrigerant controls dealt with in this unit include:

- Hand-operated needle valve
- Automatic expansion valve
- Low-pressure side float
- Thermostatic expansion valve
- High-pressure side float
- Capillary or choke tube
- Orifice

These controls depend for operation on one of the following: (1) pressure changes, (2) temperature changes, (3) volume or quantity changes, or (4) any combination of these.

Typical examples are used to illustrate the construction and principles of operation of the six types of control devices and to simplify the description of each one. The advantages, disadvantages and applications are given to develop an overall understanding of the functions of control devices under varying conditions.

Refrigerant control devices are more commonly called *metering devices*. A metering device in refrigeration work is primarily a restriction placed in the system. This restriction makes it possible for the compressor, by its pumping action, to maintain a difference of pressure. The compressor pumps the refrigerant from the low side to the high side. The refrigerant flows back from the high side to the low side through the restriction (metering device).

The compressor must have sufficient capacity to be able to pump the refrigerant as fast as it flows back through the metering device. When the compressor does this, it maintains the low pressure in the evaporator which is necessary for the boiling of the refrigerant at a low temperature. Heat can then be absorbed through the evaporator and refrigeration is achieved. Also, heat can be rejected when the compressor maintains a high-pressure and high-temperature level in the condenser. The metering device, in addition to providing a restriction, controls (meters) the rate of flow of the refrigerant from the high to the low side.

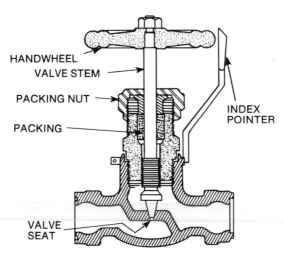

HANDWHEEL
VALVE STEM

PACKING NUT

PACKING

INDEX
POINTER

VALVE
SEAT

FIG. 12-1   Hand-operated expansion valve

HAND-OPERATED EXPANSION VALVES

While limited in application, the simplest method of feeding liquid refrigerant into the evaporator is by manually operating a shutoff valve, figure 12-1. Such a valve is adjusted by an operator to feed a definite quantity of refrigerant for the size of and conditions within the system.

The hand-operated expansion valve is used in large industrial installations, ice-making plants, and cold storage warehouses. This is practical because there is an almost constant load; therefore, the compressor operates continuously. In addition, an operator is on hand at all times.

The hand-operated expansion valve is not suitable for installations where the load varies, and the compressor runs intermittently to maintain a constant temperature. Under these conditions varying amounts of liquid refrigerant are fed into the evaporator and frequent adjustment of the hand-expansion valve is needed. Since this is not practical on small installations, other metering devices must be used.

THE LOW-SIDE FLOAT SYSTEM

The low-side float is so named because the float is located in the low-pressure side of the refrigeration system, figure 12-2. A fixed level for the liquid refrigerant is maintained in the evaporator by the position of the float, which opens, closes and controls the flow of refrigerant past a needle valve. The float and valve are connected with a lever that permits the float to control the needle valve mechanism.

The internal suction line opening is located near the top of the float chamber to permit dryer vapor to be removed at that point, and to prevent frosting in the suction line.

Float assemblies are designed with a removable or nonremovable head. Since the specific gravity of each refrigerant varies, the float selected must (1) have the correct buoyancy for the refrigerant being used; (2) be calibrated correctly to keep the refrigerant at a constant level; and (3) be in an evaporator that is level.

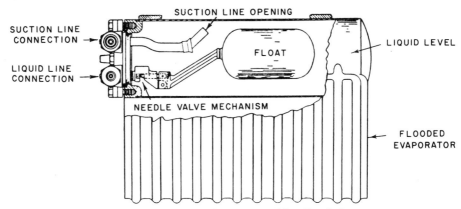

FIG. 12-2   Low-side float control for flooded evaporator

Another important consideration is the oil and refrigerant level, which must be correct to prevent suction line flooding. For those cases where oil floats on top of the refrigerant, the oil may be returned to the compressor through a hole drilled in the suction line or by a wick arrangement.

THE HIGH-SIDE FLOAT SYSTEM

The high-side float system is not as widely used as it once was. This system of controlling the flow of refrigerant gets its name from the fact that the float and needle

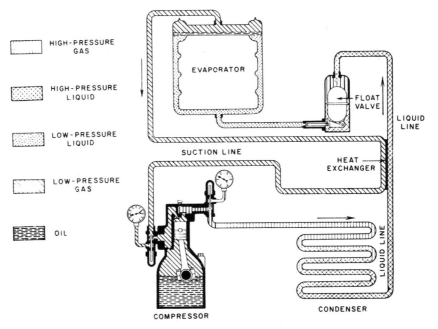

FIG. 12-3   High-side float system in refrigeration unit

valve are located on the high-pressure side. The drawing in figure 12-3 points out a few design changes from the low-side float system. Note that the float and valve are located outside the evaporator, and consequently more space is available inside for the evaporation process.

The cross-sectional drawing of a high-speed float in figure 12-4 shows that it serves as a combination refrigerant receiver and control valve. The liquefied refrigerant from the condenser flows into the float chamber. As the liquid level in the chamber rises, the float opens the needle valve. In this position the refrigerant flows into the liquid line and on to the evaporator. The level to which the float is adjusted controls the quantity of low-pressure liquid refrigerant which enters the liquid line.

Frosting or sweating of the liquid line occurs when the high-side float is used because of the low pressure. It is for this reason that the liquid line should be insulated when the float is located in the bottom of a cabinet. The sweating condition may be prevented by placing a restrictor in the liquid line near the evaporator, figure 12-5. The restrictor creates an intermediate pressure in the liquid line. This is about 25 psi greater than the low-side pressure and is high enough so that the saturation temperature of the liquid is above the dewpoint temperature of the surrounding air.

In those applications of the high-side float, where the unit is located on top of the refrigerator close to the evaporator, no intermediate pressure valve is needed. The liquid line is inside the cabinet so there is a minimum of sweating or frosting.

The refrigerant liquid level in a high-side float system is critical. Too much liquid causes flooding of the suction line; too little causes a low evaporator level with the resulting lower heat-absorbing capacity. This is in contrast with the low-side float system where the liquid level in the receiver may be at any point above the outlet to the liquid line or below the inlet from the condenser.

AUTOMATIC EXPANSION VALVES

One of the oldest types of automatic expansion valves is known as a *constant-pressure valve*. The name of the valve implies that it holds the pressure at a constant level

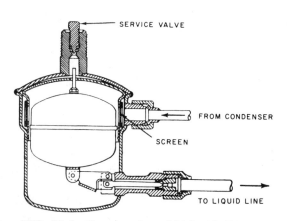

FIG. 12-4   Cross section of high-side float

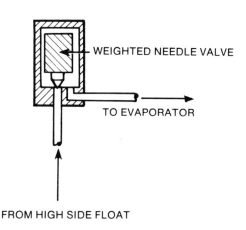

FIG. 12-5   Intermediate pressure valve

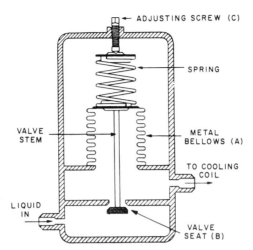

ADJUSTING SCREW (C)

SPRING

VALVE STEM

METAL BELLOWS (A)

TO COOLING COIL

LIQUID IN

VALVE SEAT (B)

FIG. 12-6 Automatic expansion valve

regardless of the load. The schematic drawing in figure 12-6 identifies the main parts of such a valve. The bellows (A) is actuated by the pressure in the evaporator. Any rise in pressure due to increased heat absorption causes the bellows or diaphragm to expand and reduce the opening between the valve stem and seat (B). This cuts down the flow of liquid refrigerant to the cooling coil and the amount of liquid that is vaporized in the coil. With less liquid being vaporized, the pressure drops because of the action of the compressor. This continues until the pressure at which the valve is set by the adjusting screw (C) is reached.

As the pressure in the coil drops, the reverse action takes place. The spring opens the valve, more liquid is admitted and the pressure inside the coil is raised. The valve regulates the rate at which liquid is admitted to the cooling coil and thus controls the amount of inside surface area in contact with the liquid refrigerant. By holding the suction pressure constant, the automatic expansion valve also keeps the compressor load the same.

This action is explained as follows. An increase in the load of a refrigerating system causes a rise in the suction pressure. This is because with an increased load more heat is absorbed, producing a more rapid boiling of the refrigerant. Vapor is formed more rapidly and this tends to increase the suction pressure. If this pressure is held at a constant level, the refrigerating capacity of the compressor is constant.

For this reason constant-pressure (automatic) expansion valves are not suited for compressors that run continuously and under varying loads. Regardless of the actual capacity required, the constant-expansion valve helps to develop a fixed capacity. As such, the entire rated capacity of the compressor is needed whenever it runs. The constant-pressure expansion valve may be used to advantage when the compressors are controlled by thermostats which are operated by room temperature.

Because this valve cannot be used when more than one coil is connected to the compressor, it has limited application in air-conditioning installations that require more than one cooling coil. If more than one coil is used, and the load on one of the coils decreases, the pressure in the coil decreases. However, the pressure equalizes in the suction line because both coils are connected to it. This equalizing of pressure interferes with the operation of the separate valves.

THERMOSTATIC EXPANSION VALVES (CONSTANT SUPERHEAT)

The control valve most commonly used in commercial refrigeration and air-conditioning systems under three tons is the *thermostatic expansion valve*. This valve maintains constant superheat in the evaporator. The difference between this valve and the automatic expansion valve is that a thermal element is connected to it by a small, sealed capillary tube. The thermal element (bulb) in liquid charged valves is partly filled with a liquid refrigerant and maintains some liquid under all conditions of temperature and load.

A typical valve is shown in the photo, figure 12-7, and in the schematic line drawing in figure 12-8. The thermal bulb is attached to the suction line so that any change in temperature of the suction line causes a corresponding change in the thermal bulb. Under an increased heat load, the refrigerant boils away faster in the evaporator. This results in a rise in temperature at the thermal bulb due to superheating. The higher temperature produces a higher pressure within the bulb and tube and this, in turn, causes the metal bellows to expand and force a wider valve opening. As a result more

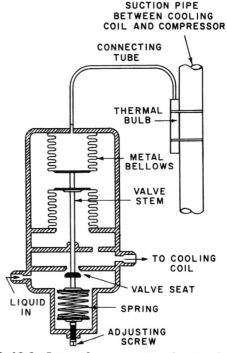

FIG. 12-7 Thermostatic expansion valve *(Courtesy of Parker Hannifin Corporation)*

FIG. 12-8 Internal construction showing function of valve

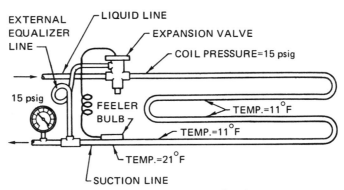

FIG. 12-9 Operating temperatures and pressures for thermostatic expansion valve

refrigerant is admitted to the evaporator to take care of the increased load. The valve opening is controlled by the temperature at which the valve is set. If the refrigeration load decreases, the valve opening admits less refrigerant to the evaporator. This continues until the valve opening becomes constant for the superheat for which the thermal bulb is set.

The refrigerant in the thermal bulb is usually the same type as the refrigerant being controlled. Ammonia is used for ammonia valves, sulfur dioxide for sulfur dioxide valves, and the like. For this reason, replacement units must be checked carefully to be certain that proper control is provided. The bulb may be attached to the side of the suction line or it may be inserted into the line. The bulb is attached at the point shown in figure 12-9 so that its temperature is the same as the superheated vapor from the evaporator.

There is always some liquid vaporized within the bulb to keep a pressure in the connecting tube and the metal bellows over the valve diaphragm. This pressure which is maintained corresponds to the refrigerant liquid temperature within the thermal bulb. Thus, the load on the evaporator coil determines the amount of liquid refrigerant that the thermostat expansion valve admits.

While any number of evaporator coils may be connected to one compressor, each coil must have its own expansion valve. Each valve then admits the quantity of refrigerant required by the coil it serves to maintain the superheat for which each valve is set. Thus, the other coils are not affected.

Equalizers

A thermostatic expansion valve with an *equalizer* connection is needed for those installations where the superheat setting of the expansion valve cannot control the amount of refrigerant that flows through the coil. An excessive pressure drop of 2 psi or more produces starved coils and high superheat.

The photo of a thermostatic expansion valve installation with an external equalizer connected to a coil, figure 12-10, gives the position of the equalizer connection between the valve and the oil return line, which is connected to the suction line. Some expansion valves have internal equalizers built into the valve. The equalizer does not eliminate the pressure drop but does make corrections for it.

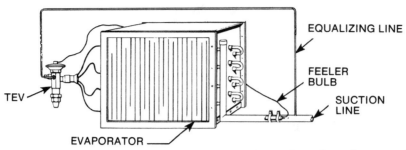

FIG. 12-10   Thermoexpansion valve hooked up to coil with equalizing line *(Courtesy of Sporlan Valve Company)*

Another design for a thermostatic expansion valve is illustrated in the cutaway section, figure 12-11. This valve combines the features of the thermal bulb and connecting tube by including them within the valve.

A power element (A) that is sensitive to the suction vapor as it flows to the compressor is found in the valve head. This produces an almost instantaneous response to changing loads.

The stem (B) at the bottom of the valve provides adjustment for superheat. No external equalizer is needed with this valve because compensation may be made for adjusting the superheat setting of the valve.

Testing Thermostatic Expansion Valves

Thermostatic expansion valves may be tested in the field with a regular service kit. A kit usually includes:

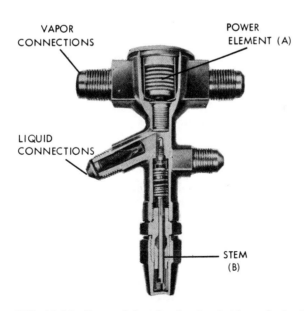

FIG. 12-11   Internal details of valve (without feeler bulb)

✓ A service drum of Refrigerant 12 to supply pressure. The refrigerant used does not have to conform to that used in the system with the valve to be tested.

✓ A low-pressure gauge.

✓ A high-pressure gauge which shows the pressure on the inlet of the valve.

✓ A quantity of finely crushed ice carried in a thermos or other container which is completely filled with ice for all testing.

✓ A power element (thermal bulb).

The illustration, figure 12-12, shows how a thermostatic expansion valve may be tested. The expansion valve outlet (A) has the low-pressure gauge screwed loosely into the adapter. This permits a small amount of leakage through the threads. The bulb (B) is placed in the crushed ice. Note that the high-pressure gauge (C) is connected in the line to the valve inlet. As the valve on the service drum (D) is opened and the warm drum builds up pressure to at least 70 pounds, the expansion valve may be adjusted.

The pressure on the gauge (E) of the thermostatic expansion valve should read:

29 lb. for Refrigerant 500
55 lb. for Refrigerant 502
22 lb. for Refrigerant 12
45 lb. for Refrigerant 22
15 lb. for methyl chloride
3 lb. for sulfur dioxide

These values were obtained by using 10° superheat. It should be noted that while methyl chloride and sulphur dioxide are used in the example, these refrigerants are not commonly used.

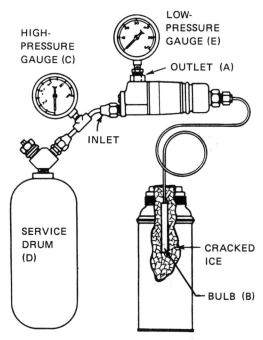

FIG. 12-12 Testing a thermostatic expansion valve

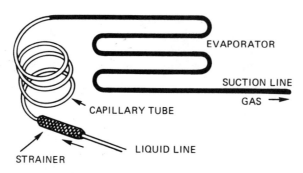

FIG. 12-13   The capillary tube

With a good valve, the pressure increases a few pounds and then either stops or else builds up gradually. A rapid pressure buildup to the point where the valve pressure equals the inlet pressure is a sign of a leaking valve.

With the new gas-charged expansion valves containing a small refrigerant charge, the pressure does not build up above the pressure marked on the valve body.

Power elements are tested by loosening the low-pressure gauge to allow some leakage through the threads, figure 12-12. The power element bulb is then removed from the crushed ice, and warmed by hand or in water at room temperature. The indicated pressure of a properly operating power element increases rapidly.

The body bellows of a thermostatic expansion valve is tested with high pressure on both gauges. The escape of gas is detected with a leak detector or by the use of oil or soapsuds. When making this test with a fairly high pressure, it is important that the body of the valve, the gauges, and other fittings are tightened securely to eliminate leakage.

Capillary Tube Controls

A *capillary* is a tube with a very small inside diameter. This diameter and the length controls the flow of the refrigerant. The capillary is the dividing point between the high side and the low side of the system.

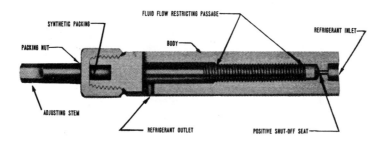

FIG. 12-14   **Application of capillary tube** *(Courtesy of Standard Refrigeration Company)*

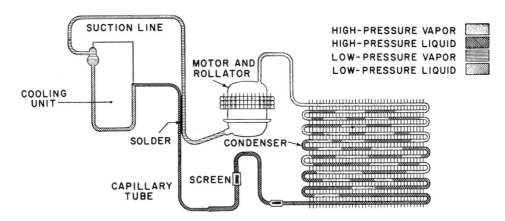

FIG. 12-15   Position of capillary tube in system

The simplified drawing in figure 12-13 shows the position of a capillary tube between the liquid line and the evaporator.

The capillary tube has the disadvantage of not being practical in commercial applications because it is not as sensitive to load changes as other control devices. This condition exits because the bore and length cannot be varied after the tube is installed.

The cutaway photo, figure 12-14, illustrates another form of capillary tube. An unusual form of tube is built into the body of the valve. A continuous helical channel is formed by the clearance between the threads on the adjusting stem and the body. The refrigerant flows through this channel and as it does so, its flow is controlled by the number of threads that are engaged. The adjusting stem provides a simple means of regulating the flow of refrigerant to match the needs of a particular system. It also provides a means for a positive shutoff.

The sketch of a complete refrigeration system with a capillary tube in the liquid line in figure 12-15 gives a better picture of the tube as a refrigerant control and metering device. When the compressor is not operating, the tube permits the pressures to equalize so that the compressor may start easily. This means that less expensive motors with low starting torque may be used. Also, since such systems use less refrigerant and the cost of a capillary tube is less than other controls, a capillary system is comparatively inexpensive and easy to produce. However, because of the tubing bore, the need to keep the refrigerant clean and dry, and the proper charge of the refrigerant, it is important that (if practical) this system be hermetically sealed.

SUMMARY

- Metering devices serve two purposes: (1) to restrict the flow of the refrigerant from the high to the low side. This makes it possible for the compressor to maintain the pressure conditions in the evaporator and the condenser that are needed for their proper functioning, and (2) to regulate the refrigerant flow according to the needs of the system.

- Metering devices may be grouped as:
  - ✓ Thermostatic expansion valves
  - ✓ Low-side floats
  - ✓ Automatic expansion valves
  - ✓ High-side floats
  - ✓ Hand-operated valves
  - ✓ Capillary tubes
  - ✓ Orifice
- Hand-operated valves are used only in large installations where operators are continuously on duty and where there is an almost constant load.
- Low-side floats maintain a constant level in the flooded evaporator. The float is usually inside the evaporator chamber and is always located on the low-pressure side of the system.
- High-side floats operate on the high-pressure side of the system.
- The automatic (constant-pressure) expansion valve permits the flow of refrigerant in response to the suction (back) pressure in the evaporator. This varies with the heat load as reflected in the rate at which the refrigerant boils. This valve operates to maintain a constant pressure in the evaporator.
- A thermostatic (constant superheat) expansion valve throttles the flow in response to the superheat of the refrigerant as it leaves the evaporator. It strives to maintain a constant superheat at that point. The controlling temperature-sensitive bulb is attached to the suction line at the end of the evaporator.
  - ✓ Equalizers on thermostatic expansion valves overcome control difficulties where there is a great pressure drop in the evaporator.
- A capillary tube control is a plain restrictor which permits the required flow and pressure equalization during an off-cycle. A less expensive motor with low-starting torque may be used with this system.

ASSIGNMENT: UNIT 12 REVIEW AND SELF-TEST

A. TYPES AND FUNCTIONS OF REFRIGERANT CONTROL DEVICES

List two metering devices for each condition given in the table.

Conditions	(1) Maintains a fixed rate of flow for a given setting	(2) The rate of flow is controlled by: The temperature or pressure of the refrigerant	(3) The rate of flow is controlled by: The level of the liquid refrigerant

B. LOW- AND HIGH-SIDE FLOATS

Check the column of the control device that fulfills each function or condition listed.

Control Device	
Low-side Float Valve	High-side Float Valve
1.	
2.	
3.	
4.	
5.	
6.	
7.	
8.	
9.	

1. Maintains constant level in the evaporator.
2. Maintains constant level in the receiver.
3. Likely to have sweating of the liquid line.
4. Located in the liquid receiver.
5. Requires an intermediate pressure valve.
6. Located in the evaporator.
7. Suction line opening near or in the top of the evaporator.
8. Float located on the low-pressure side of the valve.
9. Float located on the high-pressure side of the valve.
10. Make a simple sketch of the low-side float valve. Label the important parts.
11. Make a simple sketch of a high-side float valve. Label the important parts.

C. AUTOMATIC EXPANSION VALVES

1. Draw a simple sketch of an automatic expansion valve. Letter each important part.
2. Describe the operation of the automatic expansion valve in throttling the refrigerant as the load in the evaporator changes.

D. THERMOSTATIC EXPANSION VALVES

Add the correct condition or refrigeration term to complete statements 1 to 9.

1. The control element of the thermostatic expansion valve is fastened to the _____ at the _____ of the evaporator.
2. The control element of the thermostatic expansion valve usually contains the _____ as that used in the system.

3. The temperature of the control element is the same as that of the _____ in the suction line.
4. The temperature at the control element is _____ that of the _____ in the evaporator.
5. The refrigerant in the system is said to be _____ by the time it gets to the control bulb of the thermostatic expansion valve.
6. The primary function of the thermostatic expansion valve is to maintain _____ .
7. In order to compensate for pressure drop in the distributor and evaporator, the thermostatic expansion valve is equipped with _____ .
8. A more descriptive name for a thermostatic expansion valve is _____ .
9. As the load in the evaporator increases, the thermostatic expansion valve opens as a result of the increase in _____ of the charge within the control element.

E. CAPILLARY TUBE CONTROLS

Select the letter representing the words or terms which best complete statements 1 to 7.

1. A capillary tube system operates (a) with, (b) without a liquid receiver.
2. A capillary system provides for a (a) fixed, (b) variable capacity after it is once installed.
3. In adjusting the length of a capillary tube at the time of installation, if it is cut shorter, this (a) increases, (b) decreases, (c) does not change the rate of flow of the refrigerant to the evaporator.
4. The capillary tube (a) provides, (b) does not provide a liquid seal during the off part of the cycle.
5. The amount of refrigerant put into a capillary tube system (a) is critical, (b) is not critical.
6. The capillary system (a) permits, (b) does not permit pressure equalization during the off part of the cycle.
7. The capillary tube system requires an electric motor with a (a) high starting, (b) low starting torque.

F. METERING DEVICES AND CONTROLS FOR REFRIGERATING SYSTEMS

Determine which statements (1 to 12) are true (T); which are false (F).

1. A metering device is designed primarily to keep the evaporator as cold as possible.
2. The hand valve requires the continual attention of an operator when the load on the system fluctuates.
3. A thermostatic expansion control valve is used to turn the compressor on and off.
4. An automatic expansion valve can be used to automatically control superheat.
5. The control bulb of a thermostatic expansion valve is usually placed in cracked ice to test its operation.
6. The body bellows of a thermostatic expansion valve should be tested for leaks by applying a zero pressure to it.

7. A capillary tube is the simplest type of metering device.
8. An equalizer is used with an expansion device to equalize the pressure between the suction and discharge lines during the off part of the cycle.
9. A low-side float valve maintains a constant level of liquid refrigerant in the evaporator.
10. The earliest type of metering device was the hand valve.
11. An external equalizer makes it possible to use a motor with low starting torque.
12. A high-side float eliminates the need for a liquid receiver.

Identify the type of metering device which meets each of the conditions or functions in statements 13 to 25.

13. Permits equalization of pressure in the system during the off part of the cycle.
14. Maintains constant superheat in the evaporator.
15. Used with flooded evaporators.
16. Maintains constant pressure in the evaporator.
17. Cannot be used in a system with a receiver.
18. Located in the receiver.
19. Maintains constant refrigeration level in the receiver.
20. Located in the evaporator.
21. May be used with more than one evaporator.
22. Maintains constant level in the evaporator.
23. As the load increases the valve opens.
24. May be used with only one evaporator.
25. As the load increases the valve closes.

UNIT 13:
REFRIGERANTS AND DRIERS

Heat is removed within a refrigerating system by a *refrigerant*. Many refrigerants are known to man. In fact, any liquid that boils at a temperature somewhere near the freezing point of water can cool and preserve foods. However, a boiling point below the point at which ice forms is not in itself the only thing that makes a good refrigerant.

A refrigerant must possess other properties like being nonpoisonous, nonexplosive and noncorrosive. With a refrigerant that possesses these and other desirable qualities, the designer, craftsperson and technician can design, build and service a refrigerator where most working parts are sealed against dirt and moisture and are protected against corrosion.

Before the properties of a refrigerant may be studied, it is necessary to know its makeup. While there are some long chemical terms for refrigerants, only some of the more commonly used ones are named in this unit. A few chemical formulas are also given to explain how different refrigerants are formed.

Next, the desired properties are carefully identified and some refrigerants are described. Then, since so much depends on controlling moisture in the refrigerant, the last part of the unit covers the materials used in driers and the types and principles of operation.

FUNDAMENTALS OF CHEMISTRY APPLIED TO COMMON REFRIGERANTS

Although there are many different refrigerants and each one differs from the next, all are made from certain building blocks which may be put together in varying combinations. These building blocks are known as *elements*. Everything in this universe and in outer space is made up of one or more of 92 basic natural elements and less than a dozen man-made elements. Some of the more common natural elements are: carbon, iron, aluminum, lead, copper and hydrogen.

A few of the elements that are used in refrigeration work either as gases, liquids or solids. The first of these is *carbon*. Carbon is the main element in coal, wood, cloth, gasoline and oil. It is present in many gases like carbon dioxide, methyl chloride and other refrigerants. At normal atmospheric pressure and temperatures, carbon exists as a solid.

Iron, aluminum, lead, copper, zinc, tin, silver, and nickel are also elements. Such elements are called *metals*. These metals may be used by themselves, or they may be mixed together to form what are known as *alloys*. The metals may also be found in chemical combinations with other elements to form compounds. These metallic elements usually exist as solids.

Structure of Elements

From these few descriptions it is readily seen that the elements may form mixtures or compounds in the different states, figure 13-1. The atom, in review, is the smallest particle of a substance that can take part in a chemical reaction.

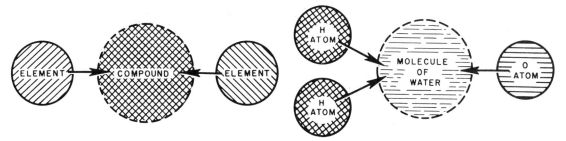

FIG. 13-1 Elements form compounds

FIG. 13-2 Water formed from hydrogen and oxygen elements

While, scientifically, the atom is composed of even finer particles (electrical charges) called *neutrons, protons* and *electrons,* for the purposes of this study the atom is considered as being indivisible (cannot be divided further) and unchangeable (retains the properties of the element). The atoms of each element are different. That is, the element lead is composed of lead atoms, zinc of zinc atoms, nitrogen of nitrogen atoms, and so on.

When the atoms of the same or of different elements combine, they form a molecule. The surprising thing about a molecule made up of atoms of more than one element is that it may have properties which are far different from any of the atoms in the individual elements. It is the energy of these molecules that makes up the heat energy of a material.

A small quantity of water contains many billions of molecules. Every molecule of water, ice or steam is made up of two atoms of hydrogen and one atom of oxygen. Both elements (hydrogen and oxygen) are gases. Hydrogen is a very light and highly flammable gas while oxygen, as a gas, supports combustion (burning). Water, which is a combination of these two gases, is a liquid having altogether different properties from either of the two elements.

Importance of Chemical Symbols in Refrigeration

Instead of using words to describe which atoms form the molecules of different compounds, letters and numbers are used to shorten the description. Each of the basic elements is always known chemically by one or two letters (called *symbols*). For instance, *O* is used instead of oxygen, *H* denotes hydrogen, and *Fe* identifies iron. The chemical formula for water is H_2O. This means that each molecule of water contains the elements hydrogen (H) and oxygen (O). It also indicates by the use of the subscript number ($_2$) how many atoms of the element have entered the combination. In the case of water, one molecule contains two atoms of hydrogen (H_2) and one atom of oxygen (O), figure 13-2.

While this description of some common chemical terms is brief, an understanding of these is important. It must be remembered from this point on that whenever refrigerants are discussed, their properties may be changed by altering the elements in the combination. The compound salt may be used as an example to emphasize this point. The common table salt (NaCl) is produced by a chemical reaction between atoms of sodium (Na) and atoms of chlorine (Cl), figure 13-3.

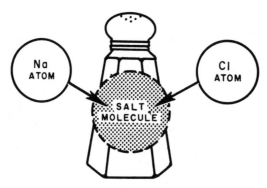

FIG. 13-3   Salt consists of sodium and chlorine atoms

Before continuing with the makeup (composition) of some of the compounds used in refrigeration systems, it is necessary to describe the qualities a refrigerant should possess.

QUALITIES OF REFRIGERANTS

The refrigerant absorbs heat and carries it through the refrigeration system, changing state from a liquid to a gas. Later in the cycle, it gives off heat, changing from a gas to a liquid. The refrigerant is effective and efficient when it has certain required properties:

Behavior of Refrigerants with Heat and Temperature

► The freezing point of the refrigerant must be lower than any temperature encountered in the system. If this were not so, the refrigerant would freeze in the evaporator. Such a condition would cause an erratic refrigeration cycle.
► The latent heat of vaporization of the refrigerant must be high. In other words, a small quantity of refrigerant must be able to absorb a large amount of heat.

Volume and Density of Refrigerants

► The volume of refrigerant vapor should be as small as practical. This makes it possible to use smaller suction lines, condensing tubes, etc., to cut down on the size and cost of parts.
► The density of the refrigerant should be high. Smaller liquid lines can be used with high density refrigerants.
► By contrast, on commercial installations where the evaporator and condensing units are on different levels, a low density liquid refrigerant is desirable. Such a refrigerant requires less pressure to force it through the liquid line.

Pressure Affects Refrigerants

► When the condensing pressure is low, lightweight equipment may be used and the chance of leakage is decreased.
► The refrigerant must be such that the difference in pressures between the high side and the low side is as low as possible.

Other Chemical Properties of Refrigerants

The refrigerant should:

- Withstand operating temperatures and pressures found in the system without decomposing (breaking down).
- Be nonflammable and nonexplosive in either gas or liquid form or when mixed with oil.
- Be noncorrosive and not attack any other substance in the system.
- Be nontoxic and noninjurious to human beings, plant and animal life.
- Have no harmful effect on the taste, color or aroma of foods and drinking water.

Other Physical Properties of Refrigerants

The refrigerant should:

- Carry oil in solution.
- Be miscible (able to mix) with oil in order to carry a sufficient quantity for lubrication.
- Have no harmful reaction with oil, even in the presence of moisture.
- Make possible the detection and location of leaks.
- Have a high rate of performance.
- Have a high resistance to electricity to prevent electrical shortages throughout the system.

Other Considerations

The refrigerant should:

- Have low initial cost.
- Be economical to maintain.
- Be easy to secure.

These are the main considerations that must be given to the selection of refrigerants. With this understanding of desirable properties, some of the common refrigerants can now be discussed.

REFRIGERANTS FOR DOMESTIC AND COMMERCIAL SYSTEMS

Ammonia as a Refrigerant

The oldest known refrigerant is ammonia. Ammonia is a compound consisting of one atom of nitrogen chemically combined with three atoms of hydrogen. Ammonia is represented by the chemical symbol NH_3. Ammonia has a very high latent heat of vaporization (565 Btu/lb.) and is widely used in industrial applications.

Equipment and fittings used with an ammonia system should be made of steel or other metals it does not attack. Ammonia is corrosive to brass and bronze wherever there is moisture.

Precautions with Ammonia

Ammonia vapor is considered flammable, explosive and toxic. Because of its sharp, penetrating odor, it can be detected quickly. Service technicians working with ammonia

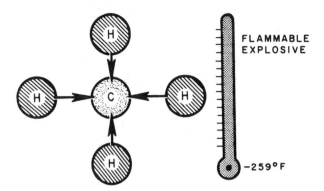

FIG. 13-4   Methane (CH4)

must be familiar with first aid rules and must be prepared to help anyone overcome by ammonia fumes or liquid. Ammonia gas is lighter than air so there is less danger near the floor. Adequate ventilation should be provided in a room where an ammonia system has developed a leak. Tests indicate that at a given pressure ammonia has over two and a half times as much tendency to leak as do the newer refrigerants. Liquid ammonia produces severe skin burns. Its vapors can also damage the eyes. For these and other reasons care must be taken when working around ammonia systems.

Hydrocarbons as Refrigerants

The name *hydrocarbon* is given to a group of chemical compounds formed from hydrogen (hydro) and carbon (carbon) atoms in different combinations. Methane (CH_4), isobutane (C_4H_{10}) and propane (C_3H_8) are three examples of hydrocarbons. The chemical composition of methane is shown in graphic form in figure 13-4.

These three hydrocarbons and others in this same family are used as fuels. They are generally known as *bottled gas*. They ignite easily and burn readily over a wide range of temperatures.

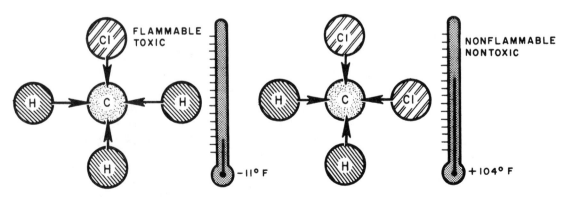

FIG. 13-5   Methyl Chloride (CH3Cl)   **FIG. 13-6   Methylene Chloride (CH2Cl2)**

Methyl chloride (CH_3Cl) is the result of chemically combining a hydrocarbon with chlorine, figure 13-5. In other words, one of the H atoms of a methane (CH_4) is replaced by a chlorine atom.

Since chlorine is one of a family of elements known as *halogens*, methyl chloride, ethyl chloride (C_2H_5Cl) and methylene chloride (CH_2CL_2) are all called *halogenated hydrocarbons*.

This group is considered to be of only limited flammability. Of the three, methylene chloride is classed with the safest refrigerants, figure 13-6. It bears the trade name *Carrene #1*.

Up to this point none of the chemicals was considered as an ideal refrigerant. So, the chemists made a fresh start with a new compound known as carbon tetrachloride (referred to as *carbon-tet*) or CCl_4. Note that this compound contains one atom of carbon, which is chemically combined with four atoms of chlorine. By substituting two fluorine atoms in place of two of the chlorine atoms in carbon-tet a new compound was formed. The result was $CCl_2 F_2$, which is chemically known as dichlorodifluoromethane.

This compound, which was originally called *Freon 12* (a company brand name), represented a major chemical achievement. Other compounds in this same family were later developed with each one having different heat-carrying abilities. This family of refrigerants, known as the R or *Refrigerant family*, did have all the desirable characteristics except high latent heat. All had in common the safe characteristics of being nonflammable and nontoxic.

Simplified Identification of Refrigerants

While *Freon* and *Genetron* are two typical trade brand names of refrigerants manufactured by different companies, the trend is toward the use of the name *Refrigerant 12*, *Refrigerant 11* and so on. The designations R-12 and R-11 are made in the manufacture and servicing of refrigeration units. Two of the most widely used of the Refrigerant family are Refrigerant 12 and Refrigerant 22. The structure of each is shown graphically in figure 13-7.

Thus, the search for a nonflammable, nontoxic, noncorrosive refrigerant has been successful. It is important, however, to point out that the Refrigerant family has one bad trait. In the presence of heat from an open flame or electric heating element, they break down into *phosgene gas* and chlorine gas. Both gases are toxic to the point of being poisonous.

Physical characteristics of common refrigerants are listed in Table 1 of the Appendix.

SAFETY PRECAUTION

It is extremely important to know the potential hazards of working with refrigerants. Information concerning flammability, toxicity and other characteristics of refrigerants is available from the *Underwriters Laboratory Reports* and/or the *Data Book* published by the American Society of Refrigeration Engineers.

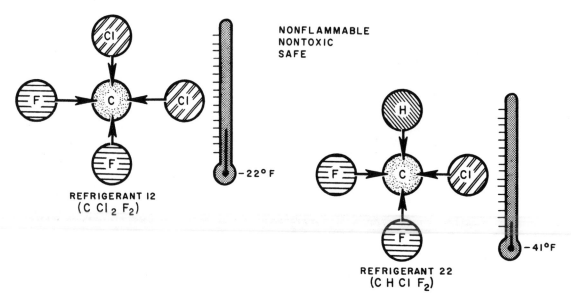

FIG. 13-7   Composition, boiling point and properties of refrigerants

Note that continuous studies are being carried on to establish the effects of R-12 and R-11 on the environment and what further regulations are needed for the use of these refrigerants as well as other products and devices using fluorocarbons.

REFRIGERANTS AND MOISTURE

Effect of Moisture

While refrigerants are normally noncorrosive, it is important to keep moisture out of systems using them. The presence of moisture in a refrigeration system can cause other troubles like a *freeze-up* at the metering device. This can take place in all systems,

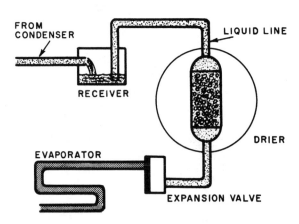

FIG. 13-8   Position of drier in liquid line

FIG. 13-9 Drier unit with cutaway view to show construction *(Courtesy of Sporlan Valve Company)*

especially those operating at the low temperatures. Moisture also encourages the formation of acids, which speed up corrosion. This same effect is found in systems where the refrigerating oil has broken down.

Drying Devices to Remove Moisture

A device designed to remove moisture from a refrigerant is called a *drier*. Driers are placed in the liquid line unless some unusual condition indiates otherwise, figure 13-8. The drier is located at approximately point T_7 on the Basic Refrigeration Cycle Diagram.

The drying material is called the *desiccant*. A desiccant is a solid substance capable of removing moisture from a gas, liquid or solid. The desiccant can be molded into a special form to fit a drier container or be used as granules or in ball form.

A small scale unit drier and core are illustrated in figure 13-9. The desiccant is a molded block designed to eliminate the disadvantages of a loose desiccant.

Another type is the angle unit, which permits the use of a replaceable desiccant core, figure 13-10. The exploded view in figure 13-11 shows the assembly of this unit.

Presently, silica gel, activated alumina and *Drierite* are the most frequently used desiccants. Drierite is the trade name for anhydrous calcium sulfate.

FIG. 13-10 Filter drier containing replaceable core *(Courtesy of Sporlan Valve Company)*

Properties of Desiccants

One of the basic properties of a desiccant is its ability to remove moisture from a refrigerant. Other properties that are important, follow.

The desiccant must be able to:

- Reduce the moisture content of the refrigerant to a low level.
- Hold moisture that may be formed in an installation while retaining efficiency.
- Act rapidly to reduce the moisture on one passage of the refrigerant through the drier unit. Unless this happens, enough moisture may be left to freeze the expansion valve or capillary tube of a low temperature system.
- Withstand the presence of oil or other substances in the refrigerating system with little or no loss to efficiency, capacity or speed.
- Hold up under increases in temperature up to 125°F without affecting efficiency, capacity or speed.
- Retain all necessary functioning properties during its storage and operation in a system.
- Dry refrigerants effectively as either gases or liquids.

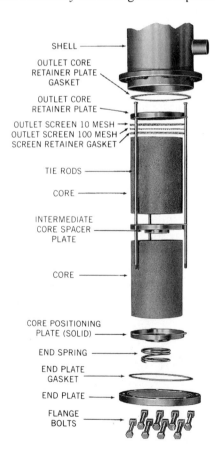

FIG. 13-11 Exploded view of filter drier *(Courtesy of Sporlan Valve Company)*

- Remove the moisture from the refrigerant to a level below that which causes corrosion. Since the corrosion level is higher than that at which freeze-ups occur, any desiccant that prevents these also prevents corrosion.
- Resist reacting chemically with the oil, refrigerant or any other material or substance used in the refrigeration system.
- Remain insoluble. This means the desiccant must not dissolve in any of the liquids found in the system.
- Stay in its original solid condition. Any change would slow up the passage of the refrigerant through the drier.
- Retain the size of its particles. Shrinkage produces unfilled areas in a drier unit.
- Be reactivated a number of times at a temperature that does not go higher than the manufacturer's recommendations. Many refrigeration and air-conditioning systems, where silica gel is used in large volumes as a desiccant for drying air, are designed so that the desiccant may be reactivated.
- Allow the steady flow of refrigerant through its grains, pellets or block with little restriction or pressure drop of the refrigerant.
- Resist reacting chemically with moisture to produce dangerous or corrosive particles.
- Work effectively and safely and be nonpoisonous and nonirritating.

How Desiccants Work

Desiccants pick up and hold moisture. This is done by either a physical process of *adsorption* or by chemical action. The adsorption process is simple when the physical structure of the desiccant is understood. The three desiccants to be studied are silica gel, activated alumina and drierite.

Silica gel is first manufactured as a jellylike material. It is then subjected to a partial drying process. As the gel begins to lose water, it develops a tremendous surface area. These surfaces are made up of fine-shaped tubes, which are called *capillaries*. Within the walls of these tubelike formations is a minute quantity of water. This water cannot be removed from the gel without ruining the desiccant. This same type structure is found in activated alumina.

In operation, the gel type desiccants pick up and hold moisture. The molecules of water stick on the inner walls of the capillaries. This physical process is referred to as the *adsorption of water molecules*. An enlarged view of the capillaries and the adsorption of water molecules is illustrated in figure 13-12. The capacity of the desiccant to pick up

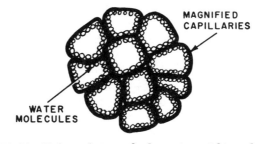

MAGNIFIED CAPILLARIES

WATER MOLECULES

FIG. 13-12 Enlarged view of adsorption within a desiccant

moisture is determined by the amount of surface area in the many capillaries in the body of the drier.

Drierite, the third of the desiccants, operates by chemical action. This desiccant also possesses an extremely large surface area with the same capillary structure. Moisture is picked up as it chemically forms another substance on the walls of the capillaries.

Drier Size

Much experimenting has been carried on with the amount of desiccant needed in a drier for different types of refrigeration systems. Differences in drying techniques, assembly operations, kinds of systems and refrigerator sizes all influence the drier size. Driers that are now on the market are rated for certain horsepower capacities.

There are a number of refillable drier units that contain the desiccant, screens and pads. Unless these are supplied in airtight factory sealed containers to keep out moisture, the cartridge must first be reactivated.

Drier Location

Driers, other than those used with sulfur dioxide, are located in the liquid line just ahead of the expansion valve or capillary tube. A sulfur dioxide drier may be placed in the suction line where it can do a better drying job than if located in the liquid line. This is practical with sulfur dioxide because there are no freeze-ups.

The Drier as a Filter

Driers are provided with screens, filter pads or porous metal cones or discs. These serve as filters to prevent the desiccant and other solids from being carried through with the refrigerant. Driers, which are equipped to carry on a filtering process, make it unnecessary, when the drier is on the high side, to use a filter in the liquid line. The only time that a filter might be needed in the suction line is when a new compressor is installed and when a need is detected in the system to replace a dirty evaporator. The filter prevents solids from reaching the compressor.

Adsorbing Coloring Matter and Acids

Acids can form as a result of action by moisture on certain refrigerants. Some manufacturers use a colored oil so if a leak occurs anywhere in the system the color helps to locate the leak.

The desiccants that operate on the adsorption principle pick up moisture, coloring matter and acids. This last action is desirable since it helps to cut down corrosion.

Driers with Moisture Indicators

One general type of moisture indicator is illustrated in figure 13-13. This model includes a *drier unit* and is designed to accept either a plain connector or one containing a *sight glass indicator*. The liquid line moisture elements turn pink when the moisture content is exessively high; blue when the system is dry. The indicator is a permanent installation and always shows a wet or dry color depending on the condition of the refrigerant. The sight glass permits easy monitoring by indicating any shortage of the refrigerant.

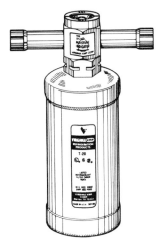

FIG. 13-13 Filter drier sight glass moisture indicator *(Courtesy Virginia KMP Corporation)*

This model would be installed at point T_7 on the Basic Refrigeration Cycle Diagram.

A *universal element moisture indicator* is shown in figure 13-14 with sweat fittings (A) and flare fittings (B and C). The cleanable glass sight permits the viewing of bubbles, which can indicate a pressure drop or a low charge in the system. Reversible pink and

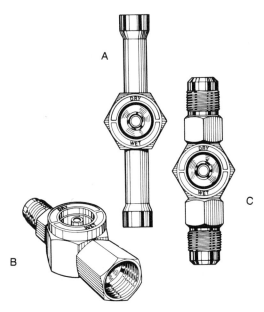

FIG. 13-14 Sight glass moisture indicator with sweat and flare fittings *(Courtesy of Virginia KMP Corporation)*

blue colors, as stated, provide sight inspection of *safely dry* and *dangerously wet* conditions of a refrigeration or air-conditioning system.

These indicators are attached at point P_2 on the Basic Refrigeration Cycle Diagram.

SUMMARY

- Heat is removed within the refrigerating system by a refrigerant which carries it from the space to be refrigerated to another place where it is not objectionable.
- A good refrigerant meets these and other conditions:
  - It reacts to changes in heat and temperature; meets certain standards of specific volume, viscosity and density; and reacts to pressure.
  - It does not decompose with use and has good insulation against electricity.
  - It is nonflammable, nontoxic, noncorrosive and noninjurious; economical and easily secured.
- A compound is formed by the chemical combination of atoms of different elements. The use of chemical symbols and formulas is a shorthand method of indicating its composition.
- Ammonia (NH_3) and sulfur dioxide (SO_2) were among the earliest compounds used as refrigerants.
- Hydrocarbons are made up of different combinations of carbon and hydrogen and are among the natural compounds used as refrigerants.
- The Refrigerant or R family is the safest group of refrigerants. These are produced by manipulating the atoms in carbon-tet with those of fluorine and hydrogen.
- Research indicates that widespread use and release of fluorocarbons cause environmental damage to the ozone layer.
- Moisture in a refrigerant can hinder the operation of the system and damage its parts by corrosion. A drier is used to control moisture.
- A desiccant is the material used in the drier to trap the moisture from the refrigerant. It must be able to:
  - Remove moisture from refrigerant and keep it below corrosive level; act rapidly; withstand the presence of oil; hold up under increased temperatures.
  - Dry refrigerants as either gases or liquids; resist chemical reactions; retain its original size and shape; and permit reactivating.
  - Allow a steady flow of refrigerant; work effectively; and be safe to work with.

ASSIGNMENT: UNIT 13 REVIEW AND SELF-TEST

A. FUNDAMENTALS OF CHEMISTRY APPLIED TO COMMON REFRIGERANTS

Provide a word or phrase to complete statements 1 to 6 correctly.

1. All refrigerants are produced by combining _____ of different elements.
2. When atoms of one element combine with atoms of another element, they form a

   _____ .
3. A compound is a combination of _____ .

4. Refrigerant 12 is a compound consisting of one atom of carbon, two atoms of chlorine, and two atoms of fluorine. Write this compound as a chemical symbol.
5. Compare the chemical makeup of Refrigerant 12 and Refrigerant 22. In Refrigerant 22, one atom of _____ has been substituted for one atom of _____ in Refrigerant 12.
6. The one change described in item 5 resulted in _____ for Refrigerant 22.

B. QUALITIES OF REFRIGERANTS

Determine which statements (1 to 7) are true (T); which are false (F).

1. Refrigerants having high density require less pressure to force them through the vertical liquid lines.
2. The freezing of a refrigerant in the evaporator is desirable.
3. A refrigerant with a high latent heat of vaporization means that a small quantity can absorb a large amount of heat.
4. High electrical resistance is unnecessary in a refrigerant.
5. The difference in pressure between the high and low sides of the system is kept as high as possible.
6. A good refrigerant should be nonflammable and nonexplosive as a gas or liquid and when mixed with oil.
7. The lower the operating temperature of the system, the lower must be the boiling point of the refrigerant used.
8. Give five physical properties of a good refrigerant.
9. List five properties required in a good refrigerant that are influenced by chemical reactions.

C. REFRIGERANTS FOR DOMESTIC AND COMMERCIAL SYSTEMS

1. Give the formula for A, B and C. Also, indicate the atoms by kind and number in each molecule of the element in the refrigerant.

	Refrigerant	Formula	Composition of Molecule
A	Ammonia R-717		
B	R-502		
C	Water R-718		

Section 4: The Heart of the Refrigeration System

2. Cite two disadvantages of using ammonia as a refrigerant.
3. Name two refrigerants consisting of hydrogen and carbon only.
4. By what name are chemical compounds of carbon and hydrogen known?
5. Name two halogens used in refrigerants.
6. List two halogenated hydrocarbon refrigerants, other than the Refrigerants.
7. Which of the halogens are added to hydrogen and carbon to form the refrigerants in problem 6?
8. List two safe operating characteristics of the Refrigerant family.
9. Which halogen is added to carbon tetrachloride in the manufacture of the Refrigerant family?
10. Refer to the formula for each of the three Refrigerants. Then name the different atoms and tell how many of each are in each compound.

	Refrigerant	Formula	Composition of Molecule
A	R-12	CCl_2F_2	
B	R-22	$CHClF_2$	
C	R-717	NH_3	

11. Name the refrigerants that fit the conditions of flammability and toxicity given in the table.

	Refrigerants		
	Highly flammable	Moderately flammable	Nonflammable
A			
B			
	Highly toxic	Moderately toxic	Nontoxic
C			

D. REFRIGERANTS AND MOISTURE

1. What is the prime purpose of a desiccant?
2. List five desirable properties of a desiccant.

248

3. State briefly, with the aid of a simple sketch, how adsorption takes place in activated alumina or silica gel.

Provide a word, term or phrase to complete statements 4 to 11.

4. A good desiccant resists reacting chemically with moisture to produce _____ .

5. A desiccant allows for the free passage of a refrigerant with as little restriction as possible so as not to cause a _____ .

6. A desiccant must not _____ in the liquids found in the system.

7. Drierite absorbs by chemically forming another substance with the moisture. State where this new substance then forms.

8. The better desiccants can be used successfully up to _____ F.

9. If an unused desiccant cartridge has had its seal broken during storage, it must be _____ .

10. Desiccant dust and other solids are prevented from being carried through the refrigerant by designing driers with _____ , or _____ .

11. When the adsorption-type desiccants pick up coloring matter, they cut out the value of the _____ used by manufacturers to help locate leaks.

12. Give the reason why driers, other than those used with sulfur dioxide, are placed ahead of the expansion valve and capillary tube.

13. State briefly what may happen when activated alumina or silica gel heats up excessively.

14. Give the two functions of a moisture indicator with a sight glass.

UNIT 14:
PRINCIPLES OF CHARGING AND TESTING REFRIGERATION SYSTEMS

The conditions under which a refrigerant is able to absorb and reject heat, and the design and functioning of the major components of small and large refrigeration systems, have been described in previous units. Building on this foundation of technical information, this unit deals with the valves, gauges, accessories and materials needed to charge and test refrigeration systems.

The unit is divided into five parts. These give the important principles and attempt to develop the technical understanding which is needed for a complete mastery of the basic skills required in the shop and laboratory. The five parts include:

- ✓ The function of valves in charging refrigeration systems.
- ✓ Service gauges and testing manifolds.
- ✓ Principles relating to the evacuation of refrigeration systems to receive a charge.
- ✓ Methods of transferring refrigerants and then adding quantities to a system.
- ✓ Tests for refrigerant leaks.

THE FUNCTION OF VALVES IN CHARGING REFRIGERATION SYSTEMS

Manifolds Shutoff Valves

Pressure gauges and valves must be installed before servicing and charging. Line valves make it possible to isolate any portion of a system or, in a multiple hookup, to separate one system from the rest, figure 14-1. They must be designed to prevent refrigerant leakage.

The term *pumpdown* refers to physically closing the liquid line service valve to remove all the refrigerant from the evaporator in order to service the system. When it is necessary to *open the system* for repairs or replacement of parts, the refrigerant is

FIG. 14-1   Two-way line shut-off valve *(Courtesy of Henry Valve Company)*

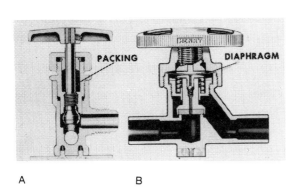
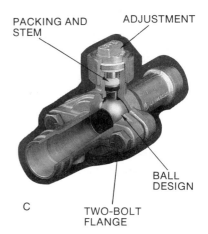

FIG. 14-2 Basic types of shut-off valves: (A) packed type, (B) diaphragm type (packless) and (C) ball valve (two-bolt model) *(Courtesy of Superior Valve Company, Division of Amcast Industrial Corporation)*

isolated in the receiver. This is accomplished by closing the liquid line service valve. The refrigerant is then drawn out of the evaporator by the compressor, condensed and stored in the receiver.

After *pumpdown*, the valves may be closed. This locks the refrigerant in the condenser or receiver so that other parts of the system may be serviced without loss of refrigerant or for winter shutdown.

The design of manual shutoff valves provides protection against leakage. Two common methods of doing this include: (1) packing or, (2) the use of a metal diaphragm, figure 14-2.

Many of these valves are also designed with a back-seating construction. When the packed valve is completely open, the reverse side of the valve seat closes against a second seat in the body of the valve. This prevents the refrigerant from leaking through the valve stem packing or the back-seat port.

The second method involves the design and construction of a flexible metal diaphragm in the body of the valve. Such valves are not provided with a back-seating port.

Some valves are built with seal caps as an added precaution against leakage at the valve stem. The seal cap must be removed with a wrench before the valve stem may be operated. The photo in figure 14-3 illustrates one type of seal cap. In some types, the hexagon end, when removed and turned upside down, is designed to be used as a wrench to operate the valve stem.

Three manual shutoff valves are usually installed as standard equipment by the manufacturers of compressors and condensers. These are the discharge and suction service (shutoff) valves on the compressor and the liquid line shutoff valve on the outlet of the receiver.

Compressor Shutoff Valves

It is essential that a suction service and a discharge service shutoff valve be installed on the suction and the discharge sides of the compressor. Both of these valves are closed

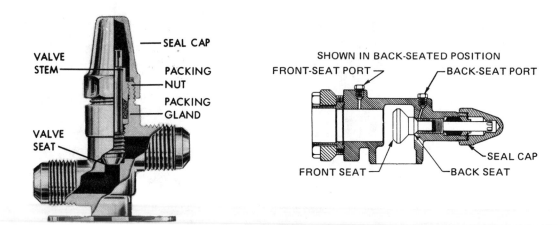

FIG. 14-3 Seal cap provides added leakage protection

FIG. 14-4 Cross section of a compressor service s.o.v. *(Courtesy of The Trane Company)*

tightly when the compressor is shipped to prevent air, dirt and moisture from entering. The expression *s.o.v.* is an abbreviation of *shutoff valve*.

The cross-section view of a compressor shutoff or service valve in figure 14-4 reveals that it is a packed valve of the back-seating type. This particular valve is designed to bolt directly to the body of the compressor. The valve is threaded and plugged with both a front- and back-seat port plug. This makes it possible to connect, disconnect and use test gauges without removing the refrigerant from the system.

One caution should be taken with the packing-type of valve. The packing nut should be loosened while the valve is being stored or before it is put in use. This prevents wearing the packing out by compressing it. When the valve is put into operation, the packing nut must be retightened.

Compressor shutoff valves usually have a front-seat port on the line side of the valve. Auxiliary safety devices, like high- or low-pressure cutout switches, can be connected at this front-seat port. Note that the front-seat port cannot be shut off from the line pressure. This prevents the operator from accidentally closing the front seat port and interrupting normal safe operation.

Liquid Line Shutoff Valves on Condensers

A third manual shutoff valve is installed in the liquid line near the condenser well. The valve is known as a *liquid line s.o.v.*, figure 14-5. It is used to shut off the flow of refrigerant between the condenser and the liquid line.

The refrigerant charge can be held between the compressor and the liquid line valve by the following procedure: (1) closing this valve, (2) pumping down the system, and (3) then closing the compressor discharge service valve.

With such controls, the remainder of the system is effectively isolated from the refrigerant charge. This is important because it permits the servicing or replacement of components on the liquid line, evaporator, suction line and compressor without removing the refrigerant charge from the system.

FIG. 14-5   Liquid line shutoff valve   FIG. 14-6   Liquid line charging valve

Liquid Line Charging Valve

Another valve that is required is the liquid line charging valve for high-side charging, figure 14-6. The initial quantity of refrigerant delivered into a system is usually charged through this valve. A system may be charged through the discharge service valve on the high-pressure side or from the suction service valve on the low-pressure side.

Suction, Oil and Discharge Gauges

The importance of using gauges properly in the start-up, testing and overall operation of a refrigeration system cannot be overemphasized. Since gauges are relatively inexpensive and give an accurate indication of performance, they should be used freely.

Some compressor manufacturers provide at least three gauges for indicating suction, discharge and oil pressures, figure 14-7. These are mounted on a gauge board. The suction pressure gauge is a compound gauge and has a range from 30 inches of vacuum to 316 psig. This gauge is connected to the low-pressure side of the system by capillary tubing.

The oil pressure gauge also has a range from 30 inches of vacuum to 150 psi and is connected to the oil pump discharge line in the compressor. Effective oil pressure is found by subtracting suction pressure from the pressure shown on the oil pressure gauge.

The discharge pressure gauge is one designed for use with R-12 and R-22 and has a range from 0 to 500 psi. It is connected to the high system.

The schematic drawing in figure 14-8 indicates the location of the three gauges which are connected to a compressor. The suction pressure (compound) gauge is connected to the back-seat port of the service s.o.v. The discharge (high-) pressure gauge is connected to the back-seat port of the discharge service s.o.v. The oil pressure gauge is connected to the oil pump discharge pressure through an opening provided in the compressor housing.

Throttling Valves

Another accessory which is usually provided whenever a gauge is installed permanently in a system is called a *throttling valve*, figure 14-9. This valve provides a method of:

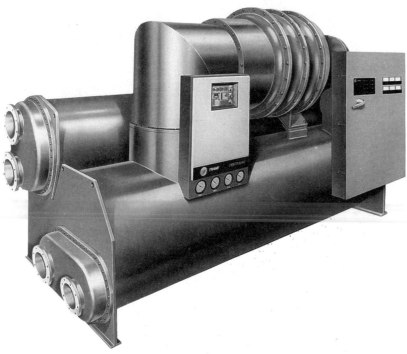

FIG. 14-7   Position of gauges on a large commercial refrigerating unit *(Courtesy of The Trane Company)*

(1) shutting off a gauge line when readings are not being taken, and (2) throttling the line to prevent fluctuation when readings are to be taken.

A throttling valve may be installed in the line to the oil pressure gauge. The other suction and discharge pressure gauges are connected to the back-seat ports of the compressor service valves. Because the service valves have only small holes through the back-seat ports, they also act as throttling valves.

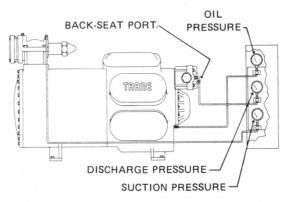

FIG. 14-8   Gauge connections *(Courtesy of The Trane Company)*

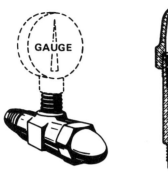

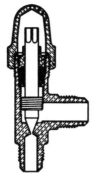

FIG. 14-10   Piercing valve with flow control for servicing a hermetic refrigerator *(Courtesy of Robinair Division, SPX Corporation)*

FIG. 14-9   Throttle valve

Special Service Valves and Adapters

A self-piercing service valve is available for hermetic refrigeration systems that may not have service valves. The type illustrated in figure 14-10 clamps onto the refrigeration line of a hermetic refrigerator. A rubber gasket is provided to make a gastight connection between the valve body and the line.

Valve attachments and valve adapters (which can be fitted to any mechanism) are made commercially. With these, it is possible to make valve connections for any refrigerator system that has fittings for valve adapters.

Pressure Relief Valves

Occasionally, the temperature of the air surrounding a refrigeration system may rise to a point where it causes the pressure of the refrigerant gas to increase to a danger point. For example, fire in a building may create such a condition. Also, if a safety control device fails to function, the internal pressure caused by compression may build up to a dangerous level.

A simple accessory known as a *pressure relief valve* can be installed to minimize the possibility of explosion, figure 14-11. When the relief valve operates, it discharges

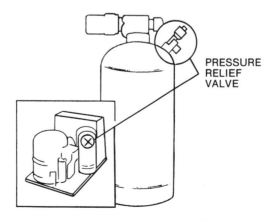

PRESSURE
RELIEF
VALVE

FIG. 14-11   Relief valve installed on liquid receiver

directly into the atmosphere, unless it is equipped with a vent line fitting to exhaust outside a building. Most building safety codes require this kind of installation. Pressure relief valves are usually mounted by the manufacturer on the liquid receiver or condenser.

The two most common pressure relief valves used in refrigeration systems are the spring-loaded and the fusible-plug types, figure 14-12.

SAFETY PRECAUTIONS

- Fusible metal plugs protect the cylinder in case of fire. Such plugs may not protect the cylinder from gradual and uniform heating. The Interstate Commerce Commission prescribes that a liquefied compressed gas container shall not be liquid-full at 131°F.
- Refrigerant drums are shipped with a heavy metal screw cap as protection for the valve and safety plug. For this reason it should be replaced after each use.

The plunger of the spring-loaded valve is held against a flat-disk seat by a spring. When the internal pressure rises above the pressure to which the valve is preset, the plunger moves away from the valve seat, permitting the release of the refrigerant. This continues until the internal pressure is reduced to the pressure at which the spring is set. The valve automatically reseats itself.

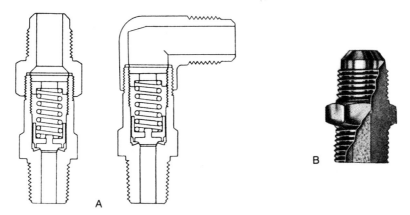

FIG. 14-12   Types of pressure relief valves: (A) spring loaded and (B) fusible plug
(Courtesy of Superior Valve Company, Division of Amcast Industrial Corporation)

Normally, this spring-loaded type of relief valve is set at 200 psi gauge pressure for Refrigerant 12 and 316 psi for Refrigerant 22. Most relief valves have a threaded flare connection. This makes it possible to attach a vent line to carry any discharge gases out-of-doors.

The fusible type of relief valve consists of a connection with a plug that is designed to melt at a desired temperature. Because temperature is related to pressure, refrigerant can be released before a dangerous internal pressure is reached. Relief valves of the fusible type cannot be automatically reset. The plug melts once the rated temperature of the fusible relief valve is reached. When this happens, the entire refrigerant charge is blown and a completely new charge of refrigerant and a new plug are required before the system may be put into operation.

Such a loss of refrigerant in a large system is costly. It is for this reason that the spring-loaded pressure relief valve is most practical and popular on high tonnage equipment.

SERVICE GAUGE AND TESTING MANIFOLDS

A service gauge and testing manifold is useful for: (1) charging and removing refrigerant, (2) checking the pressures, (3) adding oil or liquid driers, (4) bypassing the compressor, and performing many other operations without replacing gauges or trying to operate service connections in inaccessible places, figure 14-13.

A typical service gauge manifold is designed with two gauge openings, three line connections and two shutoff valves that separate the gauge openings from the center line connection. This manifold makes it possible to connect to the system for practically all service and adjustment operations without removing the gauges, figure 14-14.

The manifold valve stems may be operated by a ratchet (service) wrench unless they are equipped with hand wheels. The three-line attachment fittings are usually flared.

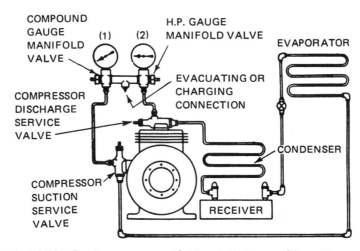

FIG. 14-13 Service gauge manifold installed in a refrigeration system

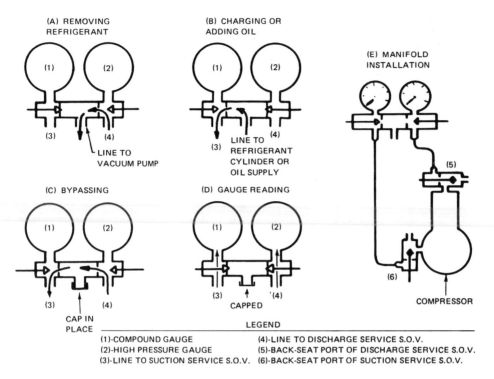

FIG. 14-14 **Testing and servicing operations possible with a service gauge manifold**

The manifold is connected to the suction service valve and the discharge suction valve with flexible lines.

PRINCIPLES RELATING TO THE EVACUATION OF REFRIGERATION SYSTEMS

There should be only two materials in a compression refrigeration system: the refrigerant and the oil which is needed to lubricate the compressor. When the major components of a complete system are assembled and connected, the system may be full of air and water vapor. In addition, there may be some foreign matter such as solder, core sand and fine metal chips.

It is good practice to add strainers to a new installation to remove dirt and other foreign matter that may have entered the system. This is important in all systems. The strainers can be left in permanently or removed after the first few hours of operation when, theoretically, all foreign matter has passed through the system. Two strainers should be used: one ahead of the expansion valve and the other in the suction line ahead of the compressor.

Air is a noncondensable gas at all temperatures and pressures encountered in refrigerating systems. As a gas it raises the normal pressures of the system. It causes excessive head pressures and excessive condensing and discharge temperatures. Air is objectionable, too, because it may combine with moisture, the refrigerant or some oils to form corrosive or other harmful compounds.

Drawing (Pumping) a Vacuum

The best way to remove air and moisture from a refrigerating system is to draw it out with a good vacuum pump, figure 14-15. Such pumps must be able to pull to within a few microns of a perfect vacuum. A *micron* is a unit of measurement that equals 1/25,000 part of one inch (.00004") or, approximately, one millionth of a meter (.0000394" or 0.001 mm). High-grade, two-stage, vacuum pumps that produce such vacuums are available commercially.

Although the system compressor is capable of pulling a vacuum, the vacuum pump instead of the compressor is used to evacuate the system, especially on a new installation. The compressor is not used because the foreign matter and the moisture would damage the valves and contaminate the oil. It is especially important that a vacuum pump instead of the compressor be used to evacuate a system with a hermetic unit.

Pulling a vacuum also lowers the boiling point of the moisture in the system. Under a 29-in. Hg vac. (about one inch Hg absolute pressure or 25,000 microns), water boils at about 77°F. The free water in the system boils into vapor at this low temperature and is carried out of the system with the air.

Pumping a vacuum is a good preliminary leak test. Unless the system is leak-proof, it is impossible to pump a good vacuum.

Purging

Purging is another method of removing air and moisture from a system. Such practice is not the most desirable because it does not remove all of the air and moisture. In purging, the refrigerant gas pushes some of the air ahead of it and out of the system. However, no matter how much the system is purged, some of the air remains.

THE TRANSFER OF REFRIGERANTS AND THEIR ADDITION TO SYSTEMS

Transferring a Refrigerant

Refrigerants are obtainable in cylinders of various sizes and in one-pound throwaway containers. A special opener and an adapter are included with the throwaway contain-

FIG. 14-15 External appearance of a high vacuum evacuation pump *(Courtesy of Robinair Division, SPX Corporation)*

ers. These small refrigerant containers are handy for such applications as water coolers and vending machines which require only limited quantities of refrigerant.

The amount of refrigerant transferred can be determined by placing the service cylinder on a weighing scale during the charging operation. As a precaution, safety goggles must be worn when working on pressurized apparatus. After the correct weight of refrigerant has passed into the service cylinder and the storage valve is closed, heat from the hand may be used to vaporize and force the refrigerant remaining in the line into the service cylinder. As soon as practical, all connections should be plugged or capped to keep air and water out of the line.

SAFETY PRECAUTIONS

- The handling of liquid refrigerant cylinders and containers is dangerous. All safety requirements recommended by the manufacturer are to be observed.
- All industrial safety codes and trade practices must be studied and carefully followed.

When the refrigerant passes into the service cylinder at a reduced pressure it becomes cool. After a period of time, the cylinder absorbs heat from the surrounding air and the refrigerant expands. For safety, the service cylinder must be filled to 80 to 85% to provide this expansion. Unless this space is allowed (and the cylinder temperature is controlled within specified safety requirements), the hydrostatic pressure resulting from the expansion of the liquid can cause the cylinder to rupture.

SAFETY PRECAUTION

- Where possible, a pressure release mechanism should be attached before any heating is attempted. Due to the expansive properties of liquid refrigerants, localized heating is prohibited.

Figure 14-16 shows the main components of one type of refrigerant transfer system for use with cylinders in an inverted or upright position. A slip-on insulating blanket covers the electric heating element which is clamped around the refrigerant supply cylinder. The blanket reduces heat loss due to radiation and sudden changes in ambient temperature.

This system provides several safety features. The heater has an automatic thermostat temperature control with a pilot light which indicates when the heater is operating. In addition, an adjustable pressure switch connected to the valve of the refrigerant supply cylinder, controls the pressure. The pressure limit is set for the type of refrigerant being used.

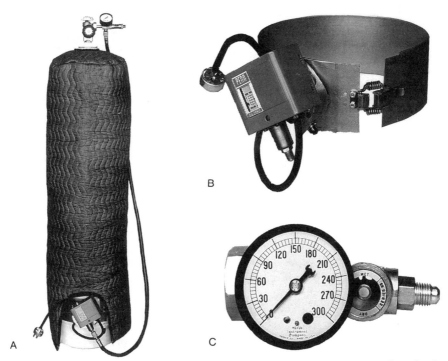

FIG. 14-16 Major components and system in operation on an upright cylinder: (A) completely protected unit, (B) heater and (C) special refrigerant indicator *(Courtesy Airserco Manufacturing Company Inc.)*

Refrigerant Measurement Using an Electronic Sight Glass Indicator

One of the newest instruments developed to indicate when a refrigerant system being charged in full is known as an *electronic sight glass indicator*. This indicator is used when charging home and commercial refrigerators/freezers, window/room packaged air-conditioning units and automotive air-conditioning systems, particularly those capillary systems that have no sight glass. The electronic sight glass indicator is also used for expansion valve systems.

Design Features and Operation of an Electronic Sight Glass Indicator

Some of the features and advantages of using an electronic sight glass indicator in charging a refrigeration system follow.

- Operates on any metal tubing in both expansion valve and CCOT systems without damage to the tubing.
- Operates on SONAR principles and emits an audible signal at full refrigeration system charge.
- Provides feedback for diagnosing refrigeration system problems such as detecting a starved evaporator and maximizing evaporator capacity; checking for refrigerant floodback; and aiding in the adjustment of temperature expansion valves.
- Ease in setting up and four-second response.

Section 4: The Heart of the Refrigeration System

The indicator depends on *ultrasonics*. A matched set of *ultrasonic transducers* are used; one transducer for receiving, the other for transmitting. Echo feedback is monitored and, when the sounds cease, an audible annunciator sounds to indicate that the system being charged is "FULL."

The electronic sight glass indicator is battery operated, portable and works within any system using hard-wall tubing. The indicator clamps onto the high-side liquid line, ahead of the TEV or capillary tube. Thus, there is no need to break a line.

Adding Refrigerant

Adapters are used to attach flexible lines to the compressor service valves. Both of these connections are left loose until the lines have been purged with refrigerant. The center connector on the manifold is for the flexible line from the refrigerant cylinder.

The lines are purged with a small amount of refrigerant. Then, all connectors are tightened. The operation of the system can be checked by the readings on the gauges. If the gauges indicate that a charge is needed, the valves are changed so that the operating compressor pulls refrigerant vapor into the system through the suction service valve. The charging cylinder must be kept upright during this operation.

This low-side method of charging has the advantage of preventing dirt, oil, scale, most of the moisture, and other contaminants from entering the system, since the cylinder is in an upright position. The refrigerant gas that is added is much cleaner and drier than a liquid refrigerant. However, the main reason for charging with vapor on the suction side is to avoid putting liquid refrigerant into the suction side, possibly causing damage to the compressor.

Whether the system has an adequate charge or not is indicated by the gauge readings and/or lack of bubbles in the sight glass. The level of refrigerant on large systems is indicated by the sight glass on the liquid receiver.

Charging a New Installation Through the Low Side

While refrigerant may be added to a new system through the low side as just described, a slightly different arrangement of the manifold connections simplifies and speeds up this operation, figure 14-17. Any new system should be evacuated thoroughly. The center hose from the manifold is connected to the vacuum pump. The compound gauge hose is connected to the compressor service suction valve. After the system is evacuated, the compound gauge valve is closed. The center hose is removed from the vacuum pump and attached to the refrigerant cylinder. Air is then purged from the center hose and the compound gauge valve is opened to allow refrigerant to flow into the system.

Charging a New System Through the High Side

A system can be charged in less time if it is possible to introduce (draw) the refrigerant as a liquid through the high side. When the system is properly evacuated, it may be charged at either one of two points: (1) through the liquid line charging valve where the liquid refrigerant is fed directly into the liquid receiver of the system, and (2) through the discharge service valve of the compressor where the liquid passes through the discharge line into the condenser and on to the liquid receiver. This second method is not recommended as it may cause damage to the compressor or valves.

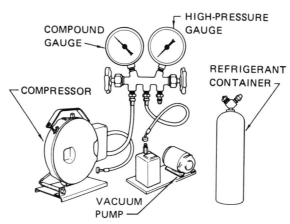

FIG. 14-17   Arrangement for evacuating and charging a system

For liquid charging the refrigerant service cylinder is held upside down with the shutoff valve at the bottom so the liquid flows out of the cylinder into the system. For obvious reasons, a good strainer and a large-size drier should be put in the line through which the refrigerant passes to prevent such contaminants as dirt, water, and air from getting into the refrigeration system. Liquid charging may also be performed through the liquid line.

SAFETY PRECAUTIONS

- Safety goggles and gloves must be worn when charging a system with liquid refrigerant. These will avoid serious personal damage caused by freezing of the skin and tissue with which any evaporating refrigerant may come in contact.
- In the event of an accidental rapid discharge of a large quantity of refrigerant in a room, it should be cleared of people until the vapor has dissipated by ventilation. The presence of an open-flame heater in the room increases the danger as the heater would change the refrigerant vapor into a dangerous toxic gas.
- Precautions for the safe disposal of pound cans of refrigerant or other disposable containers must be carefully read and followed.

Solid-State, Programmable Air-Conditioning Charging Station

The portable equipment illustrated in Figure 14-18 automatically performs two basic steps in servicing a refrigeration (air-conditioning system). The electronic programmable (solid state) charging station is used to (1) provide for complete system evacuation and (2) to meter the quantity of the refrigerant charge for the system. Programs are built in for the most common refrigerants (R-12 and R-22).

FIG. 14-18 Portable solid-state A/C charging station *(Courtesy of Robinair, SPX Corporation)*

The evacuation level and charge by weight are programmed by the air-conditioning technician. This equipment is designed with a microprocessor that is used with a front panel keypad. The vacuum level and the weight of refrigerant igned with a micropro-cessor to be delivered are preset (programmed). The high contrast screen above the keypad provides an easy-to-read digital readout.

The model shown in the illustration uses the popular industry rotary vane design pump for quick and thorough evacuation of the system and for fast charging. In place of the conventional cylinder or weight scales, the unit is designed to calculate a preset charge of refrigerant based on computer input from an electronic flow meter. The refrigerant is dispensed by weight, as programmed. Extra refrigerant may be added in increments of 0.2 pound. The one-pound time charging capacity is 30 pounds.

This type of charging system provides precise measurements of internal pressures in the system and permits easy monitoring of the charging process. An electronic scale (which is not affected by temperature or pressure changes) displays refrigerant charges within an accuracy of 0.01 pound.

The solid-state charging station is designed with solenoid controls. When the system is ready to be charged, the solenoid controls automatically stop the evacuation process. When the charge is complete, the solenoid controls close the refrigerant valve.

TESTS FOR REFRIGERANT LEAKS

All new refrigeration system installations must be carefully tested to make certain that all joints are leakproof. Leak tests are also required of parts that are installed as replacements in old systems. While the methods of testing vary with different refrigerants, they all require a positive pressure of from 150 to 300 psig.

Testing for Methyl Chloride and Isobutane Leaks

The method of finding leaks in old refrigeration systems that used methyl chloride required the use of a liquid. After all the grease and oil was removed from each joint, a solution of very thick soapsuds and water or oil was swabbed around each joint. If bubbles formed, it indicated that the joint leaked. Inaccessible places were inspected, using a flashlight and a small mirror.

In cases where the components were not hooked up to the refrigerating system, tests were made by putting the component under a pressure of 30 to 40 psi, and then submerging in a water bath. Bubbles formed at the leak.

All joints must be checked carefully. Even a minute leak can cause a complete loss of the refrigerant in a short period of time. It is important that flames be kept away from methyl chloride or other hydrocarbon systems because these gases form an explosive mixture in some concentrations.

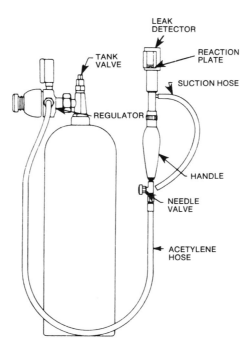

FIG. 14-19 Halide leak detector equipment setup *(Courtesy of L-Tec Welding and Cutting Systems)*

Section 4: The Heart of the Refrigeration System

Testing for Ammonia Leaks

Two common methods may be used to test for leaks in the ammonia system. The first method requires a sulfur candle. The flame of this candle gives a thick white smoke if it comes in contact with leaking ammonia. The second test is with moist pink *litmus paper*. This paper changes color the instant it comes in contact with ammonia. While both of these methods are very simple, rapid, convenient and accurate, the litmus paper method is safer.

Testing for R Refrigerant Leaks

Vapor leaks of noncombustible halide refrigerant gases (such as: "freon," "Ucon" and other R-type gases) used in refrigeration and air-conditioning systems can be detected with a *halide leak detector.* Supersensitive models of halide leak detectors and electronic leak detectors will detect the presence of as small as twenty parts of halide gas per one million parts of air.

The halide leak detector illustrated in figure 14-19 operates on acetylene. The air for combustion is drawn on through a tube from the side of the detector. If there is a leak, its presence is shown by color intensity changes in the outer, upper and lower part of the flame. The color depends on the presence and the concentration of halide gas.

SAFETY PRECAUTIONS

- Before using a halide leak detector, the technician must be knowledgeable about the principles of operation and safe hands-on practices in using air-fuel-gas equipment.
- Manufacturers' literature on Precautions and Safe Practices must be read and carefully followed.

Another similar type of leak detector burns alcohol. A third detector is of the pump-up type. This detector is not satisfactory because if there is any refrigerant vapor in the surrounding air, the flame burns green continuously.

Electronic, Solid-State Halogen Leak Detector

This *electronic halogen leak detector* is used by the service technician for leak checking refrigeration and air-conditioning systems in home, industrial and commercial applications and for automotive air-conditioning units. Electronic leak detectors are designed to detect all halogen gases, including chlorine and fluorine.

Some of the advantages of this instrument over conventional leak detectors follow.

- Capability of sensing leaks of less than one part per million (or an equivalent leak of 1/10 ounce per year), based on R-11
- Short period warm-up time (30 seconds or less)
- Instant (one second) response time

266

- Ease of operation. The instrument is hand held. The gooseneck extension for the *remote probe* permits testing leaks in and around different shapes, sizes and pieces of equipment.

Design Features and Principles of Operation. Electronic halogen leak detectors incorporate the following design features.

- Control unit consisting of *printed-circuit board amplifier, plug-in* sensor and a *power supply* (battery-type) and *speaker*.
- Control chassis on which are mounted a *calibrated reference leak bottle, leak size switch, balance control,* and *power switch;* an *input power cord* and a *sensor heat control.*
- *Probe* and *light-emitting diode (LED) leak signal lamp* that are connected to the control chassis by flexible tubing.

It should be noted that leak detectors, which include a small pump to increase the sample gathering efficiency, are called *pump-style halogen detectors.*

In operating an electronic halogen detector, the probe is held as close as possible to the area being tested. The tip is moved at about 1″ per second. As the probe approaches and reaches a leak, the detector signals the presence of the leak by an increase in the audio pitch and a faster flashing rate of the LED indicator lamp.

A heating element/catalyst decomposes freon into the elemental halogens. Halogens have extreme chemical activity. Concentrations as low as one part per million will produce a minute flow of current in the *probe circuit.* The current flow is amplified in the solid-state circuit to activate the audible and/or visual signals denoting the presence of a halogen gas leak.

WARNING: THE HALOGEN LEAK DETECTOR IS NOT TO BE USED IN A COMBUSTIBLE OR EXPLOSIVE ATMOSPHERE.

SAFETY PRECAUTIONS

- Manufacturers' warnings, health and safety precautions, and recommended safe practices must be observed at all times to prevent personal injury and to maintain the wear life of the sensor.
- There is a gradual decrease in sensitivity when the sensor heat control ages. This condition requires regular checking. Small adjustments might have to be made to correct the sensitivity and overall calibration of the leak detector.

TROUBLESHOOTING REFRIGERATION SYSTEMS

The refrigeration service technician often refers to troubleshooting charts to diagnose causes and possible solutions to system defects. The charts are sometimes prepared by manufacturers to relate to specific equipment and products. The accompanying sample troubleshooting chart identifies fourteen general problem areas, common causes of each problem and suggests remedies.

TROUBLE/PROBLEM	COMMON CAUSE	REMEDY
1. UNIT WILL NOT RUN	No power due to blown fuse or tripped breaker	Replace fuse or reset breaker.
	Defective temperature control (motor control)	Use jumper (short) across terminals of control (or check with ohm meter). If unit continues to run, replace control.
	Defective relay	Check and replace the relay, if required.
	Defective overload	Check and replace, if required.
	Defective compressor	Check the compressor. Replace, if required.
	Defective timer	Check the timer. Replace if required.
2. REFRIGERATION SECTION TOO WARM	Door not closing	Relevel or instruct user to close the door each time.
	Blocked air circulation in the cabinet	Provide for better air circulation.
	Warm or hot foods placed in the cabinet	Cool foods to room temperature before placing them in the cabinet.
	Poor door seal	Level cabinet; adjust door and/or replace seal.
	Interior light stays on	Check light switch. Replace, if defective.
	Refrigerator section airflow set too low	Check airflow. Adjust control knob to cooler position.
		Check to see that damper is opening.
		Replace control if inoperative with the door open.
	Freezer section grille not properly positioned	Reposition the grille.
	Freezer fan not running	Check for ice buildup. Replace fan, fan switch or defective wiring.
3. REFRIGERATION SECTION TOO COLD	Refrigeration section airflow control knob set to the coldest position	Turn the control knob to a warmer position.
	Airflow control defective	Replace the airflow control.
	Broken airflow heater	Replace the airflow heater.

TROUBLE/PROBLEM	COMMON CAUSE	REMEDY
4. FREEZER SECTION AND REFRIGERA-TION SECTION TOO WARM	Fan motor not running	Check fan motor. Replace, if required.
	Cold control set too low or broken	Check and adjust cold control. Replace, if required.
	Finned evaporator blocked with ice	Check defrost heater thermostat or timer. Adjust.
	Low refrigerant charge	Check for leak. Repair, evacuate and recharge system.
	Dirty condenser or obstructed condenser ducts	Clean condenser and condenser ducts.
5. FREEZER SECTION TOO COLD	Cold control incorrectly set	Reset knob to warmer position.
	Cold control capillary not properly clamped to evaporator	Tighten clamp or reposition capillary on the evaporator.
	Broken cold control	Check cold control. Replace, if necessary.
6. UNIT RUNS TOO LONG OR ALL THE TIME	Not enough air circulation around the cabinet; restricted air circulation	Relocate cabinet to remove restrictions which prevent proper clearances around cabinet.
	Poor door seal	Check seal. Make necessary adjustments.
	Improper refrigerant charge	Check for undercharge or overcharge. Evacuate and recharge with correct charge.
	Ambient temperature too warm	Ventilate area as much as possible.
	Cold control defective	Check cold control. Replace control if unit operates all the time.
7. NOISY OPERATION	Tubing vibrating against cabinet or other tubing	Relocate tubing.
	Cabinet not level	Level the cabinet.
	Drip tray vibrating	Reset tray. Place on styrofoam pad if noise continues.
	Fan blades hitting liner or mechanically grounding	Readjust fan or blade assembly.
	Compressor vibrating	Tighten or replace the compressor mounts.
8. COMPRESSOR CYCLES ON OVERLOAD	Defective relay	Replace the relay.
	Defective overload protector	Replace overload protector.
	Low voltage	Check voltage. Underload voltage should be 115 V $\pm$ 10%.
	Defective compressor	Check compressor. Replace, if necessary.

TROUBLE/PROBLEM	COMMON CAUSE	REMEDY
9. STUCK COMPRESSOR MOTOR	Valve or piston broken	Replace the compressor motor.
	Insufficent oil	Add oil to specified level. If unit is still inoperative, replace compressor motor.
	Overheated compressor	Check compressor motor. Replace if necessary.
10. ICE BUILDUP	Defective timer	Check. Replace timer, if necessary.
	Defective defrost heater	Replace the heater.
	Defective thermostat	Replace the thermostat.
11. REFRIGERATOR OR FREEZER RUNS ALL THE TIME; TEMPERATURE NORMAL	Ice buildup on the evaporator	Check door gaskets. Replace, if necessary.
	Faulty thermostat	Replace the thermostat.
	Control bulb on thermostat not in contact with evaporator	Reposition control bulb to be in contact with evaporator surface.
12. FREEZER RUNS ALL THE TIME; TEMPERATURE TOO COLD	Defective thermostat	Check thermostat. Replace, if necessary.
13. EXCESSIVE ICE BUILDUP ON THE EVAPORATOR	Leaky door gasket	Adjust door hinges. Replace cracked, brittle or worn gasket.
14. DOOR ON FREEZER COMPARTMENT FREEZES SHUT	Faulty electric gasket heater	Use alternate gasket heater or install a new one, if required.
	Faulty gasket seal	Inspect and check gasket. Replace if cracked, worn or hardened.

SUMMARY

- Manually operated valves are used in different parts of refrigerating systems. Most of these valves are back-seating and are provided with seal caps to prevent refrigerant from leaking out around the valve stem.
  - Some manual valves are just shutoff valves, such as the liquid line shutoff valve.
  - Other manual valves have both the shutoff feature and service or process ports. Pressure controls may be connected to one of the ports. When there is a second port, a pressure gauge or charging line may be connected to it at the same time.
  - Service valves are installed at the discharge and suction ports of the compressor and on the liquid line near the condenser well.

- The throttling valve dampens fluctuations of a pressure gauge and provides a way to close off the port entirely.
- The pressure relief valve provides a release for excessive pressures that might cause an explosion. These valves are usually either spring-loaded or contain a fusible plug.
- A gauge manifold, including suction pressure and discharge pressure gauges, simplifies the servicing and charging of a refrigeration system.
- Vacuum pumps evacuate systems in preparation for charging them with a refrigerant. Evacuation is the only sure way to eliminate water vapor.
  - The degree of evacuation is expressed in microns. This is an absolute pressure which indicates the residual pressure within a system.
- Refrigerants are furnished in throwaway containers, large storage tanks and small service cyclinders. Refrigerant is transferred from a storage tank or cylinder to the service cylinder, as needed.
  - A graduated charging cylinder is used for accurately measuring the charge, especially in systems using a capillary tube.
- Refrigerant vapor is charged from the low side through the suction service valve on the compressor.
- Liquid refrigerant is charged from the high side through the discharge service valve on the compressor or the liquid line charging valve.
- The electronic sight glass refrigerant charge indicator (placed ahead of the capillary tube or TEV on the high side liquid line) emits an audible signal when a refrigeration system that is being charged is full.
- The portable, automatic electronic charging station performs two major functions in servicing air-conditioning systems:
  - Evacuation of the system
  - Metering a preset quantity of refrigerant charge into the system after automatically sensing the evacuation of the complete system
- Refrigeration systems may be tested for leaks by any one of these methods:
  - Methyl chloride. The bubbling of a solution of soapsuds on a joint indicates a leak.
  - Ammonia. Moistened pink litmus paper turns blue around an ammonia leak.
  - R refrigerants. A halide torch flame turns a green color in the presence of an R refrigerant.
- An extremely sensitive electronic, solid-state halogen leak detector is designed with a heating element/catalyst for decomposing freon into elemental halogens. The presence of the halogens produces a minute current flow in the probe circuit that actuates an audible and/or visual signal.

ASSIGNMENT: UNIT 14 REVIEW AND SELF-TEST

A. THE FUNCTION OF VALVES IN CHARGING REFRIGERATION SYSTEMS

1. Identify the type of valve in Column II that is used at each location given in Column I.

Section 4: The Heart of the Refrigeration System

Column I	Column II
a. At the discharge port of compressor.	(1) Suction service shutoff
b. At the outlet of the receiver.	(2) Discharge service shutoff
c. In the liquid near the outlet of the receiver	(3) Liquid charging
d. At the end of the gauge line connected to the crankcase of the compressor.	(4) Relief
e. In a pipe fitting on top of the receiver.	(5) Liquid shutoff

Provide the terms or conditions that best complete statements 2 to 11.

2. Access to the valve stem of a service valve is obtained by removing the _____ .

3. Before turning the valve stem on a new valve, the _____ .

4. Escape of refrigerant through the outlet, when inserting a pressure gauge or charging line, is prevented by _____ .

5. The connection of a pressure control should be made to the _____ .

6. A throttling valve is used to _____

7. The type of valve that does not need packing is the _____ .

8. The three gauges that are usually connected permanently to a refrigerator compressor are the _____ , _____ , and _____ gauges.

9. Another name for the back-seat port of a service valve is the _____ port.

10. Frequently, service valves are not provided on _____ compressors.

11. Hermetic compressor systems are serviced with _____ valves.

B. SERVICE GAUGE AND TESTING MANIFOLD

Using the basic outline of a manifold as illustrated, prepare simple sketches as follows:

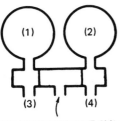

LINE TO REFRIGERANT CYLINDER

LEGEND
(1)-COMPOUND GAUGE
(2)-HIGH PRESSURE GAUGE
(3)-LINE TO SUCTION SERVICE S.O.V.
(4)-LINE TO DISCHARGE SERVICE S.O.V.

1. Insert the valves in their proper positions for operations a, b, c, and d.
    a. Charging or adding oil
    b. Gauge reading for service check
    c. Bypassing
    d. Removing refrigerant
2. Label the gauges and valves.

C. THE EVACUATION OF REFRIGERATION SYSTEMS

Select the letter representing the term or equipment which correctly completes statements 1 to 5.

1. The best method of ridding a system of moisture is by: (a) evacuating, (b) purging.
2. The best component to use for evacuation of a new system is: (a) a vacuum pump, (b) the compressor.
3. The best vacuum pumps are capable of pulling a vacuum of: (a) five microns and less, (b) fifty microns, (c) 20" of Hg.
4. Water vapor can best be removed from a system if the pressure is reduced to: (a) 29" of Hg vacuum, (b) 1" of Hg vacuum, (c) 50 microns of pressure.

D. THE TRANSFER OF RERIGERANTS AND THEIR ADDITION TO SYSTEMS

Determine which statements (1 to 10) are true (T); which are false (F).

1. Throwaway cans provide the most economical quantity in which to purchase refrigerants.
2. Service cylinders are used for convenience when transporting refrigerant to the job for servicing.
3. A throwaway can requires an adapter when connecting to a charging line.
4. The storage cylinder is set in an upright position when transferring refrigerant to the service cylinder.
5. The most common use for throwaway containers of refrigerants is in the servicing of small systems.
6. Purging increases pressure in the line and helps to add air to the line.
7. A system is charged in less time when the refrigerant is in vapor form than when in liquid form.
8. No accurate measurement is needed of the amount of refrigerant that is to be added to systems using a capillary tube as a metering device.
9. As a safety precaution, an unfilled space must be left on the top of a cylinder of refrigerant.
10. A new installation may be charged with refrigerant from either the high- or low-side.
11. Prepare a table, furnishing the required information for both high-side and low-side charging.

	Required Information	Method of Charging	
		High Side	**Low Side**
A	State of refrigerant		
B	Position of cylinder		
C	One advantage of the method		
D	One disadvantage of the method		
E	Point or points at which refrigerant is introduced into system		

12. State two advantages of using an automatic, solid-state electronic air-conditioning charging station.
13. Cite two refrigeration system problems that can be detected or diagnosed with an electronic sight glass indicator.

E.   TESTS FOR REFRIGERANT LEAKS

Add the term, material or conditions to answer statements 1 to 12.

1. a. Which has the more serious effect upon the operation of the mechanical parts of a system: a leak of refrigerant out, or a leak of air into the system?
   b. Why is this true?
2. In what part of the system are inward leaks likely to occur?
3. a. What material is used for testing leaks in an SO_2 system?
b. What is the indication of a leak when this test is made?
4. a. What is the test for a leak in an ammonia system?
b. How is a leak detected by this test?
5. Name the device that is most sensitive to leaks.
6. a. What common device is used for testing systems charged with refrigerants of the R family?
   b. Explain the operation of this device.
   c. How is a leak detected through such a test?
7. a. For safety reasons, a torch should not be used for testing systems charged with what group of refrigerants?
   b. Explain why this is so.
   c. What gas should be used for making high-pressure tests on such a system?
8. a. Name a gas that should not be used for high-pressure testing.
   b. Explain why this is so.
9. What material can be used to absorb ammonia in case of a bad leak?

10. Why is it necessary to observe great care in keeping moisture away from SO₂.
11. a. How and what material can be used to absorb SO₂ when purging it from a system?
    b. What safety precaution must be observed when using this material?
12. What device can be used safely for detecting a leak for any kind of refrigerant?
13. Cite two advantages of an electronic, solid-state halogen leak detector over a conventional leak detector.
14. Identify the upper halogen gas leak sensitivity limit of a solid-state leak detector.

F. TROUBLESHOOTING REFRIGERATION SYSTEMS

Troubleshoot the ten refrigeration system problems given in the table by indicating the corrective action to take in each case.

PROBLEM AREA	CAUSE OF ACTION	CORRECTIVE ACTION
Unit will not run	Defective motor temperature control	
Refrigeration area too warm	Freezer fan not running	
Refrigeration area too cold	Control knob set to coldest position	
Freezer section too cold	Improper clamping of cold control capillary on the evaporator	
Unit runs continuously or too long	Insufficient refrigerant charge	
Compressor cycles on overload	Low voltage	
Stuck compressor motor	Broken valve or piston	
Refrigerator or freezer runs continuously; temperature normal	Themorstat control bulb not in contact with evaporator	
Freezer runs continuously: temperature too cold	Defective thermostat	
Excessive ice buildup on evaporator	Leaky door gasket	

SECTION 5: COMPONENTS OF REFRIGERATION SYSTEMS

UNIT 15:
BASIC ELECTRICITY FOR REFRIGERATION SYSTEMS

Electricity is the major source of energy and problems encountered in operating many refrigeration systems. For this reason designers, craftspersons and technicians in the refrigeration field must have a working knowledge about applications and the nature and safe use of electricity.

The electron theory, electrical charges and static electricity serve in this unit as an introduction to electricity. With this foundation, common electrical terms are defined in a practical way. Then, factors influencing the transmission, distribution and use of electrical energy are covered. Finally, circuit requirements for refrigeration systems are analyzed in order to develop an understanding of where and how electricity is used.

THE NATURE OF ELECTRICITY

The Electron Theory

All studies of electricity and electrical effects are based on the existence of minute *charges* called *electrons*. Electrons may either build up or be moved from place to place. Electric current is produced by the action of these electrons in moving from one point to another.

What electrons are may best be described by examining the composition of an ordinary drop of water. If a single drop were divided into two smaller drops and these smaller drops were further divided, an examination under microscope reveals that each of these drops has the same characteristics and properties of water. Now if this process of dividing is continued until the smallest possible droplet is produced, still having the chemical characteristics of water, this particle is called a *molecule*, figure 15-1.

If it were possible to examine this molecule of water under high magnification, it would be found to be composed of three tiny structures. Each of these tiny structures is called an *atom*. In the water example, one molecule of water is composed of two atoms

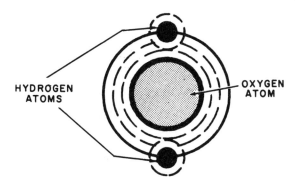

FIG. 15-1   Molecule of water

of hydrogen and one atom of oxygen. Although water is composed of two different kinds of atoms, the molecules of other materials contain many varying combinations of almost 100 different kinds of atoms. These are known as elements.

Continuing with the hydrogen atom (which is the smallest atom in the water molecule), if it were possible to view it under powerful magnification, it would look like the sun encircled by a planet, figure 15-2. The sun would represent the *nucleus;* the planet, the electron. The nucleus contains a positive charge of electricity while the electron has a negative charge. The positive charge is spoken of as a *proton.*

The electron theory states that all matter is made up of electrical charges in various combinations. The sketch shows how the parts of the copper atom are arranged according to the electron theory, figure 15-3.

It has been found that for any atom, the number of positive charges in the nucleus is equal, exactly, to the number of negative charges (electrons) in the orbits that surround it (planetary orbits). In addition to these positive charges, the nucleus also contains *neutrons.* Neutrons are electrically neutral particles thought to consist of an electron

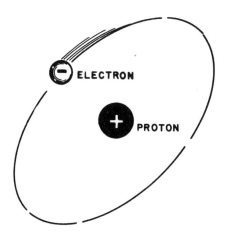

FIG. 15-2   Hydrogen atom

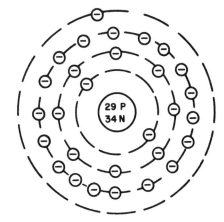

FIG. 15-3   Copper atom

and proton bonded together. Neutrons are represented on illustrations by the letter *N*. The electrons are in a constant state of motion around the protons and neutrons in the nucleus.

An electric current consists of the movement of free electrons from one atom to the next. The *free electrons* are those farthest from the nucleus which may easily be forced from their orbits. These are in contrast to the *bound electrons* which cannot be forced easily from their orbits. Stated again, electric current is the effect of too many, too few, or the movement of, electrons from place to place.

Positive and Negative Charges

A material is said to have a *negative charge* when it has an excess of electrons (⊖ charges). A material with its normal number of positive charges in the nucleus, but with a lack of electrons has a *positive charge*. When there is neither an excess nor a lack of electrons, the body is uncharged or neutral. All of these conditions are caused by moving electrons from one body to another while the positive charges remain stationary as part of the material of the body.

Attraction and Repulsion of Charges

Materials that are charged with static electricity either attract or repel each other, figure 15-4. Attraction takes place between unlike charges because the excess electrons of a negative charge seek out the charge which has a lack of electrons.

Unlike charges (⊕ and ⊖) attract; like charges (⊖ and ⊖ or ⊕ and ⊕) repel each other.

PRODUCING STATIC ELECTRICITY

Electricity may be produced by any one of six sources of energy: contact (friction), heat, pressure, light, magnetism and chemical action. Contact, as the simplest basic source, is treated at this time.

Electricity produced by friction is caused by bringing two or more parts or materials together so they are in contact with each other. During this contact, some electrons move from one material to the other. By increasing the static charges, the supply is also built up.

Materials differ in their ability to build up and transfer static electricity. Depending on which material gives up the charges more easily, the charges may be either positive or negative.

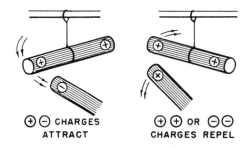

⊕ ⊖ CHARGES
ATTRACT

⊕ ⊕ OR ⊖ ⊖
CHARGES REPEL

FIG. 15-4 Attracting and repelling of charges

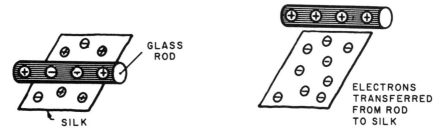

FIG. 15-5   Producing a positive charge

A glass rod, when rubbed with silk, becomes positively charged because the electrons are transferred to the silk, figure 15-5. A hard rubber rod, when rubbed with fur, becomes negatively charged because the electrons are transferred from the fur to the rod.

Similarly, the continuous moving contact between the belt and pulley of an operating compressor can build up a static charge on the compressor. This can cause a serious shock to anyone touching the compressor. This happens because when two charged (electrified) bodies are brought near each other, a potential difference exists between them. It is this force that tends to restore the charged bodies to their neutral or uncharged condition. Such a tendency can result in the movement of electrons from the (⊖) to the (⊕) body to achieve this balance.

Discharging Static Charges

Two parts with opposite static charges may discharge these charges in a number of different ways. The three most common methods include: (1) connecting together, (2) by contact, and (3) by arc (when the force or potential difference between the two charges is great enough), figure 15-6.

If a belt-driven compressor is connected to a *ground*, like a water pipe, the static electricity charges leak off harmlessly. Regardless of the method of *static discharge*, the (⊖) charges flow toward the (⊕) charges until the charges are balanced electrically. Such charges are then said to be *neutralized.*

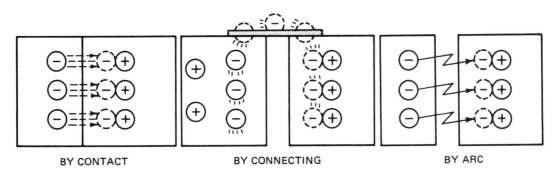

BY CONTACT   BY CONNECTING   BY ARC

FIG. 15-6   Three methods of static discharge

ELECTRIC CURRENT AND CIRCUIT

Although the exact nature of electricity cannot be defined, it is classified as *static* when the electrons are at rest and *dynamic* when they are in motion. The movement of electrons is called *current.* The term *circuit* is used when (1) a voltage source (EMF), (2) interconnecting conductors and (3) a load (usually resistive) together form a closed loop.

Before acceptance of the electron theory as the basis of electrical behavior, it was considered that current flowed from positive to negative, *conventional flow.* It has now been established that current (the flow of electrons) is actually from negative to positive, figure 15-7.

CONDUCTORS, INSULATORS AND SEMICONDUCTORS

There are three main classes of materials used in electrical work: conductors, insulators and semiconductors, figure 15-8. As the name implies, a conductor permits electrons to move through it easily. The ease and degree of speed with which this movement takes place denotes the *conductivity* of the material. Some conductors, listed in the order of their conductivity from good to poor, are: silver, copper, aluminum, tungsten, zinc, brass, iron, pure tin and lead.

On the other hand, an insulator resists the flow of electrons. Examples of insulators include: mica, rubber, Bakelite, paper and silk.

Some of these materials are used to coat or wrap wires to keep electricity from arcing from one conductor to another or to keep two wires from touching (contacting) each other. This is especially important for hermetic units in which the refrigerant flows between the wires of the motor.

A *semiconductor* is a material that conducts better in one direction than the other. Germanium, silicon, and copper oxide are examples of semiconductors. These materials are useful in building the components of thermoelectric refrigerators, such as the rectifier shown in figure 15-9.

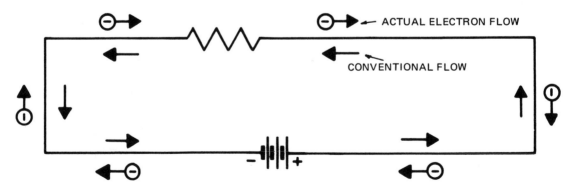

FIG. 15-7   Comparison of conventional and actual electron flow

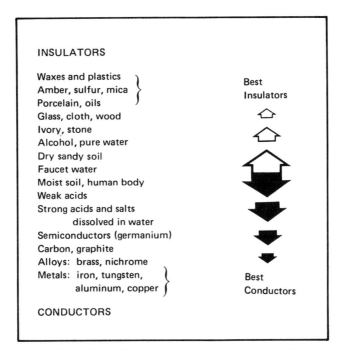

FIG. 15-8   Materials used in electrical work

POTENTIAL DIFFERENCE AND ELECTROMOTIVE FORCE

When there is a difference in the amount of an electrical charge between two points, it is called the *potential difference*. Potential difference is measured in units known as *volts*. There are a number of ways to create a potential difference between two points and, thus, force the electrons to move. This force is known as an *electromotive force* and is abbreviated *emf*. Voltage = potential difference = electromotive force.

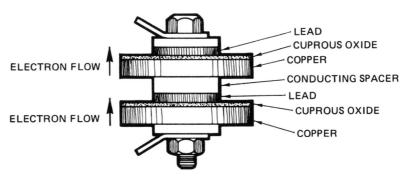

FIG. 15-9   Copper oxide rectifier, two discs in series

Section 5: Components of Refrigeration Systems

Sources of Electromotive Force

The main source for supplying electrical energy for the operation of most refrigeration units is the *generator*. A generator is a device that changes energy from one form to electrical energy (dc). An *alternator* produces alternating current (ac).

The accompanying sketch, figure 15-10, shows that an electromotive force is *induced* whenever a conductor cuts through the magnetic lines of force between two poles of a magnet. When this action takes place, a potential difference is set up between the terminals of the generator. This potential difference may be either direct current (dc) or alternating (ac).

Dc or *direct-current* potential difference means that each terminal always maintains the same identity. One terminal is always + and the other -. The size or value (called *magnitude*) may fluctuate or remain the same. The potential difference maintained between the terminals of a battery is steady. By contrast, the potential difference of a generator fluctuates, depending upon its construction and operation.

Most refrigeration motors operate on alternating-current (ac) potential difference, figure 15-11. Ac is a potential difference that changes both direction and magnitude at regular periods. This means that ac follows a pattern which repeats itself during successive fractional parts of a second.

A complete single pattern is called a *cycle*. When the pattern is repeated 60 times each second, it is called *60-hertz current* and is said to have a *frequency* of 60 hertz (Hz). This is the unit most widely used in the United States of America.

In other countries it is common to find 40-, 50-, or even 25-hertz current. For this reason, it is important that the cycle is specified when ordering electrical equipment. Generally, motors, solenoid valves and transformers which are designed to operate on one frequency cannot be operated successfully on any other frequency.

It will be remembered that thermoelectic refrigerators operate on direct current. Each unit is equipped with a rectifier which is an electrical device that changes ac to dc. Thus, the unit may be plugged into an ac line.

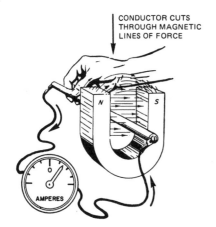

FIG. 15-10   Inducing an electromotive force

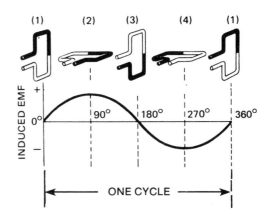

FIG. 15-11   Graph of an ac potential difference

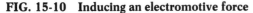

IDENTIFICATION OF ALTERNATING AND DIRECT CURRENT

A circuit may be identified in a practical way by using a neon glo-bulb in a socket with pigtail wires, figure 15-12. When the pigtails are plugged into the two openings of the receptacle, the bulb lights up. If one side only lights, the circuit is dc. When both sides of the neon bulb light, the circuit is ac.

Certain safety precautions must be observed. First, the test lamp must be designed for a voltage higher than the one being tested. Next, care must be taken to touch only the insulated part of the wire.

TRANSMITTING AND DISTRIBUTING ELECTRICITY

Step-Up Transformers for Transmission

The maximum potential difference at which an electric generator delivers energy is usually 15,000 volts. The higher this voltage can be *stepped-up* the more economically it may be transmitted across power lines to a point of operation. A standard generated pressure of 13,200 volts may be stepped up as high as 250,000 volts depending on the distance it must be transmitted and the amount of energy available.

This stepping up is done by *step-up transformers*. The transformers are found near the generating stations and look like a battery of steel tanks. The transformer has three main parts: a *primary winding* into which electrical energy is applied, a *secondary winding* which is tapped as the *output* side, and the laminated iron core around which both the primary and secondary windings are wound.

The transformer transforms electric power from one coil to another by magnetic induction. *Magnetic induction* is the phenomenon produced when a voltage applied to the primary winding sets up a voltage in the secondary winding. The required relative motion between the magnetic field and the conductor is obtained by using ac. A magnetic field surrounds the wire in the primary coil. This field fluctuates with the ac and cuts through the secondary winding.

The amount of voltage depends on the ratio of windings between the primary and secondary windings, figure 15-13. For instance, if the generator delivers electrical

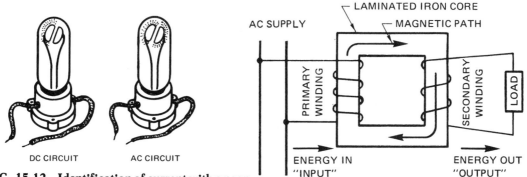

FIG. 15-12 Identification of current with a neon bulb

FIG. 15-13   Ratio of windings determines voltage

energy at 13,200 volts to the primary winding of a transformer, and there are five times as many windings around the laminated iron core at the secondary winding, the output is 13,200 × 5, or 66,000 volts. The ratio of turns in the secondary coil as compared with the number of turns in the primary coil determines the output.

Large transformer cases usually contain a high grade transformer oil. The oil tends to dissipate the heat and keep the efficiency high. It is also a very good insulator.

Electricity is distributed by parallel circuits. For example, in the above diagram, other transformers could be connected to the ac supply. Each would operate independently of the other and would be connected in parallel with each other. Also, additional loads could be connected to or *across* the transformer *in parallel* with the one shown. Again, each would operate independently of the other. Outlets in laboratories, shop or home are in parallel on the same or different circuits. Thus, lights and appliances can be operated individually or in groups.

As each load is operated its individual current is added to the others to form the total current in the secondary winding of the transformer. Therefore, the size of the transformer must be designed to carry the load without excess loss.

Line Losses

The question may be raised as to why it is necessary to transform from low to high voltage in transmission. The answer is simple. Regardless of how efficient the conductors of electricity are, they still offer resistance to the flow of current. The more current, the more heat generated, and the greater the loss in electrical energy. This is called *line loss*. Since power is equal to the potential difference × current, the greater the potential difference, the smaller the current required to transmit the same amount of power. Hence, the lower the line loss.

This loss is sometimes referred to as an $I^2 R$ (I squared R) loss. It shows clearly that reducing the current is even more effective than reducing the resistance to minimize the line loss.

Step-Down Transformers for Distribution

The longer the transmission lines, the more potential difference is required to push the current through the line itself to the place where it is used. The amount of power is, therefore, controlled by stepping up the transmission voltage. The high voltages are dangerous so in order to safely distribute power the high voltage is stepped down to 2300 or 4000 volts for use in thickly settled areas. Transformers are again used at places called substations to step down the voltages.

The common type of transformer usually mounted on a pole or in a transformer vault below surface level, further reduces the voltage to 120-240 volt three-wire service for lighter industrial and domestic use.

Three-Wire System of Distribution

The 2,300-volt current enters a step-down transformer in which the secondary coil contains fewer windings than the primary coil, figure 15-14. With this arrangement, the transformer steps down the voltage from 2,300 volts to 120 and 240 volts as needed.

The secondary coil of this step-down transformer is wired as shown so that there are three wires tapped from it. The middle wire is neutral and using this neutral wire and

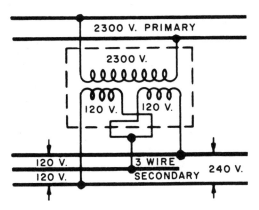

FIG. 15-14 Three-wire system

either one of the two outside wires it is possible to get 120 volts. The neutral wire is grounded at the transformer as well as at the building. When the two outside wires are tapped, the resulting voltage is 240. The combination of these three wires is known as the *three-wire system.* The values indicated are merely nominal values and they vary with the equipment and load conditions.

This three-wire service is brought into a building at a point called the *service entrance.* This service entrance is a switching center where the electric current is first sent through a meter which measures the amount of energy used. From the meter the three wires are brought into a main distribution center to branch off into the number of circuits needed. Built into this distribution center is a protective device which cuts off the service in case of any *short circuit* in electrical equipment.

CIRCUIT PROTECTION

The two basic groups of devices used for the protection of electrical circuits are: (1) fuses, and (2) circuit breakers. Each of these groups may be further classified as,

- ✓ Fuses which come in either cartridge or plug type (ordinary, super-lag, dual-element).
- ✓ Circuit breakers which may be operated by a magnetic element or thermal and magnetic element.

Fuses

Plug fuses are those which screw into sockets much the same as for electric light bulbs, figure 15-15. Most of these fuses are not renewable and must be replaced after blowing.

Cartridge-type fuses are designed to fit special sockets or clips, figure 15-16. They may be obtained in the renewable form for which replacement links are available.

- ✓ The ordinary fuse which is nonrenewable gives good protection against shorts for those circuits which are used only for lighting or heating.
- ✓ Super-lag fuses give protection to general lighting and heating circuits, allowing for temporary overloads.

(A) ORDINARY
PLUG FUSE

(B) DUAL-ELEMENT
PLUG FUSE

(C) DUAL-ELEMENT PLUG
FUSE AND ADAPTER

FIG. 15-15   Types of plug fuses

✓ The dual-element time-delay fuse gives excellent protection against short circuits and also allows for temporary overloads. Therefore, it is especially useful for motors which draw several times their operating current when they start. This type of fuse does blow when the overload extends over a long period such as when a motor encounters difficulty in starting and continues to draw its heavy starting current. Each motor should have its own dual element protection.

SAFETY PRECAUTIONS

The use of a dual-element fuse virtually eliminates fires attributed to *overheated* or *faulty* electric motors. The dual-element fuse carries the heavy current needed for starting but, if a faulty mechanism prevents the motor from dropping back to its normal operating current, the dual-element fuse will blow. Thus, the motor is protected from burning out and possibly starting a fire.

• Some motors are equipped with a *thermal overload protector.* This is essentially a bimetallic (thermostatic) switch. It opens the circuit in the event of a prolonged overload. Some protectors need to be reset manually. Others reset themselves after the motor has cooled down.

• Before changing electrical connections on refrigeration equipment, remove the fuse and throw the circuit breaker or switch controlling the circuit. Cartridge fuses should be removed with a tong-like fuse puller made of fibre or other insulating material. Attach a *DANGER* tag at the switch to tell that someone is working on the circuit and the switch should not be thrown.

• Pliers with rubber handles, screwdrivers with insulated handles, and other tools required for work on circuits should all be insulated as an added safety precaution.

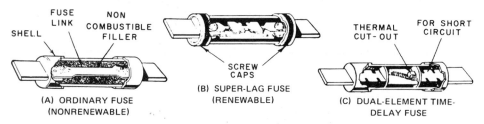

FIG. 15-16   Types of cartridge fuses

Circuit Breakers

Circuit breakers may be used to protect a circuit against overload and, also, as switches. The circuit breaker does not need to be replaced when an overload causes it to trip. After the circuit trouble is located, the circuit is restored to service by merely throwing the breaker switch knob to an on position.

Magnetic breakers provide good protection against overloads. The breaker illustrated, figure 15-17, has a time-delay feature to accommodate temporary overloads like motor starting current. On persistent overloads, an iron core is drawn into the coil increasing its flux and tripping the breaker. On heavy overloads, the time-delay feature is bypassed and the breaker trips instantaneously.

Fuses and circuit breakers are connected in *series* with the circuit or device they are to protect. Thus, if any fault develops in the circuit or operating device and the circuit becomes *overloaded* (more current than it is designed to carry without danger of overheating and causing a fire) the protecting device will blow, stopping the flow of electricity in that circuit. The fault should be found and corrected before inserting a new fuse.

CIRCUIT REQUIREMENTS FOR REFRIGERATION SYSTEMS

Three things that must be clearly understood about circuits for refrigeration systems include:

- Testing a circuit.
- Identifying the elements of an adequate circuit.
- Selecting conductors for an adequate circuit.

MAJOR CONSTRUCTION FEATURES

1. HYDRAULIC-MAGNETIC UNIT (LOAD-SENSING COIL)
2. ON-OFF, TWO-POSITION HANDLE
3. SAFETY, STRONG LATCH MECHANISM
4. TERMINALS
5. BLOWOUT GRID (MINIMIZE ARCING ON CONTACT SURFACES)
6. BALANCED ARMATURE (PREVENTS MECHANICAL TRIPPING)
7. SELF-CLEANING CONTACTS

FIG. 15-17   (Magnetic) circuit breaker with cover removed to show internal details
(Courtesy of Heinemann Electric Co.)

Testing a Circuit

Ac voltage may be checked without metal-to-metal contact. A tracer probe from an ac voltage detector is swept over specific units of the electrical circuit. Good fuses, proper voltage at terminal strips, determining which conductor is power or ground, and high voltage at load-break connectors are typical examples of voltage detector applications.

A direct reading (digital read-out) instrument may also be used for measuring volts-ohms-amperes. In such applications, the probes of the instrument are clamped around one wire of a current in a terminal box.

A combination ammeter-voltmeter instrument is used to measure the current intensity in amperes, figure 15-18. The illustration shows the unit line cord plugged into the adapter. The adapter, in turn, is plugged into the wall receptacle. The jaws of the ammeter-voltmeter are clamped through the adapter.

A test at this point measures the amount of current which the motor draws when starting and after it gets up to speed. It is normal for such an instrument to indicate an amperage of four to six times the value shown on the motor nameplate when the motor is starting. However, the amperage should drop back to the nameplate value after a few seconds as the motor gets up to speed.

The extra starting current is required to overcome the inertia of the motor and also for starting against a high pressure in the discharge side of the compressor. If the motor continues to draw a high amperage and fails to get up to speed, the trouble is either in the motor or in the line.

The potential difference (voltage) of a refrigerating unit may be read with the same combination instrument, figure 15-19. The loop of the voltmeter-ammeter is disengaged and the position of the meter switch is changed. Then, the two instrument leads are inserted into the adapter, as illustrated.

The voltage reading, under operating conditions, must fall within the limits defined in the Code governing electrical installations. The Code usually specifies that the installa-

ADAPTER

FIG. 15-18   Taking an amperage reading

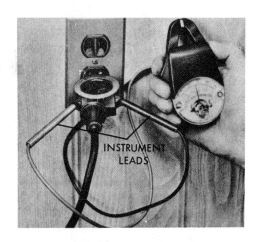

INSTRUMENT LEADS

FIG. 15-19   Setup for a voltage reading at a refrigerating unit

tion shall be adequate enough to provide the necessary voltage under the most severe operating conditions. A motor starting under load provides such a condition. This voltage should be not more than 10% lower than the stated value for the system. A maximum of two to three percent line drop provides for efficient operation and protection of the motor and the system.

Line drop refers to the potential difference required to push the current through the line from the entrance switch to the refrigeration unit itself. There is also a line drop between the entrance switch and the transformer on the pole. Line drop always results in heating the line and waste because the voltage is not made available for operating the appliance.

Line drop is sometimes referred to as *IR Drop* because its magnitude is calculated by the product of I (the current in amps) and R (the resistance of both sides of the line, expressed in ohms). Thus, line drop can be minimized by keeping I as low as possible or by using a line whose conductors are reasonably large to minimize R.

Identifying the Elements of an Adequate Circuit

When the values for the electrical conductivity of different materials are compared, it is evident that although the cross-sectional areas and lengths may be the same, one material is a better conductor than another. For instance, here are a few comparisons.

- ▶ Aluminum causes a line drop which is about 1.6 times as great as that for copper.
- ▶ Iron causes a line drop which is about 5.8 times that for copper.
- ▶ Copper causes a line drop of only 1.05 times that for silver.

Consequently, copper is widely used for electrical wires. Where weight is a problem, such as in long spans of transmission lines, aluminum may be used. Because of its value as a good conductor, silver is sometimes used to plate circuit breaker and relay contacts.

Selecting Conductors for Adequate Circuits

Three factors must be taken into account to insure that an electrical circuit is adequate for a refrigeration system, figure 15-20.

- ✓ The conductor material should provide for the least possible line drop. In most cases, copper is selected.
- ✓ The length of conductor, naturally, is determined by the location of the refrigeration unit with respect to the entrance service box or branch circuit box.
- ✓ The cross-sectional area of the conductor is the third important factor. Tables are available which give safe current limits established by the National Electrical Code for different wire sizes. Such a table appears in the Appendix.

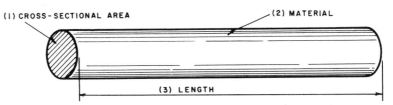

(1) CROSS-SECTIONAL AREA (2) MATERIAL (3) LENGTH

FIG. 15-20 Three factors affecting the selection of a conductor

Wire Gauge Sizes

Wire sizes are given in tables as *gauge* sizes. The standard gauges which are most commonly used in the United States are the American Wire Gage and the Brown and Sharpe (B & S) Gauge, figure 15-21.

In both systems, the same number is used to designate the size of a given wire. One common type of wire gauge that is widely used is shown by the photo. Wire that has been stripped of all insulation is tried in several slots in the gauge until the correct one is reached. The number on the slot which just slips over the wire without being forced, corresponds to the size of the wire. One side of the gauge gives the gauge number. The reverse side shows the equivalent decimal value of the wire diameter.

Wire Cross-Section Area

A system has been devised for designating the cross-section area of round conductors. The unit which identifies such areas is the *circular mil*, abbreviated *C.M.* A circular mil is a unit of area measurement of a round wire with a diameter of one mil. A mil represents 1/1000" (0.001").

A few examples are cited to show how the system works, figure 15-22. A wire of two mils diameter (0.0902") has an area of four C.M. Another wire of 102 mils diameter has an area of 10,400 C.M. In other words, the circular mils values which represent the area are merely the square of the diameter (expressed in mils).

Table 9, Wire Sizes (Appendix), indicates that a round wire of 10,400 C.M. area (Column II) is listed as a #10 wire. Examination of similar wire size tables indicates that for every three wire size numbers, the area (as the wire size numbers get smaller) approximately doubles, or is said to increase by a factor of two. This means that the area of a #7 wire is 20,800 C.M. or double that of a #10 wire. The area of a #13 wire is 5,180 C.M. or approximately half of the 10,400 C.M.

The resistance of 1,000 feet of a #10 copper conductor is 1.00 ohm at 68°F (Column III). For a similar length of #7 copper conductor wire, the resistance is only 0.5 ohms or

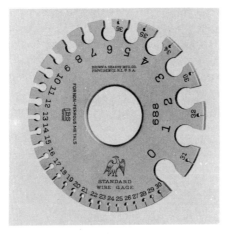

FIG. 15-21 Wire gauge for standard wire sizes *(Courtesy of Brown & Sharpe)*

DIAMETER: I MIL (.001")   DIAMETER: 2 MILS (.002")   DIAMETER: 102 MILS (0.102")
AREA: I C.M.   AREA: 4 C.M.   AREA: 10,400 C.M.

FIG. 15-22   Relationship of diameter to circular mils

half of the #10 value. On the other hand, the resistance of 1,000′ of #13 copper conductor wire is 2.0 ohms. This indicates that the resistance which a conductor offers to the flow of electricity increases as its area decreases.

Different types of insulating materials have different abilities to resist the heat formed as a result of current flowing in the wire. The capacities for copper conductors with two different insulation coverings are indicated in Columns IV and V in the wire size table. The type *RH* insulation stands higher temperatures and, therefore, can carry a heavier current (amperage), as compared with a type *T* insulation.

Sizing of Circuits

The proper size of conductor for a given installation may be conveniently selected by using tables which have been especially prepared. Figure 15-23 is a table of this type. This particular table is designed for copper conductors which may be used in a 115-volt branch circuit where the line drop is not to exceed 2%.

Such tables may be used when the approximate wattage of the unit is shown. Theoretically, one horsepower is equivalent to 746 watts. This means that an electric motor using 746 watts would do mechanical work at the rate of one horsepower. This

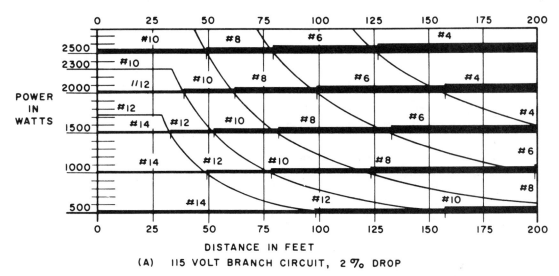

(A)   115 VOLT BRANCH CIRCUIT, 2 % DROP

FIG. 15-23   Typical information furnished in wire size and electrical characteristics tables

would be true only if there were no losses in the motor and it operated at 100% efficiency. Since no motor is 100% efficient, it takes more than 746 watts to develop 1 hp. Normally, about 900 watts is needed.

EXAMPLE:

Select the proper size of a copper conductor to meet these specifications: An installation of a 2 hp., 115-volt refrigeration system involves a branch circuit 75 feet long. The line drop must not exceed 2%.

PROCEDURE:

Step 1: Determine the wattage.　　　　$2 \times 900 = 1800$ watts

Step 2: Follow the 1800-watt line in figure 17-23 until it crosses ft. line.

75′

1800 W ---→●-- -- -- --

Step 3: Read the zone in which these two values intersect.　　1800 watts and 75′ values cross within the #8 zone

The #8 zone indicates that a copper conductor wire of this size meets the specifications.

ANSWER:

#8 copper conductor wire

Complete tables of copper conductor sizes for 115- and 230-volt branch circuits based on 2% line drop are included in the Appendix as Table 10 (A) and (B).

SUMMARY

- All matter is made up of positive (protons) and negative (electrons) electrical charges and the neutron which is a combination of a proton and electron and has no charge.
- Protection against static charges which sometimes build up on operating equipment may be provided by grounding.
- Electricity is the flow of electrons through conductors. Insulators are materials that normally deter the flow of electrons.
  - ✓ An electromotive force (emf) or potential difference (PD) is required to push electrons through a conductor.
  - ✓ Potential difference is measured in volts; current intensity (rate of flow), in amperes. A hook-on type of combination meter measures both amperes and volts.

- Mechanical energy is transformed into electrical energy by electromagnetic induction.
- Direct current is one which always flows in the same direction. It may be steady or fluctuate. An alternating-current cycle is the number of times per second the current regularly changes both direction and magnitude.
- Electricity is transmitted over long distances at stepped-up voltages. The voltage levels are reduced by transformers to safe limits near the place where the electricity is to be used.
- Distribution systems are protected by either fuses or circuit breakers.
- Line drop is the voltage needed to push the operating current through just the line. The voltage available for the operation of the motor is less than the source voltage by an amount equal to the line drop.
    - ✓ Line drop should never exceed 10% of the source voltage. Better operation results if line drop is limited to 2%.
- Wire size is measured with a wire gauge. The cross-sectional area of round wires is specified in circular-mils (C.M.) which is equal to the square of the diameter (in mils).

ASSIGNMENT: UNIT 15 REVIEW AND SELF-TEST

A. THE NATURE OF ELECTRICITY

Select the condition in Column II which describes each term in Column I.

Column I	Column II

Column I
1. An electron
2. Free electrons
3. A negatively charged part
4. A positively charged part
5. An atom

Column II
a. Composed of a heavy nucleus and one or more electrons traveling in an outer orbit.
b. Has a lack of electrons.
c. Electrons near the nucleus which cannot be forced out of their orbits.
d. A negative charge in an orbit around a nucleus.
e. Has an excess of electrons
f. Electrons that are easily forced out of their orbits.

Supply the correct terms or values to complete statements 6 to 12.

6. Electrical charges may be transferred by (a)_____, (b)_____, and (c)_____ .
7. According to the electron theory, current flows from _____ to _____ .
8. An uncharged body has neither _____ nor _____ electrons.
9. Unlike charges attract because the excess electrons in the negatively charged body seek out a positive charge which has a _____ of electrons.
10. Like charges _____ each other.
11. A dangerous static charge may be built up on a _____ compressor.
12. Electricity may be produced by _____ .
13. Give the scientific reason why some materials are good conductors.

14. a. Name three materials that are used by power companies as insulators.
   b. Explain briefly why these materials are good insulators.
15. A moving belt is producing considerable quantities of static eletricity in an industrial plant, making a safety hazard. Describe a simple device to remove the static electricity.

B. ELECTRIC CURRENT

Indicate which statements (1 to 10) are true (T); which are false (F).

1. Current electricity is the flow of electrons.
2. The intensity of current is measured in volts.
3. Electromagnetic induction is the process of changing electrical energy into mechanical energy.
4. A generator is used to change mechanical energy to electrical energy.
5. A potential difference is required to push a current through a conductor.
6. A conductor is a material that contains an abundance of free positive charges that may be moved by a potential difference.
7. A transformer changes electrical energy at one level to electrical energy at another level of potential difference.
8. A direct current flows first in one direction and then in the other.
9. A current is induced in a wire as it rests against the pole of a magnet.
10. Electricity is transmitted at high voltages in order to reduce line losses.

Add the correct term, phrase or value to complete statements 11 to 25.

11. Direct current is the kind in which the direction of the current _____ .
12. Alternating current is the kind in which the current _____ .
13. In order to induce an electromotive force there must be a magnetic field, a conductor and _____ between the two.
14. Frequency refers to the number of _____ per _____ .
15. The pattern of current flow in an ac circuit is from _____ to _____ .
16. A generator is used to convert _____ energy into _____ energy.
17. A material commonly used for insulation on a wire is _____ or _____ .
18. In the operation of the transformer it is the _____ that is moving.
19. Current intensity is measured in units called _____ .
20. Materials which allow an electric current to flow easily through them are called _____ .
21. The two classes of devices used to protect an electric circuit against overload are the _____ and the _____ .
22. Three of the best electrical conducting materials are _____ , _____ , and _____ .
23. The circuit protector that is more sensitive than a fuse is the _____ .
24. In order to provide a time lag in a circuit protector it is necessary for it to have _____ .
25. Transformers are generally used to change the _____ level.

Select the word or phrase which correctly completes statements 26 to 35.

26. A transformer changes electrical power by: (a) conduction, (b) induction, (c) neither conduction nor induction.
27. When there are more turns on the secondary of a transformer than on its primary, the transformer is a: (a) step-up, (b) step-down, (c) neither step-up nor a step-down transformer.
28. Direct current intensity: (a) never changes, (b) may sometimes change.
29. Direct current direction: (a) never changes, (b) may sometimes change.
30. Alternating current intensity changes: (a) sometimes, (b) never, (c) periodically.
31. Alternating current is: (a) always, (b) never, (c) sometimes 60 hertz.
32. Alternating current frequency is the number of: (a) cycles per second, (b) cycles per minute, (c) changes of direction per second.
33. Circuit breakers are: (a) always single element, (b) always double element, (c) may be either single or double element.
34. The electrons that flow in a circuit are: (a) created in the generator, (b) already in the wires, (c) are developed in the transformers.
35. High voltage electricity is stepped down before entering the house in order to reduce: (a) line losses, (b) danger, (c) heating of wires.

C. CIRCUIT REQUIREMENTS FOR REFRIGERATION SYSTEMS

1. Name the combination meter used for measuring electrical quantities.
2. What two electrical quantities does the combination meter measure?
3. Describe briefly how the voltmeter-ammeter is used to measure (a) amperes, and (b) volts.
4. Describe how line drop affects the voltage at the motor.
5. Identify the place where the voltage should be measured to check the operation of a motor.
6. Under what conditions should the motor voltage be measured in order to give the best check on the circuit? Why?
7. What measurements need to be taken and when in order to determine whether a line drop is excessive?
8. How is percentage line drop calculated?
9. Record the values and/or materials required to complete the accompanying table. Refer to Table 9 in the Appendix.

	Material	Gauge Size	Area in C.M.	Resistance per 1000'	Maximum Current Type RH
A	Copper	12			
B	Aluminum	10			
C			168,000	0.06	
D		14		4.14	
E			41,700	0.248	

10. Compute or select and record the missing values in the table. Refer to Table 10 (A) and Table 10 (B) in the Appendix.

	Hp.	Volts	Watts	Length of Run	Wire Size of Copper Conductors
A	1	115		100′	
B	3	230		150′	
C		115		100′	#6
D	5	230		100′	
E		230	2700		#12

UNIT 16:
ELECTRICAL COMPONENTS OF REFRIGERATION SYSTEMS

When a potential difference is applied to an electrical device, the electrons in that device move. If a direct current potential difference is applied, the electron motion (drift) is continuously in the same direction. When an alternating current is applied, the motion of the electrons is back and forth. In either case, if the potential difference is increased, the electrons move faster.

However, the amount of current for a given potential difference depends on two things: (1) the kind of device, and (2) whether the current is ac or dc.

BEHAVIOR OF DIFFERENT DEVICES IN REFRIGERATION SYSTEMS

Electrical devices usually fall into three classifications, based on their relative behavior on alternating current and direct current, figure 16-1.

- Resistors. These are devices designed primarily to produce heat; like an electric range unit or defrosting heaters.
- Inductors. Such devices make use of the magnetic effect that goes with a current. Some examples of inductors include: solenoids, relays, transformers and induction motors.
- Capacitors. Devices such as running and starting capacitors are designed for or function as a storage for electrical charges.

It is important to understand the behavior of these devices in order to know:

✓ The requirements for the successful operation of solenoids and motors.
✓ The use of running capacitors with motors to reduce line drop and to increase the number of refrigeration or air-conditioning units that may be installed on a given circuit.
✓ The application of resistive devices in refrigeration. Since Ohm's Law is limited in its usage to resistive devices (on ac or dc) and a condition of resonance for inductive or capacitive devices, and because calculations are rarely performed in refrigeration, it will not be covered in this text.

A brief consideration of their construction and operation is in order. This is followed by a comparison of their behavior on ac and dc.

FIG. 16-1 Symbols for electrical devices

Section 5: Components of Refrigeration Systems

Resistors

Electric defrosting heaters are effective in producing heat because the material through which the electric current flows is made of high-resistance wire. Resistance is the opposition to current flow provided by the length, cross-sectional area, and the material of which the conductor is made. Iron or nichrome wires are commonly used.

Electrical energy is changed into heat energy as the electrons are forced by the potential difference to move through the resistive material. This requires work and, in the process, power is consumed.

Solenoids and Relays

Inductors, as a group of devices, derive such a name from the fact that in them electrical energy is changed into a magnetic force. The magnetic force, in turn, is used to cause mechanical motion or produce electrical energy. Magnetic energy is transformed back into electrical energy by the action of a transformer.

The first of these two effects is used in the operation of refrigeration solenoids and relays. The *solenoid*, figure 16-2, opens or closes a valve. A *relay*, figure 16-3, closes or opens a set of electrical contacts.

When the current passes through the coil, it produces a strong magnetic field. This field attracts (pulls) the plunger. The pull moves the plunger up, opening a valve. In the valve as illustrated, it is closed by the weight of the plunger and stem. Some models use a spring for closing.

A similar action is required for the operation of certain types of relays. Figure 16-3 shows an ac relay with the pivoting armature removed. Thus, the shading coil (which is

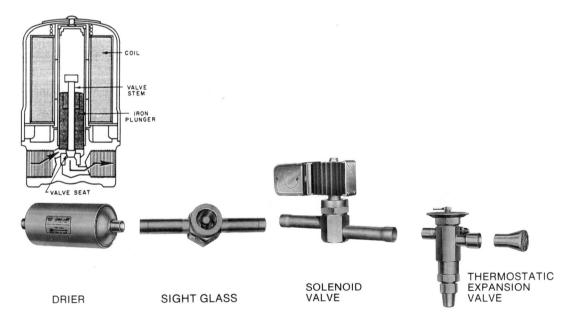

FIG. 16-2 Placement of a solenoid valve: (A) drier, (B) sight glass, (C) solenoid valve and (D) thermostatic expansion valve

298

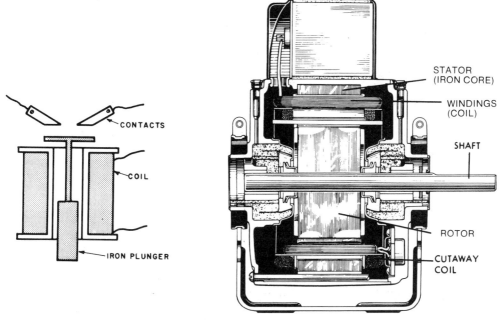

FIG. 16-3 Ac relay   **FIG. 16-4 Cutaway view of an Ac induction electric motor**

merely an oval-shaped copper ring) is exposed. The shading coil is recessed and is flush with the end of the round iron core. The shading coil acts to maintain the magnetic field between the alternations. This prevents chattering which would otherwise occur with alternating current.

The relay is designed to draw down the soft iron armature and close the contacts when the coil is energized by current flow.

Electric Motors

The induction electric motor is also an inductive device that changes electrical energy into a magnetic force. This magnetic force can be used to produce mechanical motion and to do work. In this case the motion is rotational. The cutaway view in figure 16-4 shows the coils and core which concentrate the magnetic field as the current flows in the coil. Since the principles, construction and uses of electric motors are so important to refrigeration systems, these are covered in detail in a succeeding unit.

Principles Affecting Inductive Devices

To further describe the principles underlying inductive devices, consider a wire through which a current is passed, figure 16-5. The current builds up an electromagnetic field around conductor A. In building up or expanding the electromagnetic field, the flux lines cut the conductor B and a voltage is induced in it. As soon as the electromagnetic field, caused by the current through conductor A, builds up to its maximum, it no longer changes. Because there is no change, no flux lines are cut by conductor B and no voltage is induced.

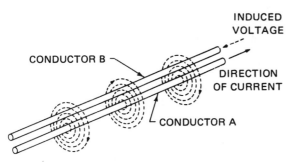

FIG. 16-5 **Fluctuating magnetic field around a conductor induces voltage**

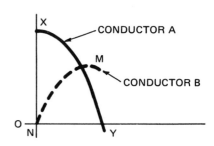

FIG. 16-6 **Inducing a current in a second conductor**

To illustrate this point, consider what happens when the current in conductor A stops. Under this condition the electromagnetic field collapses. In so doing, it again cuts through conductor B and again induces a voltage in it. Note that as long as the magnetic field changes—either expanding or collapsing—a voltage is induced in conductor B

What takes place when a current is induced by a fluctuating magnetic field is shown graphically in figures 16-6 through 16-8.

In figure 16-6, point X on the graph represents the maximum steady current in conductor A. When current in this conductor stops, it takes a short time for it to decrease to zero as shown by line XY. This condition causes the induced current as shown by dotted lines to increase (build up) from a zero value at N to a higher voltage as represented at M.

In figure 16-7, the curve Y to Z represents the negative half of an ac cycle flowing in conductor A. The corresponding changes in conductor B are shown by the dotted line from M to O. Successive changes follow along in sequence.

The peaks of the induced current curve (dotted pattern) occur after (lag behind) those of the original curve. Thus, inductive devices are said to produce a current which lags behind the voltage causing (inducing) it.

Figure 16-8 shows the pattern followed by the original current (solid black lines) and the induced current (dotted lines) for one complete cycle and the continuing production of induced current.

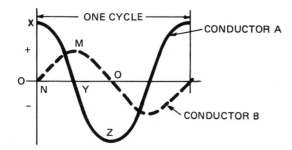

FIG. 16-7 **Induced current curve for one cycle**

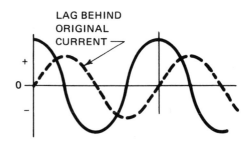

FIG. 16-8 **Pattern of original and induced current for an inductive device**

Capacitors

Construction.   Capacitors derive their name from the fact that they have a capacity for storing electrical charges (electrons). A capacitor is usually constructed of two conductor plates separated by a thin insulating material called the *dielectric*. The dielectric may be air, oil, waxed paper or mica. The conductors may be metallic plates, thin sheets of metal foil, or any other highly conductive material.

Dc Applied to a Capacitor.   When the terminals of a battery are connected to the plates of a capacitor, a flow of electrons takes place for a short time, Figure 16-9. The electrons are thus transferred through the battery from one plate to the other. This action stops when the potential difference of the battery is matched by the repelling force of the piled-up electrons. Consequently, a current cannot continue to flow in a circuit containing a capacitor because the insulation constitutes an *open circuit*.

Ac Applied to a Capacitor.   However, when a capacitor is connected to an ac circuit in which a light bulb is wired in series with the capacitor, the light glows, figure 16-10. This indicates that a flow of electrons is actually occurring in the line. This is possible because the changing potential of the alternating current merely shifts the electrons from one plate of the capacitor to the other and back again. Such movement takes place once each cycle.

It is easy to understand that as the ac potential difference starts to increase, the electrons flow easily (rapidly) into the empty capacitor, figure 16-11. The flow of electrons for even a small potential difference is great. As the change builds up, the electrons increase their repulsion of each other. The intensity of flow slows down even though the potential difference builds up as shown in (A).

As the potential difference reaches its maximum point and starts to decrease, the potential of the electrons is now greater than that of the generator. The electrons now start back toward the other plate. Since they are flowing in the opposite direction, the graph of the flow is represented as at (B) by the dotted line below the axis since it has a negative value. This negative flow increases and reaches a maximum (point c in drawing C) at the time when the potential difference of the circuit has a zero value. As the potential difference becomes negative it opposes and slows down the flow of electrons.

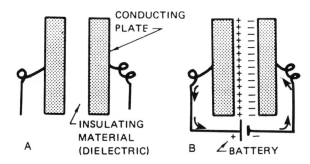

FIG. 16-9   (A) Construction and (B) charging (with DC current) of a capacitor

FIG. 16-10   Ac generator connected to a capacitor

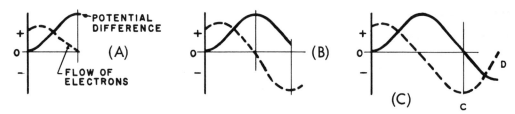

FIG. 16-11   Relationship of electron flow to a potential difference increase

A continuous graph is illustrated in figure 16-12. The action or motion takes place toward the right. The peaks of the I curve then appear to occur ahead of the peaks for the E curve. This condition gives rise to the expression that the current in a capacitive circuit *leads* the applied voltage. In refrigeration this effect has no value in itself. However, when this effect is used in combination with the one produced by a coil, it can serve a useful purpose.

The sectional view in figure 16-13 shows a starting capacitor mounted on top of a motor. Starting capacitors are intended only for intermittent use in starting a motor. They are not designed as running capacitors which need to be of much sturdier construction.

The starting and running capacitors need to match the characteristics of the motors in which they will be used to achieve the desired shift in phase between current and voltage.

ELECTRIC POWER

Electric power is defined simply as the rate at which electrical energy is used. The behavior of different devices on ac is reflected in their effect upon the power furnished by the power company. This effect is compared to the useful work performed by that power through the device. There are three kinds of power: true, apparent and reactive.

True Power

True power is measured in watts as indicated by a wattmeter, figure 16-14. This is a kind of power that does useful work on the job. For instance, true power provides the

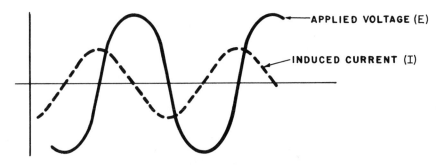

FIG. 16-12   Induced current in a capacitive circuit leads the applied voltage

STARTING CAPACITOR

FIG. 16-13   Sectional view of a 1/2 hp capacitor motor

FIG. 16-14   Meter for measuring true power *(Courtesy of General Electric Company)*

source for heating, defrosting, operating a solenoid, or the electric motor that runs a compressor. Electric meters record the amount of electrical energy used by the consumer.

Apparent Power

Apparent power is measured in volt-amperes and is calculated by multiplying volts and amperes together. There is no instrument for measuring apparent power.

Reactive Power

Reactive power is measured in *vars* as indicated on a *varometer,* figure 16-15.

Reactive power is used up in causing current to flow in a capacitor or inductor. Reactive power causes more current flow than that which does useful work. Therefore, excessive reactive power causes excessive line-drop and increases the operating costs. Reduction or elimination of this reactive power,

1. Reduces the indicated excesses, and
2. Makes it possible to increase the number of refrigeration or air-conditioning units which may operate on a given transformer installation.

There are a number of ways of expressing the relationships between the different kinds of power. The method most commonly used is that of dividing the true power by the apparent power.

Power Factor

If true power is divided by apparent power, the resulting number is called the *power factor.* A meter is available for measuring power factor (PF), figure 16-16. The importance of power factor and its measurement may best be illustrated by using data from

FIG. 16-15 Meter for measuring reactive power
(Courtesy of General Electric Company)

FIG. 16-16 Meter for measuring power factor
(Courtesy of General Electric Company)

three actual tests. In these tests a voltmeter, ammeter, wattmeter and power factor meter are used on a solenoid, a capacitor, and a refrigeration motor.

The purpose of the tests was to see how the behavior of each device affects true power calculations. Thus, the table which follows for each device indicates meter readings of current intensity (amps), watts and power factor, using a 120-volt ac source.

TEST #1 INDUCTIVE DEVICES ON AC

The following values were obtained in a test in which a solenoid was used on ac.

Volts	Amps	Apparent Power (Watts)	True Power (Watts)	Power Factor
120	2.0	240	192	0.8 lag

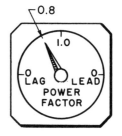

FIG. 16-17

Note that in this test the power-factor meter registers on the *lag side,* figure 16-17. This confirms the fact that when an inductive device is used on ac the current lags behind the voltage.

Secondly, the product of the 120 volts and the 2.0 amps is 240 watts. This apparent power is considerably greater than the true power of 192 watts. However, if the apparent power of 240 watts is multiplied by the power factor of 0.8, the product of 192 watts corresponds to the true power. It is the 192 watts that is converted into useful mechanical power.

$$\text{True Power} = \text{Apparent Power} \times \text{Power Factor}$$
$$\text{Watts} \quad\ = \text{Volt-Amps} \times \text{Power Factor}$$

Thus, with inductive devices like a solenoid, the product of the readings of the voltmeter and ammeter do not give the true power. This means that to determine the true power it is necessary to know the power factor produced by the particular solenoid or motor being used.

TEST #2 CAPACITOR ON AC

The following information was obtained by using a 30-microfarad (unit of capacity) running capacitor of the kind used for refrigeration motors. 120-volt, 60-hertz ac was used.

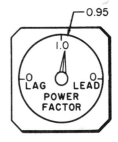

Volts	Amps	Apparent Power (Watts)	True Power (Watts)	Power Factor
120	1.25	150	142.5	0.95 lead

FIG. 16-18

This time the power-factor meter swung over to 0.95 *lead*, figure 16-18. Here, too, the power-factor reading confirms the fact that in a capacitor the current leads the voltage.

The most important thing to notice, however, is that the effect of a capacitor is just opposite from that of an inductive device. As before, the volt-amps is much greater than the true power. If the apparent power (150 watts) is multiplied by the power factor (0.95), the 142.5 watts corresponds to the true power.

TEST #3 INDUCTOR AND CAPACITOR TOGETHER ON AC

A small induction refrigeration motor was tested on 60-hertz ac (1) by itself and (2) with a 40-microfarad running capacitor connected in series with its running winding. For results see table on next page.

When the motor was operated by itself in the first part of the test, it had a lagging power factor of 0.525. This showed the typical effect of an inductive device. However, when the running capacitor was cut in, several important things happened.

	Condition	Volts	Amps	Apparent Power (Watts)	True Power (Watts)	Power Factor
#1	Motor Alone	120	4.0	480	300	0.525 lag
#2	Motor and Running Capacitor	120	2.5	300	300	1.0

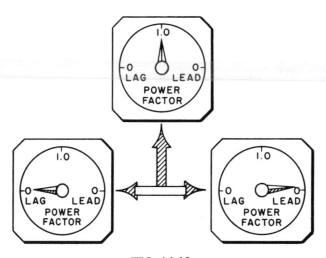

FIG. 16-19

✓ The current drawn by the motor dropped a great deal. This drop in current, resulting from the corrected power factor, is much less than before correction. This means two things: (1) less line drop and (2) more units can now operate from a given transformer installation.

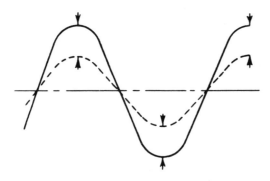

FIG. 16-20   Current in phase with voltage

✓ The power factor needle swung up to 1.0 (called *unity*). The corrected power factor resulted from balancing the value of the running capacitor with the induction of the motor. Thus, their opposing power factor effects cancelled each other out and left a 1.0 (unity) power factor, figure 16-19. This means that the current neither leads nor lags the voltage and is said to be *in phase* (in step with the voltage), figure 16-20.

✓ The true power is now numerically equal to watts. The fact that the apparent power and the true power are now numerically equal means that the reactive power has been reduced to zero by the corrected power factor.

SAFETY PRECAUTION

The use of capacitors on motors (which are inductors) can produce a condition known as *resonance*. Under resonance, the separate voltages across the capacitor and coil of the motor can actually be higher than the line voltage, causing an especially dangerous situation.

SUMMARY

- Electrical devices that are designed primarily to produce heat offer only plain resistance to the flow of the current and are known as resistors. Electric heating elements which are used for built-in defrosting of evaporator units are of this type.

- Devices that are designed to convert electrical energy to magnetic and then to mechanical energy are called inductive. Refrigeration components of this type include: solenoids, relays, transformers and induction motors. Coil-type devices offer an inductive reactance to the flow of ac in addition to plain resistance.

- Devices that function primarily to store electrical charges are called capacitive. Among these devices are the starting and running capacitors used with refrigeration motors.

- An additional opposition to the flow of ac beyond its regular resistance is called reactance. Capacitors offer a capacitive reactance to the flow of ac.

- Total opposition to the flow of ac is called *impedance*. It consists of a combination of the resistance and reactance.

- Reactance in coils is the result of an induced electromotive force. An inductive device causes the current to lag the voltage. A capacitive reactance causes the current to lead the applied voltage.

- Electric power that is converted into useful heat or mechanical power is called true power. It is measured in watts by a wattmeter.

- Apparent power is a value obtained by multiplying volts by amps. The units for apparent power are watts. Apparent power is used to calculate power factor.

- Power factor is the number by which the apparent power is multiplied to calculate true power in watts. The numerical value of the power factor is determined by dividing the watts used in the operation of a device by the volt-amperes taken at the same time.

✓ Power factor may be leading or lagging depending upon whether the device is capacitive or inductive, respectively.

- When a motor is equipped with a running capacitor of the correct value, the resulting power factor is unity (1.0). This makes it possible for the motor to do the same work with less current.

ASSIGNMENT: UNIT 16 REVIEW AND SELF-TEST

A. BEHAVIOR OF DIFFERENT DEVICES IN REFRIGERATION SYSTEMS

Add the item or condition which correctly completes statements 1 to 10.

1. A resistor is a device for converting electrical energy into _____ energy.
2. An inductor is a device for converting electrical energy into _____ energy and then into _____ energy.
3. A resistive device is used in refrigeration systems for _____ .
4. Three inductive devices used in refrigeration systems are: _____ , _____ , and _____ .
5. A capacitor is a device for _____ .
6. Two types of capacitors used in refrigeration systems are _____ and _____ capacitors.
7. Resistance is the opposition to the flow of current because of the _____ , _____ and _____ of the conductor.
8. In addition to resistance, an inductive device offers an opposition to the flow of alternating current. This is called _____ .
9. The total opposition to the flow of alternating current is called _____ .
10. A capacitor will not permit _____ to flow through it because of its _____ .

Select the letter representing the term or value which correctly completes statements 11 to 20.

11. A solenoid offers: (a) more, (b) less, (c) the same amount of opposition to the flow of dc as it does to ac.
12. A running capacitor offers: (a) more, (b) less, (c) the same opposition to the flow of dc as it does to ac.
13. A defrosting resistor offers: (a) more, (b) less, (c) the same opposition to the flow of ac as it does to dc.
14. The opposition to the flow of ac by a relay is called: (a) resistance, (b) reactance, (c) inpedance.
15. The impedance of a transformer is: (a) greater than, (b) less then, (c) the same as its resistance.
16. The impedance of a solenoid is: (a) greater than, (b) less than, (c) the same as its resistance.

17. A running capacitor causes an ac to: (a) lead, (b) lag, (c) neither lead nor lag the voltage.
18. An induction motor causes ac to: (a) lead, (b) lag, (c) neither lead nor lag the voltage.
19. A heating coil used for defrosting causes an ac to: (a) lead, (b) lag, (c) neither lead nor lag the voltage.
20. The reactance of a coil is caused by the same effect as that produced in a: (a) transformer, (b) resistor, (c) capacitor.

B. ELECTRICAL POWER AND POWER FACTOR

Determine which statement (1 to 10) are true (T); which are false (F).

1. When a solenoid is used on ac, the number of watts consumed is greater than the volt-amperes.
2. When a running capacitor operates on ac, it causes a lagging power factor.
3. When an induction motor operates on ac, the numerical value of the volt-amperes is greater than the watts consumed.
4. Volt-amperes are units of reactive power.
5. Reactive power indicates the useful work a device can do when used on ac.
6. The varmeter is the only instrument that will measure apparent power.
7. The wattmeter is the only instrument that will measure true power.
8. A motor operated on ac with the running capacitor in series with its running winding can have a true power value that is the same as the value of the apparent power.
9. When a solenoid is used on ac, the power factor is always less than one.
10. The scale on a varmeter starts at zero and shows increasing numbers the same as a wattmeter.

Add the correct term or value to complete statements 11 to 20.

11. The product of the volts and amperes is equal to _____ power.
12. True power is calculated as the product of _____ , _____ , and _____ .
13. The power that is an indication of the useful mechanical power being used is called _____ power.
14. When an induction motor operates with a running capacitor, the value of the apparent power is _____ its value when the motor operates alone.
15. When a running capacitor is added to a motor, the value of the watts consumed _____ .
16. When a motor operates with the running capacitor, the power factor meter shows _____ power factor.
17. The value of the power factor can be calculated by _____ by _____ .
18. The current flowing through the lines when the motor operates alone is _____ that when operating with the running capacitor.
19. Correction of the power factor causes a _____ in line drop.

20. Correction of the power factor _____ the number of units that can be operated on a given transformer.
21. State briefly the advantage of unity power factor to the consumer who has refrigeration equipment.
22. Explain why the correction of the power factor of existing equipment will not decrease a consumer's power.
23. Compute the missing values and name the refrigeration components to complete the accompanying table.

	Component	Voltmeter (Volts)	Ammeter (Amperes)	Apparent Power (Watts)	Power Factor	True Power (Watts)
A	Induction Motor	240		2400	0.25 lagging	
B		120	1.0		0.01 leading	
C	Solenoid	120	0.5			15
D			2.0	240	1.0	
E	Relay	120		15	0.2 lagging	

24. Tell why unity power is of interest to a power company.
25. Describe what advantage unity power factor is to the manufacturer.

UNIT 17:
ELECTRIC MOTORS FOR REFRIGERATION SYSTEMS

An electric motor is a device which transforms electrical energy (power) into the mechanical energy (power) needed to operate a refrigeration compressor. This transforming of energy is accomplished by using the forces of attraction and repulsion between magnetic poles for the operation of dc and ac motors.

DIRECT CURRENT

Direct current motors are not used very often except in places where only dc power is available or where a large installation uses exhaust steam to generate its own dc power. A dc motor consists of two sets of coil windings. The stationary coil is called the *field coil*. The rotating coil, mounted on the shaft, is called the *rotor* (ac motor) or *armature* (dc motor). Direct current flows through both the stationary and the rotating coils. The attraction and/or repulsion between the magnetic fields of these two coils causes the shaft to rotate.

Mounted on the shaft is a copper ring that is cut apart at regular intervals into small segments. Collectively, these segments are known as a *commutator*. Each segment is one terminal of an armature coil. Contact between the external circuit and the armature coils (commutator segments) is made by carbon brushes. The brushes must be properly placed to get the right interaction between the magnetic poles of the field coil and those of the rotor. Current must flow in the rotor to have the motor operate. This means the brushes must be in contact with the commutator at all times. The direction the motor rotates may be reversed by merely reversing the dc connections to the brushes.

The resistance of the armature of a dc motor is very small. Therefore, if a dc motor is connected to the line and is switched directly to its rated voltage, too much current flows to the armature. Under this condition, the current flow through the armature may reach about 25 times the full load current.

Starting Devices

The starting current may be limited to approximately 1 1/2 times the full-load current by connecting a resistance in series with the armature. A generated or counter-emf is produced when the armature starts to rotate in the magnetic field. This counter-emf gets greater as the motor speed increases. This brings about a current decrease in the armature.

The starting resistance may then be cut out gradually, as shown in the circuit diagram figure 17-1. This starter device has a crank arm that is moved slowly. As the contact is made the starter gradually cuts the starting resistance out of the armature circuit.

All resistance is cut out when the crank arm contacts the electromagnetic holding coil. The crank arm is held by the electromagnet as long as the motor is running and the voltage is maintained.

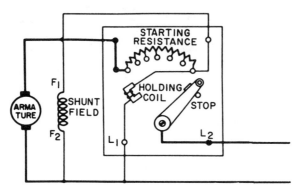

FIG. 17-1 Starting resistance for a large dc motor

An automatic controller, called a *magnetic contactor*, provides a better control of the starting operation. This type of controller changes the resistance at just the right speed.

AC DISTRIBUTION CIRCUITS

There are several different kinds of ac circuits. Single-phase and three-phase circuits are the ones most commonly used at present, figure 17-2. Either three-phase or single-phase may be distributed by three wires. It is important to be able to identify each kind of circuit because a motor designed for one will not operate on the other. The symbol ϕ is used to identify *phase*. For example, **3 ϕ** designates *three-phase*.

In general, a switch box like the one illustrated at (A) and containing two fuses with the third conductor is a single-phase circuit. Note that this three-wire, 120/240 volt entrance switch has two hot wires and a third neutral wire. The voltage is 120 volts, single-phase, when one of the hot wires is connected with the neutral wire. An entrance switch in which there are three fuses is a three-phase circuit (B).

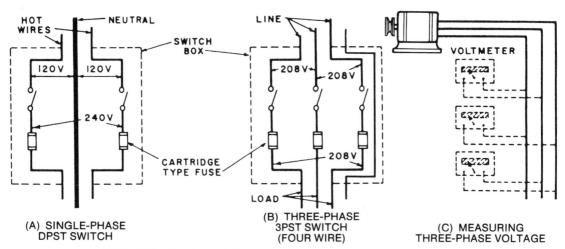

FIG. 17-2 Single- and three-phase ac circuits

312

Four-Wire System of Distribution

Electrical distribution circuits for large industrial and commercial applications are often designed for four-wire, three-phase (**3 ∅**), 480 volts, 240 volts, and 208/120 volts. The four-wire system also provides 120 volts, three-phase for lighting and low-voltage appliance requirements. The four wires consist of one neutral wire and three voltage (hot legs) wires.

A positive phase check can be made by using a voltmeter. If the same reading is obtained between any pair of conductors (as shown in the drawing [C], figure 17-2), this identifies the circuit as three phase.

If the voltage obtained between two pairs of conductors is the same, and the voltage for the third combination is approximately twice the value of the other two, then the circuit is single phase.

MEASURING POWER IN AC CIRCUITS

Single-Phase Circuits

In review it may be said that useful (true) power involves three factors: volts, amperes and power factor. In a single-phase circuit this measurement may be made very conveniently with a clamp-on wattmeter, figure 17-3(A). The potential (voltage) leads may be clamped on to the terminals in the motor junction or switch box.

This system works regardless of the type of three-phase circuit involved. Use of the clamp-on wattmeter does not require the breaking or disconnecting of lines to insert meters. This makes it an ideal method for measuring power.

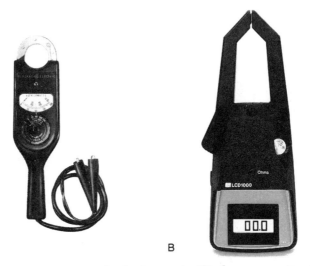

A   B

FIG. 17-3 Clamp-on power measuring instruments: (A) clamp-on wattmeter and (B) digital clamp-on volt-ohmmeter *(Courtesy of TIF Instruments, Inc.)*

Advanced Clamp-on Liquid Crystal Display (LCD) Volt-Ohm-Ammeter

Figure 17-3 (B) introduces an *advanced clamp-on liquid crystal display (LCD) volt-ohm-ammeter.* This instrument is designed with computerized electronics and is capable of sensing millisecond power surges as low as 1/10th of an amp, volt, and ohm. The instrument range is from 0 to 1,000 amps, volts, and ohms and automatically zeros on amps, volts, and ohms.

Measurements are easily visible on the high contrast readout panel. A locking switch provides for constant monitoring of amps and volts. Peak surge current is locked in.

The jaw design makes it possible to use the instrument in tight places (corners). Interchangeable jaws provide for use on large diameter cables and heavy loads. This clamp-on instrument is operated by four "AAA" batteries. A fixed probe and fused probe are provided.

Application of Clamp-on Wattmeter

The three drawings in figure 17-4 show how the clamp-on wattmeter is applied in measuring power for single- and three-phase circuits.

In a two-wire single-phase circuit (A), the potential leads are connected across the circuit. The clamp jaws are placed around either conductor depending on which one gives an upscale (positive) reading.

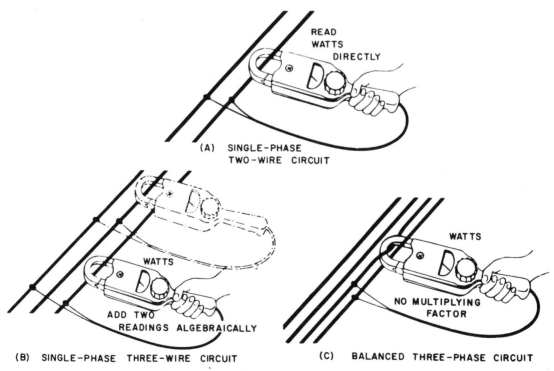

(A) SINGLE-PHASE TWO-WIRE CIRCUIT

(B) SINGLE-PHASE THREE-WIRE CIRCUIT

(C) BALANCED THREE-PHASE CIRCUIT

FIG. 17-4 Applications of clamp on wattmeter

At (B), two measurements must be taken for single-phase, three-wire circuits. The position of the clamp jaws and potential leads of the wattmeter for one measurement is shown by dark outline. The position of the meter and the leads for the second reading are indicated by broken lines.

The two readings are added unless the potential leads have to be reversed in the second measurement. In such a case, the smaller reading is subtracted from the larger. If no load is connected to the neutral wire, the power measurement may be found by connecting the potential leads across the two line conductors while the hook is placed around either line depending, again, on which gives an up-scale reading.

The third drawing (C), illustrates how a power measurement is obtained for a balanced three-phase, three- or four-wire circuit. The jaws may be placed around any two conductors with the leads attached to the same conductors (except neutral).

AC POTENTIAL DIFFERENCES

The operation of ac motors can be best understood by studying the functioning of the circuits which supply them with energy. An alternating current is one in which there is a periodic change of direction (polarity) and magnitude of the potential difference produced by the generator.

Referring to figure 17-5, single-phase ac is described graphically by a curve having one positive and one negative peak in each cycle (A). A circuit having more than one positive and negative peak per cycle is referred to as a *polyphase* circuit.

Two peaks per cycle (B) and three peaks per cycle (C) form two- and three-phase circuits, respectively. Any peak (positive or negative) which appears to be incomplete at one end of a cycle is found to be completed at the other end of the cycle. Each cycle is accompanied by a corresponding variation in the magnetism caused by the current. This is the basis for the operation of ac induction motors.

AC INDUCTION MOTORS

The induction motor receives its name from the fact that the motor requires no outside source of power. Current is induced in the rotor as the rotating field of the stator cuts through the conductors on the rotor. As rotor current is induced, a magnetic field

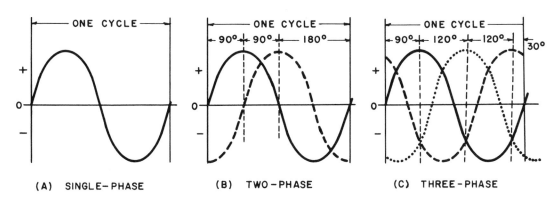

FIG. 17-5 Wave forms of one-, two-, and three-phase alternating current

develops around the conductors. These magnetic fields interact with the magnetic fields of the stator producing a torque on the rotor, causing it to turn.

Three-Phase Induction Motors

Three sets of windings, spaced 120° apart, are mounted on the stator of a three-phase induction motor. When three-phase alternating current is applied to these windings, a magnetic field is generated in each set of windings which, when combined, produces a rotating magnetic field.

Principles of Generating a Rotating Magnetic Field

Since the three sets of windings in figure 17-6 are *out-of-phase*, the maximum current is reached in line 1 first, then line 2, line 3, and the process continues. The magnetic field generated by any one of these phases at any instant depends on the current through that phase at the same movement. In one revolution the three magnetic fields shift through one complete cycle which continues over and over again.

The wiring of a three-phase stator which produces this rotating field is shown more accurately in the schematic wiring drawing, Figure 17-7.

This rotating field, which is common to all types of polyphase motors, is explained in more detail by the wave forms and drawings of stator conditions.

Wave Forms of Current and Magnetic Field of a Stator

In figure 17-8, the wave forms are lettered A, B and C to represent the three alternating magnetic fields generated on the stator during each part of a complete cycle by three-phase current. This one revolution or cycle is divided into six parts, each being equal to one-sixth of a cycle, or 60°. The wave forms also show graphically how alternating current pulsates from ($\oplus$) to ($\ominus$) values and how each phase of a three-phase current is 120° out-of-phase.

The circular drawings above the wave forms, which are lettered with the same ac input letters as the schematic wiring drawings, show three things: (1) the polarity of each phase at 60° intervals, (2) the resultant magnetic field of the stator, and (3) the direction of the magnetic field.

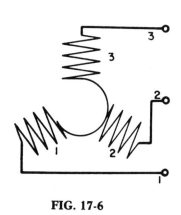

FIG. 17-6

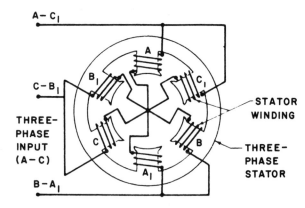

FIG. 17-7   Wiring of a three-phase stator

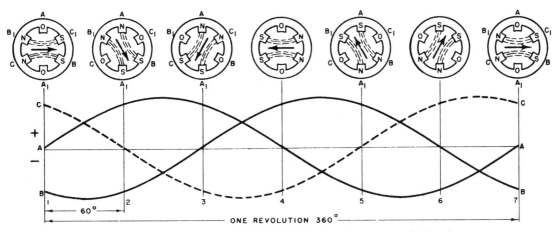

FIG. 17-8   Wave forms of three-phase current and magnetic fields of a stator

Development of a Rotating Magnetic Field

Starting at position (1) of the wave forms, the value of phase C is positive; A is zero; and B, negative – each phase being out-of-phase with the others. As a result, the polarity of the fields of the three sets of windings on the stator are shown by the letters N for north, S for south, and O for no polarity. The dotted lines show the magnetic fields created and the arrow indicates the direction of the resulting magnetic field.

At a point 60° later, phase A has a positive value; C, a zero value; and B, a negative value. Referring to the stator drawing at position (2), this means that the north pole is at phase A and the nearest south pole at phase B and, similarly, B_1 and A_1. Both of these fields, again, are equal in amplitude so the direction of the magnetic field is as shown.

At 120° the value of phase A is still positive; C, equal and negative to A; and B, zero. The north pole is A with C its nearest south pole. Similarly, C_1 is N and A_1 is S, with no polarity at windings B_1 and B.

Effect of Interaction of Magnetic Fields on Rotor

As each phase is traced at intervals of 60° through one complete cycle, the magnetic fields and the resultant direction are as illustrated. All of this action takes place in a fraction of a second so the end result of applying three-phase ac to three sets of stator windings is to create a rotating magnetic field.

This rotating magnetic field induces a magnetic field in the rotor. The interaction of the two magnetic fields causes a torque to be set up in the rotor, causing it to turn. The rotor is merely a group of rectangular loops of conductor wire mounted on the shaft. In its simplest form it resembles a squirrel-cage drum, hence its name *squirrel-case rotor*, figure 17-9.

The three-phase motor is entirely self-starting and needs no electrical connections to the rotor. Reversal is accomplished very easily by interchanging two of the three leads.

There are some compressors which should not be run backwards. The belt to the motor or the coupler may be removed to check the direction of rotation of the motor when it is first connected to the line.

FIG. 17-9   Cutaway view of induction motor with squirrel-cage rotor

A rotation tester, figure 17-10, may be employed to determine:

✓ The direction of rotation of one-, two-, or three-phase motors before they are connected to the line.
✓ The phase and polarity of unmarked motor windings.
✓ The connection for each line wire to insure proper rotation.

(A) TEST FOR MOTOR DIRECTION OF
ROTATION AND RELATIONSHIP
TO MOTOR LEAD CONNECTIONS

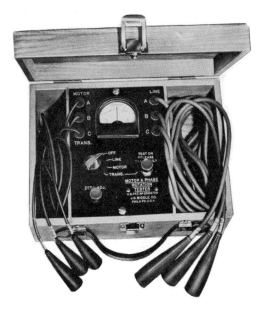

(B) TESTER WITH MOTOR
AND LINE CONNECTORS

FIG. 17-10   Motor and phase rotation tester

Synchronous Motors

Synchronus motors have

- ✓ A stationary armature winding connected to the polyphase power line.
- ✓ A moving field winding connected through slip rings to a direct current supply.
- ✓ A damper winding on the rotor which is short-circuited, similar to the rotor on a squirrel-cage motor.

Starting torque is produced by the stationary armature winding as its electrically rotating field acts on the damper winding. When the motor approaches synchronous speed, the field winding is energized with direct current. This direct current produces alternate north and south poles. These synchronize with the rotating field of the armature and continue to stay in step.

Synchronous motors are sometimes used in very large refrigeration plants. They have the advantage of maintaining constant speed, high efficiency, and they can be used to correct power factor when penalties are imposed for low power factor. By increasing the dc field current the motor can be operated with a leading power factor. Decreasing the dc field current causes the motor to operate with a lagging power factor. The synchronous motor can be adjusted to compensate exactly for the lagging power factor of induction motors. Where power factor correction is imperative but the job is too small to warrant the expense of a synchronous motor and its controls, a capacitor method is available.

However, since synchronous motors do not have especially high starting torque, care must be taken in applying them to refrigeration equipment. It is necessary to provide for unloading the compressor at its starting cycle.

Single-Phase Induction Motors

Single-phase motors operate on the same induction principle as that of a three-phase motor after they once get up to speed. However, a single-phase motor needs a supplementary device to make it self-starting.

For instance, if a single-phase motor had only a single set of windings on the stator (as does the three-phase motor) it would not start. If such a motor were connected to the single-phase line, it would merely hum but would not rotate. However, if while it were humming, the shaft were rotated by hand (in either direction) the motor would gain speed and continue to operate satisfactorily. The mechanical turning of the shaft starts the rotation of the magnetic field and establishes the induction in the squirrel-cage rotor, figure 17-11.

Thus, all that is necessary to make a single-phase motor self-starting is to equip it with a device which will start the rotation. The split-phase and the capacitor-start types are made self-starting by adding a second winding to the stator. A schematic drawing of a self-starting motor is shown in figure 17-12.

Split-Phase Motors

In the *split-phase motor* the second or *starting* windings contain a much greater number of turns than there are on the *running* winding. Because of the greater

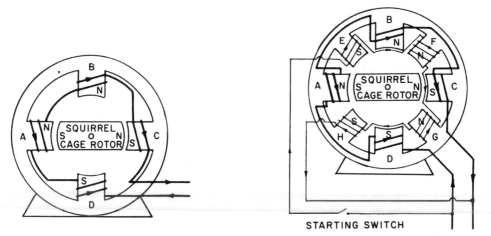

FIG. 17-11 Basic four-pole motor   **FIG. 17-12 Schematic drawing of self-starting motor**

induction in the starting winding, the current in it lags behind that in the running winding.

This, effectively, throws the two currents out-of-phase much as in the case of the three-phase motor. This condition produces the rotating magnetic field needed to start the motor. When the motor gets almost up to speed, the starting switch then automatically disconnects the starting winding. From this point on the motor continues to operate by induction, using the squirrel-cage rotor.

The use of the split-phase motor is confined to applications with light loads. This is due to the fact that the starting-torque (turning ability) of the split-phase motor is not very great. Refrigeration systems with capillary tubes that equalize pressure when off, use split-phase motors. Open motors like the one shown in figure 17-13 have a switch operated by a centrifugal mechanism to disconnect the starting circuit. A split-phase stator and frame is illustrated. The squirrel-cage rotor and housings used on this motor are not shown.

Capacitor-Start Motors

The *capacitor-start motor*, figure 17-14, has a starting winding in addition to its running winding. However, in this type of motor there is a capacitor in series with the starting winding. The capacitor produces a leading current. Therefore, the current and its magnetic field in the starting winding is out-of-phase with that of the running winding. This makes it possible for the motor to be self-starting. An automatic centrifugal switch disconnects the starting winding and capacitor when the motor gets up to speed.

The capacitor-start motor has the advantage of a starting torque several times greater than its running torque. Thus, it is able to start under heavy loads such as those encountered when starting a compressor against high pressure.

Shaded-Pole Motors

A *shaded-pole motor* is one that has only a single winding on its stator. The shift of the magnetic field and its accompanying rotation is obtained by *shading* certain pole pieces.

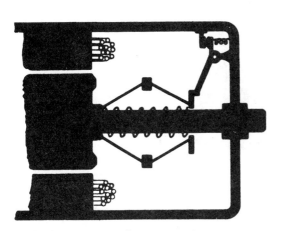

FIG. 17-13 Centrifugal switch on open-type motor

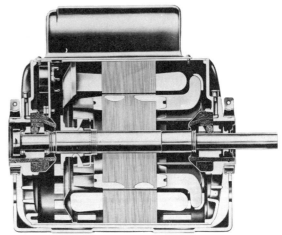

FIG. 17-14 Construction of a capactior-start motor

A shading coil is a copper loop similar to that used in the ac relay. The shading coil is recessed off-center in the pole piece. Thus, the magnetism induced in this loop is out-of-phase with the magnetism accompanying the main windings. Being out-of-phase effectively shifts or sets the magnetic field in rotation. The squirrel-cage rotor is then dragged around by induction.

Shaded-pole motors have very low starting torque. Consequently, their use is confined to operating fans and other light-starting loads.

Repulsion-Induction Motors

New models of repulsion-induction motors are no longer being designed or manufactured. However, the following information is included because many installations that were designed with repulsion-induction motors are still in operation.

A *repulsion-induction motor* should, properly, be called a *repulsion-start induction-run motor*, figure 17-15. First, the repulsion principle is used to bring the rotor up to about 75% of its running speed. Secondly, from this point on it runs as an induction motor the same as a split-phase or capacitor-type motor.

Since the armature of the repulsion-induction motor has wire coils, it is said to have a *wound-rotor*. Leads are brought out from the armature coils to the segments of the copper commutator.

The wound coils on the armature are connected by pairs of brushes which are off-center. This design feature produces a repulsion effect between the field poles and the poles of the armature coils. The effect causes the rotor and shaft to turn and brings the motor up to speed. At the proper speed, a centrifugal mechanism short circuits the commutator and lifts the brushes off the commutator. With the commutator short-circuited, the wound rotor performs the same way as a squirrel-cage rotor. The repulsion-induction motor runs from this point just as any other induction motor.

In some types of repulsion-induction motors the brushes are not lifted from the commutator. Instead, as the commutator is shorted, the brushes serve no operating function.

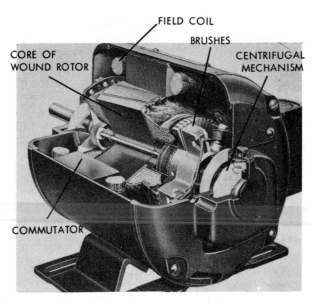

CORE OF WOUND ROTOR

FIELD COIL

BRUSHES

CENTRIFUGAL MECHANISM

COMMUTATOR

FIG. 17-15 Cutaway view of a replusion-start induction-run motor *(Courtesy of General Electric Company)*

Repulsion-start induction-run motors have starting torques which are even greater than those of capacitor-start motors. The high starting torque makes the repulsion-induction motor suitable for installations where compressors are started under very high head pressures or other high starting load conditions. The use of this type of motor is confined to very heavy starting duty and to the larger sizes of single-phase motors in which the capacitors become rather large.

Some disadvantages of the repulsion-start induction-run motors are:

√ They are more expensive to build and maintain than other motors of the split-phase or capacitor-start type.

√ Since the repulsion-induction motor has a commutator and brushes that may spark, it is not suitable for hermetic units.

√ This type of motor is not adaptable for open type units which may be located in extremely dusty or dirty places.

Dual-Voltage Motors

In order that motors may be used on either 120 or 240 volts, their field coil connections are brought out so that the coils may be put in series for 240 volts or in parallel for 120 volts, figure 17-16.

MOTORS IN HERMETIC UNITS

Hermetic units are exposed to the refrigerant. The motor, therefore, cannot be of the repulsion-induction or centrifugally operated open-contact starting switch type inside the case. Many hermetic units use a starting switch which is located outside the case.

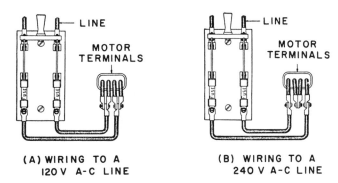

(A) WIRING TO A
120 V A-C LINE

(B) WIRING TO A
240 V A-C LINE

FIG. 17-16   Wiring an ac motor to an ac line

The switch is operated by a timing arrangement. This timer is actuated by variations in electric current during the starting period, or by the induced voltage from the starting coil. A hermetic electric motor, stator and rotor are shown in figure 17-17.

SAFETY PRECAUTION

Under certain conditions, the housing of a motor may become grounded to the wires inside the housing. While the motor will still continue to operate, anyone who touches the housing will receive a shock, especially if the person is standing on a damp floor or the ground. All refrigeration components should be grounded permanently at the time of installation to avoid this unsafe condition.

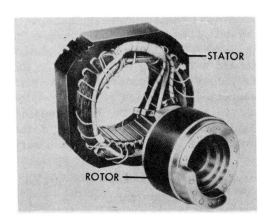

FIG. 17-17   Hermetic electric motor stator and rotor

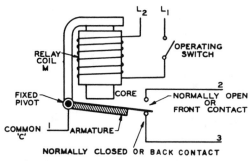

FIG. 17-18   Operation of a current relay

Current Relays

A running winding consumes more current when a rotor is not turning or is revolving slowly. As a rotor picks up speed, the magnetic fields build up and collapse in the motor. This action produces a bucking or counter-electromotive force or voltage on the running winding. This counter-emf reduces the current drawn by the running winding.

Operation of a current relay is shown in figure 17-18. A current relay is an electromagnet similar to a solenoid valve. When a system is idle, a weight or a spring may be used to hold the starting winding contact points open. When the motor control contacts close and current flows through the running winding, the magnetic switch becomes heavily magnetized, lifts the weight, and closes the contacts.

The contacts, in turn, close the starting winding circuit so that the motor may quickly reach up to 75% of its rated speed. With the increased speed, both the motor current and the magnetic strength decrease, permitting the contact points to open. Most current relays are designed with an overload cutout. Current relays are widely used on compressors of less than one horsepower capacity.

Thermal Relays

Thermal (hot wire) relays are used on certain models of older domestic refrigerators. These relays depend on three scientific principles: (1) electrical energy may be turned into heat, (2) time is needed to raise the temperature of any material, and (3) a heated metal expands.

Operating positions of a hot-wire relay are shown in figure 17-19. One type of hot-wire relay depends upon a wire under tension to operate the contact points. When cold, the wire by tension keeps both sets of contact points closed (position A). Current passing through the wire causes it to heat and expand. At a particular setting, the expanded wire opens the starting winding contact points (position B). For safe operation the relay is designed so that if the motor uses an excessive amount of current, the wire expands enough to open the running or winding contact points, (position C). The stretching of the wire permits some snap action (warp) springs to change position. In so doing, the springs open and close the contacts.

Solid-State Starting Relays

This type of relay is widely used in small motor applications. The relay depends on a *self-regulating, conducting ceramic* Any changes in current flow in the circuit produces

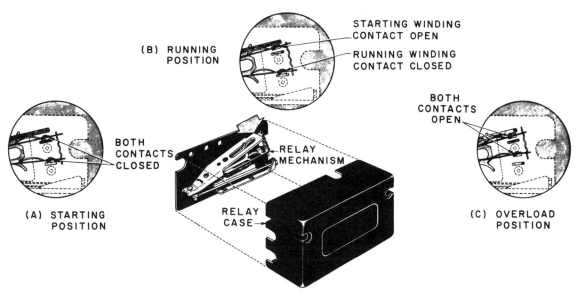

FIG. 17-19   Operating positions of an old model hot-wire (thermal) relay

resistance changes in the conducting ceramic element. On start-up, the ceramic element serves as a switch, opening the motor start circuit.

In troubleshooting a solid-state starting relay (like many other solid-state units), it is recommended that a defective relay be replaced rather than repaired. Care must be taken to establish that the replacement relay meets manufacturer's specifications.

Voltage Relays

The *voltage* or *potential relay* is used on CSCR-type compressors. A magnetic voltage relay is shown in figure 17-20. While this relay looks somewhat like a current relay, it depends for operation on the increase in voltage as a unit approaches and reaches its rated speed. The coil of the voltage relay has a large number of turns of fine wire as compared with the current relay coil which has only a few turns of heavy wire.

FIG. 17-20   Older model magnetic voltage relay

Because the voltage relay remains closed on the off cycle, this feature alone is of considerable value. Arcing is prevented because the contact points are closed as the thermostat turns on the power. As the motor speed current increases, the higher voltage creates more magnetism in the relay coil, pulling the contact points apart. By so doing the starting circuit is opened. The heating of the relay coil and core is kept at a minimum because the relay coil is (1) connected across the starting winding and (2) is made of small wire so very little current passes through it.

OVERLOAD PROTECTORS

Electric motors may deliver more power than the nameplate rating. The limit of the power developed depends largely upon the temperatue of the motor. Heavy or continuous overloading causes: an excessive rise in temperature, the insulation to be destroyed, and the motor to burn up., Although some motors, in cool locations, may be overloaded as much as 20% without damage, this is a poor practice. Under usual operating conditions, motors should not be overloaded by any more than a few percent.

Inherent Type of Overload Protector

The inherent type overload protector is affected by both motor current and temperature. The elements of this device are designed to carry full motor current and must always be connected in the main line to be effective.

Inherent protectors are of two different types:

- ✓ The two-terminal type consisting of a thermal disc and heater connected in series. This type is always connected between the common terminal and the supply line.
- ✓ The three-terminal type which provides extra-fast operation under stalled or breakdown conditions. The heater is connected in the starting winding; the disc, in the running winding. Any extra heating caused by the starting winding current trips the disc if the motor fails to start. The three-terminal type is used primarily in welded and other compressors which are designed for high back pressure operation.

Thermal Type of Compressor Overheating Protector

A motor temperature and current overlay device is attached to the outside case of the compressor, figure 17-21A. The function of the protective device is to sense any unusual rise in temperature or excess electric current draw.

The compressor overheating device, as illustrated, is actuated by a bimetal disc. Under normal operation, the electrical circuit within the overload protector is "open," figure 17-21B. If there is either a surge in the amount of current drawn or there is an unusual increase in temperature, the bimetal disc springs downward to a "closed" position, figure 17-21C. This action disconnects the compressor from the power source, providing protection from overheating.

The thermal-type protector should be connected in the pilot circuit of a motor starter, which is designed with approved heater elements for overcurrent protection.

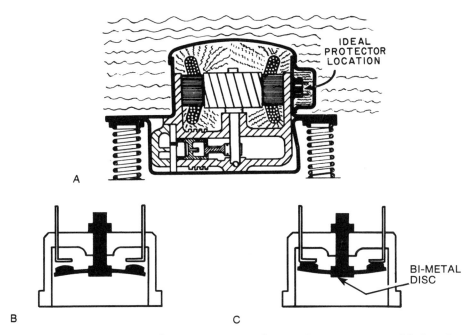

FIG. 17-21 Compressor overheating protective device: (A) protector assembled on the outside of the compressor housing, (B) open circuit and (C) closed (disconnected circuit)

Thermal Plug-on Motor Protector

A *thermal plug-on protector*, figure 17-22A, is used on refrigeration compressor motors. This device protects the compressor from overheating caused by running overload or locked rotor conditions by sensing the current and the motor temperatures.

The thermal protector line drawings at (B) show a noncarrying current snap-acting disc ⑥ located above the heater element ⑤ . The disc ⑥ senses radiant heat generated from the heater, compressor housing, and the metal pin that carries line current through the hermetic terminals. The movable spring arm ⑦ (located above the disc) is actuated as the disc flexes. This movement opens the circuit contacts ⑧ and ⑨ , thereby shutting off the compressor motor from the power source.

Thermal motor protectors for compressors are used in such refrigeration applications as refrigerators, freezers, water coolers, dehumidifiers, and other installations.

TRIAC: Bidirectional Current Flow Control

The standard relay is designed to provide electrical isolation between the control circuit and armature/contact assembly. While serving a similar electrical isolation function, a solid-state relay (SSR) uses a photoelectric effect that is activated by a light-emitting diode (LED).

The presence of the control signal turns the LED "ON." The light emitted by the LED is received by a photocell, which activates a *triac*. A triac is a solid-state bidirectional

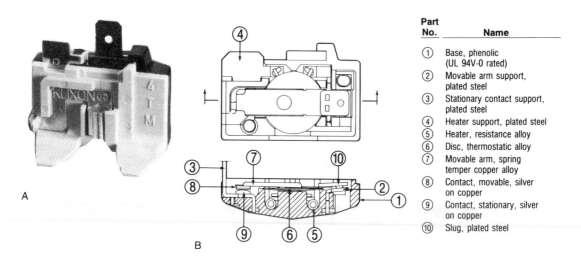

Part No.	Name
①	Base, phenolic (UL 94V-0 rated)
②	Movable arm support, plated steel
③	Stationary contact support, plated steel
④	Heater support, plated steel
⑤	Heater, resistance alloy
⑥	Disc, thermostatic alloy
⑦	Movable arm, spring temper copper alloy
⑧	Contact, movable, silver on copper
⑨	Contact, stationary, silver on copper
⑩	Slug, plated steel

FIG. 17-22 Thermal plug-on motor compressor protector: (A) plug-on protector and (B) section view and parts identification *(Courtesy of Motor Controls Department, Texas Instruments Incorporated)*

switch. The switch duplicates the action of a regular open and closed relay circuit.

Three advantages of the solid-state relay are as follows:

- There are no moving parts.
- SSR provide an additional boost to the compressor on some compressor applications.
- Greater reliability.

SSRs are designed for direct plug-in replacement for many conventional models. A *fin clip* style SSR is available for installations that are wire-for-wire compatible but may not fit mechanically.

SSRs may be used to replace such individual components as a relay, overload protection and start capacitor unit, or a combination. Standard SSRs are available for residential and commercial air-conditioning systems and heat pumps ranging from 4,000 to 60,000 Btu (1/2 Hp through 10 HP). SSRs are designed for use on 120- through 288-volt units. Applications vary with use from ordinary low voltage start problems to low voltage, hard starting compressors, to severe low voltage or hard starting compressors.

The solid-state relay shown in figure 17-23 contains a specially designed relay and larger start capacitor to handle severe voltage and starting compressor problems.

AC MOTOR CONTROLS

Switches that are used to connect and disconnect motors and their power supply are referred to as *motor starters.* These starters range in design and operation from simple switches to elaborate devices that provide motor protection, current limits, time delay, and sequence operation. Motor starters have overloads incorporated in their design and normally are used on **3 ∅** motors. Most single-phase motors have *contactors* (definite purpose relays).

FIG. 17-23 SSR for severe, hard starting compressor problems *(Courtesy of Sealed Unit Parts Company, Inc.)*

Manual Motor Starters

As the name implies, *manual motor starters* are used wherever a control is required that is not automatic and where the motor is often stopped and started at infrequent intervals. A manual motor starter is generally used with a fan or pump motor. They usually operate continuously and with the starter located near the motor.

Manual motor starters are provided with overload protection in the form of melting alloy latches. These permit the starter to open as the current load becomes too heavy. Protection against undervoltage is not provided. Manual starters are available for two-speed separate winding motors. They are also designed with resistors or auto transformers which limit starting current.

Magnetic Across-the-Line-Starters

A magnetic-type motor starter, figure 17-24, must be used in installations requiring an automatic control. The *magnetic-motor starter* consists of a set of heavy contacts for carrying the main current. These contacts are held whenever the magnetic *holding* coil is energized. Since only a very light current is needed to energize the holding coil, this circuit can be controlled by a pressure controller, thermostat, or some other device having relatively light contacts.

The magnetic across-the-line starter illustrated is widely used on practically all medium sized refrigeration and air-conditioning work. The motor is connected directly to the line so that full voltage is impressed through the motor windings instantly. Overload protection is provided through melting alloy relays.

When used with three-wire control, under-voltage protection is provided. The starter has a reset button in the cover. This button resets the starter and makes it operative after the switch has "kicked-out" due to an overload. This particular model requires a thermostat or push button station as an outside pilot control. Other designs have a push button instead of a reset button station in the cover, the starting button serving also as a reset.

FIG. 17-24 Magnetic motor combination starter (mounted in NEMA Type I, general purpose enclosure) *(Courtesy of Allen-Bradley, Rockwell International Company)*

Magnetic across-the-line starters are designed for motors of two or more speeds. When necessary, multispeed starters may be fitted with automatic time delay devices which provide the necessary delay in switching from high to low speed. These starters may be manufactured to take care of special applications.

Magnetic Reduced Voltage Starters

Magnetic starters are available that provide for a starting voltage reduction on the motor. Magnetic starters are required by many power companies for all motors that call for some specified minimum starting current. Starters that provide reduced voltage use either resistances or auto transformers which are placed in the circuit by a relay. The impedance usually is held in the line for a definite time by a timing device connected to the relay. Reduced line voltages up to 35 percent may be used with fans, centrifugal pumps or compressors which start unloaded. A reduction of not more than 20 percent is recommended for compressors that start fully loaded. The compressor is equipped with unloaders.

Two- and Three-Wire Control. Magnetic starters are often connected with other controls to provide electrical interlocks which prevent operation of the starter when other conditions are not right. These electrical interlock connections are made to be the starter holding coil circuit.

A circuit diagram of a typical magnetic starter under three-wire control is shown at (A) in figure 17-25. The three heavy lines represent the power circuit entering from the three-phase line, and leaving for the motor at T_1, T_2, and T_3. The holding coil circuit is represented by the lighter lines. At the push button station the stop button is a normally closed contact; the start button the open contact.

When the start button is pushed, current flows from L_1 through 1, across the closed contacts of the stop-and-start button, through 3, energizes the holding coil, and proceeds to the power supply at L_2. The energized holding coil closes the main contacts and the auxiliary contact. The holding coil remains energized even though the start button is released because current flows from L_1 across the closed contacts of the stop button to 2, across the auxiliary contact to 3, and through the holding coil back to L_2. The main current flows through the starter to the motor as long as the holding coil is energized.

Overload control is provided by an overload relay and heater coils in the main circuit. When the motor is running, current passes through the heaters. If there is an overload, the heater temperature rises until the alloy trips the overload relay and de-energizes the holding coil. When the main contacts open, the heaters cool. However, the motor cannot be started until the overload relay is closed manually by pushing the reset button.

The second schematic drawing (B) in figure 17-25 shows the same magnetic starter arranged for a two-wire control. Note that a single maintaining contact is used between 1 and 3. This may be a thermostat, a pressure controller or a switch. The holding coil remains energized only as long as the control switch is closed at 2, and the auxiliary contact is not used. This is the usual connection for compressor control by means of a low-pressure switch.

The two-wire control system provides current overload protection by using heaters. Undervoltage protection is also provided because the holding coil opens the main contacts whenever the voltage becomes too low. Upon return to normal voltage, the holding coil again closes the main contacts and restarts the motor.

Starter Enclosures

Many types of coverings (enclosures) are commercially available for manual and magnetic starters. Besides the general purpose enclosures for normal indoor duty, dustproof enclosures are used for dusty locations; waterproof and explosion-proof enclosures, for hazardous locations. Conditions affecting installation should be considered for specifying the size, kind and type of starter and its enclosures.

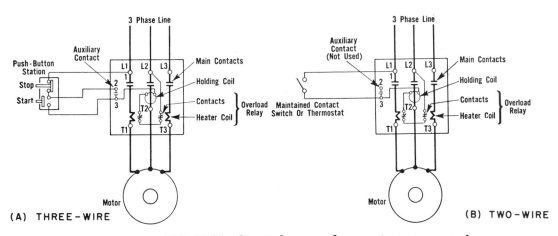

FIG. 17-25   Circuit diagram of magnetic starter control

Disconnect Switches

Most electrical codes specify that each electric motor circuit shall be provided with a disconnect switch. When disconnect switches are installed, the system can be completely disconnected from the source of power for emergency, repair, maintenance or other work. As a safety precaution, the location of the switch must be known.

A fusible disconnect switch is generally used on air-conditioning systems. The fuses can be removed and the enclosure locked to prevent anyone from attempting to operate the system when it has been pumped down for seasonal shutdown. In some cases, the fusible disconnect switch is combined with the magnetic starter.

Reversing Starters

The direction of rotation of a polyphase ac motor may be reversed by interchanging two of the incoming leads. Two line contactors are required: one for forward direction and the other, for reverse direction of rotation. The circuit diagram in figure 17-26 shows standard control connections. Mechanically actuated limit switches connected in the coil circuits as shown, stop the motor automatically at the ends of the forward and reverse travel.

The controller must be so designed as to prevent simultaneous closing of both reversing contactors that would result in short circuiting the line. Mechanical and

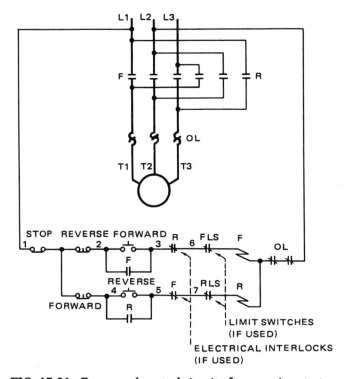

FIG. 17-26 Power and control circuit of a reversing starter

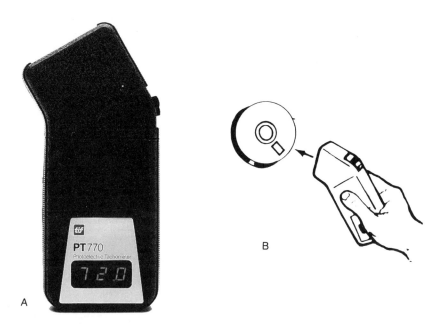

FIG. 17-27 Solid-state photoelectric digital tachometer: (A) portable tachometer and (B) measuring RPM *(Courtesy of TIF Instruments Inc.)*

electrical interlocking of reversing contactors must be included for safe operation. The circuit diagram indicates how electrical interlocking may be provided through back contacts of the reversing push buttons.

SOLID-STATE PHOTOELECTRIC DIGITAL TACHOMETER

The *digital, photoelectric tachometer* is used in air-conditioning and refrigeration work to electronically measure speeds of motors, compressors, or any other type of rotating equipment.

The model shown in figure 17-27 has a range from 0 to 10,000 revolutions per minute (RPM), with an accuracy of ± 2%. This particular tachometer is adaptable for shop or field work.

Design Features and Principles of Operation

The digital, tachometer depends on a photoelectric cell (electronic "eye") and light source, solid-state circuitry to electronically decode impulses, and a digital readout to display the RPM. This measuring instrument combines the concept of an electronic "eye" with the physical science principle of reflected light. Impulses are produced by variations of the angle of incidence of the reflected light (produced from a reflective sticker placed on the rotating part whose RPM are to be determined) to the photoelectric cell. The impulses produced are, in turn, decoded in the electronic circuit into RPM. The RPM are displayed as a bright digital readout.

Some of the major advantages of this tachometer design are the following:

- The instrument is not sensitive to fluorescent lighting.
- Dependability, accuracy, and sturdiness of solid-state electronic circuitry
- Electronic measurement of RPM without torque loss
- Flexibility of use on any type of moving equipment (there are no electrical connections)
- Safe, noncontact measurement
- Portable, hand-held, easy-to-use

SUMMARY

- DC motors operate entirely on the principle of magnetic repulsion. There must be current in the armature while operating. Thus the brushes are always in contact with the commutator.
- A starter consisting of a variable resistance must be used with a dc motor to prevent excessive currents from flowing.
- The most common ac circuits are three-phase and single-phase. Either may be distributed over a three-wire system.
  - ✓ Single-phase is usually 120/240 volts with 120 volts between the neutral (ground) wire and either of the other two. The 240 volts is between the two hot wires. A single-phase current contains one positive and one negative peak per cycle.
  - ✓ Three-phase is usually 208 or 230 volts with the same voltage across any one of three pairs of wires. Three-phase current contains three positive and three negative peaks per cycle.
- A hook-on wattmeter which operates on induction can be used to measure true power (in watts) on either the single-phase or three-phase circuits.
- Three-phase current effectively produces a rotating magnetic field in sets of coils connected to the circuit.
- Induction motors operate on the induction principle with no wires as such on the rotor. These rotors are called squirrel-cage rotors. The magnetic fields induced in the loops of these rotors by the rotating magnetic field of the stator drag the rotor around causing the torque or turning-effect of the motor.
- Three-phase induction motors are self-starting and can be reversed simply by changing two of the three connections.
- Single-phase motors operate by induction, having squirrel-cage rotors. The repulsion-start, induction-run motor, however, has a wound rotor and starts by magnetic repulsion.
- Single-phase induction motors require a device to start rotation of the magnetic field in order to start the motor by induction.
- Centrifugal switches on open-type motors are used to disconnect the starting winding once the motor gets up to speed.
- The split-phase motor uses a starting winding with a large number of turns in addition to the running winding. This throws the field of the starting winding out-of-phase with that of the running winding so rotation starts. This motor has only fair starting torque but is adequate for capillary-tube systems.
- The capacitor-start motor uses a starting capacitor in series with the starting winding to cause the phase difference and start the rotation. This motor has a strong starting torque.

- The repulsion-induction motor has the strongest starting torque of any of the single-phase motors. The shaded-pole single-phase motor has the least starting torque.
- Hermetic motors need external starting switches. These may be:
  - ✓ Current relays which operate as electromagnets to close the starting contacts.
  - ✓ Potential (voltage) relays which operate on the induced back-emf to open the starting contacts.
  - ✓ Hot-wire relays which operate on the change of tension of a heated wire and the snap action of springs.
- Synchronous motors are a combination of the repulsion-induction ac motor together with some added dc excitation on the field. Such motors have a very low starting torque but have the advantage of having an adjustable power factor to offset the lagging power factor of other induction motors.
- Overload protectors are provided on motors to prevent burnout, fires and other safety hazards.
- The thermal type of overload allows primarily for the amount of current that the motor draws. It protects against overloads that result in excess currents only.
- Solid-state relays achieve electrical isolation by using a photocell effect to actuate a bidirectional triac switch, thus duplicating an open and closed relay contact.
- The digital photoelectric tachometer uses a photocell and light source and reflective tape on a rotating part to electronically count and display the RPM.
- Motor controls are classified as manual, magnetic across-the-line, or magnetic reduced voltage. Automatic controls require the use of magnetic across-the-line and reduced-voltage starters. Across-the-line starters may be used when the starting current of the motor is not excessive.
- Reduced voltage, either auto transformer, resistance or reactance, must be used to keep the starting current within reasonable limits. Disconect switches and interlocking systems provide for convenience and safety of operation of the controls.

ASSIGNMENT: UNIT 17 REVIEW AND SELF-TEST

A. DIRECT CURRENT MOTORS

Add the term that correctly completes statements 1 to 10.

1. An electric motor is a device for changing _____ energy into _____ energy.
2. A dc motor rotates as a result of _____ forces.
3. Ac motors use rotating magnetic fields. The second set of magnetic fields is produced from the first by a process known as _____.
4. The stationary coil of the dc motor is known as the _____ or _____.
5. The magnetic forces responsible for the rotation of the dc motor are called forces of _____.
6. A dc motor requires a starting device which employs a _____ to _____ in the rotor and the line.
7. The magnetic field of the armature in a dc motor is produced by _____.

8. The segments of the rotor that are the terminal connections of each coil, collectively, are called the _____ .
9. Connection is made from the external circuit to the coils of the rotor of a dc motor through _____ .
10. The process of induction involves _____ or _____ magnetic fields.

B. ALTERNATING CURRENT CIRCUITS

Select the letter representing the phrase or value which correctly completes statements 1 to 5.

1. Three wires are used for the distibution of: (a) single phase only, (b) three phase only, (c) either single or three-phase circuit.
2. A switch box containing three fuses is used for: (a) single phase only, (b) three phase only, (c) either single or three-phase circuit.
3. A switch box containing two fuses and a solid conductor is used for: (a) single phase only, (b) three phase only, (c) either single or three-phase circuit.
4. A voltage test on the three different pairs of wires in a switch box showed two pairs with the same voltage and one pair different from the others. This identified the circuit as (a) single phase, (b) three phase, (c) neither single nor three phase.
5. The voltage between the solid conductor and either of the two fused lines in a switch box is generally: (a) one-half, (b) two times, (c) neither one-half nor two times that between the two fused conductors (lines).

Determine which statements (6 to 11) are true (T); which are false (F).

6. A hook-on wattmeter indicates true or useful power.
7. When using a hook-on wattmeter to measure power in a two-wire, single-phase line the potential leads are clipped on to the same two wires passing through the loop.
8. The needle of a wattmeter sometimes reads to the left of the zero on the scale.
9. When using a hook-on wattmeter on a four-wire (one is neutral) circuit, the potential leads are clipped to the same two wires that pass through the loop.
10. The hook-on wattmeter cannot be used to measure power on a three-wire, single-phase circuit.
11. When using a hook-on wattmeter to measure the power in a three-wire, three-phase circuit, the two conductors need to be crossed as they pass through the loop.

Supply the correct value in answer to statements 12 to 16.

12. Give the number of positive and negative peaks in one cycle of single-phase ac.
13. State how many degrees there are between the peaks of the graph of a two-phase circuit.
14. State how many degrees there are in the graph on one complete cycle of single-phase ac.
15. What length of time is represented by one cycle of 60-hertz, three-phase ac?
16. How many positive and negative peaks per second are there for one cycle of a 50-hertz, three-phase circuit?

C. INDUCTION MOTORS

Identify from the following information whether each motor is single phase or three phase. Use (1) for single phase and (3) for three phase.

1. Self-starting but no starting switch
2. Self-starting but has starting relay
3. Shaded pole
4. Brushes and commutator
5. Synchronous
6. Insert the number of the motor listed which fills conditions (a) through (j).

   (1) Synchronous
   (2) Three-phase
   (3) Split-phase
   (4) Capacitor
   (5) Shaded-pole
   (6) Dual-voltage
   (7) Repulsion-induction

   a. Cylindrical shaped container mounted on top of the motor.
   b. Only one winding on the stator; two wires connected to the motor.
   c. Commutator and brushes that lift when motor gets up to speed.
   d. Centrifugal switch, but no capacitor.
   e. Four terminals with instructions on the plate as to how to make the connections in pairs.
   f. Three wires, but no starting switch.
   g. Driving a small condenser fan.
   h. Used on a system with capillary tube restrictor.
   i. Connected to three-phase line, but with dc supply.
   j. Can be reversed by changing any two of the three connections.
7. Define the term induction.
8. Give one advantage of a three-phase over a single-phase motor.
9. Tell how a repulsion-induction motor can be recognized at a glance.
10. List three parts that are common to both capacitor and split-phase motors.
11. Describe the operation of a centrifugal switch.
12. State how many sets of circuits there are in the field of a three-phase motor and identify the purpose of the circuits.
13. Give the number of degrees there are between the different phases of a three-phase circuit.
14. Describe how the rotating field is produced in a split-phase motor.
15. State how the rotating field is produced in the capacitor motor.
16. Tell how the shaded-pole motor produces a rotating magnetic field.
17. List four refrigeration motors in the order of increasing torque.

D. HERMETIC MOTORS

Add the term or condition which correctly completes statements 1 to 9.

1. a. The type of starting switch that cannot be used with hermetic units is the _____ switch.
   b. This is true because the contacts on opening may _____ .
2. a. The type of starting device that is used with hermetic units is called a _____ .
   b. These types of starting devices are generally classified as the: _____ , _____ , and _____ types.

337

3. Both current and voltage starting devices operate on the principle of _____ .

4. The primary difference between the current and voltage type starting device is that the voltage type has a _____ .

5. The contact points of the current relay are normally _____ .

6. The contact points of the voltage type are normally _____ .

7. The current type starting device operates because of the difference between the _____ and _____ currents.

8. The voltage type operates because of the increase in _____ as the motor picks up speed.

9. The hot-wire type of starting device operates because the tension of the wire changes as it is _____ by the current.

10. A triac is a _____ control switch.

E. OVERLOAD PROTECTORS

1. Describe briefly the operation of (a) the inherent type and (b) the thermal type of protectors.

2. Identify three air-conditioning electrical components that can be replaced by solid-state relay controls.

F. ALTERNATING CURRENT MOTOR CONTROLS

Add the term or phrase to correctly complete statements 1 to 9.

1. a. The simplest and least expensive type of motor starter is the _____ .
   b. This type has the disadvantage in that it requires _____ .

2. Automatic control requires a _____ or a _____ starter.

3. The type of automatic control that must be used when the motor can have the full voltage applied to it immediately is the _____ starter.

4. The type of automatic control that must be used when it is necessary to limit the starting current is the _____ starter.

5. The type of automatic control that limits the starting current employs either an _____ , _____ or _____ .

6. Magnetic starters sometimes include _____ in the same box.

7. Added safety is afforded to the service technician if a _____ is used in addition to the starting device.

8. Starters, other than manual, usually operate on the principle of the _____ .

9. When a controller operates on a sequence of events it is provided with an interlock that may be either _____ or _____ .

G. SOLID-STATE DIGITAL PHOTOELECTRIC TACHOMETER

1. State the purpose served by a photoelectric digital tachometer.

2. Name two major components of a solid-state digital tachometer.

UNIT 18:
BASIC REFRIGERATION CONTROLS

The primary function of a refrigeration system, as has been stated many times, is to remove heat from a place where it is not needed and is objectionable to a place where it is not objectionable. This is accomplished by circulating the refrigerant through the system and subjecting it to certain conditions.

A low-pressure condition is necessary in order for the refrigerant to pick up heat by boiling. This produces a desired cooling effect. A high-pressure and temperature condition is required for the refrigerant to effectively eliminate heat from the system as it condenses.

The control system must operate in such a way as to maintain the required temperature as closely as possible. In addition, temperature control should be accomplished with a minimum of operation of the components and maximum of safety to the equipment and/or materials and people occupying the conditioned space.

CLASSIFICATION OF CONTROLS

The term *basic refrigeration* control refers to a device that starts, stops, regulates and/or protects the refrigeration system and its components.

Operating or Primary Controls

The first group of controls is sometimes called *operating* or *primary controls*. It is their job to be sensitive to changes in the desired conditions such as temperature (or its related pressure) and humidity. Operating controls translate such changes into motion or mechanical force so as to initiate whatever action is necessary to correct the change and maintain the proper level of the desired condition.

In general, control devices act merely as electric switches to turn the proper component on or off. Some electric switches work in direct response to changes in temperature. Others operate as a result of the effect of temperature changes on air pressure (pneumatic controls) or on liquid pressure (remote-bulb controls).

Typical of these operating controls are the thermostat, humidistat and pressurestat. The name indicates the type of actuating force upon which operation of the control depends. Each can be used to control the refrigeration system. When the temperature in a household refrigerator rises too high for food storage, the sending element in the thermostat responds by starting the compressor. When the thermostat is satisfied, it stops the compressor.

Temperature in a display case can be indirectly controlled by a pressurestat connected to the evaporator or system suction line. This is possible because the temperature of the evaporator depends upon the saturation pressure of the refrigerant. As heat is absorbed by the refrigerant its pressure and temperature rise. This increase in pressure is used to start the compressor. As the temperature and pressure are lowered to the right level, the pressurestat operates to stop the compressor.

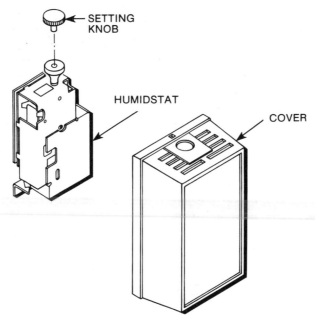

FIG. 18-1   Humidistat mechanism

Humidity Controllers

Specific humidity levels are important in industrial/commercial storage rooms, in many manufacturing processes and products, and for climate control of businesses and homes. Correct humidity is maintained by a humidity controller, figure 18-1, sometimes called a humidistat.

Humidity controllers are designed with sensing elements that respond to minute changes in relative temperature over a wide ambient temperature range. Humidity controllers are used with humidifiers and dehumidifiers to compensate for humidity changes, to control mildew problems, and as components of air-conditioning systems.

Actuating or Secondary Controls

The second group of controllers is called *actuators* or *secondary controls* because they indirectly control the operation which is to carry out the corrective measure. Among this group of controls are the relays (or contactors), solenoid valves, water valves, four-way valves and back-pressure valves.

Limiting and Safety Controls

The third group of controllers is the *limit and safety* group. Refrigeration systems sometimes approach limits of temperature or pressure for which they are not designed. It is the function of this group of controls to exercise judgment and stop the system before excessive limits are reached.

The controllers do this by taking over the control of the actuators, regardless of what the operating controls call for. Among the safety controllers are: the safety thermostat, high-pressure cutoff, low-pressure cutoff and oil safety switch. Others of the safety group include the freeze controls, voltage controls, electrical overloads, fusible plug and rupture disc.

One of the purposes of this unit is to describe how these three groups of controls function in relation to the successful and safe operation of the refrigeration system. These controls are connected in series with the operating control so that an abnormal condition at any one of the controls will stop the operation of the system.

OPERATING CONTROLS

Manual Operation

If the load on a refrigeration system were constant, hand-operated switches would be adequate because the system would run continuously until the power was turned off.

Manual switches are used with automatic controls so that in servicing the units they can be turned on and off as needed. Usually, the fans on the evaporators are circuited separately from the condensing units and should have their own control.

Separate manual switches for the fan motor and compressor are used in some installations, figure 18-2. Two switches are necessary. Even after the compressor is turned off, the evaporator fans operate separately as they continue to run during the off cycle.

Simple Automatic Control

More often than not the heat load on a refrigeration system is not constant. The varying heat load may be caused by any one or a combination of three things:

✓ Variations in ambient temperature around the refrigerated space. The leakage of heat into the refrigerated space is affected.

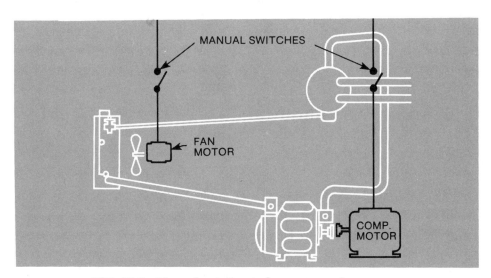

FIG. 18-2   Manual switches to fan motor and compressor

✓ Variations of the amount of warm material placed in the refrigerated space. The heat load is very heavy until this material is cooled down to the temperature of the refrigerated space.

✓ Variations in the frequency with which the doors of the refrigerated space are opened and closed. Quantities of warm, humid air and heat from lights affect the heat load.

All of these factors tend to justify mandatory automatic control for modern equipment applications.

Automatic controls are designed to maintain a constant temperature in the refrigerated space by allowing the equipment to operate *intermittently* (off and on as needed). The automatic controls are in addition to the different methods of capacity control and the use of multiple evaporators or compressors.

Automatic controls may be operated in either of two ways. They may

- Be located directly in the refrigerated space itself, or
- Have a thermal bulb located in the refrigerated space at a place where the control is desired.

Manual Opening of Solenoid Valve

Solenoids should be designed for manually opening the valve. These valves are usually built so that the operation of a screw in the bottom lifts the plunger. This makes it possible to hold the valve open for long periods. Such operation is necessary for initial evacuation and to pump down the system for servicing.

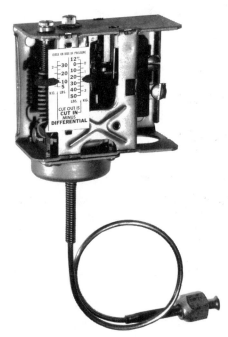

FIG. 18-3 Pressure control switch *(Courtesy of Ranco North America)*

Low-Pressure Cutoff Switch

The lower-than-normal pressure provides the way for stopping the compressor. A pressure-sensitive switch called a *pressure control* or *low-pressure cutoff* is connected to the suction side of the compressor, figure 18-3. When the compressor pumps the system down to the pressure for which the valve is set, the control breaks the connection and deenergizes the magnetic starter, stopping the compressor. The model as illustrated has a dual scale reading pressure in either pounds or kilograms.

The low-pressure cutoff switch is usually one-half of a combination low-pressure control and high-pressure safety cutout. For simplicity, the single control is covered at this time. The high pressure cutout is discussed under **Safety Controls.**

Restarting a Pump-Down System

After a compressor has been off for a period of time, the thermostat calls for the operation of the system. The thermostat contacts close, energizing the solenoid. The plunger is lifted, opening the valve. Liquid refrigerant flows into the evaporator where it boils. This raises the vapor pressure in the evaporator and suction line to which the low-(suction) pressure control is connected. The pressure-sensitive element closes the contacts, energizing the magnetic starter. The compressor starts and operates until the thermostat is satisfied.

SECONDARY OR ACTUATING CONTROLS

Direct-Coupled Control System

The dotted outline on the drawing in figure 18-4 traces the location of a thermostat and magnetic contactor and shows how such a control adds the automatic feature to

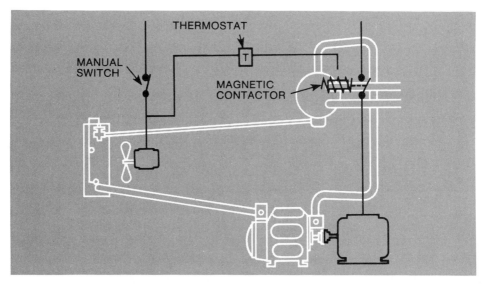

FIG. 18-4 Location of thermostat for magnetic contactor for automatic operation

the system. This system provides for the continuous operation of the fans as long as the manual switch is closed. The compressor operates under the control of the thermostat as needed. The magnetic contactor, with its heavy contact points, is used to turn the heavy current on and off.

This type of control adequately provides for automatic starting and stopping. However, it leaves the system with an undersirable operating condition. When the compressor stops, there is liquid refrigerant left in the evaporator. In spite of the fact that the fan is left running, not much of the liquid evaporates because the compressor is not operating to carry away the vapor formed by evaporation. This causes the pressure to build up and stops the boiling. This unsatisfactory condition may be corrected by changing the control system so that the thermostat operates a solenoid valve instead of the compressor.

Pump-Down Cycle System

The purpose of this system is for liquid control and to prevent compressor overload on initial start-up.

In the pump-down cycle system, figure 18-5, the solenoid is placed in the liquid line just ahead of the thermostatic expansion valve. When the temperature of the refrigerated space is reduced to its desired value, the thermostat is said to be satisfied and its contacts are open. Current no longer continues to flow through the solenoid coil. It loses its magnetism. This permits the plunger and needle valve to drop, closing the valve.

Liquid refrigerant is no longer permitted to flow to the evaporator although the compressor keeps running. This continuing action pumps the refrigerant vapor out of the evaporator and back through the condenser to the receiver. This is the reason the system is called the *pump-down system*. The low side is pumped down to a pressure lower than that at which it normally operates.

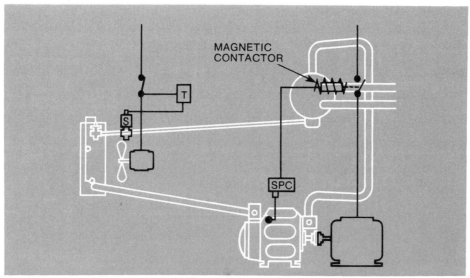

FIG. 18-5 Pump-down cycle system

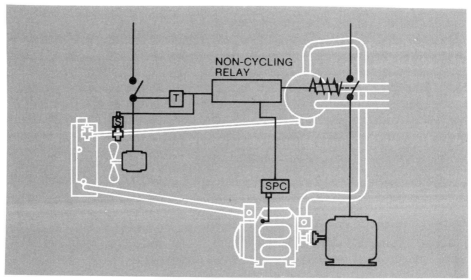

FIG. 18-6   Noncycling system

Noncycling System

Refrigerant is quite apt to leak back into the low-pressure side of the system during the off part of the cycle. The low-pressure control responds to this increase in pressure by starting the compressor. The operation continues until the pressure is again reduced to the preset value.

The unnecessary on-off operation of a system is called *short cycling*. Correcting this condition is a very simple matter. A *noncycling relay* connected to the thermostat permits the suction-pressure switch to start the compressor only when refrigeration is required, as shown in figure 18-6.

The relay is essentially a switch controlling current from another source. In this case, the source of current is through the suction-pressure control. The suction-pressure control can, therefore, stop the compressor even though it cannot of itself (without the help of the thermostat controlled relay) start the compressor. Short cycling is thus prevented. The system can be stopped either by the operation of the thermostat or the low-pressure control.

Solenoid Pilot Valve

The liquid line on huge systems must of necessity be large, requiring a correspondingly expensive solendoid. The same control effect may be produced with a smaller unit called a *solenoid pilot valve*, figure 18-7. When the thermostat calls for the system to operate, the equalizer line of the thermostat expansion valve is connected through the pilot solenoid to the suction line. The expansion valve operates normally under these conditions.

When the thermostat is satisfied, it permits the pilot solenoid to close and stop the pressure from the suction line to the thermostatic expansion valve, figure 18-8. At the same time another internal solenoid port opens and allows high pressure from the liquid

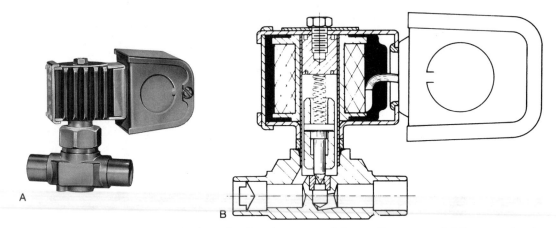

FIG. 18-7   Electrically operated solenoid valve: (A) external design features and (B) internal valve construction features *(Courtesy of Sporlan Valve Company)*

line to pass through to the thermostatic expansion valve. This expansion valve is closed tightly by this action. It thus serves not only as a metering device but as a shutoff valve as well.

LIMITING OR SAFETY CONTROLS

All of the primary or operating controls up to this point have been discussed in relation to normal operating conditions. It is necessary to have other controls which will keep the system within reasonable and/or safe limits of pressure and temperature. This group is

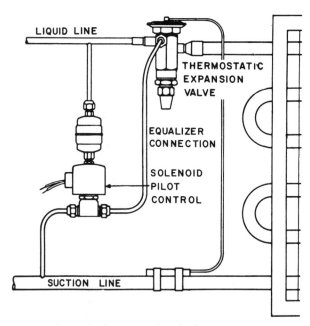

FIG. 18-8   Solenoid pilot control with thermostatic expansion valve

called *safety controls.* Excluded from the treatment are those controls designed specifically to protect the electrical circuit from overloading or shorting and to prevent motors from overheating.

Low-Pressure Control

The mechanical operation of a low-pressure (safety) control is the same as when it is used as an on-off switch to stop and start the system

The low-pressure control stops the compressor at a preset minimum operating pressure. As a safety control, it protects against:

- Extreme compression ratios.
- Freeze-up of the evaporator.
- Entrance of air and water vapor resulting from low-side leaks.

High-Pressure Cutoff

Sometimes operating conditions can cause excessively high head pressures. Conditions such as elevated ambient condensing temperatures or the pressure of noncondensable gases in the system will do this. Also, high head pressure can result from any condition that restricts or stops the flow of water through a water-cooled condenser.

A high-pressure cutoff control is used to prevent the buildup of pressures that may cause damage to the equipment and unsafe conditions. The high-pressure control is similar in construction to the low-pressure control. The bellows or diaphragm which is connected with a small tube to the compressor, actuates an electric switch. The discharge pressure affects this control. A combination low-pressure control and high-pressure safety cut-out (graduated to read in pounds or kilograms pressure) is shown in figure 18-9.

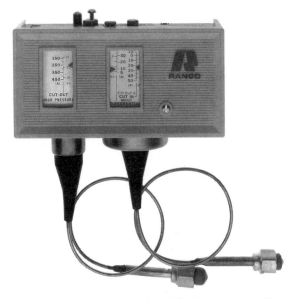

FIG. 18-9   Combination low-pressure control and high-pressure safety cutout *(Courtesy of Ranco North America)*

In addition, the electric switch operates in reverse manner from the low-pressure control. If the pressure rises beyond a specific setting for a maximum safe operating pressure, the bellows or diaphragm in the high-pressure cutout opens an electric switch and thus stops the condensing unit.

The switch stays open and the condensing unit is not permitted to run until either the condition is corrected or the discharge pressure is reduced and causes the discharge pressure to lower considerably. The lowered pressure causes the diaphragm or bellows to close the electric switch and start the condensing unit.

Oil Pressure Failure Control

An oil pressure failure control stops the operation of pressure-lubricated refrigeration equipment when the oil pressure falls below a safe limit for a specified time. This control serves as a combination differential pressure control and a time-delay relay. The differential pressure control measures the useful oil pressure. The time-delay relay prevents the compressor from cutting out when it should be operating.

The useful oil pressure in a pressure lubricated compressor is the difference between the oil pump discharge pressure and the suction pressure. The oil pressure is related to the suction pressure of the oil pump as protection against oil failure. The oil and suction pressures are related in the differential pressure control by using two pressure bellows which are opposed to each other. The oil pump discharge pressure is exerted on one bellows; the suction pressure on the other. Repeating, the usable oil pressure is the difference between the oil pump discharge and the suction pressure.

The normal usable oil pressure in reciprocating compressors may range between 20 to 45 psi. The differential pressure control provides adjustable cut-in and cutout points in the usable oil pressure range. This control is usually adjusted at a cut-in point of 18 pounds above suction pressure and a safe operating cutout point as specified by the manufacturer.

The time-delay relay permits the compressor to operate for about two minutes to establish the correct oil pressure differential. If the pressure differential does not come up to the cut-in point within this predetermined time, the compressor motor shuts off. Also, whenever the usable oil pressure drops below the cutout point during operation, the oil pressure switch closes, causing the time-delay relay to shut off the compressor in a given time.

An oil pressure failure control is illustrated at (A) in figure 18-10. The wiring diagram (B) identifies the main parts and shows how the control works. The contacts of both the pressure differential switch and the timer switch are closed as the compressor is started. If the usable oil pressure does not build up to the pressure at which the control is set, the crankcase pressure keeps the contact of the differential pressure switch closed. The position of the oil pressure failure control in the control circuit is also illustrated in this diagram.

The setpoints of pressure differential controls depend on the model of compressor and the manufacturer's specifications. A general range of factory fixed settings is from 5 to 60 psi.

The time delay switch illustrated in figure 18-10 operates by a thermal heater. This heater is energized by a control switch. The control switch senses a low pressure differential. There is a built-in ambient temperature compensating device to permit accurate timing to meet varying ambient temperature changes. If the pressure is not corrected within the allowable time, the heater opens a bimetal switch.

The photo at (A) shows the manual reset button on the time delay switch. The reset button is used to restart the compressor after shutdown due to loss of oil pressure. Time delay settings are usually 60, 90, or 120 seconds, depending on the compressor manufacturer's specifications.

Another feature of the time delay switch (figure 18-10) is that when one circuit opens to stop the compressor (resulting from oil pressure failure), a second circuit closes an alarm circuit.

When the usable oil pressure builds up to the cut-in point within the specified time, the contacts of the differential pressure switch open. This de-energizes the heat circuit before the bimetal bends and opens the timer switch contacts. The compressor, thus continues to operate normally.

By contrast, if the usable oil pressure fails to build up to the cut-in point within the allotted time, heat from the energized heater acts to bend the bimetal strip. The contacts of the timer switch are broken and the compressor stops.

The compressor cannot be restarted until

- The heater and bimetal have cooled, and
- The timer switch contacts on the control are reset manually by the button.

If during operation the oil pressure drops below the cutout point, the crankcase pressure forces the contacts of the differential pressure switch to close. This causes the

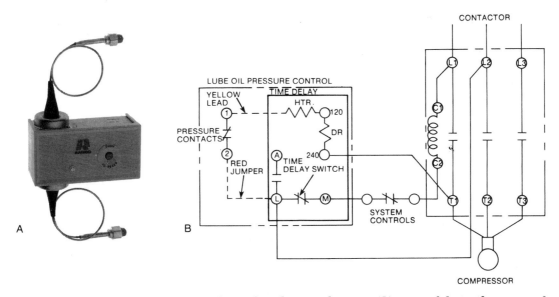

FIG. 18-10   Oil pressure failure control switch and wiring diagram: (A) external design features and (B) typical three-wire diagram of control switch *(Courtesy of Ranco North America)*

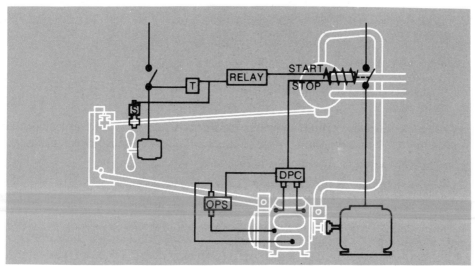

FIG. 18-11   Oil pressure failure control in relation to other controls

heater circuit to become energized and, unless the oil pressure returns to normal within the time delay period, the compressor shuts down. With such safeguards, the compressor never runs more than a predetermined time on subnormal oil pressure. Figure 18-11 shows the position of this device in the control circuit.

Differential

A *differential* between the cut-in and cutout settings is built into most temperature, humidity and pressure-sensitive controls. When applied to a thermostat having a two-degree differential, it means that the contacts close (cut in) if the temperature setting is 70° and the temperature drops to this reading. The contacts remain closed until the temperature rises to 72°, the cutout setting.

In another example, the low-pressure control on a compressor might close to start the compressor when the suction pressure rises to 37 pounds and continues to run until the suction pressure is pumped down to a cutout pressure of 25 pounds.

Controls are designed with differentials to prevent flutter, short cycling and almost continuous operation due to minute changes of temperature or pressure. The differential of heating or cooling equipment must be set to operate over as long periods as possible and less frequently than would be the case in short cycling (short periods of operation at frequent intervals).

Suction Line Regulators

A thermostat or pressure switch is inadequate as a simple compressor control for many refrigeration systems. Instead, suction line regulators provide a very effective method of balancing the capacity of the refrigeration system to the load requirements. Such regulators control the operating pressure of the evaporator or the suction pressure of the compressor. Suction line regulators enable the system to maintain maximum operating efficiency by meeting the requirements of a wide range of load.

Suction line regulators may be divided into two distinct groups:

- The evaporator pressure regulator regulates the upstream or evaporator pressure.
- The suction pressure regulator or hold-back valve, which regulates the downstream or suction pressure at the compressor.

The smaller suction line regulators are of the direct acting or internally pilot operated type. The large suction line regulators are externally pilot operated for high sensitivity and accuracy of control.

Evaporator Pressure Regulators. The *evaporator pressure regulator* is the oldest and most widely known suction line control. This regulator prevents the evaporator pressure from falling below a predetermined pressure setting. The regulator may be used on a single evaporator like a water chiller or several may be installed in a multiple system to maintain certain minimum pressures on individual evaporators.

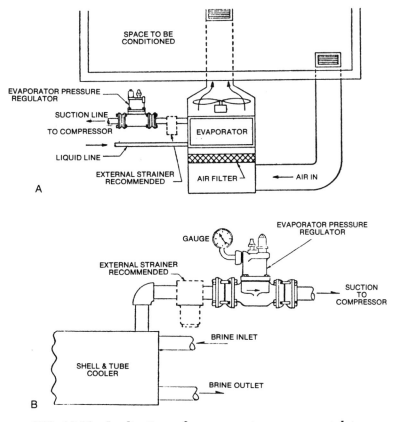

FIG. 18-12 Applications of an evaporator pressure regulator

Evaporator pressure regulators may be used on any refrigeration system to maintain a minimum evaporator pressure or temperature. The regulators may be fed by thermo expansion valves, high-side floats, low-side floats or solenoid liquid valve and float switch.combinations.

The schematic drawing at (A) in figure 18-12 shows how an evaporator pressure regulator is installed on a blower-type finned evaporator in order to prevent or minimize dehumidification. A humidified condition is essential in vegetable storage. The regulator prevents the evaporator pressure and the corresponding saturation temperature from falling below a predetermined pressure setting at low loads. This prevents large temperature splits between evaporator and air temperature, with resulting dehumidification.

Evaporator pressure regulators may also be used on brine or water chillers (B). The regulators prevent freeze-up of the chiller at a low load. As a load increases, an evaporator pressure in excess of the regulator setting is produced. This causes the regulator to open until it reaches the wide-open position at full load.

Suction Pressure Hold-Back Valves. A compressor may be overloaded when heavy loads are thrown periodically on the system unless an oversized motor is selected. This condition may exist in a liquid chiller system where a quantity of liquid is cooled from a high to a low temperature. Such a chiller requires a large surface at low temperature while at high temperature it is able to evaporate an excess amount of refrigerant.

A *suction pressure hold-back valve* is used to prevent the suction pressure from running too high at the compressor. The suction pressure hold-back valve permits only a predetermined maximum amount of gas to reach the compressor. The suction pressure at the compressor is thus prevented from becoming higher than the setting of the valve. The capacity of the system is effectively limited because the gas held back by the valve increases the pressure in the evaporator and reduces its capacity.

SUMMARY

- A refrigeration control system accurately and safely maintains the temperature of the refrigerated space within a minimum of operation.
- Operating or primary controls function as an electric switch to turn the refrigeration system on and off. Thermostats, humidistats and pressurestats are examples.
- Actuating controls function in response to the requirements of the primary group in actually controlling some part of the refrigeration system. This group consists of relays, solenoid valves, pilot solenoids and four-way valves.
- Safety or limiting controls operate to keep the refrigeration system within safe limits of temperature and pressure. This group consists of the safety thermostat, the high- and low-pressure cutoff and the oil safety switch.
- The thermostat provides automatic control by responding to the temperature in the refrigerated space. The thermostat gives on-off operation by energizing the compressor through a magnetic starter.
- When a thermostat operates a solenoid which shuts off the supply of refrigerant to the evaporator, this permits the compressor to continue operating until the evapora-

tor is pumped down clear of refrigerant. Any leakage of refrigerant causes the compressor to start on cycle unnecessarily even though additional refrigeration is not required.

- The disadvantage of the pump-down system is overcome by using a relay controlled by a thermostat.
- Solenoids are used in the liquid or suction line to stop the flow of refrigerant. They may be used as a pilot solenoid to control the flow of refrigerant by controlling the pressure to the equalizer of the thermostatic expansion valve.
- Excessively low pressures are prevented by low-pressure cutoff controls.
- Damage to the compressor, due to lack of lubrication, is prevented by the oil pressure failure control.
- Differential is the difference of temperature or pressure between the on and off operation of the control.
- A limiting control which prevents fluctuations in pressure in the suction line (and hence the evaporator) is called an evaporator pressure regulator. It is used in conjunction with a blower coil or water chiller.
- A suction pressure hold-back valve limits the suction pressure at the compressor caused by overloading.

ASSIGNMENT: UNIT 18 REVIEW AND SELF-TEST

A. CLASSIFICATION OF CONTROLS

Controllers may be classed as (a) operational, (b) actuator, and (c) safety or limit. Indicate by appropriate letter the group to which each controller belongs.

1. Thermostat
2. Solenoid
3. High-pressure cutoff
4. Pressurestat
5. Suction line regulator
6. Humidistat
7. Oil pressure failure control
8. Relay or contactor
9. Manual switch
10. Low-pressure cutoff

B. OPERATING CONTROLS AND ACTUATORS

1. Name the components or conditions necessary to complete the control systems given in the table on the next page at A, B, C and D. *Note:* If there is no component or disadvantage, write *none.*
2. Explain briefly why it is necessary or desirable to have a manually operated feature on a solenoid valve.
3. State (a) the need for and (b) the operation of the solenoid pilot valve.
4. Describe briefly the operation of the noncycling relay.

Section 5: Components of Refrigeration Systems

	Type of System	Operating Control	Actuator	Motor Control	Safety Control	Disadvantage
A	Manual					
B	Simple Automatic					
C	Pump-down					
D	Noncycling					

C. SAFETY AND LIMITING CONTROLS

Give the letter of the device in Column II which matches each operation or condition in Column I.

Column I

1. Minimizes dehumidification of vegetables.
2. Prevents excessively high head pressure at the compressor.
3. Prevents excessively high suction pressure at the compressor.
4. Prevents damage to the compressor from lack of proper lubrication.
5. Prevents freeze-up of the water chiller.
6. Two controls in a single case.
7. Avoids excessively high compression ratio.
8. Maintains a constant minimum pressure in the evaporator.
9. Prevents excessively low pressure in the evaporator.
10. Maintains constant suction pressure at the compressor.

Column II

(a) Low-pressure cutoff
(b) High-pressure cutoff
(c) Oil pressure failure control
(d) Evaporator pressure regulator
(e) Suction pressure hold-back valve

UNIT 19:
TEMPERATURE INDICATORS, MEASUREMENT, AND CONTROLS

Temperature has been defined up to this time as the hotness of a material as compared to a fixed point on a temperature measuring scale. While the terms *Fahrenheit, Celsius, fixed points,* and *scale* have been used, no explanation was given as to their exact meaning.

This unit identifies three fixed points for different temperature measuring scales. Then, since the common temperature measuring systems are different, a few examples are used to show how to change temperature from one scale or system to another.

This is the foundation upon which liquid- and pressure-operated temperature-measuring instruments are described and illustrated. The function of different types of temperature controls and recording indicators used in refrigeration systems are also given.

TEMPERATURE MEASURING SYSTEMS

The word *scale* is used with temperature measuring devices to identify a system of measurement. The scale must have definite *fixed points* or standards that always have the same value and may be reproduced easily.

Fixed Reference Points on Temperature Scales

Certain standard temperatures are used that depend upon the physical conditions of a material, figure 19-1. For example, the temperature at which water freezes was selected as one of the standard temperatures that can always be reproduced. This freezing temperature is influenced by extreme changes in pressure and only minutely by ordinary changes in atmospheric pressure. So, the freezing point of water at atmospheric pressure is the first of the standard temperatures that is used as a fixed reference point.

(A) FREEZING POINT   (B) BOILING POINT   (C) ABSOLUTE ZERO

FIG. 19-1   Fixed reference points on temperature scales

Section 5: Components of Refrigeration Systems

Another common temperature is the boiling point of water under normal conditions. Unlike freezing, the temperature at which water boils is greatly affected by atmospheric pressure. For this reason it is important to know what the pressure conditions are when boiling takes place.

The third fixed reference point is known as *absolute zero*. This is the temperature at which scientists believe all movement of molecules ceases. Since their movement indicates heat energy, it follows that if there were no movement there would no longer be any heat. Experimenters have been able to produce temperatures within a few hundredths of a degree of this absolute zero.

Ambient and Operating Temperatures

The temperature of the surrounding air is known in refrigeration work as *ambient temperature*. This ambient temperature is usually expressed in degrees Fahrenheit and is used whenever an *operating temperature* is required. The operating temperature is equal to the sum of the ambient temperature and the rise in temperature of the unit itself.

Suppose for instance that the temperature rise as marked on an electric motor specification plate reads 55°C. Under normal operating conditions the motor can be expected to run hot within 55°C above the temperature of the surrounding air. If the ambient temperature is 30°C, the operating temperature is:

$$55°C + 30°C = 85°C$$

Temperature Measuring Scales

Changes of temperature are measured on either the *Fahrenheit* or *Celsius* scale, figure 19-2. These two basic scales are known as the *normal scales*. There are two fixed points on both scales that indicate:

1. The temperature at which ice melts.
2. The temperature at which pure water boils under a standard atmospheric pressure.

These fixed reference points may be duplicated exactly in any part of the world.

Measurements are made on either the Fahrenheit or the Celsius scales in units called *degrees*. The degree is written ° and is followed by a letter to show to what scale it applies. A reading like 25°F means that the temperature is 25 degrees as measured on the Fahrenheit scale. The letter C is used for the Celsius scale.

Comparing Values on Normal Temperature Scales

The point at which ice melts is known as zero degrees on the Celsius (C) scale and 32 degrees on the Fahrenheit (F) scale. The boiling point of pure water at standard pressure is 100°C and 212°F. The locations of the freezing and the boiling points are shown in figure 19-3.

There are 180 Fahrenheit degrees between the freezing and boiling points of water, but only 100 Celsius degrees. Thus, the difference between the melting point of ice and the boiling point of water is 100 degrees in the Celsius system and 180 degrees in the Fahrenheit system. This means that each degree change in temperature on the Fahrenheit scale is equal to five-ninths of a degree on the Celsius scale.

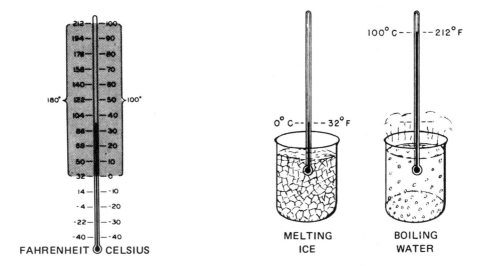

FIG. 19-2   C and F scales   **FIG. 19-3   Starting points on C and F scales**

Note from the drawing, figure 19-3, that the freezing point on the Celsius scale starts at zero while the Fahrenheit scale begins with the number 32. These two numbers must be remembered when a value on one scale is changed to its equivalent value on the other scale.

Knowing that a degree on the Fahrenheit scale is smaller (5/9's) than a degree on the Celsius scale, and that the freezing point of water is 0° Celsius and 32° Fahrenheit, it becomes a simple matter to change readings from one scale to the other.

To change a Celsius reading to Fahrenheit

1. Multiply the reading by 9/5
2. Add 32°

To change a Fahrenheit reading to Celsius

1. Subtract 32° (or add a -32°)
2. Multiply by 5/9

Three examples are used to show how temperatures are converted (changed) from one normal scale to the other. The reason for saying "add a -32°" will be shown when below zero readings are to be converted.

EXAMPLE 1:

Find the Fahrenheit temperature equal to 35°C.

PROCEDURE:

Step 1: Multiply the reading by 9/5

$$35° \times \frac{9}{5} = 63°$$

Step 2: Add 32°

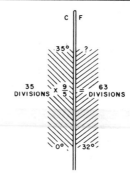

ANSWER:

63° + 32° = 95°F

EXAMPLE 2:

Find the Celsius equivalent of 77°F.

PROCEDURE:

Step 1: Subtract 32° from the reading

77° − 32° = 45°

Step 2: Multiply the remainder by 5/9

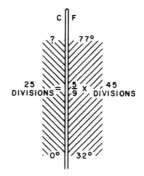

ANSWER:

$$45° \times \frac{5}{9} = 25°C$$

EXAMPLE 3:

Find the C temperature equal to –58°F.

PROCEDURE:

Step 1: Subtract 32° from the reading. Note: Because this is a *minus* reading, subtraction actually increases the *minus* quantity because a minus 32 is added.

$$-58° + (-32°) = -90°$$

Step 2: Multiply by 5/9

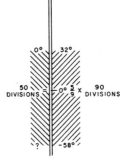

ANSWER:

$$-90° \times \frac{5}{9} = -50°C$$

ABSOLUTE TEMPERATURE SCALES

The Celsius and the Fahrenheit scales indicate relative temperature. *Absolute temperature* may be measured only when the reading scale starts at a true zero temperature where, because there is no heat, there is no degree of heat. This *absolute zero* is the basis of two other temperature scales. These are called the *Kelvin* and the *Rankine* scales.

The Kelvin (K) scale is referred to as the *absolute Celsius scale* and is used in scientific work.

The Rankine (R) scale is the *absolute Fahrenheit scale.*

These are the scales that are applied in refrigeration and air conditioning. A comparison between the normal scales and the absolute scales for boiling and freezing points is illustrated in figure 19-4.

The Kelvin Temperature Scale

Scientifically, it has been established that a gas decreases 1/273 of its 0°C pressure for each degree that it is cooled. This condition exists for all gases. Thus, the temperature at which no pressure is exerted is at the same point as the temperature at which there is no heat, or –273°C.

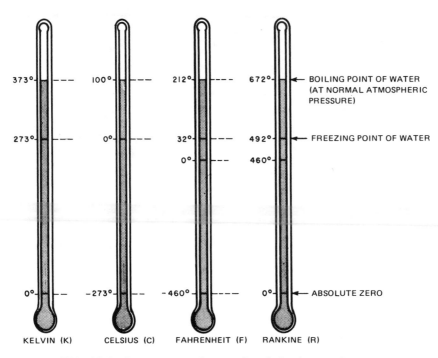

FIG. 19-4 Comparison of normal and absolute scales

Before gases change state, it is also true that at a constant pressure there is an expansion (or contraction) of 1/273 of their volume for each degree Celsius increase (or decrease). Referring to the illustration on which the temperature scales are compared, note that the unit degree on the Kelvin absolute scale is equal in value to the Celsius degree.

To convert from the Celsius to the Kelvin scale when the reading is 0°C or higher, merely add 273°. This is so because the zero reading on the Kelvin scale is located 273° below 0°C. For example, 35°C is equal to 273 + 35, or 308°K. The 273 represents the number of Celsius degrees between absolute zero and the melting point of ice. Since the reading is 35° above 0°C, 35 is added to 273 to get the 308°K value.

On readings which are below 0°C, subtract the reading from 273 in order to get the Kelvin scale reading. For instance, –15°C is equal to 273 – 15, or 258°K.

The Rankine Temperature Scale

On the F scale, absolute zero is 460 Fahrenheit degrees below 0°F or –460°F. This is the place at which there would be no heat whatsoever and no motion of the molecules. Since the size of the units on the F and R scales is the same, ordinary thermometer temperatures in Fahrenheit may be easily changed to R temperatures. If the temperature reading is above 0°F, merely add 460 to the reading. When a temperature reading is below 0°F, subtract the reading from 460. The answer will then be in degrees on the R scale.

EXAMPLE 4:

A. Change a 50°F thermometer reading to R absolute temperature.

   460 + 50 = 510

B. Convert –20°F to its equivalent R value.

   460 – 20 = 440

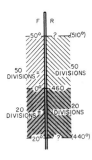

ANSWER:

A. The 50°F reading is 510°R (Rankine absolute).

B. The –20°F reading is 440°R.

The absolute scales must be used when calculating the changes in pressure that a refrigerant gas experiences when being superheated in the evaporator or in a cylinder left out in the hot sun.

TEMPERATURE INDICATORS

Two general groups of temperature indicators which are used in refrigeration work are briefly described and illustrated. These indicators are either (a) liquid or (b) pressure-operated.

Liquid-Operated Temperature Indicators

There are many different types of indicators used for temperature measurement. A number of these have been selected in order to show the range that is available.

The liquid thermometer is the most common among the temperature indicators. The liquid may be mercury, alcohol, or any other colored fluid, depending upon the range to be covered. This thermometer may be conveniently used to measure the temperature of the evaporator. The location of the thermometer, indicated by the letter T, is shown on the Basic Refrigeration Cycle Diagram at the front of the text. Clamps are available to attach this type of thermometer to tubing lines.

Stem-Type Refrigeration Thermometers

A thermometer is also used to check the refrigerator cabinet temperature and the condensing unit temperature, figure 19-5. (Refer to point T$_4$ on the Basic Refrigeration

FIG. 19-5 Stem-type refrigeration thermometer and case *(Courtesy of Thermometer Corporation of America)*

FIG. 19-6 Checking accuracy

Diagram.) A mercury-in-glass thermometer is the type used in industrial refrigeration service. The range of this thermometer is from –42°F to 110°F in graduations of two F degrees. This thermometer has a triple lens front, which magnifies the mercury in the tube. Unless there is a special magnifying lens, the mercury-filled thermometer is hard to read.

The thermometer can be checked for accuracy by using an ice and water bath, figure 19-6. It should check within one degree of accuracy at 32°F. Usually the thermometer carried by refrigerator service technicians will not go up to 212°F. Therefore, the ice bath is the only check that can be easily used.

Occasionally, the fluid (liquid) in the column of a thermometer separates. The best way to make the liquid column unite again is to cool the bulb by spraying it with a small quantity of Refrigerant 12. The cooling effect from the refrigerant causes the fluid in the column to shrink into the bulb. When the liquid re-expands it will not be separated. Care must be taken to make certain that the liquid is not frozen solid so as to damage the thermometer.

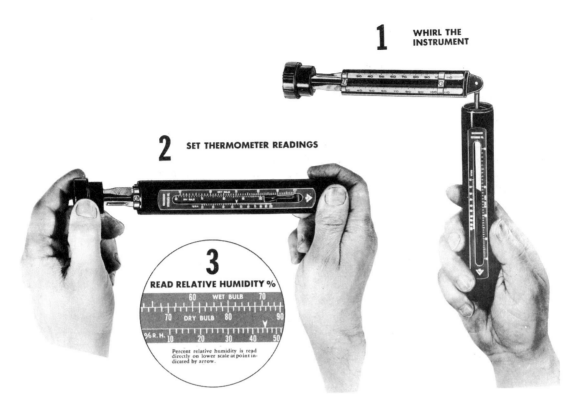

FIG. 19-7 **Humidity measuring instrument (sling psychometer)** *(Courtesy of Bacharach Inc.)*

Humidity Measurement

Inasmuch as refrigeration is the main part of an air conditioning system, one type of instrument which is used for measuring humidity (moisture) is shown in figure 19-7. This instrument is known as a *sling psychrometer* and uses mercury-filled thermometers. The recent model illustrated was designed to be conveniently carried by refrigeration service technicians.

Superheat

The term *superheat* refers to the temperature difference between the inlet and outlet of the evaporator. The normal superheat setting should be 10°F (5.6°C). This temperature can be obtained by adjusting the thermostatic expansion valve.

OTHER TEMPERATURE INDICATORS

Metallic Thermometers

Dial thermometers are also used extensively for indicating temperatures. They may be operated by either a single metal or a bimetal strip.

The bimetallic type operates on the principle of unequal expansion of two metals in response to changes of temperature. The illustration in figure 19-8 shows the effect of applying heat to a bimetal strip. The movement of the bimetal strip causes a corresponding movement on the indicator. The amount of movement is controlled by the change in temperature.

Pressure-Operated Thermometers

The movement on some dial thermometers is controlled by a bellows, a capillary, and a remote sensitive *bulb*. The *capillary* is a small hair-fine tube used to transmit pressure. The pressure is produced by changes of temperature of the fluid in the bulb.

The range of the dial-type thermometer, which is shown in figure 19-9 with capillary and remote sensitive bulb, is from –40°F to 120°F, in either °F or °F/°C. The range of other models in this series is from 40°F to 240°F and equivalent degrees celsius. Some thermometers are provided with an adjusting screw to permit resetting or to correct inaccuracies.

Dial thermometers are sometimes used in pairs to measure superheat as in figure 19-10. Note that the bulb (known also as *temperature sensing element*) for each instrument is clamped to one of two different lines: the evaporator (liquid) inlet or the outlet (suction) line. The points of measurement correspond to the points marked T_2 and T_3 on the Basic Refrigeration Cycle Diagram at the front of the text. The difference in readings between the two thermometers is designated as superheat even though it is a temperature difference.

Many authorities insist that the proper way to measure superheat (on mechanical systems which use compressors) is to place only one thermometer on the suction line at the point indicated. The other temperature is obtained indirectly from the pressure of the suction gas. This pressure is indicated on a gauge in the suction-service valve of the

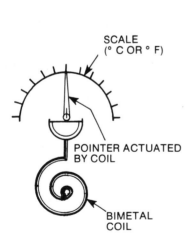

FIG. 19-8   Heating effect on bimetal strip

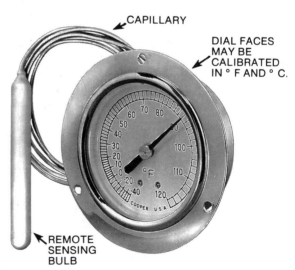

FIG. 19-9   Pressure-operated thermometer *(Courtesy of Cooper Corporation)*

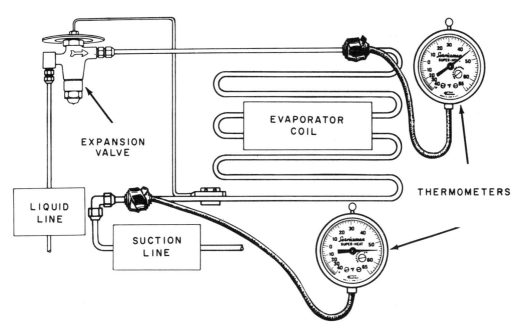

FIG. 19-10 Measuring superheat

compressor. (This point is indicated as P_1 on the Basic Refrigeration Cycle Diagram; P_2 measures discharge pressure.) The difference between the thermometer reading and the temperature obtained indirectly from the gauge reading (called *inferred temperature*) is the superheat in degrees. Since absorption-type refrigerating units contain hydrogen gas, they are not recharged in the field and thus do not require these measurements.

Thermoelectric Temperature Indicator

The *thermoelectric temperature indicator*, figure 19-11, is another type of temperature measuring instrument. Such an indicator gives a quick, reliable response and is convenient to use for refrigeration work. This indicator operates on the principle that minute quantities of electric current may be produced by heating two dissimilar metals which are joined at one end. The instrument is so named because a thermoelectric (heat-electric) effect is produced.

The lead wires for the instrument shown are insulated from one another except where they are joined together and welded at the tip. As such, the tip forms a *thermocouple* or *thermojunction*. Any two dissimilar metals like iron and constantan may be used. The metals are selected on the basis of the range of temperatures which must be measured.

When the wires are heated at the tip (known as the *hot junction*) different electron vapor pressures are set up within the two metals. This electron vapor pressure refers to the movement of atoms and their electrons. As heat is applied, the electron vapor pressure increases. The difference in this pressure between the two metals causes the electrons to flow from one metal to the other, producing an *electromotive force*.

365

FIG. 19-11 Thermoelectric temperature indicator *(Courtesy of Simpson Electric Company)*

This electromotive force is an electrical pressure which moves through the circuit of the instrument. The electrical effect produced is transmitted to the pointer of the thermoelectric indicator. In back of the pointer is a scale graduated in degrees of temperature. Thus, changes in temperature produce changes in electrical energy. This electrical energy, in turn, operates the thermoelectric indicator to measure degrees of temperature.

Electronic Solid-State Thermistor Sensors

Thermistor sensors are small electronic solid-state semiconductor devices. As a *semiconductor,* a thermistor permits small quantities of current to flow (depending on its temperature). There is less resistance to current flow as the temperature of a thermistor rises; greater resistance as the temperature drops.

Changes in current flow are monitored through special electronic circuits. These circuits are designed to provide temperature readouts and/or to control functions such as starting, stopping, accelerating, or decelerating machine components or equipment or processes within a system.

Digital, Battery-Operated (Electro-) Test Thermometers

Mechanical thermometers of conventional bimetal or vapor tension types may be *retrofitted* (converted) with *digital temperature test instruments.* Digital thermometers are designed in pocket models, figure 19-12A, and for flush panel mountings as shown at (B). These instruments are designed for through-wall and for *remote readings* where the probes may be extended up to 1,000 feet from the instrument.

Digital temperature testers (thermometers) are quick reading and accurate to one-tenth of one degree. A few examples of applications include heating, ventilating, and air-conditioning systems; freezers and coolers; energy management and solar systems; laboratory work, food storing and processing; heating and curing ovens; and others.

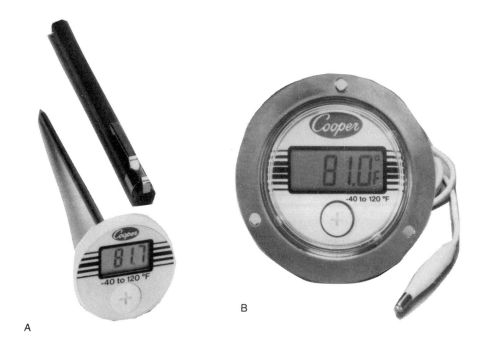

A

B

FIG. 19-12 Examples of battery-operated (electro-) digital test thermometers: (A) pocket type with carrying tube and (B) flush mounting panel type *(Courtesy of Cooper Instrument Corporation)*

Solid-State Digital Thermometer/Pyrometer

The *solid-state digital thermometer/pyrometer* utilizes fast-response thermocouple probes to collect temperature data from different locations in air-conditioning and refrigeration systems. The input data is essential in analyzing and solving performance problems in superheat, subcooling, temperature differentials (TDs) in evaporator and condenser coil applications and to obtain other temperature readings for air-conditioned areas, piping, etc.

This new instrument, figure 19-13, extends the service mechanic's ability to quickly and accurately take temperature measurements over a wide range of 2,000°F. As the product name implies, the digital thermometer/pyrometer combines measurements of conventional thermometers with an extended pyrometer range from –40°F to 1999°F (or –40°C to 1100°C).

Features and Principles of Operation

Measurement accuracies of the instrument illustrated are ± 1% in the 32° to 212°F range and ± 3% in the full range up to 1999°F (1100°C scale). Temperature probes are available in some models for air, surface, immersion, and superheat applications.

The *microprocessor (digital circuit)* used provides for the conversion of the thermocouple input voltage to the four-digit display (°F). The instrument is designed with memory storage and arithmetic problem-solving capability. This permits the storage of on-the-job test results and the establishing of data to establish temperature differentials.

FIG. 19-13   Digital thermometer/pyrometer with measuring probes *(Courtesy of TIF Instruments, Inc.)*

Other advantages of this instrument include the following:

- Rapid measurement display update of three times per second
- Easy-to-read, accurate digital display of each measurement

TEMPERATURE RESPONSIVE CONTROLS AND RECORDERS

Thermostatic Expansion Valves

The thermostatic expansion valve, figure 19-14, is used for refrigerant flow control and operates on increased pressure resulting from a rise in temperature. The bulb on the end of the small tubing is the sensing element which responds to the temperature of the part to which it is attached.

The sketch in figure 19-15 demonstrates the proper way to fasten the sensing element to a suction line. (Referring again to the Basic Refrigeration Cycle Diagram, such an element would be attached to the suction line at T_3). The response of the element to the suction temperature (really its superheat) controls the flow of the refrigerant.

MOTOR CONTROL THERMOSTATS

The heart of a mechanical system may be operated through its electric motor. The motor can be controlled by a remote-bulb temperature-sensing thermostat. A typical control is shown in the cutaway view, figure 19-16.

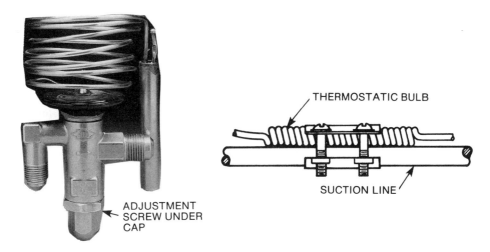

FIG. 19-14 Thermostatic expansion valve *(Courtesy Alco Controls Division, Emerson Electric Company)*

FIG. 19-15 Fastening the sensing element to a suction line

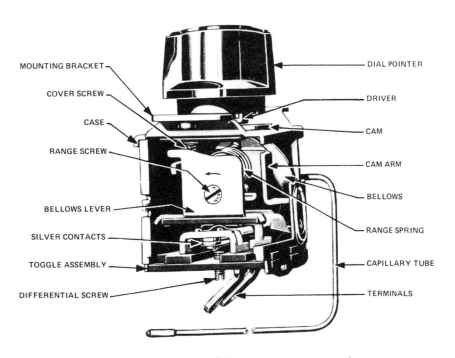

FIG. 19-16 Motor control thermostat *(Courtesy of Ranco)*

The pressure in the sensing element responds to the temperature in the refrigerated space. This pressure moves (actuates) certain parts of the control unit to turn the motor on or off as required. This kind of a control is used in small commercial units and for household units.

Thermostatic Water Valves

The last type of temperature controller to be described is the water valve. One that operates as a result of the temperature-pressure effect is shown in figure 19-17. This valve is controlled by the temperature of the outlet water from the condenser. A thermostatic element is connected to the bellows, which operates the valve. The sensing bulb is charged with a volatile liquid.

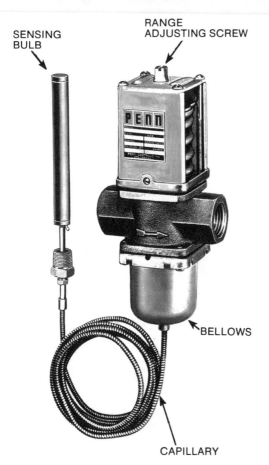

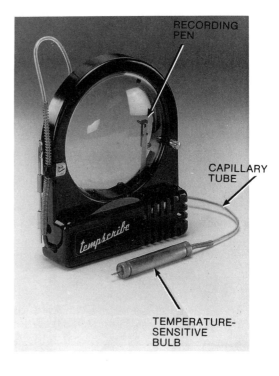

FIG. 19-17 Thermostatic (temperature-actuated) modulating water valve *(Courtesy of Control Product Division, Johnson Controls, Inc.)*

FIG. 19-18 Front case assembly of temperature recorder instrument *(Courtesy of Bacharach Inc.)*

The sensing bulb is clamped to the condenser cooling water outlet line. This point is represented as Tw on the Basic Refrigeration Cycle Diagram. The pressure created by the volatile liquid in the bulb opens the water valve when the temperature of the outlet water becomes warm and closes it as the line cools. Thus, the temperature of the gaseous refrigerant in the discharge line is controlled.

Control For Absorption-Type Refrigerators

The thermostat on an absorption-type refrigerator is actuated by temperature. The response of the sensing element to the temperature in the evaporator regulates the flame. The flame, in turn, controls the operation of the unit to maintain the desired temperature in the refrigerator box.

Temperature Recorders

A *temperature recorder* (as its name suggests) is a device for automatically recording temperature over a definite period of time. One of the latest designs of a temperature recorder consists of two basic assemblies. The first assembly includes a temperature chart drive unit. A temperature chart is mounted on the front face of the drive so that temperatures can be recorded over various time spans.

The chart drive assembly is hinged to the front case (second assembly), which contains the measuring element, figure 19-18. The temperature-sensitive bulb is placed in the area where temperature is to be measured. The temperature impulse is transmitted through the capillary tube from the bulb to the recorder. Increases in temperature at the temperature-sensitive bulb cause an expansion of the gas and an increase in pressure in the system.

Pressure changes are transmitted through the capillary tube to a pressure-controlled bellows. The bellows actuates a pen arm, which records temperature changes on a time-rotating temperature chart mounted on the face of the chart drive assembly.

When it is possible to place the recorder directly into a refrigerated space, another model of the same recorder may be used. In this case the temperature-sensitive element is a spiral-shaped metal spring of either one or two different metals. Changes of temperature cause expansion or contraction of this spring. The motion produced is transferred to an arm which holds a recording pen.

Heating-Cooling Thermostats

Thermostats operate on the principle of expansion and contraction due to temperature change. Illustrated in figure 19-19 is a combination heating-cooling thermostat of a type used in year-round air conditioning.

Programmable Thermostats

Programmable thermostats provide automatic control of single-stage heating or heating/cooling air-conditioning systems. These thermostats are designed to be powered through heating/cooling system controls, direct from a transformer, or by batteries.

Programmable means that different schedules during a day, weekdays and weekends may be set up to meet specific heating/cooling needs. The digital clock area shown in

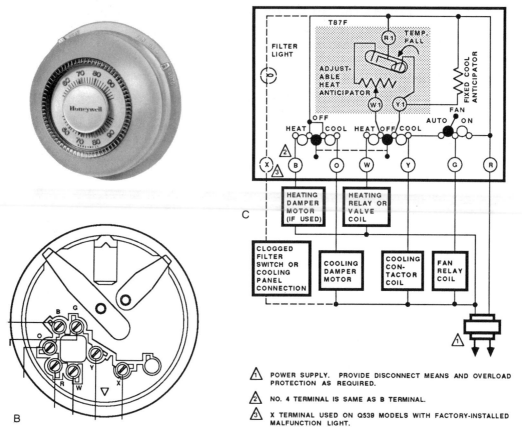

FIG. 19-19 **(A)** Heating-cooling temperature control thermostat and **(B)** wiring hookup thermostat and **(C)** schematic wiring diagram for thermostat *(Courtesy of Residential Division, Honeywell Inc.)*

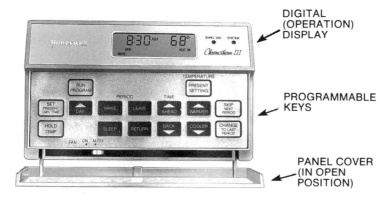

FIG. 19-20 **Programmable thermostat** *(Courtesy of Residential Division, Honeywell Inc.)*

figure 19-20 displays the time of day, day of the week, room temperature and operating mode. In addition, program times and temperature set points are displayed on request.

The *keys* (inside the thermostat case cover) are programmed for time and temperature heating or heating/cooling requirements in air-conditioning systems. The programs may be temporarily or permanently changed.

Fuel savings are realized by setting the thermostat for lower (heating) *setback* or higher (cooling) *setup temperatures* for those periods of time when heating/cooling needs might be reduced. A small light on the digital clock panel shows when the system is in the energy saving mode.

SUMMARY

- Absolute zero refers to the temperature at which all molecular motion ceases, according to the kinetic theory of heat.
- Ambient temperature is the temperature (usually of the air) surrounding operating equipment.
- The C and F scales are known as normal temperature scales.
  Freezing point: 0°C, 32°F   Boiling point: 100°C, 212°F
  F = C × 9/5 + 32°   C = (F −32°) × 5/9
- Absolute temperature scales are used for calculating changes in refrigerant vapor pressures.
- The Kelvin scale measures absolute Celsius temperature.
  K − C (0° or above) + 273 *or* 273 − C (below 0°)
- The Rankine scale measures absolute Fahrenheit temperature.
  R = F(0° or above) + 460 *or* 460 − F (below 0°)
- The effect of expansion and contraction caused by temperature changes is used in thermostats and thermometers.
- The effect of pressure changes is used in remote bulb indicators and refrigeration controls.
- Liquid thermometers of mercury, alcohol or other fluid are used to check refrigeration cabinet and condensing unit temperatures.
- Pressure-operated temperature indicators usually operate on the effect of temperature on a bimetal strip, or bellows filled with volatile fluid.
- Temperature recorders combine the design features of a temperature recording unit and a coordinated drive assembly, which identifies specific periods of time over which the temperature measurements are taken.
- The solid-state digital thermometer/pyrometer is used in superheat, supercooling, evaporator and condenser coil, and other applications that require measurement capability over an extended range (−40°F to 1999°F). Temperature data, collected by thermocouple probes in different locations, are converted through a digital circuit (microprocessor) where memory is used to store data and to calculate temperature differentials for four-digit display in degrees F.
- Thermostatic expansion valves depend on increased pressure from a rise in temperature to operate as a refrigerant flow control device.
- Motor control devices depend on the response of a sensing element to temperature change.

- Thermostatic water valves are used to control the temperature of the refrigerant in the condenser.

ASSIGNMENT: UNIT 19 REVIEW AND SELF-TEST

A. TEMPERATURE MEASURING SYSTEMS

1. Match each temperature term in Column I with the correct condition in Column II.

Column I	Column II
a. The Celsius and Fahrenheit systems of temperature measurement.	(1) Multiply by 9/5.
	(2) Subtract 32° and multiply by 5/9.
	(3) Add 460 to the temperature reading.
b. The Kelvin temperature scale.	(4) This system measures absolute temperature starting with –273°C.
c. Changing a temperature from °C to °F.	(5) There are two fixed points on the temperature scale: (a) to indicate the melting point of ice, and (b) to show the boiling point of pure water at normal atmospheric pressure.
d. Changing an above zero degree reading on the F scale to an absolute reading on the R scale.	
e. The Rankine scale.	(6) Multiply by 9/5 and add 32.
f. Changing a temperature from F to C.	(7) Starts at approximately –460°F.
	(8) Multiply by 5/9.

2. Give the boiling point of water at standard atmospheric pressure on the C, F, K, and R temperature scales.
3. Give the freezing point of water on these same temperature scales.
4. Name the absolute Fahrenheit scale.
5. Name the absolute Celsius scale.
6. Change the following normal temperatures:
   a. 77°F to C   b. 35°C to F
7. Change the following normal temperatures to absolute:
   a. 40°F to absolute F   b. 27°C to absolute C
8. Is 20°C higher, lower or the same temperature as 20°F?
9. Is 80°F higher, lower or the same temperature as 100°C?
10. Is a change of temperature of 50°C greater, smaller or equal to a change of temperature of 90°F?
11. Compute the missing equivalent values on the F, C, K, or R scales as indicated in the table on page 375 for the temperatures given.
12. Pure copper wire melts at approximately 1080°C. What is its melting point in degrees Fahrenheit?
13. Rubber covered wire may be used in all installations except those where the temperature exceeds 122°F. What is the C temperature limit?
14. A certain transformer has a 55 C degrees temperature when operating. What is the temperature rise in F degrees?

Temperature Readings				
	°F	°C	°K	°R
A	14			
B	95			
C		0		
D		100		
E			0	
F			100	

15. What is a transformer's operating temperature in F degrees if the ambient temperature is 70°F and it has a 55 C degrees temperature rise when operating?

16. If the ambient temperature is 68°F and motor temperature is 140°F, what is the temperature rise of the motor in C degrees?

17. A motor has a rated temperature rise of 50 C degrees. If the ambient temperature is 86°F, what is the continuous running temperature of the motor in degrees C?

B. TEMPERATURE MEASUREMENT PROBLEMS ON BASIC REFRIGERATION CYCLE DIAGRAM

Refer to the Basic Refrigeration Cycle Diagram to determine the temperature locations and conditions described in statements 1 to 5.

1. At what two points on the diagram are temperatures taken to determine superheat.

2. Is the temperature of the refrigerant at T_5 on the diagram higher, lower, or the same as that at T_6?

3. Is the temperature of the liquid refrigerant at T_2 higher, lower or the same temperature as that of the liquid at T_7?

4. Is the temperature of the liquid refrigerant in the evaporator higher, lower or the same as that of the gaseous refrigerant at the same place?

5. What is the difference in temperature between T_2 and T_5 called?

C. TEMPERATURE INDICATORS

Supply the correct refrigeration terms to complete statements 1 to 6.

1. The temperature in the evaporator may be conveniently checked with a _____ .

2. A refrigeration cabinet temperature and condensing unit temperature may be checked with a _____ .

3. Periodically a thermometer should be checked for accuracy by using _____ bath.

4. Pressure-operated temperature indicators depend for operation on a _____ .

5. The effect of change of temperature on a bimetal strip is used on _____ .

6. Superheat may be measured by attaching a bulb from each of two instruments on these two different lines: a. _____ b. _____ .

Describe briefly how to perform the following operations (7 and 8):

7. How to connect a broken liquid column in a thermometer.
8. How to determine superheat using a second method.
9. Name the kind of instrument or device which:
    a. Records changes in temperature by changes in electron vapor pressure of two dissimilar metals.
    b. Makes a graphic record of the temperature in a refrigerated space.
    c. Consists of two dissimilar metals joined and welded together at one end.
10. State two advantages of using a digital thermometer/pyrometer in temperature measurement applications in comparison with the use of a standard pressure-actuated thermometer.

D. TEMPERATURE CONTROL AND RECORDERS

Indicate whether each of the following instruments or devices operates as a result of change of (a) pressure or (b) size when subjected to change of temperature.

1. Remote-bulb recorder.
2. Service technician's thermometer used to check the evaporator.
3. Cooling and heating thermostat.
4. Thermometers used for checking superheat.
5. The spiral coil that actuates a motor control thermostat.
6. Thermometers used in the sling psychrometer.
7. Recorder to be placed directly in a refrigerated space.
8. The bulb (sensing element) on a thermostatic expansion valve.
9. The sensing element on an absorption-type refrigerator.
10. Thermostatic expansion valve.

UNIT 20:
SUPPLEMENTARY
REFRIGERATION CONTROLS

Many different types of devices, mechanisms and refrigeration systems have been treated thus far. These were intended to show the components which, when combined, may be designed for and assembled into a refrigeration system to take care of most common applications.

In this unit, controls, which are grouped as supplementary types, are covered. These show newly described types of modulating controls, timers and recorders, pneumatic control systems and finally, controls for multiple installations.

MODULATING CONTROLS

Modulating Thermostats

The thermostats that have been discussed up to this point were temperature-operated switches. As such, the circuits they control are either closed or open. Another important type of thermostat is a *modulating temperature controller*, which requires a special winding instead of switch contacts.

A modulating thermostat must be used with a *companion* motor having characteristics matching those of the thermostat. The modulating thermostat works on the principle that as a temperature change is sensed, a contact moves to change the resistances in the winding. This change in resistance unbalances the current flowing through the two branches of the control circuit. The companion motor immediately reacts to restore electrical balance. Thus, for every change made by the thermostat, a corresponding change is produced by action of the motor.

Modulating type remote-bulb thermostats, pressure controls, and humidity controls are a few other examples of modulating controls. Modulating thermostats, when used with modulating damper motors, control the operation of dampers for bypassing the air around cooling coils. An example of this system is shown in figure 20-1.

These devices are also used for controlling the flow of secondary refrigerants in remote refrigeration systems, using water chillers.

Thermostatic Expansion Valve Capacity

Superheating the gas before it leaves the evaporator makes part of the evaporator ineffective for cooling. The greater the amount of superheating required, the less evaporator surface is available for the evaporation of a liquid refrigerant.

By controlling the equalization pressure it is possible to control the superheat necessary to operate a valve. The superheat (and the capacity of the evaporator) may be controlled by regulating the pressure in the equalizer line to the expansion valve.

An old application of this principle of controlling evaporator capacity by varying pressure in an external equalizer line is illustrated graphically in figure 20-2. A bulb or temperature sensitive power element actuates a thermostat expansion valve which

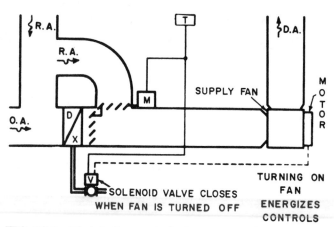

FIG. 20-1   Modulating control of a return air bypass damper

controls the pressure in the expansion valve equalizer line. Pressure is obtained by bleeding a little liquid from the liquid line to the equalizer line. Modern thermostatic expansion valves have an equalizing line connected to the suction line, without any pilot control.

For full cooling capacity the very small bleed is vented to the suction line and the expansion valve operates in its normal manner.

For a reduction in capacity, the pilot valve partially closes the vent allowing some pressure to build up in the equalizer.

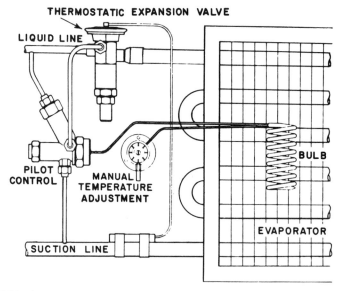

FIG. 20-2   Old application of controlling evaporator capacity by varying external equalizer line pressure

For complete shutoff, the vent is closed securely allowing liquid line pressure to build up in the equalizer and to close the expansion valve tightly.

A reduction in capacity is accompanied by a reduction in suction temperature. This condition exists unless provision is made to regulate suction pressure or to control the compressor capacity. Some means must be provided to prevent too great a suction temperature drop when the flow of refrigerant is greatly reduced. Suction pressure regulators accomplish this and are discussed later with multiple evaporators.

WATER REGULATING VALVES

A considerable amount of water is wasted unless the flow of water to a water-cooled condenser is automatically regulated. Water flow may be controlled by any one of a number of automatic water regulating valves. A cutaway section of a typical automatic water regulating valve showing the major operating parts is illustrated in figure 20-3.

Such a water regulating valve is installed in the water supply line to the condenser. The capillary control line (1) from the valve bellows is connected to the compressor discharge manifold, or the hot gas line, or the top of the condenser.

Any pressure buildup in the high side of the refrigeration system is transmitted to the valve bellows (2). There the pressure works against the spring which forces the water regulating valve to close. As the pressure in the capillary tube becomes greater than the spring tension, the valve (3) opens and passes more water. This action lowers the discharge pressure.

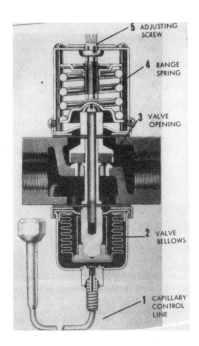

FIG. 20-3 Cutaway view of pressure-operated water regulating valve *(Courtesy of Controls Products Division, Johnson Controls, Inc.)*

The point at which the valve begins to open is regulated by the tension of the range spring (4). The tension is regulated by turning the adjusting screw (5). The valve opens and closes automatically in response to condensing pressure and maintains the set or desired condensing pressure.

The water regulating valve remains closed until the condenser pressure rises to valve setting. Pressure increases in the condenser cause the water regulating valve to open until enough water passes to maintain the condensing pressure for which the valve is adjusted. When pressure in the condenser decreases, the force of the range spring (4) closes the valve.

In installations where condenser water is supplied by cooling towers or where a well pump serves only the condenser, water regulating valves are seldom used. This is due to the fact that the cooling tower fan and the circulating pump are interlocked with the compressor motor control circuit. This means that the compressor motor circuit, pump and fan operate only when the compressor is running.

REFRIGERATION TIMERS AND RECORDERS

Automatic Timer Controls

Timer controls have been applied to the defrosting of systems. They may also be used to schedule the operation of refrigeration and air-conditioning equipment during hours or days when it is not needed. Timer controls can also be set to start a refrigeration unit to meet anticipated needs. The buttons on the outside of a timer wheel are set in positions corresponding to the time during or at which a particular operation is to be controlled.

Operation Recorders

Recorders may be connected into an electrical circuit to check the on-off operation of a system. A 24-hour timer record may be made without requiring the presence of an operator. The record may then be analyzed and, if necessary, corrective measures may be taken.

PNEUMATIC CONTROL SYSTEMS

Controls that are *actuated* (caused to operate) by compressed air are known as *pneumatic controls*. The main applications of pneumatic control systems are in commercial air conditioning. The thermostat, humidistat, or other controllers may be used to vary the pressure to a branch line containing a valve, damper motor or relay to be controlled. The layout sketch in figure 20-4 shows a very simple typical system. The air compressor (A) supplies air to a tank at between 80 to 100 psig. The air from the storage tank (B) is filtered (C) and passed through a reducing valve (D) where its pressure is reduced to 15 psig. This air goes to a preset thermostat (E) which changes the pressure in the line to the control valve according to specific requirements.

There are two types of pneumatic control systems. In the conventional system, a thermostat may be used to change the pressure on the diaphragm or bellows of the control valve. The motion of the valve is opposed by a spring. When the forces of the bellows and spring balance each other, the valve or air motor remains stationary. Any

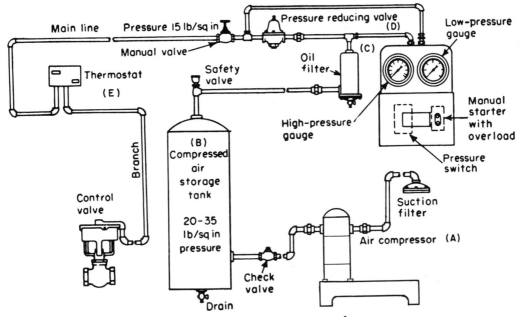

FIG. 20-4 Basic pneumatic control system

change in the branch-line pressure due to action of the thermostat unbalances the forces. The valve or motor operates to assume a new position where the forces of the spring and bellows are again in balance.

There are some disadvantages to the conventional control system. The principal drawbacks are the variable forces created by bearing or valve stem friction or air velocity against dampers. Due to these forces there is no fixed motor position for each unit of branch-line pressure. This makes it almost impossible to very accurately position dampers or valves with a conventional system.

The second type of pneumatic control is the positive-positioning system which depends on the branch-line pressure only. The operation of this system is shown by a diagram in figure 20-5. Note that a relay is installed between the thermostat and the air

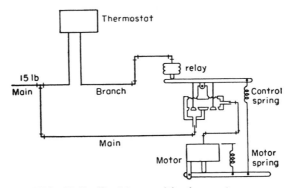

FIG. 20-5 Positive positioning system

motor. A relay with two ports operates on the line pressure due to a demand upon the thermostat. Piped to one port is a main-line pressure of 15 psig. The second port is open to the atmosphere. The port which is open determines whether air is fed to the motor or bled from it.

A control spring which is positioned by the motor arm balances the relay diaphragm. The diaphragm is subject to small variations in branch-line pressure. This system has the advantage over the conventional system because the motor position is independent of friction. This positive-positioning system permits close control of dampers and other units in air-conditioning applications.

MULTIPLE UNIT INSTALLATIONS

The trend is toward using multiple installations in commercial and air-conditioning applications, when possible. A *multiple system* is one in which two or more evaporators in different refrigerators are operated from one compressor, or vice versa. Multiple unit installations are common in restaurants, soda fountains, and in other places where there are several single refrigeration units.

Capacity control is obtained in large installations by using two or more compressors with one evaporator. A good example is an ice plant where the compressor may be started or stopped according to the load demand.

Multiple Evaporators

There are two basic groups of multiple-unit systems. The first and simplest group is the one in which all the evaporators operate at the same temperature. In the second group the temperatures vary in each different refrigerator. The temperature of individual evaporators in such a multitemperature installation is controlled by various combinations of valves and controls. The correct selection and installation of these valves and controls determines the success of the installation.

Control Valves for Multiple Evaporator Units

No special valves are needed for two or more evaporators operating from the same compressor where the temperature of the warmer refrigerator is not more than 5°F higher than the colder refrigerator. However, for temperature differences greater than 5°F, some sort of valve or control is required for the warmer refrigerator.

Suction-Pressure Regulating Valves

A *suction-pressure regulating valve* is also known as an *evaporator-regulating valve*. When it is placed in the suction line of the warmer evaporator, the valve serves to control the suction line pressure and, consequently, the saturation temperature in the evaporator. With such a control valve (although the pressure might vary), it will not drop lower than a predetermined setting of the valve. Thus, when two or more evaporators are operated with one compressor, the desired temperature in the warmer evaporators can be maintained by the proper setting of the evaporator-regulating valve. The location of such valves in a system is shown in figure 20-6.

One model of an evaporator-regulating valve, as shown in figure 20-7, has an inlet connection (A) from the evaporator and an outlet connection (B) to the compressor. The

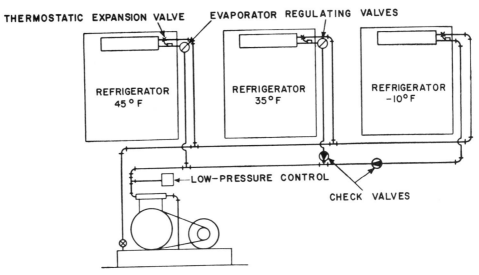

FIG. 20-6   Single compressor with multiple evaporator units

evaporator pressure acting on the valve seat of the bellows (C) and the force of a small spring under the valve are opposed by the heavier spring (D).

When the forces are balanced, the valve is in equilibrium and maintains a definite opening. Any reduction in evaporator pressure produces an unbalancing of forces,

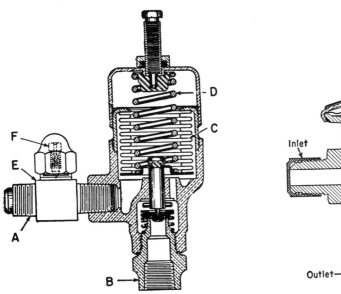

FIG. 20-7   Cross section of evaporator regulating valve

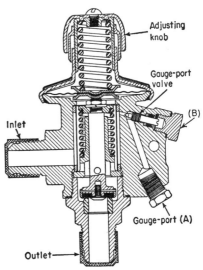

FIG. 20-8   Diaphragm type of evaporator regulating valve

causing the valve to throttle. The decrease in the flow of vapor which results prevents the pressure in that evaporator from dropping too low. The condensing unit continues to provide reduced suction pressure at the other evaporators.

The evaporator regulating valve is provided with a fitting (E) at the place where the evaporator pressure may be taken. A gauge adapter valve must be used. The cap is removed and the adapter is screwed into the fitting with the key, engaging the needle valve (F).

The fitting is also used for connecting the vacuum pump when pumping down the evaporator. Unless the regulating valve is bypassed when pumping down, it closes on a reduction of suction pressure. A line is run from the adapter to the suction valve gauge port of the compressor to bypass the regulating valve when necessary.

The diaphragm evaporator-regulating valve, figure 20-8, is still another type of valve. This valve uses a diaphragm instead of a bellows. Pressure graduations are marked on the collar under the adjusting knob. This means that a correct setting can be made with the adjusting knob. This saves the time that might otherwise be spent taking a pressure or temperature reading.

The diaphragm valve operates in a manner similar to the evaporator-regulating valve previously described. The valve remains open when the warmer evaporator pressure is high. It then throttles as the compressor lowers the pressure. The valve is designed with a gauge port for checking pressure and for bypassing. In order to get a reading, the gauge is attached to the port, the cap is removed and the gauge shutoff valve is opened.

The adjustment in another type of two-temperature valve, shown in figure 20-9, is made by nut (A). The handwheel (B) is used to close the valve without affecting the setting. An auxiliary valve (C) is provided for attaching a gauge and to bypass the main valve when the coil is to be pumped down.

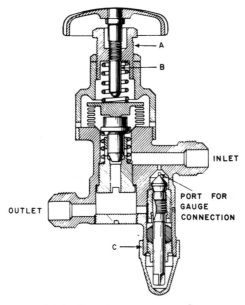

FIG. 20-9   Two-temperature valve

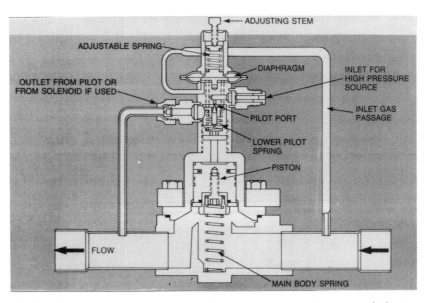

FIG. 20-10   Evaporator regulating valve (snap-action type) *(Courtesy of Alco Controls Division, Emerson Electric Company)*

The auxiliary valve is closed during normal operation and turned to mid-position when the coil is being evacuated.

The evaporator-regulating and two-temperature valves should be located where frosting will not occur and reasonably close to the refrigerator to be controlled. These valves may be used on flooded or direct-expansion evaporators, providing no defrosting is required.

Evaporator Regulator Valves (Snap-Action Type)

A suction-pressure regulating valve of the snap-action type, shown in figure 20-10, may be used to operate an evaporator on a defrosting cycle or when a shorter operating time is required than that provided by the condensing unit. This valve is not a throttling type and can be set to cut in and out at specified pressures. This valve is either wide open or closed securely. A snap-action valve on an evaporator in a multiple system produces the same effect as if it were connected to a separate compressor. The snap-action valve should be used only with a low-side float or with a thermostatic expansion valve. A gauge port is provided in order that a gauge may be attached to simplify the proper setting.

Thermostatic Suction-Pressure Valve

A *thermostatic suction-pressure valve*, shown in figure 20-11, is used for close and nonelectrical control of single evaporators. Applications where this valve is used include: sweet-water baths, water coolers, beverage coolers, soda fountains, and similar installations. In multiple installations such valves are placed in the suction line from the warmer evaporator. The thermal bulb is located in the refrigerated space or liquid.

385

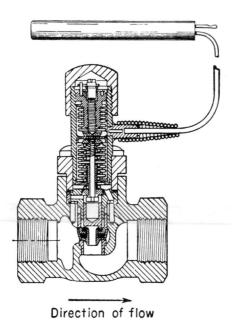

Direction of flow

FIG. 20-11   Thermostatic suction-pressure valve

The thermostatic suction-pressure valve shows it is of the snap-action type. However, the thermostatic type is also available in a valve that has throttling action. The valve is wide open and the coil is subject to refrigeration from the condensing unit when the refrigerator temperature is higher than is required. At the desired temperature the valve snaps shut, isolating the coil from the rest of the system.

The coil defrosts while the valve is closed if the refrigerator temperature is above frosting. In installing this valve it must be located in a horizontal part of the suction line, with a strainer placed ahead of it, and the bulb positioned where it reports the average conditions. In water bath installations the bulb should be placed in the liquid but not too close to the coils.

Check Valves

The pressure in the warmer evaporator in a multiple installation is higher than that in a colder evaporator. Unless a check is provided, when the control valve opens the high-pressure vapor in the warmer evaporator backs up into the colder evaporator. Such a condition would cause a warming up of the colder evaporator and cut down on its efficiency. Figure 20-12 shows two check valves which permit the vapor to flow in only one direction. These check valves are installed in the lines of individual evaporators.

Solenoid Valves in Multiple Evaporator Installations

Solenoid valves are widely used in multiple evaporator installations. These valves may be placed in the liquid line or the suction line. Figure 20-13 shows a multiple hookup with solenoid valves installed in the liquid line. A thermostat is located in each refrigerator. In

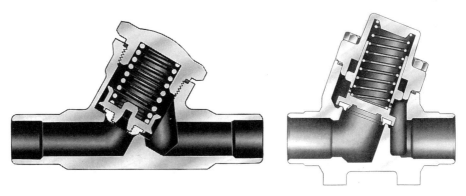

FIG. 20-12   Internal design features of two basic check valves (*Courtesy of Standard Valve Company Division of Amcast Industrial Corporation*)

turn, each thermostat is connected to the solenoid valve that is in the liquid line leading to its refrigerator. The thermostat-solenoid valves are connected to a separate electrical circuit.

The operation of a multiple installation is simple. For instance, suppose all the thermostats call for cooling and the compressor is operating on all of the evaporators. When one thermostat is satisfied, it opens the circuit. The solenoid valve closes the liquid line to the evaporator which that thermostat controls. The compressor pumps the refrigerant out of the particular evaporator and continues to operate on the other evaporators. Each time a thermostat is satisfied, its solenoid valve closes. When all of the valves are closed, the compressor is stopped by a low-pressure cutout.

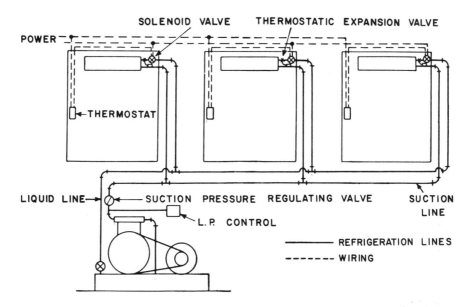

FIG. 20-13   Solenoid valves in liquid line of multiple evaporator installation

Here are some important factors to consider in designing or maintaining multiple systems, using solenoid valves.

- When each solenoid valve closes, its evaporator is pumped down, returning the refrigerant to the receiver.
- The size of the receiver must be large enough to hold the entire charge in the system.
- When the condensing unit is operating on all the evaporators, it operates at full load and at a high suction pressure.
- As refrigeration stops in one evaporator after another, the suction pressure drops progressively lower.

This action continually upsets the balance between the refrigerator temperature and the coil temperature, affects the humidity and, in some cases, prevents the defrosting of the coil.

Some of the disadvantages may be offset by placing an evaporator-pressure regulating valve in the suction line near the compressor. This will hold pressure at the desired point in the evaporators. However, when only one evaporator is operating from the compressor, the crankcase pressure will be low. The low-pressure control, which shuts the compressor off after the last solenoid valve closes, must be set lower than the suction pressure when the compressor operates on only one evaporator. The cut-in point should be set so that the opening of one solenoid valve starts the compressor.

MULTIPLE CONDENSING UNITS

Condensing units are sometimes connected in parallel for greater flexibility and for capacity control. The practice of operating multiple compressors in units is quite common in ammonia plants. The main problem in interconnecting condensing units is in the oil return to the crankcase. Since oil and ammonia are not miscible, it is possible to use multiple condensing units.

By contrast, the R's, methyl chloride and other oil-miscible refrigerants present some difficulties when condensing units using such refrigerants are interconnected. For these reasons, multiple installations of condensing units should be made only when it is not possible to split up the units so that each evaporator has its own condensing unit.

As far as practical, the individual condenser unit in a multiple unit should be of the same manufacture and size. A good installation is one in which each condensing unit carries its share of the load. The oil return must be planned so that the proper level is maintained. This requires the interconnection of liquid, suction and discharge lines. Oil-equalizer and gas-equalizer lines are installed between the crankcases. The condensing units must be located close to each other in order to keep the connecting lines as short as possible. The piping sketch in figure 20-14 shows the arrangement and location of lines for interconnecting two condenser units

Control of Multiple Condensing Units

After condensing units are interconnected, some way must be provided for starting and stopping the compressors according to load demands. Manual operation may be used on large installations where an operator is on hand or when load changes can be

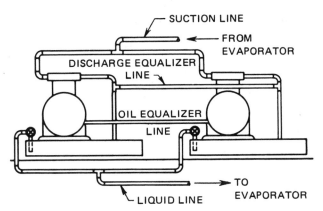

FIG. 20-14   Arrangement of lines and piping for interconnecting two condenser units

determined in advance. When close temperature control is not required for liquid or air cooling, two temperature controls may be used on large installations where an operator is on hand or when load changes can be determined in advance. When close temperature control is not required for liquid or air cooling, two temperature controls may be used, one set a degree or two higher than the other.

A common method of starting and stopping compressors is to use pressure controls on the common suction line. These pressure controls are set in sequence to start and stop the compressors according to changes in suction pressure. Thermostats and solenoid valves are nearly always installed with such a system.

When the installation of multiple compressor units is such that all compressors start at the same time, this puts an exceedingly heavy load on the circuits and the power system. It is desirable in such instances to install a timer which delays the starting of each compressor. The timer provides a delay of at least 10 to 15 seconds from the time when the contactor for the first compressor is closed, before the timer closes the control circuit of the second compressor, allowing it to start.

The timer can take the form of a *step controller* operated by a *modulating motor* (M) as identified in figure 20-15. This motor is, in turn, operated by a *modulating space thermostat* as indicated at (T).

The compressors are staged (1 to 6) in sequence to give the required total cooling as measured by the modulating space thermostat. When refrigeration is still required after a power interruption, one compressor unit is turned on at a time until the required number are operating.

The same system of control is sometimes applied to cylinder unloading. Figure 20-16 shows the system used to stage the compressors (or unloaders) to give the required chilled-water temperature.

The *recycling relay* (R) in the diagram holds the compressors inoperative when power is resumed after an interruption. During this time, the relay returns the step controller to a starting position.

At this point the chilled water temperature controller (T) operates the step controller to turn on stages as the temperature of the chilled water rises. It also turns off stages as the temperature of the chilled water drops.

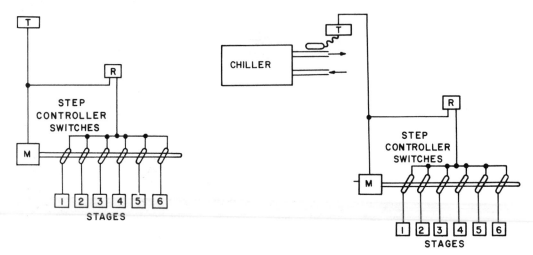

FIG. 20-15 Staging for compressors

FIG. 20-16 System for staging compressors to provide a required chilled water temperature

The floating motor (M) of the step controller actuates switches to turn the stages (1 to 6) on or off in sequence. The step controller first returns to a starting position after a power interruption and then goes to the position required by the chilled water temperature controller.

SUMMARY

- Modulating controls provide for variations by steps as contrasted to the off and on operation of refrigeration systems with ordinary controls. Modulating thermostats need a matching motor (actuator) to produce increasing steps of operation.
- Modulating thermostats are used to operate dampers on DX coils and valves for varying the flow of chilled water.
- A modulating thermostatic expansion valve varies the capacity of the valve in response to variations in load on the system.
- A water regulator valve, operated by the head pressure of a compressor, varies the flow of cooling water to the condenser in response to variations in the need for cooling.
- An operation recorder makes a graphic record for diagnosis of equipment performance in a given operating time.
- Pneumatic control systems employ changes of air pressure between operating and actuating controls.
- The positive-positioning pneumatic-control system provides positive operation of the damper or valve independently of wear, friction or other operating conditions.
- Evaporator-pressure regulating valves provide independent temperature control for each evaporator. They are equipped for (1) setting at a desired temperature and

(2) for connecting a gauge to measure the pressure in the evaporator or for making a temporary bypass around the regulating valve.

- One or more check valves must be used with evaporator-pressure regulators to prevent the higher pressure in one evaporator from backing up into another evaporator of lower pressure when a regulator is open.
- Multiple condensing installations, which provide for capacity control, present problems like the distribution of oil to the different compressors and the proper distribution of load between the compressors.
- A step controller (timer) that is operated by a modulating motor and controlled by a modulating thermostat controls multiple condensing units.

ASSIGNMENT: UNIT 20 REVIEW AND SELF-TEST

A. MODULATING CONTROLS

Select the letter corresponding to the term which correctly completes statements 1 to 10.

1. A modulating thermostat is an: (a) operating control, (b) actuator, (c) safety control.
2. A modulating thermostat provides operation that is: (a) on-off, (b) stepped, (c) neither on-off nor stepped.
3. A modulating damper motor will operate with: (a) an ordinary thermostat only, (b) a special thermostat only, (e) either an ordinary or special thermostat.
4. A modulating valve motor controls: (a) the secondary refrigerant only, (b) the primary refrigerant only, (c) either the secondary or primary refrigerant.
5. A modulating control system provides capacity that is: (a) variable, (b) fixed, (c) neither variable nor fixed.
6. A modulating thermostatic control is used to directly control: (a) secondary refrigerant, (b) primary refrigerant, (c) neither secondary nor primary refrigerant.
7. A thermostatic expansion valve is controlled by: (a) solenoid pilot valve, (b) modulating pilot valve, (c) either solenoid or modulating pilot valve.
8. A modulating thermostat may be used with: (a) a damper motor only, (b) a liquid valve motor only, (c) either a damper or liquid valve motor.
9. A water regulator valve is directly activated by: (a) change of temperature, (b) change of pressure, (c) either change of temperature or pressure.
10. The modulation of the capacity of a thermostatic expansion valve is through: (a) its equalizer, (b) its thermal bulb, (c) either its equalizer or thermal bulb.

B. REFRIGERATION TIMERS AND RECORDERS

1. Explain briefly the operation and use of the refrigeration control timer.
2. Describe how the operation recorder works.

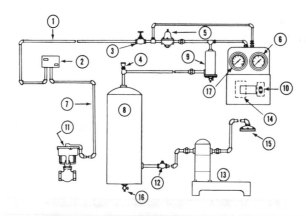

C. PNEUMATIC CONTROL SYSTEM

1. Identify each component listed in the appropriate circle on the basic pneumatic control system drawing.
2. Trace the operation of a pneumatic system of controls, explaining the medium used, the nature of the control device and how the system operates.

D. MULTIPLE INSTALLATIONS

Add the correct term or value to complete statements 1 to 20.

1. Multiple unit installations refer to multiple _____ or multiple _____ .
2. Suction pressure regulators are used on the multiple _____ system.
3. Another control device that may also be used on the same system as the suction pressure regulator is a _____ .
4. When suction pressure regulators are used, they also require _____ to keep the units operating independently.
5. The only unit that does not require a suction pressure regulator is the one operating at the _____ temperature.
6. Evacúation of a multiple evaporator system equipped with a suction pressure regulator requires a _____ .
7. When a gauge is connected to a suction pressure regulator, loss of refrigerant is prevented by a _____ .
8. Solenoids may be used in the _____ or _____ lines of a multiple system.
9. The use of solenoids with a multiple evaporator system requires that each evaporator be equipped with a _____ .
10. When a multiple evaporator system uses a solenoid, it requires a suction pressure regulator to prevent excessively _____ pressure in the evaporator(s).

11. The pressure in the suction line at the compressor is _____ the pressure in the evaporators when the solenoids are all closed.

12. The compressor in multiple evaporator systems is stopped by the action of the _____ .

13. The solenoids in multiple evaporator systems are controlled by _____ .

14. With multiple compressors connected in parallel, the _____ , _____ , and _____ lines of one compressor are connected to similar lines on the other.

15. When using halogen refrigerants in a multiple compressor system it is necessary to provide _____ and _____ equalizer lines between the compressor crankcases.

16. Equalizer lines are not needed when the refrigerant in a multiple compressor system is _____ .

17. Oil return problems are encountered in multiple systems when the oil and refrigerant are _____ .

18. A common method (besides multiple temperature control) for controlling multiple condensing units is to use _____ on the _____ .

19. Simultaneous starting of multiple condensing units is prevented by using a timer which sometimes is operated by a _____ and _____ .

20. A system of control using various modulating devices, when used for the capacity control of a single compressor, is then known as a _____ control.

SECTION 6:
ALL-WEATHER AIR-CONDITIONING SYSTEMS

UNIT 21:
HEAT PUMPS

The *heat pump* is a refrigeration system designed to remove heat from one place and pump it to another. With capability to pump heat *two ways*, the heat pump is used for heating, cooling and space conditioners. The distinction between the functions of an air conditioner and a heat pump lies in the fact that an air conditioner pumps heat laden vapor one way.

COMPRESSION CYCLE REFRIGERATING MACHINES

In review, the four major components of a refrigerating machine include an evaporator, condenser, compressor, and metering device. Although these components are the same for heat pumps, there is an additional component known as a *four-way valve*.

The Four-Way Valve

The *four-way valve*, figure 21-1, is designed to divert discharge gas and heat in a particular direction, depending on whether a conditioned area is to be heated or cooled. The control of the four-way valve is by means of a space-temperature thermostat, which positions the valve for heating or cooling.

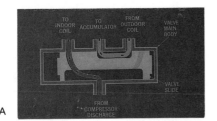 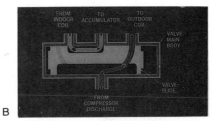

FIG. 21-1   Schematics of reversing valve function: (A) heating mode and (B) cooling mode *(Courtesy of William Buynum Training Center, Carrier Corporation)*

The typical heat pump in winter removes (absorbs) heat from outdoor air and deposits it in an indoor area that is to be heated. The process is reversed in summer: Indoor heat is removed from an interior area and is deposited outdoors.

DESIGN FEATURES OF HEAT PUMPS

In terms of appearance, the exterior case (cabinet) of a heat pump is similar to a summer air conditioner, as illustrated in figure 21-2. The identification and location of each major design feature are shown.

Solid-State Electronic Defrost Control

The *solid-state electronic defrost control* in the model displayed is shown as item 9. This type of control is designed to replace conventional electro-mechanical defrost timer controls. The solid-state digital countdown timer has no moving parts and serves the function of a synchronous clock motor and cam-drop defrost switch. A solid-state switch serves to energize the reversing valve. A *thermistor* is used as a *defrost sensor* in place of the conventional *remote sensor bulb.*

PRINCIPLES OF OPERATION OF A HEAT PUMP

A heat pump operates on the same principle as a refrigeration system. Heat is moved from a higher temperature to a lower temperature. Heat is absorbed into the system

CODE   DESIGN FEATURE

①   EXHAUST FAN MOTOR
②   CONTINUOUS SPIRAL COIL
③   HOUSING
④   SERVICE VALVES WITH LEAK-
      FREE FITTINGS
⑤   OPERATING CONTROLS/WIRING
      SERVICE ACCESS
⑥   4-WAY VALVE
⑦   HEAVY-DUTY COMPRESSOR
⑧   CRANKCASE HEATER
⑨   SOLID-STATE AUTOMATIC DEFROST
      SYSTEM
⑩   ACCUMULATOR
⑪   START-ASSIST SYSTEM
⑫   LOW-PRESSURE SWITCH

FIG. 21-2   Internal design features of year-round heat pump *(Courtesy of Carrier Corporation)*

through the evaporator. Heat is rejected from the system through the condenser. The heat-laden vapor is absorbed into the refrigeration system through the compressor. The flow of refrigerant is controlled by a metering device. A thermostat-controlled, four-way valve diverts the discharge gas in a required direction for either heating or cooling a designated area.

In a heat pump, both the condenser and the evaporator are used to transfer heat and cold. Although the heat pump is often called a reverse-cycle mechanism, it should be noted that the cycle is not reversed. Instead, the condenser and evaporator functions are interchanged.

Since the operation of heat pumps and compression units or systems is similar, both contain the following components:

- Compressor
- Indoor Coil (Evaporator)
- Motor Controls
- Outdoor Coil (Condenser)
- Liquid Line
- Suction Line (Vapor)
- Refrigerant Controls
- Reversing Valve
- Check Valves

The reversing valve, check valves and refrigerant controls are optional depending on the metering devices used in smaller systems to change from heating to cooling.

Heat pumps are designed according to basic sources of energy: air, water, and earth. The three categories of heat pumps include:

- Air to air
- Air to water
- Air to ground coil

INDOOR AND OUTDOOR HEAT PUMP COILS

The *indoor coil* of a heat pump serves inside a structure. The *outdoor coil* serves the outdoor unit. In the *heating mode,* the indoor coil is a *condenser.* In the *cooling mode,* the indoor coil works as an *evaporator.*

Reciprocally, the *outdoor coil* is an *evaporator* in the heating mode; a *condenser* in the *cooling mode.* These functions are shown graphically in figure 21-3 at (A) and (B).

The heating cycle (A) shows hot gas flowing into the inside condenser to give up heat in the conditioned area. At (B), during the cooling cycle (mode) the hot refrigerant gas flows from the indoor evaporator coil to the outside condenser coil. Thus, heat is removed from an indoor area, resulting in a cooling effect.

HEAT PUMP METERING DEVICES

Thermostatic Expansion Valve and Check Valves

Heat pump systems require two metering devices: one for the indoor unit; the other, the outdoor unit. Two locations of the thermostatic expansion valve (as metering

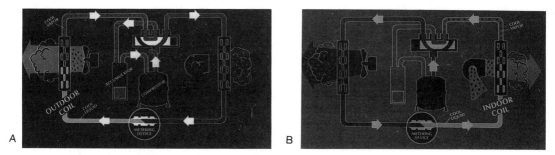

FIG. 21-3   Heating and cooling cycles of a heat pump and direction of refrigerant gas flow: (A) to heating cycle and (B) cooling cycle *(Courtesy of Carrier Corporation)*

devices) in a heat pump system are shown in figure 21-4. Note in this instance that the system is in the cooling mode (summer cycle).

As illustrated, the thermostatic expansion valve metering device meters refrigerant flow in the direction of the indoor unit. A check valve forces the refrigerant to flow through the thermostatic expansion valve.

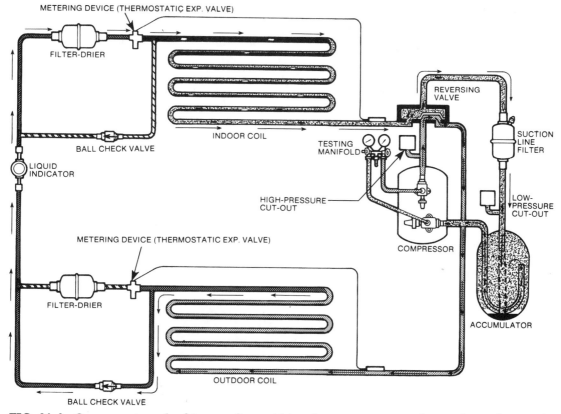

FIG. 21-4   Components and refrigerant flow within a heat pump system during the cooling mode *(Courtesy of Mueller Brass Company)*

When the summer cooling mode is changed to a winter mode, the refrigerant liquid flows in the direction of the outdoor unit.

SAFETY PRECAUTIONS

- Since refrigerant gas lines sometimes reach temperatures of 200°F, care must be exercised to avoid touching a hot gas line.
- Exposure of an expansion valve sensing bulb on the hot gas line to temperatures of 200°F might cause the expansion valve to rupture.

LIQUID LINE FILTER-DRIERS

A liquid line filter-drier is used to control the flow through a metering device. The filter-drier is installed in series with an expansion device (in a system with check valves) to control the flow direction. One filter-drier is installed for heating in the indoor unit. The second filter-drier for cooling is located in the outdoor unit.

Biflow Liquid Drier

Biflow liquid driers, figure 21-5, combine the operation of single driers into one component. Check valves inside the shell of the liquid drier provide direction for liquid flow, as required.

Fixed-Bore (Orifice) and Check Valve

This type of metering device is a combination check valve and flow device. It permits full flow in one direction; restricted flow in the opposite direction, figure 21-6. In split system installations, two fixed-bore metering devices are required. The bore size for the indoor coil is larger than the bore size on the outdoor coil. More refrigerant is generally used in the summer (cooling) cycle.

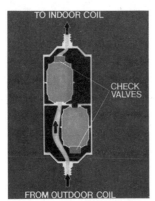 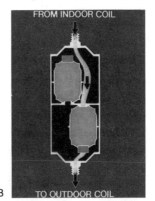

FIG. 21-5 Functions of a biflow drier for cooling and heating modes: (A) to cooling and (B) to heating *(Courtesy of Carrier Corporation)*

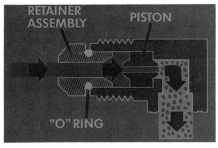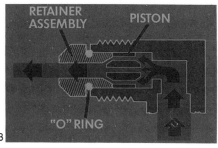

FIG. 21-6   Cross section of a combination fixed-bore (orifice) and check valve metering device *(Courtesy of Carrier Corporation)*

AIR-TO-AIR HEAT PUMPS

Smaller models of air-to-air heat pumps are used in residential, commercial and small industrial applications. Since these heat pumps include a reversing valve, they provide a practical system for heating and cooling.

Heating Cycle of the Air-to-Air Heat Pump

Figure 21-7 indicates the position of the reversing valve during the heating mode and the direction of refrigerant flow. The heated outdoor coil produces heat in a desired area. Compression-type refrigeration systems are normally used for heat pumps. Heat is transferred using the condenser located outdoors and the evaporator located indoors.

During the heat cycle the outdoor coil (the condenser) becomes the evaporator. Liquid refrigerant in the system enters the outdoor coil, picks up heat from outdoors and is vaporized. The vapor is compressed to a higher temperature and pumped into the indoor coil. Since the compressed refrigerant temperature is higher than the indoor temperature, heat is released in the required area. The process (cycle) continues until the thermostatic control device indicates that the required temperature is reached.

For the cooling mode, the reversing valve causes the indoor coil to become an evaporator. In this mode, the system is basically a capillary tube system.

Air-to-air heat pumps are ideally suited to locations where (1) the ambient air temperature is almost always above freezing temperature and (2) the summer cooling and winter heating loads are about equal. Heat pump efficiency is greatly reduced when the outdoor temperature drops below freezing. When such conditions prevail, the difference between the heat capacity of the heat pump and the required Btu heating needs must be made up by *auxiliary heat.*

In general, the primary heat of the heat pump is supplemented by heat produced by electric, oil, gas, or solar energy sources. An auxiliary energy unit is fitted into the indoor section to produce the required indoor temperature. Electrical heaters installed in an air-to-air heat pump system are illustrated in figure 21-8.

GROUND OR WATER-COIL HEAT PUMPS

Since there are great daily temperature fluctuations, the efficiency of a heat pump can be increased by some method of removing heat from the ground. In some heat

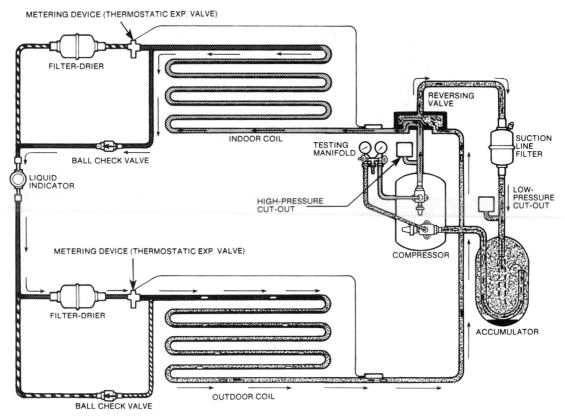

FIG. 21-7 Components and refrigerant flow within a heat pump during the heating mode *(Courtesy of Mueller Brass Company)*

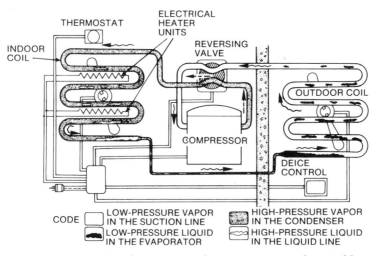

FIG. 21-8 Heating cycle of an air-to-air heat pump using electrical heater units

pump installations, a *pipe loop (coil)* or a great length of pipe is buried outdoors in the earth below the frost line. The *ground coil* replaces the outdoor air coil.

To review, a refrigerator, air conditioner or heat pump all serve to concentrate existing heat and to move it from a low temperature location to a higher temperature location. Cooling functions in refrigerators and air conditioners depend on the removal of heat from colder interior spaces to warmer exterior spaces.

In the case of heat pumps, for purposes of heating interior spaces, heat within the system is moved from a low temperature outdoor source to a higher temperature indoor space. In other words, an air-source heat pump extracts heat from outdoor air and pumps it into a building.

Ground-source heat pumps operate in a similar manner. Heat (thermal) energy that is stored in the earth or in ground water is absorbed. Since this source of heat energy has a more constant temperature in comparison with the temperature of air, ground-source heat pumps have an advantage over air-source units. For example, in some states, ground water/earth temperatures average 50°F (10°C). By contrast, air temperatures in these same geographic areas average 40°F (5°C), with outdoor temperatures dropping below freezing in the winter heating season.

A cycle of evaporation, compression, condensation, and expansion is involved in heat pumps in the process of elevating low temperature heat to over 100°F (38°C) and transferring this heat to indoor air. A refrigerant, circulating through the heat pump system, serves as the heat transfer medium.

ENERGY SOURCES FOR GROUND-SOURCE HEAT PUMPS

Ground-water aquifiers are used in ground water heat pump systems. Well water or a pond provide the low temperature energy source. Water (removed from the aquifier) passes through to a liquid refrigerant heat exchanger in the heat pump, is returned to the aquifier and discharged into a stream or pond or reused. Ground-water heat pump systems are also known as *open-loop systems.*

Earth-coupled heat pump systems require pipe loops or long lengths of pipe buried in the earth. Basic loop configuration systems include *horizontal* and *vertical.* Antifreeze solution circulates through the ground coil in the horizontal system. The trench in which the ground coil is buried (and from which it absorbs heat from the earth) ranges in depth from 2 1/2 feet to 6 feet. The depth depends on weather conditions (climate) in a geographic location and operating characteristics of the heat pump.

Horizontal loops that use heat stored from ground water may also be laid on the bottom of a lake or pond, figure 21-9. In this type of installation, the antifreeze enters a liquid-to-refrigerant heat exchanger in the heat pump where it exchanges heat with the original refrigerant.

The same results of a ground-coil system can be accomplished by the *vertical-earth loop system.* Since such a system can be installed in vertical holes, site area needs are minimized. An *earth-coupled heat-pump system* is also identified as a *closed-loop system.*

A common component of both the open-loop and closed-loop heat pump systems is the water-source heat pump. Care must be taken in selecting equipment for a closed-loop application that will operate at colder source temperatures, usually ranging from 25°F to 30°F (approximately –5°C to –1°C).

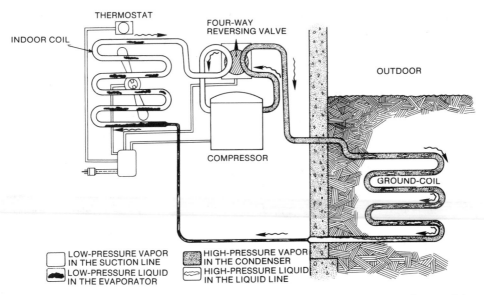

FIG. 21-9 Application of horizontal loops in a ground-water system (cooling cycle)

Water-to-air, ground-source heat pumps are available as *single-package* or *multi-component package units*. The single-package unit illustrated in figure 21-10 is designed with the refrigerant-cycle components and blower mounted in one piece of equipment. This unit serves as a heat extractor and an air handler.

Multicomponent packages generally house the compressor, liquid-to-refrigerant heat exchanger, blower, and refrigerant-to-air heat exchanger in two or more separate units.

WATER-TO-WATER HEAT PUMPS

This type of ground-source heat pump requires two liquid-to-refrigerant heat exchangers, a compressor and an expansion device. Water-to-water heat pumps are used in *hydronic heating system* applications. Heat is distributed via a secondary closed-loop water system with a small circulating pump instead of an air handler and a duct system.

AIR-CONDITIONING OPTIONS

There are two common methods of providing air conditioning by a ground-source heat pump. The first method (as discussed earlier) involves the heat pump refrigeration cycle. When the heat pump is equipped with a reversing valve, the cycle can be reversed. Heat from air inside a building is absorbed and rejected into the water.

A second method of air conditioning involves direct water-to-air cooling. In this type of system, a liquid-to-air heat exchanger is added to the heat pump system. Direct water-to-air cooling is feasible with open-loop heat pump systems because ground-water temperatures are constant (about 50°F, 10°C). This second method, however, is not applicable to earth-coupled heat pump systems due to the fact that during the summer months earth-loop temperatures are too high.

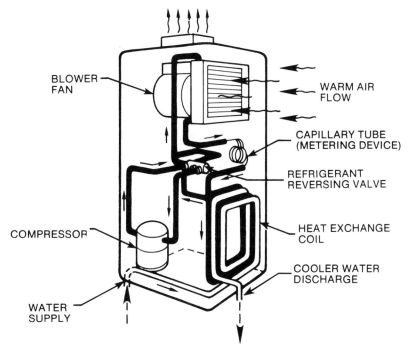

BLOWER FAN

WARM AIR FLOW

CAPILLARY TUBE (METERING DEVICE)

REFRIGERANT REVERSING VALVE

COMPRESSOR

HEAT EXCHANGE COIL

COOLER WATER DISCHARGE

WATER SUPPLY

FIG. 21-10   Single-package, water-to-air ground-source heat pump

In summary, direct water-to-air cooling is only applicable to ground-water systems. Unfortunately, humidity control is compromised for lower energy consumption.

THERMOSTATIC CONTROLS

A *thermostat* generally provides the primary control for a ground-source heat pump. A typical heat pump thermostat provides for two-stage heating and one-stage cooling. Supplemental electric resistance, fossil fuel or other energy heating can be switched on when the heat pump cannot handle the load.

One-stage cooling can be controlled with either manual or automatic changeover. Also, separate switches are provided for emergency heat and fan operation.

SUMMARY

- Compression cycle refrigeration systems that have capability to pump heat in either direction are considered heat pumps.
- The four-way valve is designed to permit heat pumps to reverse the refrigeration cycle and to reject heat in either direction.
- The four-way valve serves in the cooling mode in one position; the heating mode in the other position.
- Heat pump space-temperature thermostats control the position of the four-way valve.
- Indoor and outdoor thermostatic expansion valves are required as metering devices in a heat pump system.

- The thermostatic expansion valve meters refrigerant flow in the summer mode in the direction of the indoor unit. A check valve forces the refrigerant to flow through the thermostatic expansion valve.
- A solid-state electronic defrost control may be used to replace and to serve the functions of a conventional electromechanical defrost control timer.
- One liquid line filter-drier is installed in the indoor unit for heating; another filter-drier in the outdoor unit for cooling.
- The earth, lakes, deep wells, and water from industrial cooling systems may be used to provide energy sources for ground-coil or water-coil heat pumps.
- Earth-coupled heat pump systems use horizontal or vertical earth loops or pipe buried in the ground.
- Water may be used in industrial/commercial buildings to absorb heat in one location; rejecting it in another area of a building.
- During the heating cycle (mode) of an air-to-air heat pump, outdoor heat is vaporized, pulled into the compressor and compressed to a higher temperature, and pumped into the indoor coil; releasing heat to the surrounding area.
- The indoor coil within a building serves as a condenser for winter operation and as an evaporator for summer air conditioning.
- The outdoor unit of a heat pump serves as a condenser for summer air conditioning and as an evaporator for winter.
- When a heat pump cannot handle the entire load, auxiliary heat from electrical resistance, oil or gas energy may be required.
- Multicomponent water-to-air ground-source heat pumps usually contain the compressor, liquid-to-refrigerant and refrigerant-to-air heat exchangers and the blower.

ASSIGNMENT: UNIT 21 REVIEW AND SELF-TEST

Supply the missing heat pump terms, conditions or components as required to complete statements 1 through 11.

1. Heat pump systems have the capability to pump heat-laden vapor _____ .
2. The _____ and _____ in a heat pump system are used to _____ _____ .
3. A reversing valve permits the transfer and release of heat during hot weather from an _____ heated area to _____ .
4. Heat absorbed in the _____ of a heat pump from one heated area is released through a _____ in another location.
5. A solid-state _____ _____ may be used to replace a regular electromechanical defrost timer control on a heat pump.
6. A reversing valve is used for _____ and _____ .
7. The direction of heat flow in the compressor is _____ .
8. Water-source heat pumps are common to _____ _____ and _____ _____ systems.
9. Auxiliary heat energy in the form of _____ , _____ or _____ is used to supplement the capacity of heat pumps.

10. Refrigerators and air conditioners for cooling move heat from _____ areas to _____ _____ areas.

11. The indoor heat pump coil serves as the _____ during the heating cycle of an air-to-air heat pump.

Circle the letter of the basic refrigeration principle or condition that correctly completes statements 12 through 18.

12. Coils or pipes for heat pumps that are buried in the ground are called _____ .
    a. ground water aquifiers
    b. water-source heat pumps
    c. earth-coupled heat pumps
    d. horizontal or vertical loops

13. A water-to-water heat pump consists of _____ .
    a. vertical and horizontal air loop systems
    b. two liquid-to-refrigerant heat exchangers, a compressor and an expansion device
    c. an air handler and duct system
    d. a compressor, a refrigerant-to-air heat exchanger and a blower

14. The heat pump thermostat is designed _____ .
    a. for one-stage heating
    b. to compromise humidity control
    c. for one-stage cooling
    d. for two-stage heating and one-stage cooling

15. Air-to-air heat pump systems are ideally suited to geographic areas where the ambient air temperature _____ .
    a. is almost always above freezing
    b. drops and remains below freezing
    c. fluctuates continuously above and below freezing temperatures
    d. produces a great variation in heat load summer and winter

16. During the *heating cycle* of an air-to-air heat pump, the refrigerant flows as a _____ .
    a. low pressure/high pressure liquid in the evaporator
    b. low pressure vapor in the condenser
    c. high pressure liquid in the liquid line
    d. high pressure vapor in the suction line

17. During the *cooling cycle* of an air-to-air heat pump, the refrigerant flows as a _____ .
    a. low pressure liquid in the evaporator
    b. low pressure liquid in the liquid line
    c. low pressure vapor in the condenser
    d. high pressure vapor in the suction line

18. During the *cooling cycle* of a ground-coil heat pump in which a four-way reversing valve is used, _____ .
    a. low pressure is produced in the liquid line

     b. low pressure liquid is produced in the evaporator

     c. low pressure vapor is produced in the condenser

     d. high pressure vapor is produced in the suction line

19. List five principal components of a heat pump system.
20. Cite four basic refrigeration processes that are essential to the operation of a ground-water heat-pump system.
21. State one advantage of ground-source heat pumps over air-source units.
22. Identify four refrigeration process cycles required to elevate heat at a low temperature and to transfer heat over 100°F (38°C) to an indoor area.
23. Describe briefly the heating cycle operation of an air-to-air heat pump.
24. Name four major components of a water-to-air ground-source heat pump.

UNIT 22:
RESIDENTIAL AND COMMERCIAL AIR-CONDITIONING SYSTEMS

Two important factors must be considered during the designing of residential, commercial and industrial air-conditioning systems. The first factor deals with *load estimating;* the second, determining the required *capacity of the duct system.*

FACTORS AFFECTING SYSTEM DESIGN

The American Refrigeration Institute and many manufacturers provide checklists for recording data, construction features and other technical information relating to systems design. Figure 22-1 shows a portion of a heating/cooling *survey checklist.*

Answers are obtained from building floor plans, elevation drawings, sections, and other drawings and specifications. For example, consideration is given to building size, materials used in construction; other structural design features, shading and sunlight; exhausts and vents. If an add-on system is planned, sizes of ducts, registers and grilles existing in the air distribution system must be identified and included. Computations must be made for sources of heat such as windows, walls and roof; other heat-producing appliances, equipment and processes; and other factors, such as the number of people to be accommodated and outside sources of air.

Estimating the Cooling Load

A form similar to the one illustrated in figure 22-2A is used to estimate *cooling load.* The *total* residential or commercial *installation load* is based on the individual room load (air quantity) requirements in *cfm.* Load depends on the room area in square feet.

Other information requires the use of technical reference tables, figure 22-2B, related to such factors as *heat gain* per square foot through walls, doors, roofs, and windows. The grand total of Btu/hr is calculated. The total cfm is then determined by dividing the Btu value by 30.

Estimating the Heating Load

The steps, procedures and forms used to establish the heating load are similar to those used for cooling load. The total cfm for the system is equal to the maximum heat loss divided by 108. The section of a heating estimate form illustrated in figure 22-3 shows the total of the separate room heating requirements (losses). Duct losses are added to establish the maximum heat loss in Btu/hr and to compute the cfm.

The values from the cooling load estimate and the heating load estimate in Btu/hr are used to determine the heating and cooling equipment needs. It is good practice to select equipment that has a slightly higher capacity than the needs computed for the system.

DESIGN CONDITIONS

	Summer	Winter
Outside Temp. (F)		
Daily Temp. Range (F)		
Inside Temp. (F)		

ORIENTATION

House Faces: N NE E SE S SW W NW

CONSTRUCTION

Type

Single Story ☐   Two story ☐   Split level ☐

Walls

Frame ☐   Heavy masonry (over 10" thick) ☐

Light masonry (10" thick or under) ☐

Insulation: None, 1", 2", 3⅝", 6" ⸺⸺⸺⸺

Ceiling heights: 1st fl. ⸺⸺'⸺⸺", 2nd fl. ⸺⸺'⸺⸺"

Roof & Ceiling

Pitched roof ☐   Flat roof ☐

Studio ceiling ☐   Rafters covered ☐   Rafters exposed ☐

Insulation: None, 2", 4", 6", 8", 12" ⸺⸺⸺⸺

Insulated Sheathing ☐

White roof ☐   Dark roof ☐   Attic Fan ☐

Floor

Slab on ground ☐   Edge insulation: None, 1", 2"

Over crawl space ☐

    Open ☐   Closed: Vented ☐   Unvented ☐

    Vapor seal on ground ☐

    Insulation: Underside of floor ☐ ⸺⸺⸺⸺"

               Crawl space walls ☐ ⸺⸺⸺⸺"

Over Basement ☐   Over garage ☐

    Insulation: None, 1", 2", 4" ⸺⸺⸺⸺

Windows

Movable ☐   Fixed ☐

Single pane ☐   Double pane ☐   Glass block ☐

Double hung ☐   Casement ☐

Weatherstripped ☐   Plain ☐

SERVICE AND UTILITIES

Electrical service

Current characteristics ⸺⸺ volts, ⸺⸺ phase, ⸺⸺ cycles

Capacity of service ⸺⸺⸺⸺ amps.

Water service (water source equipment only)

Source:

City ☐   Well ☐   Other ⸺⸺⸺⸺⸺⸺⸺⸺

Maximum temperature: Summer ⸺⸺⸺ F.   Winter ⸺⸺⸺F.

Pressure ⸺⸺⸺⸺ PSI.

Meter location ⸺⸺⸺⸺⸺⸺, Meter size ⸺⸺⸺"

Service pipe size ⸺⸺⸺"

Available flow (wells) ⸺⸺⸺ gpm.

FIG. 22-1   Abstracted portions of survey and checklist of building details, services and conditions *(Courtesy of Carrier Corporation)*

ROOF FACTORS

Construction	Base Factor	Insulation Factor			
		R7	R11	R13	R19
Light No Ceiling 10#, U= .20	8	.41	.31	.28	.21
Light Ceiling 10#, U= .13	5	.53	.41	.37	.29
Medium No Ceiling 40#, U= .51	24	.22	.15	.13	.09
Medium Ceiling 40#, U= .21	10	.41	.30	.27	.20

1. Factors based on 95 F outdoor design temperature. For each 5 F higher design temperature add 1 to base factor for light roofs, and medium roofs with ceiling. Add 3 to medium roof with no ceiling.

2. Overall roof factor - Base factor x insulation factor.

3. If ceiling space is ventilated by a fan, multiply factor by .75.

4. R-7 approximates 2" insulation, R-11 approximates 3", R-13 approximates 4", R-19 approximates 6".

OUTDOOR AIR FACTORS

Room Conditions	Total					Sensible	
	Outdoor Wet Bulb					Outdoor	
	65	70	75	78	80	Dry Bulb	Factor
75 F, 50%	8	26	46	59	69	85	11
						95	22
75 %, 55%	3	21	41	55	64	105	32
						115	43
75 F, 60%	–	17	37	50	60		

In determining the outdoor air quantity for calculating the outdoor air load be guided as follows:

1. Outdoor air through the unit.

   a) No exhaust fans, use value from Table 9.
   b) Exhaust fans, use exhaust fan air quantity or value from Table 9, whichever is greater. In the absence of exhaust air information, base air quantity for 20 air changes per hour for toilet room and 10 air changes for other ventilated rooms.

2. No outdoor air through the unit.

   a) No exhaust fans, use 7 cfm per person.
   b) Exhaust fans, use exhaust air quantity.

A

COOLING ESTIMATE

ITEM	Exposure	Quantity	X Factor	=	BTUH
WINDOWS (Table 1)		sq ft			
		sq ft			
		sq ft			
		sq ft			
WALLS (Table 2)		sq ft			
		sq ft			
		sq ft			
Partitions (Table 2)		sq ft			
Roof (Table 3)		sq ft			
Ceiling (Table 4)		sq ft			
Floor (Table 5)		sq ft			
Electrical and Appliances (Table 6)					
People (Total) (Table 7)		persons			
Room Total Heat (RTH)					
Outdoor Air (Total) (Table 8)		Cfm			
Grand Total Heat (GTH)					

1. Room Sensible Heat (RSH)
   = (RTH) - [People X Latent Factor (Table 7)*]

2. Sensible Heat Factor (SHF) = RSH/RTH

3. Total Sensible Heat (TSH)
   = RSH + CFM** X outdoor air factor (sensible) (Table 8)
* Also deduct 50% of the load of moisture producing appliances.
** Cfm is outdoor air quantity.

B

FIG. 22-2 Factors and forms for estimating cooling load requirements: (A) examples of tables used to establish heat gain factor and (B) portion of form for estimating cooling load *(Courtesy of Carrier Corporation)*

HEATING ESTIMATE

Room Temperature (occupied) _____ F (unoccupied) _____ F   Outdoor Temperature _____ F

Items	Description	Quantity	X Factor						= BTUH/° F
			R0	R4	R7	R11	R13	R19	
Glass Windows/doors	Single Pane	sq ft	1.13	.....	.....	.....	.....	.....	
	Double Pane	sq ft	.61	.....	.....	.....	.....	.....	
Walls-Light	8" lt. wt. agg. concrete block or frame	sq ft	.34	.15	.10	.07	.06	.....	
Walls-Medium	4" concrete block with 4" brick facing plastered	sq ft	.39	.15	.11	.07	.06	.....	
Walls-Heavy	8" brick-plaster finish	sq ft	.45	.16	.11	.08	.07	.....	
Roofs-Light	Preformed slab NO ceiling	sq ft	.22	.....	.09	.....	.06	.04	
Roofs-Light	Same w/suspended acoustical tile ceiling	sq ft	.14	.....	.07	.....	.05	.04	
Roofs-Medium	4" concrete NO ceiling	sq ft	.56	.....	.11	.....	.07	.05	
Roofs-Medium	Same w/suspended acoustical tile ceiling	sq ft	.23	.....	.09	.....	.06	.04	
Floors	2" concrete over vented crawl space	sq ft	.48	.....	.11	.08	.07	.....	
	Same over enclosed space or unheated basement	sq ft	.24	.....	.05	.04	.03	.....	
	Hardwood floor over vented crawl space	sq ft	.33	.....	.10	.07	.06	.....	
	Same over enclosed space or unheated basement	sq ft	.16	.....	.05	.03	.03	.....	
	Concrete slab on grade (perimeter) (Note 1)	lin ft	.85	.....	.....	.....	.....	.....	
Basement	6" Masonry wall (perimeter)	lin ft	.05	.....	.....	.....	.....	.....	
Infiltration (See Note 4)	1/2 air change (floor area)	sq ft	.10	.....	.....	.....	.....	.....	
	3/4 air change (floor area)	sq ft	.15	.....	.....	.....	.....	.....	
	1 air change (floor area)	sq ft	.20	.....	.....	.....	.....	.....	
	Sub Total (1)								
Ventilation	Outdoor air thru apparatus	cfm	1.1	.....	.....	.....	.....	.....	
	Sub Total (2)								

Unoccupied Heating Load	Occupied Heating Load
Sub Total (1) x Unoccupied Rm. Temp-Outdoor Temp. = BTUH	Sub Total (2) x Occupied Rm. Temp.-Outdoor Temp. = BTUH.
_____ x _____ T.D. = _____ * BTUH	_____ x _____ T.D. = _____ BTUH.
If Unoccupied or Setback Temp. is 10 F below Occupied Temp. Set Back Capacity Equals:	Less Credit for Lights = _____ BTUH. (Note 4)
Unocc. Htg. Load _____ * x 1.20 = _____ BTUH.	Occupied Heating Capacity = _____ BTUH.

FIG. 22-3 Sample form/worksheet for estimating heating load *(Courtesy of Carrier Corporation)*

DUCT SIZING AND LOCATION

Air-conditioning equipment and duct systems are located as centrally as practical. The heating and cooling load estimates are used for *sizing* the supply and return ducts, registers and grilles. The size of the plenum(s) surrounding the heating/cooling central unit, branch supply ducts, risers and outlets, and return outlets and ducts is determined from the survey information and technical data.

AIR-CONDITIONING CONTROL TERMS

Some of the commonly used terms for controlling air-conditioning systems include set point, control point, cycling, and deviation.

- *Set point* is the temperature, humidity or pressure value that is set on an air-conditioning device. If the temperature setting of a thermostat is 68°F, the *set point* is *68°F.*
- The *control point* represents the actual measurement of a temperature or humidity value that is recorded by the instrument. For instance, although the set point on a thermostat may be 68°F, the control point temperature (recorded on the thermostat) may differ; like, 70°F.
- The term *cycling* signifies a repetition of a planned series of events. As an example, if the thermostat control is set at 68°F and the space temperature drops two or more degrees below this set point, the system responds to the thermostat call for more heat. However, the thermostat shuts off the burner before the set point of 68°F is reached.
- When the space temperature cools slightly, the thermostat again calls for more heat. Heat is supplied until the space temperature is two or more degrees above the set point when the thermostat shuts off the burner. This patterned sequence of events is called cycling.
- *Deviation* means the difference between the set point and the control point. The deviation between a set point of 68°F and a control point of 70°F is 2°F.

SINGLE-PACKAGE AND SPLIT-PACKAGE AIR-CONDITIONING UNITS

A *single-package air-conditioning system* is self contained and includes a heating and cooling coil, condenser, compressor, fan, and fan motor. Supply plenums, registers, return plenums, and grilles are generally built into single-package units. Figure 22-4 identifies a self-contained water-cooled air-conditioning system.

In many commercial applications of single-package air-conditioning units, a supply register moves (throws) air from the unit directly into a required space. The return air is moved through the return grille and the unit itself.

SPLIT-PACKAGE HEATING/COOLING UNITS

A *split-package air-conditioning system* consists generally of two major units, figure 22-5. The system includes a cooling coil that has been added to a hot (warm) air furnace. The cooling coil is mounted above the furnace. The blower circulates return air around the furnace and cooling coil and feeds the cooled or heated air (depending on the operating mode) to the required room areas.

ACCESSORY
PLENUM

INDOOR
AIR FAN

THERMOSTATIC
EXPANSION VALVE

EVAPORATOR

RETURN
AIR FILTER

SELECTOR
SWITCH AND
THERMOSTAT

REFRIGERANT
FILTER-DRIER

CONTROL
BOX

WATER-COOLED
CONDENSER

COMPRESSOR

FIG. 22-4 Self-contained, water-cooled air-conditioning unit *(Courtesy of Carrier Corporation)*

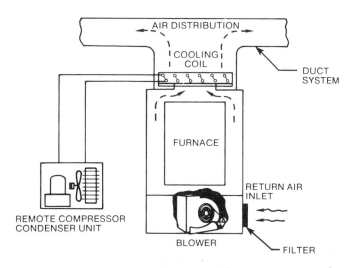

AIR DISTRIBUTION

COOLING
COIL

DUCT
SYSTEM

FURNACE

RETURN AIR
INLET

REMOTE COMPRESSOR
CONDENSER UNIT

BLOWER

FILTER

FIG. 22-5 Split-package, heating and cooling air-conditioning unit

The compressor and condenser in this kind of installation may be placed in a remote indoor or outdoor location. Although condensers are generally air cooled, many commercial applications require water-cooled condensers. Also, forced-draft and natural-draft cooling towers are used, particularly where water supply and water drainage are not a problem.

WINDOW AND THROUGH-THE-WALL-MOUNTED AIR-CONDITIONING COOLING UNITS

The window-type, single-room air-conditioning cooling unit, figure 22-6, is popular. This unit is designed to cool a specific area, usually a single room. In multiple dwellings and structures such as apartments and offices, it is common practice to plan the building construction to accommodate window or through-the-wall air conditioners.

Such air conditioners have the following advantages:

- Each unit may be controlled (set) to air-condition each room at a different temperature.
- Duct systems and plumbing systems may be eliminated in geographic locations where the climate is mild.
- Installation and maintenance are simple, and special construction features are not required.
- Units of varying capacities are commercially available to meet particular needs.

However, there are a number of disadvantages.

- Unless the unit works properly, there may be a continuous dripping of condensate from the unit.

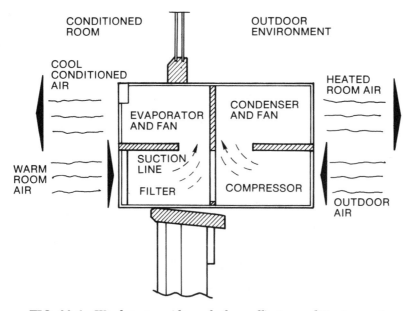

FIG. 22-6   Window-type (through-the-wall) air-conditioning unit

- The room lighting for which the window was designed is partially blocked off. Window units also extend beyond the sill into the room.
- The inside and outside appearance of a unit may not be pleasing or may conflict with the architectural design of the building.
- The sound level within the room may be objectionable.
- The flow of air may not be as precisely controlled as a central air-conditioning system.

Principles of Operation of Air-Conditioning Window-Type Cooling Systems

Window/through-the-wall mounted units fall within the 5,000 to 36,000 Btu range. There are four basic components in the outdoor compartment: ① hermetic compressor, ② motor relay and terminal box, ③ condenser and ④ condenser fan. The schematic drawing, figure 22-7, shows the position of each component.

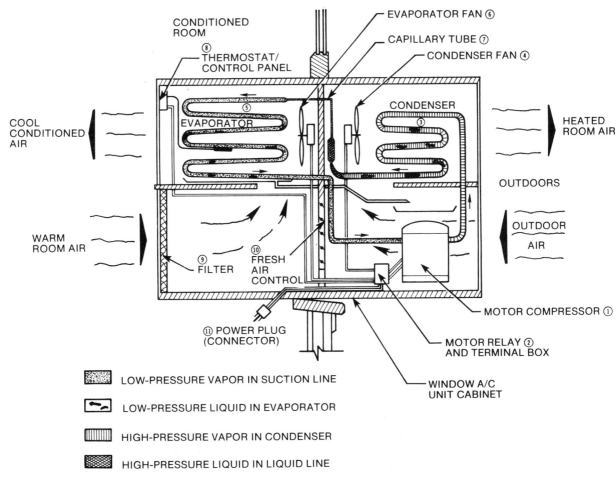

FIG. 22-7 Principles of operation of a window-type air-conditioning cooling unit

The components in the air-conditioned room compartment of the unit include the evaporator ⑤ , evaporator fan ⑥ , capillary tube ⑦ , thermostat/control panel ⑧ , air filter ⑨ , fresh air control ⑩ , and power connector plug and cord ⑪ .

When the set point on the thermostat calls for cooling, the compressor turns "ON." Low pressure vapor from the evaporator is drawn and compressed to a high pressure vapor. In turn, this high pressure vapor is forced into the condenser where the heat of compression and the heat picked up by the evaporator is given off to the atmosphere.

The removal of this heat changes the high pressure vapor to a high pressure liquid. This liquid flows through a metering device (capillary tube or TEV) into the evaporator.

During the compressor operation, the evaporator (which is under low pressure) causes the liquid refrigerant to boil. This condition allows the refrigerant to pick up the heat that has accumulated on the surface of the evaporator.

As shown in Figure 22-7, the evaporator fan pulls the warm air over the evaporator. Heat is removed from the room air and cooled air is pushed back into the room. This replacing of the warm air with cool air continues until the set temperature is reached and the thermostat control shuts down the operation. Cooled air flowing over the evaporator serves as a dehumidifier. Also, moisture, pollen and dust (which may accumulate on the evaporator) flow off. Drip pans are provided for water removal.

The control panel, which is included in the design of window units, usually includes a temperature control, fan speed control for regulating airflow and air selection control.

The condenser fan serves to draw outside air, circulate it over the compressor and condenser and then discharges the air to the atmosphere.

CENTRAL AIR-CONDITIONING SYSTEMS

Central air-conditioning systems have the following advantages over smaller single-room units:

- The entire air-conditioning requirements of a residence, apartment or commercial building are provided from a central automatic control point.
- Air distribution is better.
- System maintenance is easier as compared with maintaining a number of individual units in which many components are duplicated.

The schematic drawing, figure 22-8, of a dual heating/cooling unit in a central air-conditioning installation is used to describe the components and operation during summer (cooling) and winter (heating) modes. The evaporator ① is mounted in the plenum; the condensing unit ② (for cooling), outdoors. These two units are connected by tubing (usually precharged), which runs through the wall. The metering device is generally located near the evaporator.

Heating, cooling and humidity are controlled by a thermostat and humidistat. These are mounted in the area to be air conditioned. Cooling is provided by using a blower ③ to force air through cooled evaporator coils ④ . Cooled air is distributed throughout the duct system ⑤ . Warm air is pulled through the return ducts ⑥ to be recirculated around the air-cooling evaporator.

Humidity is controlled by removing the moisture on the surface of the evaporator coils. The condensed moisture is carried away by a drain pipe (tube) ⑦ .

The blower fan in this kind of an installation serves the dual function of circulating cool air during the summer (cooling) season and warm air throughout the heating cycle.

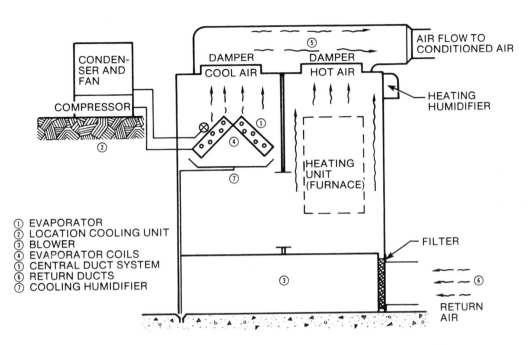

FIG. 22-8   Dual heating/cooling units in a central all-year air-conditioning system

The blower fan is often used independently to circulate air in particular areas when no heating or cooling may be needed.

EVAPORATIVE CONDENSER SYSTEMS

This system, figure 22-9, uses the same major components as most air-conditioning systems: motor compressor, evaporator, thermostatic expansion valve, sight glass, drier, shut-off valve and receiver, and motor and other controls.

The condenser, fan, water pump, holding tank, and float mechanism are located away from the conditioned space.

Compressed (hot) high pressure refrigerant is pumped to the evaporative condenser. Water is piped to a holding tank and maintained at a required level. Water is moved out of the holding tank and sprayed over the condenser. The fan in the evaporative condenser area forces air over the condenser producing droplets of cool water that cool and speed up the evaporative process.

The condenser is cooled by the flow of cool water over it. The evaporative condenser reuses the same water plus additional water to replace water losses due to evaporation.

COOLING TOWERS AND WATER-COOLED CONDENSERS

Cooling towers and *water-cooled condensers* are used in combination with natural draft, forced draft and evaporative cooling condenser systems. The cooling towers serve to remove heat from the water that returns back to a condenser. In turn, the water-cooled condenser absorbs heat from the refrigerant.

415

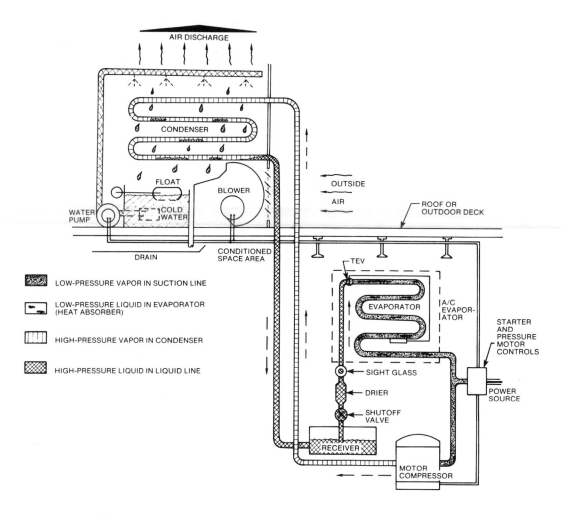

FIG. 22-9 Evaporative condenser used in an air-cooling installation

CHILLER/HEATER UNITS: GAS FIRED

Chillers/heaters are designed to supply on demand either chilled water for cooling or hot water for heating residences, apartments and industrial air-conditioning systems. A typical residential installation of a gas-fired (natural or propane gas) system is illustrated in figure 22-10.

Circulating water connections at the outdoor chiller/heater unit are shown in the sketch, figure 22-11. Note the location of return and supply water thermometer wells in each line.

The *throttle valve* for regulating chilled/heated water flow is installed in the return water line from the coils. All water circulating lines are completely insulated, including a vapor barrier. One chiller/heater unit may be used in a number of zones, providing the overall load is within the rated capacity of the unit.

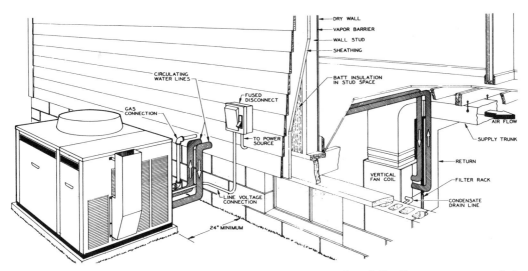

FIG. 22-10 Installation details for the major components of a chiller/heater, water-cooled air-conditioning system *(Courtesy of The Dometic Corporation)*

SAFETY PRECAUTIONS

- The model chiller/heater unit as described uses an ammonia refrigerant. Safety codes require outdoor venting of the water circulating system.
- All gas piping, connections and shut-off valve must comply with applicable codes and standards.

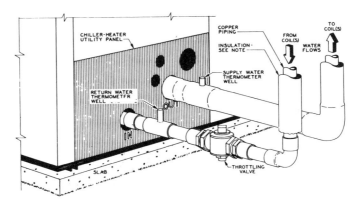

FIG. 22-11 Piping and control details for an outdoor chiller/heater, water-cooled air-conditioning unit *(Courtesy of The Dometic Corporation)*

417

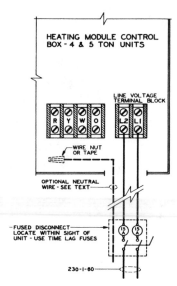

FIG. 22-12   Typical electrical wiring diagram of a 4- and a 5-ton chiller/heater installation *(Courtesy of The Dometic Corporation)*

Electrical Circuitry

Figure 22-12 is a typical electrical circuit diagram. All wiring connections, wiring protection devices, switch locations, and fuse circuits are shown.

SAFETY PRECAUTIONS

- The fused disconnect switch should be installed within sight of the unit, *but not on the unit.*
- Unit grounding must conform to applicable local and national electrical industry codes.

Low Voltage Control Wiring

Low voltage wiring is required between the chiller/heater (A), room thermostat (B), and fan coil (C). Figure 22-13 is a typical wiring diagram of low voltage control wiring and line voltage wiring for the fan coil unit.

Temperature Control Thermostats

Heating-cooling requirements for a single zone or multiple zones for gas air conditioners are controlled by special thermostats. These automatic heating-cooling, multiple-zone, thermostats generally operate on 24-volt control current power.

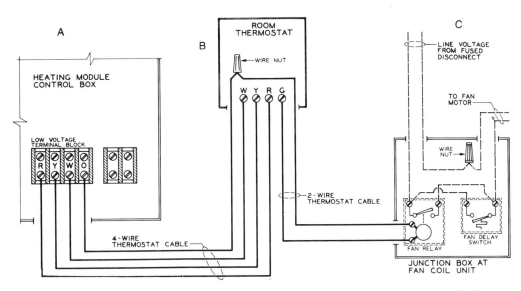

FIG. 22-13 **Typical low voltage control and line voltage wiring diagram** *(Courtesy of The Dometic Corporation)*

COOLING CYCLE OF A CHILLER/HEATER UNIT

The refrigerant solution in the chiller/heater unit illustrated in figure 22-14 consists of water (absorbent) and ammonia (refrigerant). The cooling cycle begins in the generator ① with the refrigerant solution. A burner causes the solution to boil and release

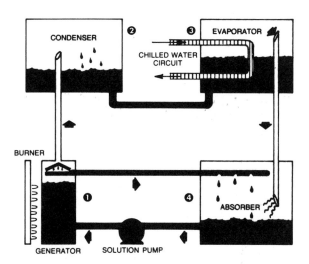

FIG. 22-14 **Principles of operation: cooling cycle of a chiller/heater, water-cooling system** *(Courtesy of The Dometic Corporation)*

refrigerant vapor. The vapor (under pressure) is pushed to the air cooled condenser ② . In the condenser, the vapor is cooled, condensed back to a liquid refrigerant and moved to the evaporator ③ .

In the evaporator, water that contains heat from the air-conditioned space flows over the outside of the tube. As heat is transferred to the liquid refrigerant, the refrigerant vaporizes, is drawn to the absorber ④ , and is absorbed back into the solution. The solution pump pumps the refrigerant solution of water and ammonia back to the generator. The cycle continues until the cooling demand is met.

Producing the Cooling Effect

The cooling effect part of the air-conditioning system is produced by circulating water that has been chilled through coils. Return air from the conditioned space flows across the coil unit. In this process, some of the heat from the air is transferred to the chilled water. The air is cooled and some of the moisture is cooled and condensed. This combination of conditions produces drier, cooler air for distribution in conditioned areas.

Heat is dissipated to the inside air from the water, which contains the heat that has been removed from the air. The water flows back to the chiller to be rechilled.

HEATING (HOT WATER) CYCLE OF A CHILLER/HEATER UNIT

When the thermostat is set in the heating mode, the chiller water pump is "OFF," figure 22-15. In this position, a ball valve in the chiller tank seals off any flow of water.

Hot water is circulated by the hot water pump through the hot water generator and air handler, back to the generator (through the water reservoir). The reservoir permits air to be released from the system.

CONSTRUCTION OF GAS CHILLER/HEATER AIR CONDITIONERS

An example of an all-season, gas energy, heating/cooling air-conditioning unit is illustrated in figures 22-16 and 17. Units in this particular series are designed for 3-, 4-

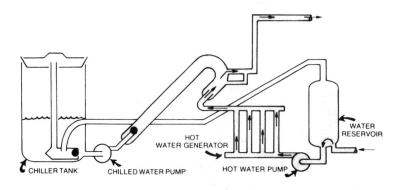

FIG. 22-15 Heating cycle of a chiller/heater, water-heating system (*Courtesy of The Dometic Corporation*)

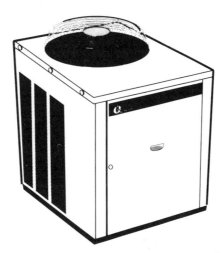

FIG. 22-16 External appearance of all-season, gas energy, chiller/heater air-conditioning unit *(Courtesy of The Dometic Corporation)*

and 5-ton (36,000 to 60,000 Btu) cooling capacities and 120,000 to 180,000 Btu/hr heating capacities.

Natural or propane gas is used as fuel. The units are all-season and are adaptable to residential, light commercial and industrial applications.

BUILT-UP WINTER-SUMMER AIR-CONDITIONING SYSTEMS

A common example of a built-up air-conditioning system for winter heating and summer cooling is shown graphically in figure 22-18. The components for the heating

FIG. 22-17 Internal operating features and construction of an all-season, gas energy, chiller/heater air-conditioning unit *(Courtesy of The Dometic Corporation)*

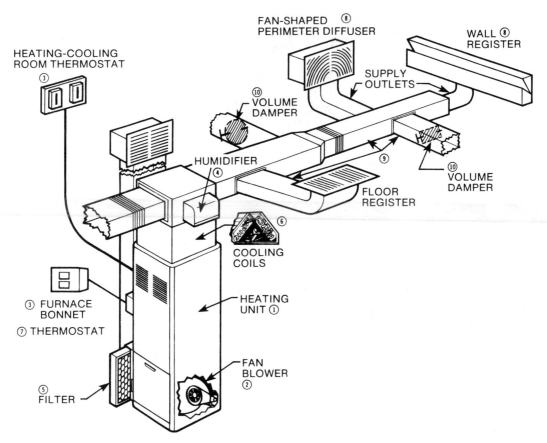

FIG. 22-18   Typical central air-conditioning system duct work and components (excluding remote location cooling unit and connections)

mode consist of a heater ①, blower ②, controls ③, humidifier ④, and air filter ⑤. The additional components for the cooling mode include the compressor, condenser, expansion valve or capillary tube assembly, cooling coil, and interconnecting tubing. The cooling coil ⑥ is located in the supply plenum. A condensate drain pan is included.

A room thermostat is used to control winter-summer temperature requirements. A temperature actuated switch ⑦ (bonnet [also called fan and limit] thermostat) controls the blower operation and limits the maximum bonnet temperature.

The air distribution system consists of supply and return air ducts, supply outlets, return air intakes (grilles), and volume control dampers. Supply outlets (located in the conditioned rooms) are also known as *registers* ⑧ or *diffusers*. Diffusers are of different designs like fan shape to flow the air in a fan pattern or parallel design to flow the air parallel to a wall or ceiling.

The blower fan distributes (circulates) air through the air-conditioning unit components and throughout the air distribution (duct) system ⑨. The air conditioning system

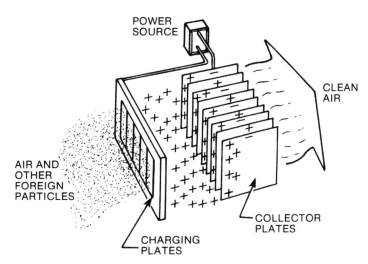

FIG. 22-19   Principle of operation of electronic filters

is *balanced* by adjusting *volume dampers* ⑩ (that are built into various ducts) to regulate the quantity of air flowing through each duct.

The cooling coil serves to remove heat and moisture. In this particular application, the "A-shape" cooling coil is contained in the supply plenum. The cooling coil may also be located in the supply duct system. Cooling coils are used where there is either an upflow or a downflow of air.

MECHANICAL, ELECTROSTATIC AND ELECTRONIC FILTERS

Filters serve to remove foreign particles such as dirt, pollen, lint, and smoke and to produce a healthier, cleaner air. Mechanical (dry-type) filters are designed with a fibrous, filtering medium that is encased in a wide-mesh filter frame. Large foreign particles are entrapped in the filter. Disposable-type filters are inexpensive and are replaced when they become dirty. Washable-type filters are cleaned and reused.

A plastic filtering material is used in another filter design, which depends upon an electrostatic charge. The action of moving air as it passes through the filtering material generates an electrostatic charge. This charge attracts airborne dirt and other foreign particles to the surfaces of the filtering material. Generally, dirty electrostatic (self-charging) filters are washable and reusable.

Charged-Media Electronic Filters

These filters consist of a series of alternate positive and negative electronically charged *collecting plates*, figure 22-19, and a number of *charging plates*.

As foreign particles in the air stream flow through the charging plates, they receive a strong charge. As the air stream continues to move through the bank of collecting cells (collector plates) that have an opposite charge, the dirt particles are attached electronically to the collector plates. These plates are removed for recleaning and reuse.

SUMMARY

- Cooling load estimates are based on overall dimensions of residential or commercial structures, floor area (square feet), room height, and living/working areas, and other survey data.
    - Btu/hr values are established for each room using appropriate tables of factors for window, floor, and roof areas, linear feet of walls, etc.
    - Consideration is given to heat gain produced by equipment, processes and numbers of people within the conditioned area.
    - Duct losses are added to obtain load requirements in Btu/hr. The grand total cooling load is calculated by multiplying the total load by 0.75.
- Heating load estimates require similar calculations based on the sum of the heat losses per room, adjusted for actual temperature differences.
    - The total heating load losses = 20% of the duct heat losses added to the heat losses of the room.
    - The heating load cfm requirement equals the total Btu/hr heat loss divided by 108.
- Room/window air-conditioning units provide temperature and air control conditions for a room or compartment without requiring duct work.
- Cooling equipment may be added to or combined with a residential or industrial central heating system to permit all-season heating/cooling within a building.
- Single-package fan coil units are designed for water- or air-cooled condensers.
- Single- or split-package air-conditioning units are available as self-contained for use in commercial and industrial applications. These units may be floor or ceiling mounted either inside or outside a controlled area.
- Duct systems include the plenum and one or more branch ducts, supply and return outlets; appropriate air distribution grilles, radiators and registers; and damper controls.
- The terms set point, control point, deviation, and cycling are applied to the thermostat and conditions of temperature control.
- Window (through-the-wall) cooling units permit economical installation and practical air-conditioning (cooling) applications.
    - The outdoor compartment of a window air-conditioning unit includes a hermetic compressor, condenser, fan, and motor relay and terminal box.
    - The indoor compartment includes an evaporator, fan, suction line, thermostat, control panel, air filter, fresh air intake, and power source.
- Chiller/heater units are designed for all-season chilled water cooling (summer) and hot water (heating) applications.
    - Safety precautions for operation and other national/state/local industry codes must be observed.
- The chiller produces cooled water that is circulated through evaporator coils. Room air is cooled as it is passed over the coils and redistributed to conditioned air areas. In this process, some of the moisture cools and condenses to produce drier, cooler air.
- The heater generates hot water that is circulated through the hot water generator and air handler.

- Smaller 3-, 4- and 5-ton chiller units are commercially available for residential or light commercial applications that require 36,000 to 60,000 Btu cooling capacities and from 120,000 to 180,000 Btu/hr heating capacities. Larger units are available for heavier load industrial plant applications.
- Water-cooled condensers are used in evaporative condenser systems to absorb heat and to cool a compressed high-pressure refrigerant.
- Natural draft, forced draft and evaporative condensers are used in combination with cooling towers to absorb heat from the refrigerant in a system.
- Built-up systems provide flexibility by combining all essential components into a single central system from which branching may take place through special duct systems.
- Humidifiers reduce humidity from moist air within a controlled space by evaporating the fine particles during summer cooling. Humidifiers also serve the reverse process of supplying moisture when called for in winter heating periods.
- Mechanical, electrostatic charge and electronic filters serve to trap and collect dust, dirt, lint, and other undesirable foreign airborne particles (dirty air), preventing them from recirculating throughout an air-conditioning system.

ASSIGNMENT: UNIT 22 REVIEW AND SELF-TEST

A. ESTIMATING HEATING AND COOLING WORK LOADS

1. Name four prime sources of information that are used to estimate heating and cooling requirements for residential and industrial air-conditioning systems.
2. a. Secure a set of the *survey forms* furnished by an air-conditioning unit manufacturer for estimating loads.
   b. List three different load factors that relate to (1) construction details and (2) three others that deal with geographic location conditions.
3. a. Refer to a handbook or manufacturer's table of factors for *roof areas*.
   b. Determine the heat factors for the following two conditions:

	Range of Daily Temperature	Roof Insulation	Outside Design Temperature
(1)	15°F	1" Thick	90°F
(2)	25°F	4" Thick	105°F

4. Compute the CFM capacity required for a residential heating unit where the room heat losses = 116,750 Btu/hr. Consider duct losses at 20% additional. Use the formula: $\text{CFM} = \dfrac{\text{Maximum Heat Loss}}{108}$

Cooling Loads	
Windows	5,750 Btu/hr
Walls	3,056 Btu/hr
Roof	6,924 Btu/hr
Electrical/	
Appliances	7,216 Btu/hr
Individuals	15,120 Btu/hr
Outside Air	20,240 Btu/hr

5. a. Use the Btu/hr cooling loads in the table.
   b. Determine the total cooling load.
   c. Compute the actual tons of refrigeration that are required.
      Note: Use 1 Ton = 12,000 Btu/hr.
   d. Select the capacity of the unit.
   b. Determine the total cooling load.
   c. Compute the actual tons of refrigeration that are required.
      Note: Use 1 Ton = 12,000 Btu/hr.
   d. Select the capacity of the unit.

B. INDIVIDUAL WINDOW-ROOM AIR CONDITIONER COOLING UNITS

1. State three advantages of window, single-unit air conditioner cooling units.
2. List the basic components of a window-type air-conditioning cooling unit that are located in the outdoor compartment.
3. Insert the number of the *definition or condition* that correctly relates to each of the air-conditioning temperature control *terms*.

Temperature Control Term   Definition or Condition

( ) a. Set Point   (1) Plenum chamber temperature
( ) b. Control Point   (2) Patterned sequence of temperature events
( ) c. Deviation   (3) Position of a damper in a duct
( ) d. Cycling   (4) Actual recorded temperature
                          (5) Compressor temperature relay
                          (6) Difference between set point and point control
                          (7) Fixed thermostat temperature

C. CENTRAL AIR-CONDITIONING SYSTEMS

1. State two advantages of self-contained, central air-conditioning systems.
2. Explain briefly the two following conditions:
   a. How cooling is produced in a central air-condition unit and
   b. How summer humidity is controlled.

D. EVAPORATIVE CONDENSER AIR-CONDITIONING SYSTEMS

1. List the components of an evaporative condenser air-conditioning system that are remotely located.
2. Explain briefly how hot, compressed, high-pressure refrigerant in an evaporative condenser is cooled.

E. CHILLER/HEATER AIR-CONDITIONING UNITS

Complete statements 1 through 3 by adding the correct term or condition.

1. Chiller/heater units are designed to supply either _____ and/or _____ .
2. The _____ regulates chiller/heater water flow.
3. During the cooling cycle of a chiller/heater unit,
   a. _____ is cooled and condenses back to a _____ _____ .
   b. as _____ is transferred to the _____ _____, it becomes a _____ and is absorbed back into solution.
   c. cooling is produced by circulating _____ _____ through coils, and _____ return air (heated from the conditioned space) across the coil unit.
   d. _____ is controlled by removing moisture from the air as it passes over the cooling coils.

F. HEATING AND COOLING AIR-CONDITIONING SYSTEMS

Select an actual or a manufacturer's built-up, year-round, residential air-conditioning system.

1. Name the system: manufacturer, type model, capacity.
2. Identify the major components for (a) heating, (b) cooling, (c) distributing, and (d) cleaning the controlled air.
3. Give the heating and cooling work loads for which the system is designed.

G. FILTERS FOR AIR-CONDITIONING SYSTEMS

1. List three different types of air-conditioning system filters that are commercially available.
2. State the function served by air-conditioning system filters.

UNIT 23:
NONMECHANICAL REFRIGERATION SYSTEMS

The high and the low pressures needed in a refrigeration system may be produced by either mechanical or nonmechanical means in what is called the "heart" of the system. When a compressor is used for this purpose, the heart is said to be mechanical. The heart in nonmechanical refrigeration systems produces the required pressures without the use of a compressor or other similar device.

This unit begins with a review of the principles upon which the nonmechanical absorption systems depend. Complete cycles of both domestic and commercial absorption installations are illustrated and described. The steam jet, as a second system of nonmechanical refrigeration, is covered in a great deal of detail because of its importance in large industrial applications. This is followed by examples of new developments in thermoelectric refrigeration systems. These show the tremendous possibilities of this method of refrigeration.

The important magnetic system of reaching the near absolute temperatues necessary in research is the last of the nonmechanical systems to be covered. Throughout the unit, comparisons are made of the values of mechanical and nonmechanical systems to point but where each may best be employed.

ABSORPTION SYSTEMS

As a refrigerant boils in the evaporator it sbsorbs heat. For the boiling to take place at a low enough temperature to refrigerate, the pressure in the evaporator must be very low. This low pressure may be produced by removing the refrigerant vapor as fast as it is formed by the boiling liquid. Th vapors can be absorbed readily by a second liquid to bring about this condition. One of the required properties of such a liquid is that it must be able to absorb the vapor when it is cool and release it when heated.

The most commonly used refrigerant in absorption refrigeration systems is ammonia. Vapors from ammonia may be absorbed in large quantities by cool water at temperatures between 90° and 100°. Water has the ability to absorb ammonia vapor so fast that it is as effective as a mechanical compressor.

For example, refrigeration is produced when a tank of pure liquid ammonia under high pressure is fed from the high side through a metering device and into an evaporator, figure 23-1. This refrigeration continues until either the high-side liquid ammonia is exhausted or the water in the absorber tank is so saturated that it can no longer absorb ammonia. The size and design of the absorber and ammonia tank can be such as to make this system practical.

Refrigerated trucks use a similar system. During the summer, the ammonia tank is filled daily. At the same time the absorbers are drained and refilled with pure water. The ammonia, which is absorbed by the water, is reclaimed from the absorber solution and

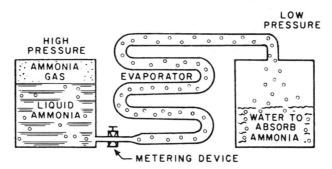

FIG. 23-1   Absorption refrigeration system principle

is used again. In the wintertime, this servicing operation is carried on every few days. While this kind of absorption system is not a continuous one because of the servicing required, a refrigerating plant may be built to include an absorber for reclaiming the ammonia from the water.

Continuous Operating Absorption Systems

Like the mechanical systems, continuous operation of an absorption system requires a condenser, liquid receiver, expansion valve and an evaporator. The schematic drawing in figure 23-2 shows the absorber, generator and ammonia solution pump for such an ammonia absorption system. In this system most of the ammonia is removed from the water. Since the generator does not remove all of the ammonia, a weak water solution of ammonia flows by gravity to the absorber.

The water in the absorber (A) absorbs the ammonia until the ammonia and water solution contains a concentration of almost 30 percent ammonia. This strong solution, called *strong aqua*, is pumped by the *strong aqua pump* (B) up to the generator (C). This

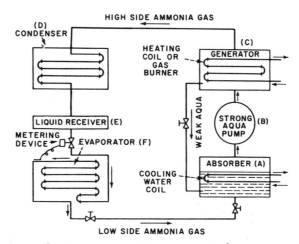

FIG. 23-2   Continuous operating ammonia absorption system

is necessary because the absorber operates at low-side evaporator pressure as contrasted with the generator (high-side) pressure.

In the generator, ammonia vapor is driven out of the strong aqua in much the same manner as air is driven out of water when heated. As the ammonia vapor which is at a high temperature and pressure level rises, it travels to the condenser (D). The weak aqua liquid drops in the separator and then flows back to the absorber by gravity.

The condenser removes the latent heat from the ammonia vapor and condenses it. The condensed ammonia liquid flows through the liquid receiver (E) to the evaporator (F). Here it boils at reduced pressure. The latent heat is absorbed while the liquid ammonia changes into a vapor and thus, the refrigerating effect is produced.

Large Commercial Absorption Refrigeration Systems

The absorption flow diagram in figure 23-3 shows a large installation in which steam is used to heat the generator. In addition, the heat from the steam accomplishes two things:

- It separates the ammonia refrigerant from the absorbent (water).
- It raises the ammonia vapor to a high level of temperature and pressure as it enters the condensers.

The cooling water is also circulated through the absorbers. This keeps the water at a temperature level where it has a high absorption rate for the ammonia. The strong aqua

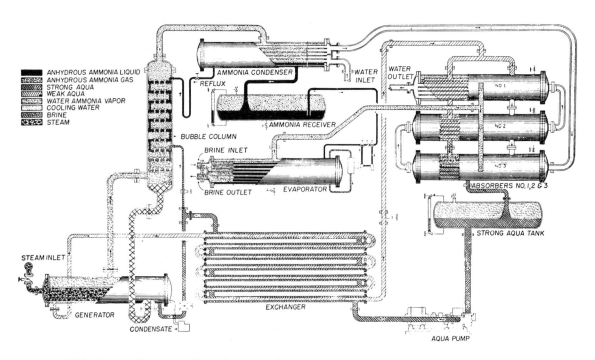

FIG. 23-3 Absorption flow diagram of a large installation (*Courtesy of York Corporation*)

is pumped back to the generator and it moves through the exchanger and cools the weak aqua, which is on its way to the absorbers.

In this large system the boiling ammonia is used to chill the brine solution in the evaporator. The brine solution, in turn, is circulated through insulated pipes to the refrigerated space. While ammonia and water have been discussed, other combinations may be used in absorption systems. For instance, the absorbent salt lithium bromide may be combined with water as the refrigerant.

Domestic Absorption Refrigerating Systems

The absorption cycle of the domestic unit (illustrated in figure 23-4) may be easily traced through the numbers which identify the components and their operation. The refrigerant in this case is ammonia because liquid ammonia has a high latent heat of vaporization. Water is used as the absorbent.

At ordinary temperatures and pressures water absorbs large quantities of ammonia. In the reverse process, the ammonia (A) absorbed by the water may be driven off by adding heat. Hydrogen gas (H) is present in this system to increase the rate of evaporation of the ammonia and to provide the necessary pressure balance in the system. Heat applied from the flame through the generator (1) causes ammonia vapor to be released from the solution. This hot ammonia vapor moves upward through the percolator tube (10), carrying the solution over the upper level in the separator (11). The liquid solution, which settles in the bottom of the separator, flows through the liquid heat exchanger (9) to the absorber (4).

The hot ammonia vapor in the separator moves down the center tube to the analyzer (6). From the analyzer the hot ammonia vapor rises into the rectifier (7) where any water vapor that is still in the ammonia condenses and drains back to the analyzer. As shown, the rectifier has a series of baffle plates around the tube. From the analyzer the pure ammonia vapor moves upward to the finned condenser (2). Air circulating around the fins in the condenser (2a) takes out heat and condenses some of the vapor into a liquid. At this point the heat has completed its work and circulation for the rest of the cycle depends on gravity. The pure ammonia flows into the evaporator (3b). On the other hand, the ammonia vapor that does not condense in the first part of the condenser (2a) moves into the upper trap (2b).

The liquid ammonia, which builds up to a predetermined level in the U-tube storage and receiving compartment, flows by gravity into the cooling coil (3a). Liquid ammonia forms a series of large shallow pools on the baffle plates as it falls into the cooling unit (3a, 3b). Since hydrogen is fed to the cooling unit, its presence makes it possible for the liquid ammonia to evaporate more rapidly at a low temperature. During the evaporation process, the ammonia absorbs heat from the refrigerator and causes water in the ice-cube containers to freeze.

When the ammomia evaporates, it mixes with the hydrogen and, since the mixture is heavier than hydrogen, it travels down through the gas heat exchanger (8) to the absorber (4). It is the absorption of ammonia which reduces the pressure in the evaporator. As the ammonia and hydrogen mixture passes through the center tube of the gas heat exchanger, it cools the hydrogen in the outside tube.

A weak water and ammonia solution in the separator (11) flows through the liquid heat exchanger (9) and the precooler (12) to the top of the absorber (4). At this point the

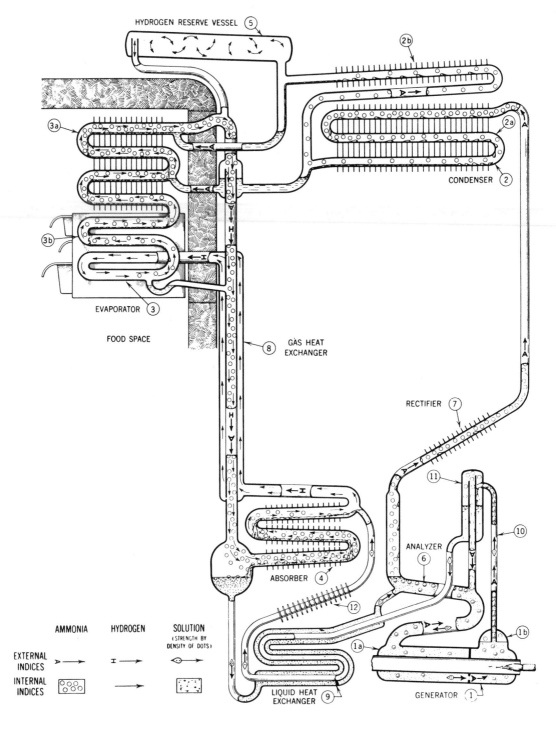

FIG. 23-4   Domestic absorption unit

solution meets the ammonia and hydrogen vapor which has passed through the gas heat exchanger. Since it is a weak solution, it absorbs the ammonia vapor and frees the hydrogen vapor. The hydrogen vapor is extremely light and insoluble in water and, therefore, it rises to the top of the absorber and passes around the outside of the gas heat exchanger up to the cooling unit.

The absorber illustrated is air cooled through the use of fins. As it cools, the weak solution of water and ammonia absorbs the ammonia gas out of the mixture of ammonia and hydrogen. In this process the sensible heat of the mixture is released through the fins. The resulting strong solution of ammonia and water flows to the absorber and returns to the liquid heat exchanger. From this point, the strong liquid refrigerant returns to the analyzer (6) and back to the generator to begin the cycle again.

The preceding type of a domestic absorption refrigeration system is welded and there are no moving parts to be adjusted or to wear out. The construction must be rugged because a total pressure of 200 psig throughout the cycle must be maintained. To produce a 0°F refrigerant in the evaporator, ammonia must boil at 15.7 psig. The 184.3 psig balance of pressure must be made up by the hydrogen.

Gas Volume Control. The domestic absorption automatic control refrigerator has both a gas volume and a safety control. A pressure-temperature control bulb located within the cooling unit regulates the amount of gas needed to produce certain desired temperatures and conditions. The larger the flame, the quicker the cycle is completed in the system. As the cooling unit cools, the power element cools. This reduces the pressure on the gas valve diaphragm and closes the gas opening. The size of the flame is regulated in this manner. The flame can be manually adjusted by turning the two adjusting screws clockwise to reduce the gas supply; counterclockwise, to increase the supply, figure 23-5.

Usually, absorption refrigerators have an automatic pressure regulator in the gas line for maintaining a constant gas pressure as fluctuations must be reduced to a minimum.

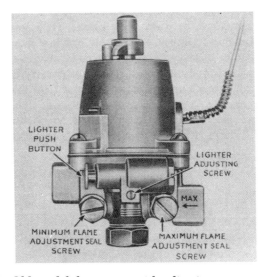

FIG. 23-5 Old model thermostat with adjusting screws and lighter

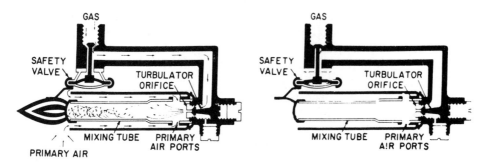

FIG. 23-6 Principle of operation of a safety shutoff valve

Safety Controls. A principle of operation of a safety control valve that is provided on absorption refrigerating systems (to prevent the waste of gas and to be protected against other dangers which may be present if the gas flame is extinguished) is shown in figure 23-6. Note that the valve is temperature controlled and is of a *dish-button type.* Since this safety control is located next to the flame, the disk remains hot as long as the gas is ignited. When the flame is extinguished, the disk cools and snaps the dish-shaped button out. This pulls the rod, connected to the valve, so that it cuts off the supply of gas to the burner.

The burner can be automatically ignited by depressing a special push button or by turning a gas control knob on other models. Once the burner is "ON," the thermostat is set to the desired temperature.

The burner is turned "OFF" by first moving the thermostat to the lowest setting. The gas control knob (inside the access panel) is then turned to the "OFF" position.

SAFETY PRECAUTIONS

- There is always the possibility of a fire or explosion unless safety precautions are strictly followed. Safety instructions are provided by manufacturers to protect against personal injury, possible loss of life, or property damage.
- Manufacturer's instructions deal with safe practices in installation and maintenance and safety procedures in operation.

Automatic Defroster Devices on Absorption Refrigerators. Semiautomatic and automatic defrosters are provided for domestic absorption refrigeration systems. The thermostat on a semiautomatic defroster may be set to a minimum flame, figure 23-7. In this defrost position the thermostat will automatically start the refrigerating cycle after the cooling coil is defrosted.

The electric timer in the wiring diagram of an automatic defroster is controlled by the defrost switch, figure 23-8. The thermostat operation stops at the hour at which the timer is set. At the same time an electrical connection is made with the defrost heater.

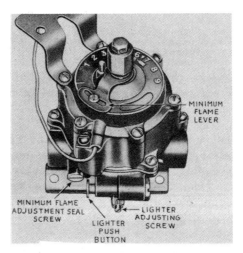

FIG. 23-7   Semiautomatic defrosting thermostat

Electrical energy is sent through resistance wires which are positioned under the cooling coil. The heat defrosts the outside of the cooling coil without disturbing any frozen foods within the coil. Normal operation starts automatically when the coil is defrosted.

Ice Cube Makers.   Ice cubes may be manufactured automatically in an ice cube maker. The ice cubes are held in a container. When the container is partially full, an electrical circuit opens a water valve. This valve is controlled by the level of the water in the ice cube freezer trays. As the trays fill, the water flow stops and the refrigeration

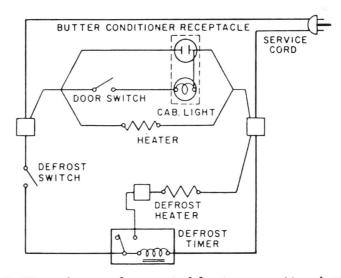

FIG. 23-8   Wiring diagram of automatic defrosting system (time shutdown)

cycle starts and continues until the water is frozen into ice cubes. At this point, a thermostat closes a heating circuit which frees the ice cubes. A small electric motor then turns the ice cube device, unloading the cubes into the basket.

COMPARISON OF NONMECHANICAL AND MECHANICAL SYSTEMS

Before proceeding with other types of nonmechanical systems, a comparison of the advantages of the absorption over the compression systems is made. Some ideas may then be gained from this summary as to the places where each one may work best:

- The absorption system is quiet and subject to limited wear because the only moving part is the aqua pump. In comparison with a mechanical refrigerating system of the same capacity, the engine, pump motor, or turbine is small.
- Where the absorption system is designed to operate on high- or low-pressure steam, the exhausts from other equipment may be used. The system does not need to depend on electrical power for the pump motors.
- Multiple absorption units of over 1,000-ton capacity may be built. By comparison, single compressor units are now being built to exceed 5,000 tons.
- While absorption units take up more space, they can be located outdoors as a vertical unit to cut down on the ground area. No housing is needed.
- The absorption system is favored for low evaporator temperatures and for automatic control requirements.
- There is only a limited decrease in capacity of an absorption unit when the evaporator pressure and temperature are reduced. This may be compensated for by increasing the steam pressure to the generator. By contrast, the capacity of a compression system is decreased rapidly as the evaporator pressure is lowered.
- The absorption unit is almost as efficient at reduced loads as at full capacity. The quantity of aqua circulated and the steam supplied to the generator may be changed to meet different load conditions.
- Any liquid refrigerant in the absorption system merely unbalances the system. By comparison, preventive measures must be taken in the compression system in order to avoid damage to the compressor.

STEAM-JET REFRIGERATION SYSTEMS

The steam-jet refrigeration unit has many advantages over other systems in those locations where water and high pressure steam are both available at moderate cost. Because it has no moving parts, the steam-jet unit has limited vibration and can be located anywhere. It is also especially adapted for use where hazards due to toxicity or fire must be minimized. Steam-jet refrigeration systems are used when low temperatures are required: for example, industrial processes such as the separation of paraffin wax from lubricating oils.

Principles of Operation of Steam-Jet Systems

Water remains liquid up to the temperature at which its boiling point corresponds to its pressure. When water at a temperature higher than 45°F enters a tank having a pressure of 0.148 psia, the water cools to 45°F as it surrenders heat. The water itself is vaporized because there is no other medium at a temperature low enough to receive this heat.

Thus, the sensible heat given up by the water in cooling is used as latent heat. The latent heat vaporizes a small quantity of the water coming into a tank. The warm water, which is returned to the tank, is cooled to a temperature corresponding to the pressure in the tank. The low pressure determines the temperature to which the water may be cooled. The chilled water is removed by a pump, which increases its pressure as it is delivered to a desired place.

The tank in which the water continuously flashes into steam is known as a *flash tank*. Unless the vapor produced is removed as quickly as it forms, the required low pressure within the tank will not be held. This means the vapor must be removed and then compressed to the point where it condenses at a temperature above the temperature of the medium available for such purpose. A water-cooled condenser is usually used here, so water is the cooling medium.

Usually, the pressure of the vapor must be increased to the point where it condenses at approximately 100°F. From tables, the pressure which corresponds to this 100°F temperature is 0.95 psia. Referring back to the chilled water at 45°F, in order to keep it at that temperature in the flash tank, a pressure of 0.15 psia is needed. This means that to condense the flash vapor, its pressure must be raised from 0.15 to 0.95 psia.

Primary Ejector and Condenser

One of the methods commonly used to compress the flash vapor is the *steam-jet ejector*. The steam-jet ejector uses the energy of a fast moving jet of steam to *entrain* (capture) the flash tank vapor and compress it.

In figure 23-9, high-pressure steam expands while flowing through the nozzle ①. The expansion causes a drop in pressure and an enormous increase in velocity (to about 4,000 feet per second). Because of this high velocity, flash vapor from the tank ② is drawn into the swiftly moving steam and the mixture enters the diffuser ③. The velocity is reduced gradually because of the increasing cross-sectional area of the diffuser.

Interestingly enough, in this process the pressure of the steam increases from 0.15 psia at the entrance of the diffuser to around 0.95 psia at the condenser ④. This pressure corresponds to the condensing temperature of 100°F. This means that the mixture of high-pressure steam and the flash vapor may be liquefied in the primary condenser at this temperature. The latent heat of condensation is transferred to the condenser water which may be at 80°F. The condensate ⑤ is pumped out and is returned to the boiler from which it may again be vaporized at a high pressure.

Flash Tank

The warm water that is returned to the flash tank for cooling is sprayed so that a large surface area is exposed. As a result of cooling, some of the flash vapor is withdrawn from the flash tank by the primary ejector and delivered to the condenser. It is important that the tank be insulated to prevent heat conduction through the walls. Such heat would add to the weight of the vapor which must be compressed and delivered to the con-

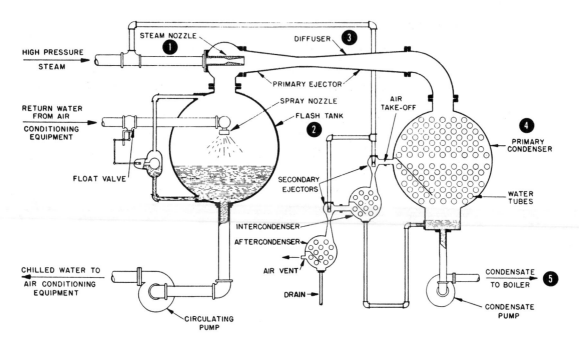

FIG. 23-9   Steam-jet water cooling sytem

denser. There must also be a constant return of water from the refrigerated space to replace the loss of several pounds of water/hr/ton of refrigerating capacity.

Open and Closed Systems for Cooling Air

The term *closed system* indicates that the chilled water from the flash tank is pumped through a coil to cool air and is then returned to the flash tank. In an *open system* the chilled water is sprayed into the air to be cooled and it is then collected in the air washer tank and returned to the flash tank and is again cooled.

Condenser

The *surface type condenser* is common in air-conditioning installations. This condenser is similar to the shell-and-tube condenser used in regular refrigeration installations. Water flows through the tubes while steam condenses on the outside of the tubes.

Jet condensers are also built in two other types: *barometric* and *low-level*. The schematic drawing in figure 23-10 shows a barometric condenser raised a distance of 34 feet above the water level in the hot well. This height is needed because of the high vacuum (28 in. Hg) carried in the condenser and also to adequately drain the condensate water and condensate by gravity.

In the low-level jet type, the condenser water and condensate are pumped under a 28 in. Hg vacuum to a hot well, which is at atmospheric pressure.

THERMOELECTRIC REFRIGERATION SYSTEMS

Before discussing principles of thermoelectric systems, it is necessary to review the effect of heat on certain metals and materials that are electrical conductors. Electricity may be produced by heating the junction of two dissimilar materials. This fact is important in refrigeration work because a number of control devices and testing instruments depend on this action for operation. A pilot light flame may produce enough current to operate a solenoid valve, regulating it by the flame.

An opposite effect to this is called the *Peltier effect*. This provides new ways of producing refrigeration. In the Peltier effect, changes in the direction of an electrical current at a junction produces either heat or cold. The greatest refrigerating effect is produced by the ideal thermoelectric material which meets three sets of conditions.

Condition 1	The thermoelectric material must be an excellent conductor of electricity, to minimize resistance losses.
Condition 2	The thermoelectric material must be a very poor conductor of heat because the heat must be absorbed at one end, rejected at the other. It must also have limited flow from the hot to the cold end.
Condition 3	The thermoelectric material must have high thermoelectric power. This means it must have a high rate of change in voltage with temperature.

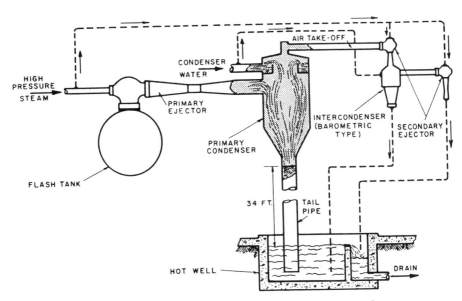

FIG. 23-10   Schematic drawing of a barometric condenser

Metals are usually good conductors of heat and electricity. On the other hand, insulators are poor conductors. This means that an altogether different material must be used as a thermoelectrtic material. *Semiconductor* is the name used for a material that combines the properties of a conductor with those of an insulator.

While some semiconductors have a thermoelectric power which is hundreds of times greater than that of metals, the quantity is still exceedingly small. This thermoelectric effect is increased by varying the composition of the semiconductor. To summarize, the overall efficiency of a thermoelectric material depends on its electrical and thermal conductivity and its thermoelectric power.

The two line drawings in figure 23-11 show how a single junction may be assembled using two different types of semiconductors *(N* and *P)* and the effect of changing the direction of current. The junction has a comparatively large thermoelectric power when two different semiconductors are used.

The basic parts of a thermoelectric system are shown in greater detail in figure 23-12. The thermoelectric elements (A) are connected by the *couple* (B) which is a metallic bar. In actual practice, a full-sized thermoelectric refrigerator would contain a large number of these couples because of the limited capacity of a single couple.

The couple is connected to a source of electrical energy (C). The thermometer (D) indicates the temperature of the metallic bar. The direction of current flow is controlled by the position of the switch (E)

In the cool position, heat is absorbed at the metal bar (B) and is thermoelectrically pumped into the base in which the other ends of the elements are secured. The bar becomes colder than ambient temperature and may even become frosted. When the current flow is reversed, the elements absorb heat from the base and again pump it to the junction formed by the metal bar. The bar then reaches such a high temperature as to melt the frost and even boil a drop of water.

While the Peltier effect was known for many years, it had limited application in refrigeration work because ideal semiconductors were not developed until recently.

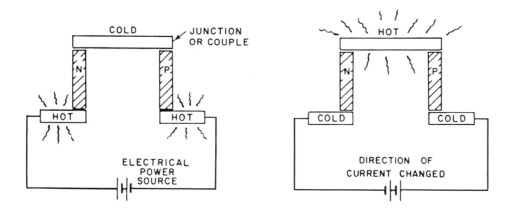

FIG. 23-11   Assembly of a single junction

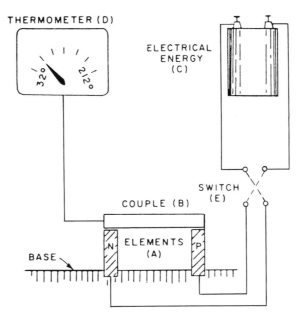

FIG. 23-12   Schematic of a simple thermoelectric refrigeration unit

Applications of Thermoelectric Refrigeration

An early *cooler-warmer unit* that was capable of cooling a small container of liquid to 40°F or of heating it to 100°F, is shown in figure 23-13. These temperatures were controlled automatically. This particular product operated from a regular electrical power source. There were about 50 Peltier junctions in contact on one end with the wall of the well. The other end of the couples was in contact with the outside finned heat exchanger.

Since direct current was used, a rectifier was built into this unit. The liquid in the container could be kept at the low temperature or a high temperature, with the time for temperature change controlled automatically. A simplified unit was also available for use within automobiles, which already had a source of direct current.

Thermoelectric refrigerators include compartments for ice and frozen foods, a crisper, and conventional food refrigeration area. The thermoelectric dehumidifier to keep foods dry and crisp is another application. One of the chief advantages of thermoelectric refrigerators and freezers is the feature of noiseless operation.

MAGNETIC SYSTEM OF PRODUCING LOW TEMPERATURES

Extremely low temperatures are obtained by evaporating liquefied gases in insulated containers. Temperatures of 90°K may be reached with liquid air, 54.3°K with liquid oxygen, 35.6°K with liquid nitrogen, 14°K with liquid hydrogen, and below 1°K with liquid helium. The lowest temperature reached by boiling liquid helium under the smallest pressure obtainable is 0.71°K.

FIG. 23-13 **Early design of automatic small cooler-warmer** *(Courtesy of Westinghouse Corporation)*

All substances fall within one of three classes according to their magnetic properties. Those materials that are strongly attracted by a magnet are classed as *ferromagnetic.* Those that are repelled by a magnetic pole are *diamagnetic;* those that are weakly attracted, *paramagnetic.*

When paramagnetic salts are precooled to a very low temperature, the heat motion of the molecules is reduced to a minimum. These molecules may then be thought of as elementary magnets when they are subjected to a strong magnetic field. If the paramagnetic material is then demagnetized, it becomes cooled because there is neither a heat gain nor loss from outside of the system itself. This is said to be *adiabatic cooling.*

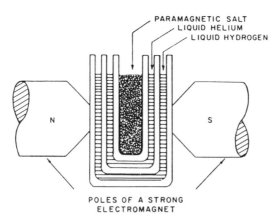

FIG. 23-14 **Magnetic cooling of a paramagnetic substance**

While it has not been possible to measure the lowest temperature reached by magnetic cooling with any instrument, it has been determined that temperatures as low as 0.001°K and even lower are possible by this method. The schematic drawing in figure 23-14 shows the arrangement of the materials and the power source for the magnetic cooling of a paramagetic substance.

The paramagnetic salt is first cooled. This is done by surrounding it with liquid helium which is boiling under a reduced pressure.

The salt (surrounded by the boiling helium) is next subjected to a very strong magnetic field. The heat given off is absorbed without a temperature change. This brings the temperature down to below 1°K. Finally, demagnetization without heat gain or loss from outside the system causes the temperature to drop to 0.001°K and lower.

Magnetic cooling is used primarily for research. Further development of this method will lead to new materials, products and processes for industry, agriculture, business, medicine, and the home.

SUMMARY

- Nonmechanical refrigeration systems are those that obtain the required high and low pressures by some method other than a mechanical compressor.
- An absorption system uses the ability of a liquid to absorb a vapor to produce the necessary conditions.
- The main components in a continuous operating absorption refrigeration system include:
  - ✓ Generator
  - ✓ Separator
  - ✓ Gas and liquid heat exchanger
  - ✓ Absorber
  - ✓ Analyzer
  - ✓ Rectifier
  - ✓ Condenser
  - ✓ Evaporator
  - ✓ Volume and safety controls
  - ✓ Defroster
- In large commercial installations steam is used to heat the generator, separate the refrigerant from the solvent, and raise the level of the refrigerant vapor and pressure.
- Some of the major advantages of absorption refrigeration systems over mechanical include:
  - ✓ Quietness, limited wear, efficiency at any load.
  - ✓ Possibility of using hot exhausts from other equipment.
  - ✓ Outdoor location without need for housing.
  - ✓ Lower evaporator temperature and control requirements.
  - ✓ Limited decrease in capacity when the evaporator pressure and temperature are reduced.
- A steam-jet system uses a device in which the extremely rapid flow of a vapor through a narrow tube reduces the pressure and permits evaporation of a liquid. This produces a cooling effect.

443

- Thermoelectric refrigeration depends upon passing electrical energy to a couple through two dissimilar semiconductors.
- Temperatures lower than those possible by ordinary methods may be produced by magnetic cooling. The relationship between magnetism and the energy of the molecules of certain materials is used to produce near absolute temperatures.

ASSIGNMENT: UNIT 23 REVIEW AND SELF-TEST

A. ABSORPTION SYSTEMS

Determine which statements (1 to 10) are true (T); which are false (F).

1. All absorption systems are capable only of intermittent operation.
2. The refrigerant pressure in the evaporator of an absorption system is low.
3. The refrigerant boils in the cold evaporator of an absorption system.
4. Ammonia dissolves in water better at temperatures below 100°F.
5. Hydrogen is condensed in the condenser of the absorption system.
6. The term *strong aqua* refers to heavy hydrogen.
7. The absorption system of refrigeration is noted for its quiet operation.
8. The absorption system cannot freeze ice cubes.
9. The pressure in the evaporator at low temperatures is lower than the pressure in the condenser of the absorption system.
10. Absorption systems need defrosting occasionally.

Provide the correct term or process to complete statements 11 to 20.

11. Two refrigerants commonly used in the absorption systems are _____ and _____ .
12. Two materials that remove refrigerant vapor from the evaporator are _____ and _____ .
13. The name of the material that is used to remove the refrigerant vapor from the evaporator is called _____ .
14. The energy for operating an absorption system is _____ .
15. The part of the system in which the refrigerant is separated from the absorbent is called the _____ or _____ .
16. Two sources of energy for operating an absorption system are _____ and _____ .
17. The refrigerant in the condenser changes from the _____ state to the _____ state.
18. The component in the refrigeration system in which the level of the temperature and pressure are raised is the _____ .
19. The device used to prevent the loss of raw gas if the flame goes out is called _____ .
20. The automatic defrost system used with domestic absorption systems is the _____ .

Select the letter representing the conditions or terms which best complete statements 21 to 25.

21. The pressure of the refrigerant in the evaporator of the absorption system is (a) higher than, (b) lower than, (c) the same as it is in the condenser.
22. The temperature of the refrigerant in the evaporator of absorption systems is (a) greater than, (b) less than, (c) the same as it is in the condenser.
23. The temperature of the refrigerant in the generator of absorption systems is (a) greater than, (b) less than, (c) the same as that of the absorbent.
24. The pressure of the hydrogen gas in the evaporator of absorption systems is (a) greater than, (b) less than, (c) the same as that of the refrigerant.
25. The efficiency of an absorption system (a) increases, (b) decreases, (c) remains the same as the load increases.

B. COMPARISON OF NONMECHANICAL AND MECHANICAL SYSTEMS

1. Indicate whether the condition refers to the mechanical (M) or nonmechanical (N) system.

	Condition	Type of System	
		M	N
A	Quietness of operation		
B	Considerable wear due to many moving parts		
C	Utilizes exhaust from other equipment		
D	Capacity almost unlimited beyond 1,000 tons		
E	Operates at higher evaporator temperatures		
F	Capacity unaffected by decrease in evaporator pressure		
G	Good operating efficiency at reduced loads		
H	Liquid in the compression part of system must be removed		

C. STEAM-JET REFRIGERATION SYSTEMS

1. Name the refrigerant used in the steam-jet refrigeration system.
2. In what respect is the steam-jet system comparable to the absorption system?
3. What change takes place in the refrigerant of the steam-jet system that makes it possible for it to absorb heat?
4. In what part of the steam-jet system does the change in the refrigerant take place?
5. What is the pressure condition in the part of the system in which there is a change in the refrigerant?

6. Prepare a table containing information for the absorption, mechanical, and steam-jet systems pertaining to the following six conditions:

Condition		Type of Refrigeration System		
		Absorption	Mechanical	Steam-Jet
A	Refrigerant (R) used			
B	Source of energy			
C	Component used to raise R temp. and pressure to high level			
D	Component used to produce low evaporator pressure			
E	Method of producing the low pressure			
F	Method of producing the high pressure			

7. State the condition that produces the change in the refrigerant making it possible to absorb heat.
8. What pressure condition exists in the flash tank of a steam-jet system?
9. State briefly how the pressure condition is accomplished in the condenser?
10. What change takes place in the refrigerant within the condenser?
11. Does the steam and refrigerant vapor absorb or give up heat within the condenser?
12. What is the usual condensing medium for the steam-jet refrigeration system?
13. To what part of the system does the condensate return?

D. THERMOELECTRIC AND MAGNETIC REFRIGERATION SYSTEMS

1. Describe briefly how heating and cooling may be produced thermoelectrically.
2. Name the man after whom the thermoelectric effect is named.
3. How are materials classified that are readily affected thermoelectrically?
4. What system is used to obtain temperatures below that of liquid helium?
5. What term is used to classify materials that (as they become demagnetized) produce temperatures below 1°K?

UNIT 24:
NONMECHANICAL TRANSPORT REFRIGERATION

Four nonmechanical transport refrigeration systems are in common use. These systems include: (1) water-ice refrigerant, (2) dry-ice (carbon dioxide) refrigerant, (3) eutectic solution refrigeration, and (4) chemical refrigeration. Principles relating to each system and the advantages and disadvantages are treated in this unit. It concludes with a section on combinations of systems.

WATER-ICE NONMECHANICAL TRANSPORT REFRIGERATION SYSTEM

The water-ice nonmechanical refrigeration system for in-transit use depends on the absorption of heat that accompanies the change from the solid to the liquid state. This change involves the latent heat of fusion and occurs usually at a definite temperature, depending on the material.

Historically, the Greeks and Romans used snow and ice in insulated pits to produce a cooling effect. Interestingly, transport refrigeration, as first used by the railroads in 1851, applied the same principle. Several tons of butter were shipped from Ogdensburg, New York to Boston in a wooden boxcar that was insulated with sawdust and stocked with ice. One hundred years later the industry still used water-ice as the standard refrigerant in freight and express refrigerator cars.

No important changes in refrigerator car design were made until 1919 when new United States standards called for a minimum thickness of 2 to 2 1/2 inches of insulation, basket bunkers, insulated bulkheads, floor racks at least 3 3/4 inches high, and a five-ton ice capacity. By 1945, the standards for all new general service forty-foot refrigerator cars included 4 to 4 1/2-inch insulation, sidewall flues, easier riding trucks, end bunkers with water-ice, and the important addition of air circulation fans to give uniform temperature throughout the lading. Figure 24-1 shows the construction and components of a typical ice bunker refrigerated freight car of the 1950s and 1960s.

The operation of this system was simple. The circulating fan was driven by belt from a drive shaft pulley, which was connected to the wheel shaft. Other systems used a flexible shaft to drive the fan. When the freight car was on a siding, the fan was driven by a gasoline engine or electric motor. As air was drawn from under the flooring and forced over the ice, it cooled. The ice remained at a constant temperature of 32°F even as it melted and absorbed heat. At normal speed, the fan operated to circulate air through the car and over the ice several times per minute.

Even with this equipment, control of temperature was not possible for all cargoes. For instance, for certain leafy products it was found that proper cooling could be maintained only by completely immersing the product packages in crushed or snow ice. In this way the temperature could be brought down close to 32°F and held there throughout the transit time.

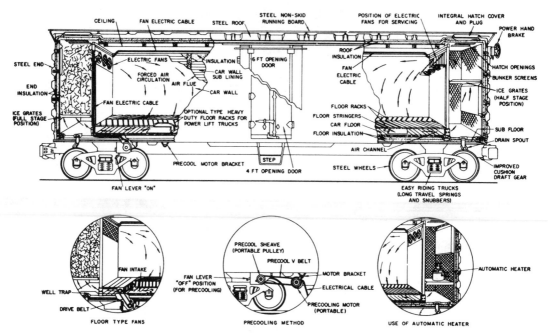

FIG. 24-1 Major components of an early ice bunker refrigerated freight car *(Reprinted by permission from ASHRAE Guide and Data Book)*

When shipping other produce that would freeze around 30°F, care had to be exercised not to use new, fresh ice from an ice plant because it was likely to be *supercooled.* Supercooling occurs during the freezing process when a container and its contents are left in the brine after the ice has completely solidified (by the removal of latent heat). The removal of sensible heat causes the ice to be cooled below its freezing point.

Sample Problem in Refrigerating Freight Cars

A sample problem is included to illustrate the principles and some computations required in establishing needs in the old ice bunker refrigerated freight car system.

PROBLEM:

Assume a 40-foot freight car meets the requirements of four inches of solution, five tons of bunker ice, and forced air circulation. The average design temperature difference between the outside and inside of the car is 40°F. Under these conditions, the heat leakage is known to be 6,000 Btu per hour. The problem is to calculate how long the ice will last.

SOLUTION:

It is known that in melting, one pound of ice absorbes 144 Btu. Further, the heat lost in cooling the cargo and compartment equals the heat gained by the ice in melting.

$$\begin{matrix} \text{Cargo} \\ \text{loss} \end{matrix} \, (6{,}000 \text{ Btu}) \times (N) \, \begin{matrix} \text{Hours in} \\ \text{transit} \end{matrix} = \begin{matrix} \text{Ice} \\ \text{load} \end{matrix} \, (5 \text{ tons} \times 2{,}000 \text{ lb.}) \times (144) \, \begin{matrix} \text{Btu/lb.} \\ \text{of ice} \end{matrix}$$

$$6{,}000(N) = 1{,}440{,}000$$
$$N = 240 \text{ hours or } 10 \text{ days}$$

Variations in temperature, heat losses and other conditions made it necessary to consider these variable factors to determine the required refrigeration.

Cushioned refrigerated cars are equipped with mechanical refrigeration units and shock-absorbing couplers. The refrigeration equipment, installed in one end of the car, is similar to an electromechanical system used in trucks, figure 24-2. In brief, a diesel-driven generator provides electricity for the semihermetic compressor, the condenser and the evaporator fans. Cold air is circulated in the interior car compartment through openings in the false ceiling and returned under the gridwork floor.

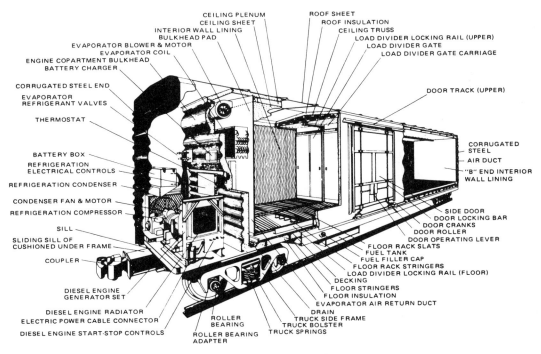

FIG. 24-2 Refrigeration components for mechanical refrigerator cars (early design)
(Reprinted by permission from ASHRAE Guide and Data Book)

Refrigeration cars are entirely automatic in their operation and require only periodic refueling and service checks. The service time is minimal since two 500-gallon fuel tanks are mounted under the cars. A sliding door provides access to the complete system and makes it easy to check the oil level in the prime mover. A temperature indicator, mounted at eye level on the outside of the car, permits the inspector to make a quick check while standing on the ground.

Application of the Water-Ice System in Refrigerated Trucks

It was natural that refrigeration systems for truck transport should be designed after those of railroad cars that had been in use for many years and were satisfactorily moving large quantities of perishable goods. The development of mechanical, water-ice units that would stand the rigors of over-the-road applications rose rapidly, beginning in the 1950s.

Figure 24-3 shows the construction of one of the early refrigerated trailers, called a *high-temperature transport*. Ice provided satisfactory refrigeration for produce that needed to be kept in a temperature range from 35° to 40°F.

DRY ICE (CARBON DIOXIDE) REFRIGERANT

Dry ice, or carbon dioxide in the solid state, has been used in trucks to produce temperatures close to 0°F. Since the temperature of dry ice is –110°F, gloves should be used for safety in handling bare blocks. Dry ice is normally supplied in multilayer paper or cloth bags to insure safe handling and to control the rate of sublimation. One great advantage of dry ice is that as a refrigerant it *sublimes*. That is, as the dry ice absorbs heat it changes state. It passes directly from the solid to the gaseous state without going through the liquid state.

The *heat of sublimation* of dry ice at normal atmospheric pressure is approximately 246 Btu/lb, figure 24-4. Thus, one pound of dry ice needs to absorb 246 Btu in order to completely sublime or change into the gaseous state. If, for example, only one half pound of dry ice is sublimed and changed to a gas, 123 Btu would be needed. It follows

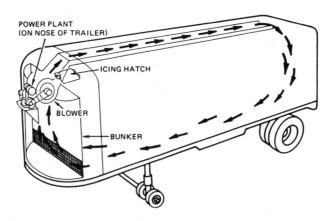

FIG. 24-3 Early model of a high-temperature transport trailer based on the water-ice railroad refrigeration system

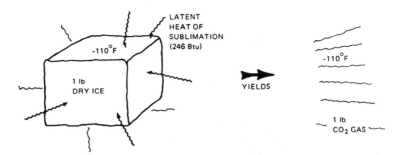

FIG. 24-4   Sublimation of dry ice

that a ten-pound cake of dry ice placed in a truck compartment would cool the air by absorbing 2,460 Btu of heat by the time it sublimed. The dry ice would offset 2,460 Btu of heat leakage through the insulation of the truck, cargo temperature or other heat losses.

Since the temperature of the carbon dioxide gas is also at –110°F after sublimation, further cooling results. Another example is used, this time with dry ice, to explain the principle of refrigeration that is involved.

PROBLEM:

A low-temperature truck body has a heat gain of approximately 1,500 Btu/hr. The required temperature is 5°F. The temperature difference between the outside and inside of the body is 100°F. The truck is to be used for local delivery with the doors to be opened an estimated 50 times per ten-hour trip. The problem is to determine how many pounds of dry ice are needed to sublimate to offset the heat gain and to maintain the required temperature at 5°F.

(Note: It is standard practice to add a *service heat gain* of 50% of the body heat gain to the body heat gain when the doors are to be opened and closed continuously.)

SOLUTION:

In this problem the heat is absorbed in two stages.

Stage One—When the dry ice sublimes, each pound absorbs 246 Btu.

Stage Two—When the resulting carbon dioxide (CO_2) gas warms from –110°F to 5°F, there is a rise of 115°F. The sensible heat is calculated by the formula: $(m) \times (c) \times (\Delta t)$; where m is the number of pounds, c is the specific heat, and Δt the temperature rise.

Using one pound and substituting values for specific heat and the temperature rise in the formula:

$$1 \times 0.184 \times 115 = 21.16 \text{ Btu}$$

The 246 Btu + 21.16 Btu = 267 Btu as the total heat absorbed by one pound of dry ice. The total heat gain per hour is 1,500 + 750, or 2,250 Btu/hr. For the ten-hour trip the total heat gain would be 22,500 Btu.

If M equals the number of pounds of dry ice required, then
$$267 \times M = 22,500$$
$$M = 84.3 \text{ lb. (rounded off) of dry ice}$$

As the dry ice sublimes, the resulting carbon dioxide gas drives air out of the vehicle. If the cargo is produce, the deterioration due to oxidation is reduced because of the removal of air. This system, which is free from mechanical failure, is also desirable because it is clean.

EUTECTIC SOLUTION REFRIGERATION

As the demand for refrigerated transport increased, so did the need for a wider range of temperatures and refrigerants other than pure ice. For example, a frozen water-salt mixture will drop the freezing temperature of ice to below 32°F. Such mixtures were employed in insulated truck bodies to produce temperatures in the vicinity of 0°F and below. The name *low-temperature transport* was given to refrigerated trucks that were used for this purpose.

With the need for lower temperatures came the requirement for improved insulation of the bodies. However, the water-salt mixture was not used very long because of its destructive effects on the truck body or railroad freight cars. In addition, it was not possible to control the temperature of the melting ice with the required degree of accuracy.

A satisfactory way of using a nonmechanical system to obtain a wide range of temperatures that could be controlled was developed after 1925 by placing the refrigerant in a container. This provided refrigeration and also avoided the harmful effects accompanying melting ice. The production of ice-making machines at about the same time reinforced this development of containing the refrigerant.

Water (or any other liquid), placed in a closed container may be frozen by immersing it in a brine that is chilled by an *ammonia machine*. The colder the brine, the quicker the liquid in the container freezes. The container enclosing the ice may then be placed in an insulated enclosure. As the container and ice absorb heat, the ice melts in the container. Since ice changes state, either freezes or melts at the same temperature of 32°F, its application is limited to that of a high temperature body around 40°F.

However, a mixture called a *eutectic* will freeze and melt at a specified temperature, depending on the proportions of chemicals in the eutectic. Consider a salt solution like calcium chloride. The chart in figure 24-5 shows a few sample freezing and melting temperatures. These differ according to the concentrations of calcium chloride in water. According to practice, a refrigerant should be provided at a temperature of about 15° below the desired temperature in the cargo space. This provides for reasonably fast cooling.

Therefore, if a cargo space temperature of about 30°F is needed, a closed container with a frozen eutectic of 14% calcium chloride can be used. The frozen solution melts

CaCl$_2$ in Solution Actual %	Freezing Point of Eutectic °F
0.	+32
8.0	+24.6
11.4	+20.3
14.0	+15.5
20.0	− 0.4
22.4	− 9.4
25.0	−21.0
27.0	−31.2
29.0	−49.4
29.6	−59.8

FIG. 24-5   Sample eutectic concentrations with 0°F freezing points

near 15°F as it absorbs the latent heat. The temperature remains at this point as long as there is some frozen solution in the container. The solution may then be refrozen and made ready to be reused.

The containers that hold the eutectic and provide refrigeration are called *holdover plates*. The eutectic is generally one of the refrigerants. A single holdover plate and a cutaway section to show the internal construction are illustrated in figure 24-6. The size

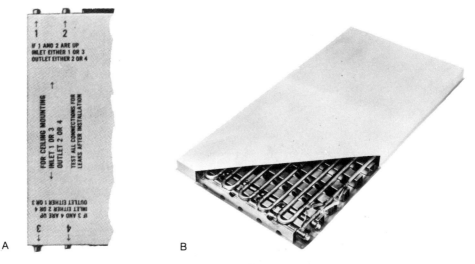

A   B

FIG. 24-6   Holdover plate details: (A) portion of pan side showing connections and (B) cutaway showing internal construction *(Courtesy of Dole Refrigerating Company)*

453

of the plate is determined by the quantity of eutectic required to provide adequate heat absorption capacity for the length of service anticipated. The area of the holdover plate must be sufficiently large to provide the proper rate of cooling to meet required operating conditions.

To facilitate refreezing, refrigerant tubing is built into the plate. Chilling brine or refrigerant from a mechanical system is circulated through the tubing during the layover period. Numbered arrows on the holdover plate, figure 24-6A, point to the refrigerant connections. Usually, these are extended to the side of the vehicle where they terminate in quick-coupler valves to facilitate connecting and disconnecting. Thin holdover plates have the advantage of occupying a minimum of area in the cargo compartment.

The interior space of the plates contains a network of cells formed by metal webs which also contact the surface, figure 24-6B. This construction provides much better transfer of heat by conduction between the eutectic solution and the face of the plate. The simplicity of construction and the limited need for maintenance makes this a very dependable system.

Holdover plates are designed for ceiling or wall mounting, figure 24-7A and B. A fan is sometimes installed in the truck body to provide a uniform temperature. The air

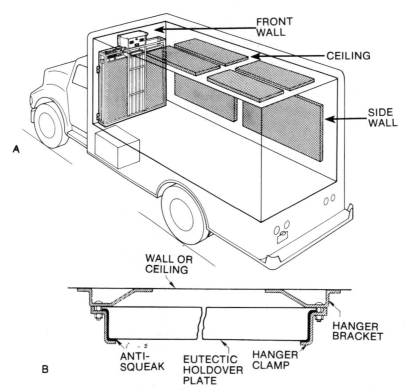

FIG. 24-7 Holdover plate mountings and locations: (A) ceiling and wall installation and (B) example of plate mounting *(Adapted from drawings of Dole Refrigerating Company)*

454

circulates over the holdover plates to insure the even distribution of refrigeration around the product being transported.

Compact eutectic blower systems are available to provide uniform temperatures in the medium temperature range (34°F to 40°F) truck bodies. Two or more holdover plates are mounted in a treadplate case, ready for connecting to the refrigerating system. A fan, mounted at the top of the unit, draws air over the plates and discharges it through an adjustable grille for circulation throughout the body.

A newer holdover plate blower unit is illustrated in figure 24-8. This unit provides increased Btu/hr cooling capacity for faster body temperature recovery after door openings. The increased capacity results from (1) the development of a more effective higher velocity airflow path through the unit, (2) increasing the cubic feet per minute air volume, (3) adding extended secondary surfaces to transfer heat from air to the plates, and (4) changing from +18°F to 0°F holdover plates.

The dual blowers are powered by the truck battery during over-the-road operation. When the truck is garaged, a built-in conversion unit which utilizes a regular 115- or 230-volt alternating-current power supply provides power for the dual blowers. The

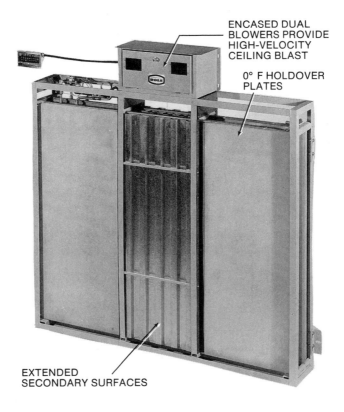

ENCASED DUAL
BLOWERS PROVIDE
HIGH-VELOCITY
CEILING BLAST

0° F HOLDOVER
PLATES

EXTENDED
SECONDARY SURFACES

FIG. 24-8 Increased capacity eutectic holdover plate blower unit *(Courtesy of Dole Refrigerating Company)*

blowers operate at all times, except when a door is open. A door switch and a dash-mounted snap switch and pilot light are available for controlling the operation of the fan. A thermostat may be installed in the control circuit if accurate temperature control is required. This type of system gives better control of excess moisture in the cargo space and concentrates the defrosting in the blower unit.

Refreezing the Eutectic Solutions

Holdover eutectic plates provide refrigeration for a limited time, usually one day of service. This requires that the eutectic solution must be frozen for service the next day. Refreezing may be accomplished in one of two ways. (1) The holdover plates may be connected with quick couplers to a remote system as shown in figure 24-9. (2) The holdover plates may be refrozen by a self-contained condensing unit mounted on the truck. This is sometimes called a complete holdover type and is usually an electrically operated hermetic unit.

EXPENDABLE REFRIGERANT REFRIGERATION (LIQUID NITROGEN SYSTEM)

The last of the nonmechanical refrigeration systems for transport use to be discussed is referred to as a *chemical* or *liquid nitrogen* (LN$_2$) system. Liquid nitrogen systems were popular and widely used at one period of time. These systems were displaced by other systems that were more economical to operate. Liquid nitrogen systems are included because the principles of refrigeration and an understanding of sample transport applications are important to a study about refrigeration.

Like other refrigerants, liquid nitrogen has a very low boiling temperature at normal atmospheric pressure. This characteristic makes it a very good refrigerant. Actually, when exposed to normal atmospheric pressure, liquid nitrogen boils at –320°F. Consequently, when used for in-transit refrigeration, the container must be specially constructed with super insulation and pressure-release valves.

One or more containers of liquid nitrogen may be used in each vehicle depending on compartment size; amount of insulation in the walls, roof, and floor of the trailer; the

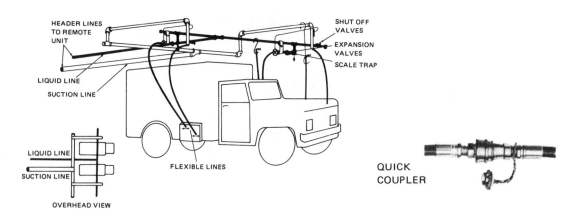

FIG. 24-9 Remote system for refreezing eutectic plates

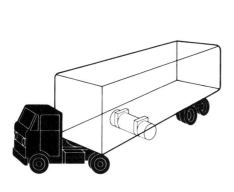

FIG. 24-10 External mounting of liquid nitrogen tank (former system) *(Courtesy Linde Division, Union Carbide Corp.)*

FIG. 24-11 Formerly used liquid nitrogen storage tank for external mounting *(Courtesy of Linde Division, Union Carbide Corp.)*

temperature at which the cargo has to be maintained; and the number of times the doors are to be opened.

The location of the tank is flexible. One or more tanks were installed in a *penthouse* in the nose of the trailer which often extended out over the cab of the tractor. Another location is shown in figure 24-10. External mounting provided maximum cargo space.

Figure 24-11 shows a formerly used liquid nitrogen storage tank that was designed for underbody installation. Its capacity was 425 pounds. The internal design provided superinsulation and rugged construction to stand up under over-the-road operation.

When cell-like, modular refrigerated containers were used for multiple-unit transportation, figure 24-12, and several containers were placed on a trailer bed, safe operation required that there be no projection on the shipping container or controls. Thus, the cylinders were placed inside and the controls were flush mounted.

This system was especially adaptable to the refrigeration of multiple compartments. Cargoes with different temperature requirements were shipped in one vehicle as shown in the illustration. Today, the operation of multiple refrigerated compartments (using movable dividers) is similar to that of former nitrogen systems.

Operation of the Liquid Nitrogen System

When the tanks were filled with liquid nitrogen and the cargo space was loaded, the desired temperature was selected at the main control. The temperature-sensing element in the cargo space sensed any rise in temperature resulting in heat absorption through the sides or other parts of the vehicle.

The temperature controller opened the solenoid valve in the liquid line. This permitted the liquid nitrogen to issue from the spray header. The *spray header* was a

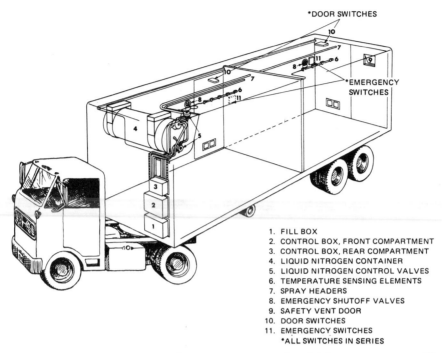

1. FILL BOX
2. CONTROL BOX, FRONT COMPARTMENT
3. CONTROL BOX, REAR COMPARTMENT
4. LIQUID NITROGEN CONTAINER
5. LIQUID NITROGEN CONTROL VALVES
6. TEMPERATURE SENSING ELEMENTS
7. SPRAY HEADERS
8. EMERGENCY SHUTOFF VALVES
9. SAFETY VENT DOOR
10. DOOR SWITCHES
11. EMERGENCY SWITCHES
    *ALL SWITCHES IN SERIES

FIG. 24-12 Old liquid nitrogen refrigeration system of multiple compartments *(Courtesy Linde Division, Union Carbide Corp.)*

perforated pipe mounted along the ceiling of the cargo compartment. The liquid nitrogen vaporized immediately. In so doing, it absorbed heat from the surrounding air in the top of the cargo space. As the air cooled its density increased and it settled toward the floor, setting up convection currents to keep the entire cargo space cooled. When the temperature at which the sensing element was set was reached, the element actuated the controller to close the solenoid and shut off the flow of liquid nitrogen.

The system was flexible in that the controller could be set with a minimum of effort on the part of the driver to match the temperature requirement of any given load. Since there was a minimum of moving parts, the performance was dependable. The extremely low temperature of -320°F of the vaporizing nitrogen made this system particularly suitable for quick drawdown of temperature after each door opening. Although the liquid container required special superinsulation and construction, the liquid lasted a long time as it was used only to provide cooling to overcome heat which came into the cargo container. The cargo was prefrozen and/or chilled to the required temperature so the nitrogen was not called upon to perform this function.

The use of quick-disconnect couplings speeded the refilling of the tanks as needed. The storage tanks were equipped with safety valves which vent nitrogen to the outside if the pressure in the tank rose above 22 psig. All vehicles were further equipped with a safety vent which was spring loaded. This device permitted gas to exhaust to the atmosphere when the pressure inside the cargo space exceeded atmospheric pressure.

Advantages of the Liquid Nitrogen Spray Cooling System

The former spray cooling system, which used nitrogen, had other advantages to those of simplicity of equipment, minimum of moving parts, ease of maintenance, ruggedness, and dependability. The system provided an inert atmosphere which not only reduced the respiration rate for fruits and vegetables but also protected them from spoilage by the exclusion of oxygen. Furthermore, the system prevented the dehydration of produce in transit. To repeat, another important advantage was that the system provided extremely rapid recovery of the desired temperature in the cargo space.

SAFETY PRECAUTIONS

- In the spray cooling system, safety switches were connected to each door. Whenever a door was opened, the safety switch automatically stopped the operation of the unit. There were three reasons for this design feature.
  - ✓ First, it prevented waste of liquid nitrogen.
  - ✓ Second, it protected the operator from coming in contact with the dangerously cold vapor as it issued from the spray head. Since the liquid issued from the nozzles and immediately changed to a vapor at -320°F, any contact with flesh would freeze it instantly, even before the person felt the cold.
  - ✓ Third, safety precautions were needed to prevent possible asphyxiation due to any condition where nitrogen (N_2) might displace the oxygen in the compartment.
- Warning signs giving the following instructions were usually placed on each cargo compartment door and on the outside truck door of a liquid nitrogen system.

WARNING
BEFORE ENTERING
1. **TURN MAIN SWITCH OFF**
2. **BEFORE ENTERING, WAIT THREE MINUTES WITH THE DOORS WIDE OPEN "IF" ONE HOUR OR MORE HAS PASSED SINCE LAST ENTRY.**

Personal safety was required to insure that no person was in the refrigerated space when the liquid nitrogen system was in operation. As an added safety precaution, an emergency door release was provided on the inside of each door.

Other safety measures required that in addition to making certain the safety doors were opened, the truck body was ventilated, and no refrigerant passed through the system, service people always wore goggles and knew the required pressures and the nature of the safety valves and controls used in the system. Safety precautions were also followed in the driver's cab where upholstery materials had to comply with the Department of Transportation's *burn-rate regulations*.

COMBINATIONS OF SYSTEMS: LIQUID NITROGEN, HOLDOVER PLATE, BLOWER COIL SYSTEMS

Combinations of systems like the liquid nitrogen, holdover plate, and low-temperature blower coil system provided a desirable and less expensive yearly cost system than liquid nitrogen, a more reliable and economical system having advantages not obtained with the auxiliary engine-powered mechanical blower-type when used alone, and a better system in terms of maintaining product quality. Combination systems also provided for continuous product refrigeration and the added protection of two independent refrigeration systems.

Two common combination systems (that were used before the LN_2 system was replaced) include: (1) the electric plug-in blower coil and liquid nitrogen system and (2) a combination holdover truck plate and liquid nitrogen system.

Low-Temperature Blower Coil and Liquid Nitrogen Combination System

The former combination of liquid nitrogen and plug-in blower coil system, figure 24-13, was practical for product quality and reduced total yearly refrigeration costs for peddle-delivery service requiring the continuous opening and closing of the refrigeration compartment. The electric plug-in low-temperature blower coil system took care of body heat leak, reduced return-product temperature during dockside plug-in operation, and provided refrigeration for overnight, weekend and other standby hours.

The liquid nitrogen refrigeration system handled body leak and door-opening service during the drop-delivery run. This system provided the most rapid vehicle body air temperature recovery after each door opening and was comparatively trouble free and quiet.

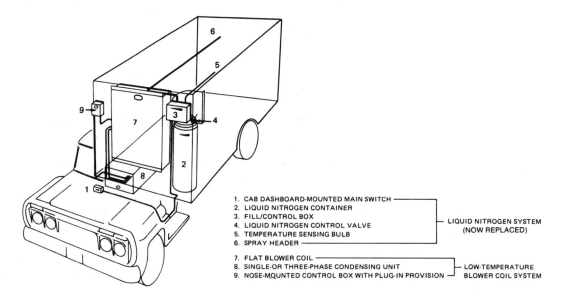

1. CAB DASHBOARD-MOUNTED MAIN SWITCH
2. LIQUID NITROGEN CONTAINER
3. FILL/CONTROL BOX
4. LIQUID NITROGEN CONTROL VALVE
5. TEMPERATURE SENSING BULB
6. SPRAY HEADER
— LIQUID NITROGEN SYSTEM (NOW REPLACED)

7. FLAT BLOWER COIL
8. SINGLE-OR THREE-PHASE CONDENSING UNIT
9. NOSE-MOUNTED CONTROL BOX WITH PLUG-IN PROVISION
— LOW-TEMPERATURE BLOWER COIL SYSTEM

FIG. 24-13   Major components of a former combination liquid nitrogen and low-temperature blower coil system *(Courtesy of Linde Division, Union Carbide Corp.)*

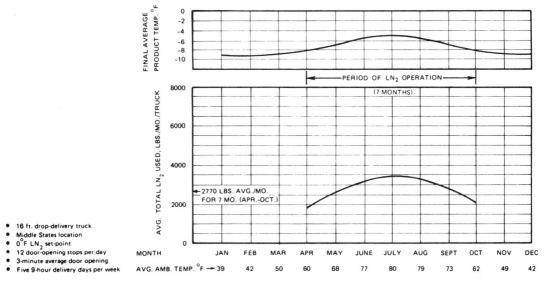

FINAL AVERAGE PRODUCT TEMP. °F

AVG. TOTAL LN₂ USED, LBS./MO./TRUCK

PERIOD OF LN₂ OPERATION
(7 MONTHS)

2770 LBS. AVG./MO. FOR 7 MO. (APR.-OCT.)

- 16 ft. drop-delivery truck
- Middle States location
- 0°F LN₂ set-point
- 12 door-opening stops per day
- 3-minute average door opening
- Five 9-hour delivery days per week

MONTH	JAN	FEB	MAR	APR	MAY	JUNE	JULY	AUG	SEPT	OCT	NOV	DEC
AVG. AMB. TEMP. °F →	39	42	50	60	68	77	80	79	73	62	49	42

FIG. 24-14   Sample graph of performance and liquid nitrogen consumption of a holdover truck plate and liquid nitrogen system (7 months)

Holdover Truck Plate and Liquid Nitrogen System Combination

In the former holdover plate/liquid nitrogen system, figure 24-14, that was used for drop-delivery service, holdover truck plates handled the body leak, a portion of the door-opening service load at 100°F ambient temperature, and all of the door-opening service load at 60°F ambient temperature. Also, the system met refrigeration requirements during dockside plug-in operation.

The liquid nitrogen phase handled most of the door-opening service at temperatures above 60°F in order to rapidly recover the required air temperature.

The advantages of this combination of systems included:

- Dependable, economical refrigeration with holdover truck plates on low ambient temperature days and effective peak-demand refrigeration supplied by liquid nitrogen on high ambient temperature days.
- Quick, efficient body air temperature recovery by the liquid nitrogen system after door openings.
- Protection at dockside and on-the-road of a holdover plate system.
- Low maintenance, and trouble-free, quiet operation.
- The protection of two independent refrigeration systems.

The holdover plate part of the system was run from dockside station electric power to freeze the eutectic solution in the holdover plates. When outside average daily temperatures were below 60°F, product protection was provided by the plates alone. As average daily temperatures rose above 60°F, the liquid nitrogen system was turned on to recover the required temperature.

461

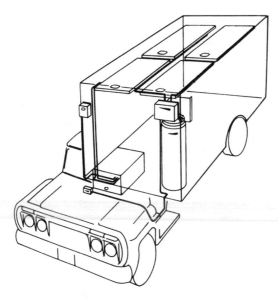

FIG. 24-15   Former liquid nitorgen and holdover plate combination refrigeration system *(Courtesy of Linde Division, Union Carbide Corp.)*

The major components of the liquid nitrogen system were the same as those described earlier. The combination system shown in figure 24-15 had four ceiling-mounted holdover plates.

The holdover truck plates were designed to:

- Handle body leak and a portion of the door-opening service load at 100°F ambient temperature, and to take care of all body heat leak and door-opening service loads at 60°F ambient temperature.
- Provide extra protection for safe return of product during any over-the-road failure on a high ambient temperature day.
- Provide low-cost plug-in dockside operation.

The liquid nitrogen refrigeration system was designed to:

- Handle the remaining door-opening service loads at ambient temperatures greater than 60°F.
- Provide protection in handling the entire body heat leak and door-opening service load during any holdover plate system failure.
- Provide rapid recovery of compartment air temperature after each door opening.

SUMMARY

- One of the earliest commercial applications of refrigeration for transport use depended on a water-ice refrigeration system. Heat was absorbed as a result of the change of state of the refrigerant.
- Dry ice passes directly from the solid state (–110°F) to the gaseous state at normal atmospheric pressure. The heat of sublimation of one pound of dry ice is 246 Btu. The dry-ice system reduces product deterioration due to oxidation.

- Sensible heat is equal to (m) number of lbs × (c) specific heat × (△t) temperature rise.
- Wide ranges of temperature in a nonmechanical system are obtained by enclosing a eutectic low-freezing mixture in closed containers, known as holdover plates.
- Holdover plates are located on side walls and/or the ceiling and are designed with adequate area and capacity to meet rigid cargo cooling requirements.
- Better control in the holdover plate refrigeration system and product quality are provided by the additional use of a fan for over-the-road operation.
- The eutectic solution in holdover plates may be refrozen by connecting with quick couplers to a remote system or by a self-contained condensing unit mounted on the truck.
- The liquid nitrogen system of transport refrigeration (that has been replaced) was easily adapted to a single vehicle as well as multiple compartments with different temperature requirements. Liquid nitrogen at –320°F vaporizes instantly as it issues under pressure from a spray header.
- In the old liquid nitrogen system, heat was absorbed from the surrounding air, convection currents were set up, and cool air was circulated to keep the entire cargo space cooled.
- The liquid nitrogen system was especially desirable for quick drawdown of temperature to meet particular requirements.
- Safety precautions had to be strictly observed in the liquid nitrogen system during installation, operation, refilling of the tanks, and in exhausting of inside compartment gas to the outside before entering the cargo space.
  - ✓ Safety door switches, emergency door design, safety valves and vents, pressure controls, low burn-rate materials used in construction of the vehicle, and other safety features, were essential to the safety of the operator and the system.
- By combining the advantages of two or more refrigeration systems, economies of operation and maintenance were achieved with the added product quality protection of two independent systems.
- The liquid nitrogen system provided the most rapid temperature recovery from truck body heat leak, door-opening losses, and other leak sources, under trouble-free, quiet conditions of operation, and with limited maintenance costs.
- Former combinations of the liquid nitrogen and holdover truck plate systems provided for the holdover plate system to handle all of the door-opening service load at 60°F ambient temperature, a portion of the load at 100°F ambient temperature, with the peak-demand refrigeration supplied by the liquid nitrogen system on high ambient temperature days. Plug-in operation was provided at dockside.
- Product protection, when the outside daily temperature was below 60°F, was provided by the holdover plates alone.

ASSIGNMENT: UNIT 24 REVIEW AND SELF-TEST

A. EARLY WATER-ICE TRANSPORT REFRIGERATION SYSTEMS

1. Refer to a trade handbook containing a formerly used mechanical refrigeration system for railway, automotive or other transport installations.
   a. Identify and name the major components of the system.

    b. Describe briefly how the system worked.

2. State two disadvantages of the old, water-ice transport refrigeration system.
3. Compute the quantity of ice required to maintain an average temperature difference for a 40-foot transport vehicle equipped for refrigeration operation. The heat leakage is known to be 4,500 Btu/hr. The transit time is 96 hours.

B. DRY-ICE REFRIGERANT

1. State (a) two advantages to using dry ice as a refrigerant as compared with a water-ice solution and (b) one safety precaution to be followed.
2. Describe heat of sublimation as applied to dry ice.
3. Compute the quantity of dry ice required to cool a low-temperature truck compartment to meet the following specifications (rounded to nearest pound).

• Heat gain of compartment. .1,250 Btu/hr
• Required compartment temperature . . . . 10°F
• Differential between internal and external temperatures. . . . . 80°F
• Door openings per 10-hour trip 25

C. EUTECTIC SOLUTION REFRIGERATION

1. Secure a table of eutectic concentrations with freezing points. Then, select a eutectic solution and its freezing point that may be used to provide the following temperatures:
    a. 30°F   b. 15°F   c. −36°F
2. Describe briefly the function of holdover plates.
3. Cite two advantages of a holdover plate system in comparison to the dry-ice and water-ice systems.
4. Describe the function of a blower system when used with the holdover plate system.
5. State (a) two methods by which holdover plates may be refrozen at dockside and (b) describe one of the methods for refreezing the eutectic solution.

D. EXPENDABLE REFRIGERATION: FORMER LIQUID NITROGEN SYSTEM

1. Locate a technical manual with a schematic drawing and circuit diagram of the electrical components for one of the former liquid nitrogen refrigeration systems.
    a. Make a freehand sketch by tracing the major components of the whole refrigeration system.
    b. Identify the major components by name.
    c. Describe the function of each major component in the former system.
2. Use the now replaced liquid nitrogen refrigeration system and drawings as in the previous problem. Identify and name the safety components and features that were incorporated in the system design.

3. Explain the operation of the earlier liquid nitrogen system.
4. State two advantages of the liquid nitrogen system over other single refrigeration systems.
5. Give three safety precautions that the vehicle operator and the maintenance persons had to observe with a liquid nitrogen refrigeration system.

E. COMBINATION FORMER LIQUID NITROGEN, HOLDOVER PLATE AND BLOWER COIL SYSTEMS

1. Locate a manufacturer's technical manual which shows a combination of a former liquid nitrogen and either a holdover plate or blower coil system used in combination with LN_2.
   a. Make a simple sketch of the system by tracing the major components.
   b. State briefly the function performed by each major component in the total system.
2. Give four advantages of earlier combination liquid nitrogen/holdover truck plate systems.

UNIT 25:
MECHANICAL TRANSPORT REFRIGERATION

Mechanical transport refrigeration systems operate on the same principle as most stationary commercial refrigeration. However, added consideration must be given in the design, installation and maintenance of the system and the individual components. Each must meet a wide range of ambient and operating temperatures and other severe conditions and stresses found in over-the-road operation during many continuous hours of service.

Components must be positioned to provide maximum efficiency by avoiding obstructions to free airflow and the damaging effects of weather. Other consideration must be given to the use of shock mounts to reduce vibration that causes fatigue failure. The systems must be designed with a minimum number of soldered joints and with corrosion-resistant materials.

Each unit must have its own dependable prime mover. Where refrigeration is required, as for the layover of perishable goods, a supplementary system (compatible with the facilities available) may be required. Valuable cargoes with critical temperature requirements call for a backup system to insure against any possible mechanical failure in the primary system.

Special cargoes sometimes require a special suspension system on the vehicle. Economical and effective operation dictate that attention must be given to details of body construction in order to provide ruggedness, maximum insulation, and even special colors for the transport roof and sides. An added design factor is the need for flexibility in the system for long distance hauling to permit vehicular return with a different type of cargo.

In this unit the basic principles of refrigeration are reviewed. This review is followed by treating mechanical transport refrigeration according to the following classifications: (A) self-contained refrigeration systems, (B) systems operated by the vehicle engine, (C) systems where the compressor is driven by power take-off and (D) combination mechanical and nonmechanical systems.

PRINCIPLES OF MECHANICAL REFRIGERATION

While truck and trailer body designs have been improved and better insulation materials are used, it still is not possible to create the conditions of a perfectly insulated compartment. With high ambient temperatures there is heat leakage through the walls, floor and roof, and at body seams, door seals, and drain holes. Still another factor in maintaining a fixed temperature and other climatic conditions is the generation of heat by cargo. This heat is known as the *heat of respiration* or *product heat.*

Mechanical transport refrigeration systems follow the principle that a liquid refrigerant absorbs heat as it evaporates to a gas and gives off heat as it reliquefies. As the refrigerant boils, the evaporator becomes cold. The blower in the evaporator section pulls air through the coil to dissipate the heat in the air which is picked up by the metal

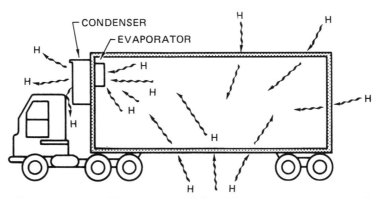

FIG. 25-1 Heat entering and being removed from the cargo compartment of a transport vehicle

fins surrounding the coil. The air, having given up its heat, is now cold and is circulated through the trailer to cool the load, figure 25-1.

The heated refrigerant gas is pumped out of the evaporator and into the condenser. Here the gas reliquefies, giving up its heat to the air which is passing through the coil (outside the compartment). The liquid refrigerant then flows into the receiver tank to be recirculated through the system. The mechanical transport refrigeration system maintains the same or lower desired temperature by replacing the quantities of heat units being removed with the same or a greater number of cool air units.

Obviously, refrigeration is needed to maintain or to lower the temperature. This is done by removing the number of Btu's (heat units) to equal the number of Btu resulting from the product load, leakage, and other causes. Figure 25-2 shows a typical installation with capacity to balance the heat load.

SELF-CONTAINED REFRIGERATION SYSTEMS

Self-contained refrigeration units are, as the name implies, mounted on a single chassis. The design and size of the unit depend on the truck body size and the required

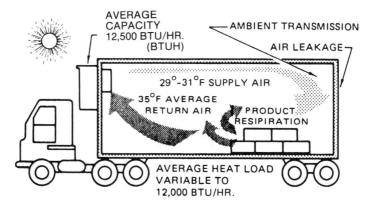

FIG. 25-2 Balancing the heat load in a transport vehicle

467

and auxiliary temperature range requirements. The evaporator is located inside the body or in a penthouse extension of the nose.

In this type of unit the high-side components extend in front of the body as illustrated in the phantom in figure 25-3A. The electric motor behind the fan is used for standby and layover operations. Having all of the components on the same chassis permits the high side to be connected directly to the evaporator, eliminating the need for long or flexible refrigerant lines.

Gasoline-powered units like the one illustrated are designed for run-stop When cargo space has been drawn down to the set temperature, the engine drops down to a fast idle. The centrifugal clutch then disengages the compressor which stops until the thermostat calls for cooling. At that point the compressor starts again. A governor maintains a single, steady number of rpm's during the on cycle.

When the compressor is driven by the gasoline engine, the electromagnetic clutch disengages the electric motor from its pulley. The drive pulley on the electric motor acts only as an idler between the gasoline engine and the compressor. When the compressor is driven by the electric motor on standby, the electromagnetic clutch on the gasoline engine disengages to permit the freewheeling of its drive pulley.

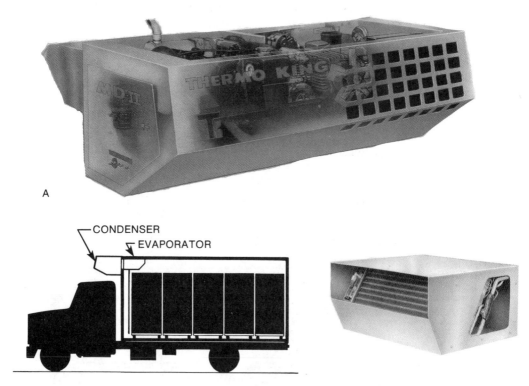

FIG. 25-3 High- and low-side components of a typical self-contained unit: (A) phantom view of high side nosemount condenser and (B) low side standard ceiling-mounted evaporator (inside the body) *(Courtesy of Thermo King Corporation)*

An electric motor drives the evaporator fan which eliminates a solenoid and damper mechanism and permits the use of a door switch to prevent blowing out cold air when the doors are open. Although automatic defrost is used, some medium temperature models (above 35°F) have compressor off-cycle defrost. Low-temperature (down to 0°F) models have reverse-cycle defrost. These models heat as well as cool. Cargo space temperature is selected by just setting the thermostat. An engine hour meter, which is standard with all manufacturers, facilitates regular maintenance of the equipment.

Capacity of Mechanical Transport Refrigeration Units

The Air Conditioning and Refrigeration Institute (ARI) maintains a standards and certification program. Manufacturers who ate and qualify may use the ARI symbol and other data related to their series of units. An example of the high and low temperature ranges and other certified characteristics for this type of unit of one manufacturer are shown by the symbol and accompanying data in figure 25-4.

This arrangement indicates that the manufacturer's product conforms to the design specifications of ARI when the capacity is determined according to a table of standard rating conditions shown in figure 25-5. The display of the ARI certification symbol indicates that the models and sizes, certified according to ARI standards, will have a cooling capacity of not less than 95% of the established rating.

The data accompanying figure 25-4 shows considerable differences in the capacity of the unit between the high-temperature operation (13,000 Btuh or Btu per hour) and the low-temperature (6,000 Btuh) ratings, both at 1075 rpm. This is explained by the fact that a greater number of Btuh can be removed at the higher temperature. Remember, the capacity of a substance to pick up heat is calculated by the product of its mass (m), its specific heat (c), and the difference in temperature (Δt) between the mass and the environment; (m) × (c) × (Δt).

In both the high (35°F) and low (0°F) return-air temperature ratings, the evaporator fan circulates cold air at the same rate, the mass is the same for both, and the specific

RATED IN ACCORDANCE WITH ARI
STANDARD 1110-69
STANDARD RATING: 35°-13,000 BTU/HR.
STANDARD RATING: 0°-6,000 BTU/HR.
R-12 REFRIGERANT
COMPRESSOR SPEED 1075 RPM
ELECTRIC STANDBY RATING:
STANDARD RATING: 35°-12,500 BTU/HR.
STANDARD RATING: 0°-6,000 BTU/HR.

FIG. 25-4   ARI symbol and manufacturer's certified capacities for a specific unit
(Courtesy Air Conditioning and Refrigeration Institute)

ARI STANDARD RATING CONDITIONS *						
Service	Return-Air to Air-Cooler	Ambient Air		Vehicle Engine Speed †		
	Dry-Bulb Temp	Dry-Bulb Temp	Barometric Pressure	Low	Medium	
	F	F	In. Hg	rpm	rpm	
Refrig Temp — High	35	100	29.92	600	1800	
Refrig Temp — Low	0	100	29.92	600	1800	
* Standard Ratings shall include compressor speed.						
† For variable-speed units only. Speed-governed units shall be rated at governed speed, to be shown on the unit nameplate.						

FIG. 25-5   ARI standard 1110-69 rating conditions *(Courtesy Air Conditioning and Refrigeration Institute)*

heat (for practical purposes), is the same. Since the rpm's of the compressor are the same in both cases, the chilled air issuing from the evaporator is the same, for example, –10°F. Obviously, there is a much greater difference in temperature (45°F) for the high temperature operation than that of only 10°F for the low temperature operation. However, a given unit can remove a greater number of Btu per hour at the higher temperature.

Large Capacity Nosemount Units

Mechanical transport refrigeration units for large trailers such as the one in figure 25-6 have components arranged vertically. The high temperature unit illustrated has a certified capacity of 40,500 Btuh at 1900 rpm. This series is designed to continuously and automatically maintain the cargo space temperature at the desired thermostat setting over a range from –20°F (–28.9°C) at 80°F (26.7°C). Four steps of operation include: high-speed cool, low-speed cool, low-speed heat, and high-speed heat. The required operating mode is automatically selected by the thermostat in response to evaporator air temperature. The diesel engine runs continuously when the unit is needed. The placement of the components in this unit is shown in figure 25-7.

Heating is accomplished by circulating hot gas directly from the compressor to the evaporator coil. Two electric solenoid valves control the refrigerant circuit. In the event of failure from the 12-volt dc power source, these valves return to their normal position and permit the unit to be manually operated in the cooling mode until control power is restored.

Normally, when a prechilled load is being transported, a diesel unit will alternate between low-speed cool and low-speed heat. This causes some *breathing* of the atmosphere within the cargo compartment. Furthermore, under these conditions, the thermal load from the outside plus any *heat of respiration* from the cargo may produce (for example) between 12,000 to 13,000 Btuh. These quantities are below the low-speed cooling capacity for this unit which, when it drops back, has a range from 28,000 to 30,000 Btuh.

In order to avoid *top freezing*, excessive trailer breathing, and cycling between low cooling and low heating, one manufacturer has a cylinder-unloading feature option. When less cooling is required, four of the six compressor cylinders stop pumping refrigerant gas and the cooling capacity is reduced to a 10,000-14,000 Btuh range. This range is close to the load requirement in the previous example. Also, there is a rise in evaporator coil temperature, tending to reduce both top freezing and excessive dehydration of the cargo. Cylinder unloading also reduces the diesel load, requiring less fuel.

On one model of a nosemount unit, the diesel is coupled directly to the six cylinder compressor. The compressor is fitted with two-ring pistons which reduce blow-by and high oil circulation, especially under operating conditions requiring very low pressure on the suction stroke.

Additional refrigeration components are mounted in the upper half of the housing directly over the engine and compressor. The evaporator extends through the mounting

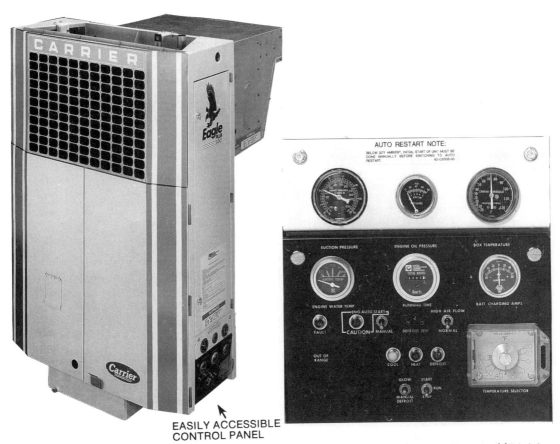

EASILY ACCESSIBLE
CONTROL PANEL

FIG. 25-6 Nosemount unit components mounted vertically *(Courtesy of Carrier Transicold Division, Carrier Corporation)*

471

FIG. 25-7 Location of components for large capacity nosemount units in the lower section provide easy access *(Courtesy of Carrier Transicold Division, Carrier Corporation)*

port into the cargo compartment. A portion of the condenser is shown in front of the tubular radiator that serves the diesel engine.

Automatic evaporator coil defrosting is initiated by sensing the pressure drop across the coil with a differential air switch. Manual defrosting may be started by closing a switch on the control panel. Both the condenser and evaporator fans are mounted on the same shaft and are driven by a belt.

Figure 25-8 (A) shows a high-capacity compressor and direct-injection engine components for a transport refrigeration system. The temperature and process control panel at (B) provides an instant visual *digital readout* on unit "power-up." A *keypad* is used to provide information and to operate all functions such as *temperature setpoint, defrost time interval, cumulative engine running hours, coolant temperature, engine oil pressure,* and *probe (controlling) temperature.* A *manual defrost key* and an *alternating display key* (for box temperature and setpoint temperature) serve these specific functions.

Another Vertical-Profile Self-Contained Nosemount Unit

Another line of reefers operates in much the same manner as the previously identified units. However, there are other significant mechanical details that must be understood.

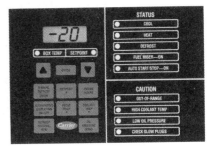

FIG. 25-8 Compressor, engine and control panel: **(A)** engine compartment showing high-capacity compressor and direct-injection engine and **(B)** digital temperature and process control panel *(Courtesy of Carrier Transicold Division, Carrier Corporation)*

The model illustrated in figure 25-9 has a certified cooling capacity of 40,000 Btuh at 2,200 rpm for high temperature operation. The evaporator section extends to the rear with one end of the damper showing (closed) in the discharge opening.

The header on the evaporator is clearly visible. A heated drop pan runs the entire length under the evaporator. The damper of this model has flexible lips. Thus, when the damper is closed, it seals the evaporator compartment completely. Under this condition, when the blower fan is operating, its compartment is pressurized during defrost and the water is actually forced out through the drains. Reverse cycle refrigeration is used for the defrost and heating modes. A three-way valve channels the hot gas directly from the compressor to the evaporator and drip pan.

The finned-tube engine radiator is shown behind the grille in the lower portion. The expansion tank for the coolant appears at the top of the housing. The condenser is located directly above the radiator in the upper portion.

The gauges and control panel are clearly visible on the side of the mounting. The engine exhaust muffler extends through the housing. Other refrigeration components such as the drier, expansion valve and accumulator are also housed in the upper part of the housing.

In this model the accumulator uses engine coolant to supplement the heating and defrost cycles. The U-tube inside the accumulator prevents the liquid refrigerant from slugging the compressor and meters oil to the suction line.

The engine, which is directly coupled to the compressor, the sight glass on the compressor for visual check of oil level, and other features, are contained in the lower section behind the doors.

Figure 25-10 shows the electrical circuit for the unit. Such a circuit is either 12 or 24 volts depending on the design requirements. Naturally, all of the components must match the voltage of the circuit.

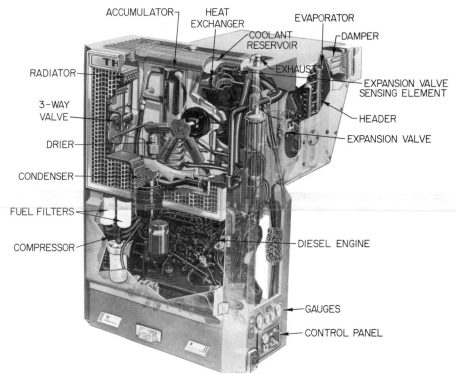

FIG. 25-9 Cutaway view of vertical-profile self-contained nosemount unit *(Courtesy of Thermo King Corporation)*

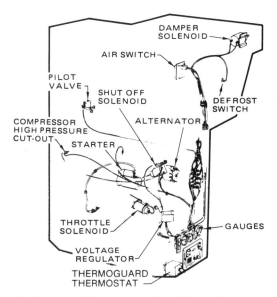

FIG. 25-10 Electrical circuit for a vertical-profile self-contained nosemount unit *(Courtesy of Thermo King Corporation)*

When a gasoline engine is used as the prime mover, it usually cycles between off and on, depending on the refrigeration requirements. The engine starts and runs at a fixed speed when the thermostat in the refrigerated body calls for cooling.

By contrast, diesel-powered units usually run continuously when refrigeration is required. They either idle or run at low-cool speed if refrigeration is not needed. This arrangement is designed to provide low and high heat as well. Under normal operating conditions, the unit cycles between low-heat and low-cool to maintain a uniform temperature in the cargo compartment.

COOLING CYCLE OF A NOSEMOUNT UNIT

The cooling cycle of a diesel nosemount unit is diagrammed in figure 25-11. Major components such as compressor, condenser coil, evaporator coil, valves, lines, solenoids, etc., are identified by encircled numbers from ① through ㉘. Conditions within the refrigeration system during the cooling cycle can be traced by following the code symbols that are used for high and low pressure liquid and high and low pressure vapor.

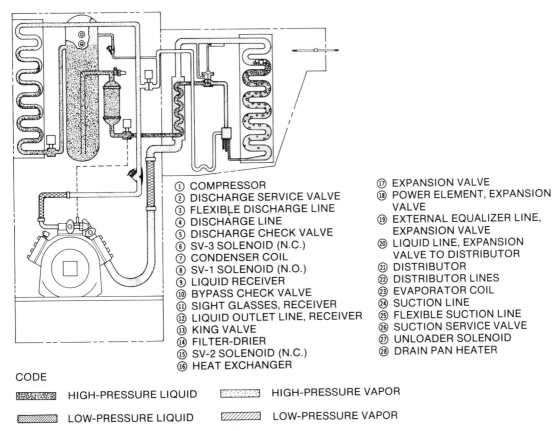

① COMPRESSOR
② DISCHARGE SERVICE VALVE
③ FLEXIBLE DISCHARGE LINE
④ DISCHARGE LINE
⑤ DISCHARGE CHECK VALVE
⑥ SV-3 SOLENOID (N.C.)
⑦ CONDENSER COIL
⑧ SV-1 SOLENOID (N.O.)
⑨ LIQUID RECEIVER
⑩ BYPASS CHECK VALVE
⑪ SIGHT GLASSES, RECEIVER
⑫ LIQUID OUTLET LINE, RECEIVER
⑬ KING VALVE
⑭ FILTER-DRIER
⑮ SV-2 SOLENOID (N.C.)
⑯ HEAT EXCHANGER
⑰ EXPANSION VALVE
⑱ POWER ELEMENT, EXPANSION VALVE
⑲ EXTERNAL EQUALIZER LINE, EXPANSION VALVE
⑳ LIQUID LINE, EXPANSION VALVE TO DISTRIBUTOR
㉑ DISTRIBUTOR
㉒ DISTRIBUTOR LINES
㉓ EVAPORATOR COIL
㉔ SUCTION LINE
㉕ FLEXIBLE SUCTION LINE
㉖ SUCTION SERVICE VALVE
㉗ UNLOADER SOLENOID
㉘ DRAIN PAN HEATER

CODE

▨ HIGH-PRESSURE LIQUID   ▨ HIGH-PRESSURE VAPOR

▨ LOW-PRESSURE LIQUID   ▨ LOW-PRESSURE VAPOR

FIG. 25-11 **Cooling cycle of a diesel nosemount unit** *(Courtesy of Carrier Transicold Division, Carrier Corporation)*

FIG. 25-12 **Engine-drive generator set for center-of-gravity mounting under tractor-trailer chassis** (*Courtesy of Carrier Transicold Division, Carrier Corporation*)

ELECTROMECHANICAL SYSTEMS

The electromechanical system is another type of self-contained unit. The nosemount unit is a complete refrigeration system consisting of both the high and low sides, usually with a hermetic or serviceable hermetic compressor. The source for over-the-road operation is the diesel driven alternator which is mounted under the tractor-trailer chassis just ahead of the rear wheels. Electric cables provide the only connection between the nose-mounted unit and the alternator.

Figure 25-12 shows an engine-drive generator unit that slides out on rails for ease of maintenance or replacement. This system has the added advantage that the hermetic unit can be operated by simply plugging into a dockside outlet of matching characteristics of phase and voltage.

Evaporator and condenser fans are driven by electric motors. Some generator systems have capacity enough to operate lift gates, loading booms, and other equipment.

OTHER APPLICATIONS OF SELF-CONTAINED UNITS

Another type of self-contained unit, complete with electric standby, now provides refrigeration for insulated bodies that are designed for mounting in a pickup truck.

Refrigerated containers for piggyback use are another application of self-contained units. Several units of this type can be placed on an open trailer bed. Each container has its own refrigeration unit with the evaporator on the inside and the condenser on the outside. The piggyback system requires pneumatic-hydraulic jacks on the trailer and truck for positioning the container legs for loading and unloading.

Piggyback applications are used to good advantage in pre-ordered drop deliveries. Each container is loaded in the delivery sequence. The view through a typical installation of piggyback containers, figure 25-13, shows the forward door of the last two cells opened. Thus, the containers can be loaded right on the trailer as a single compartment. This feature also makes for ease of cleaning.

The piggyback system has these advantages.

- It eliminates the need for transferring lading from trailer to truck.
- The lading travels from plant to store in the same refrigerated atmosphere.
- The transfer of container to truck is accomplished by the driver in a matter of minutes as contrasted with the time required to load and unload the lading.
- A large number of drops can be handled by the driver.
- The chances of loading and unloading errors at the terminal are reduced.

COMPRESSOR DRIVEN BY THE VEHICLE ENGINE

A truck refrigeration system in which the compressor is driven by the vehicle engine has all of the usual components. The truck engine is used to belt drive the compressor when it is mounted in the engine compartment. An extra pulley is added to the crankshaft to drive the compressor in addition to the fan, generator, and other engine components. Such an arrangement provides for flexibility in placing all of the components except the evaporator, figure 25-14. The evaporator is always within the insulated body.

Flexible suction and hot-gas lines run all the way from the engine compartment to the condenser, which is mounted either on the cab roof or nose of the truck body. Both the evaporator and condenser have electric fans driven from the truck battery. The unit has plenty of capacity for precooling the body and over-the-road operation.

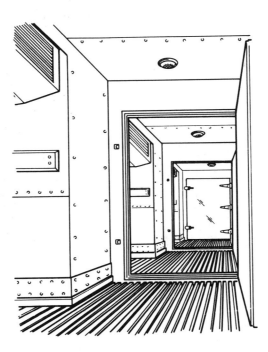

FIG. 25-13   Refrigerated container units arranged piggyback facilitate cargo handling

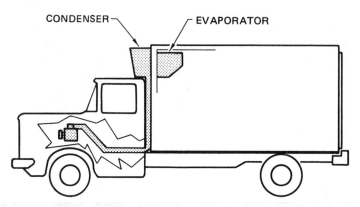

CONDENSER — EVAPORATOR

FIG. 25-14   Location of major components in a system driven by the vehicle engine *(Courtesy of the Trane Company)*

If an electric standby is used for backup protection against mechanical failure or to provide for overnight storage, the unit is usually mounted in the truck or under the body. An electromagnetic clutch between the continuously rotating pulley and the compressor shaft is used to put the compressor in service.

COMPRESSOR DRIVEN BY POWER TAKEOFF

Like the preceding system, the *power takeoff (PTO)* driven refrigeration system for trucks provides for flexibility in the location of the components and does not require an extra engine to be purchased and maintained. The system is driven during the over-the-road operation from the vehicle engine through the power takeoff which is attached to the transmission.

One type of PTO unit locates the condenser on the nose of the body. In other designs, the PTO unit which drives the compressor and the condenser are on the same frame, requiring one line to carry high pressure liquid refrigerant from the condenser to the expansion valve and evaporator inside the insulated body. The return line is the usual suction line.

This unit includes an optional standby electric system for dockside use. In the PTO-driven units both the condenser and evaporator have fans driven from the battery of the vehicle. An electromagnetic clutch is used between the compressor and the PTO drive.

COMBINATION MECHANICAL AND NONMECHANICAL TRANSPORT REFRIGERATION SYSTEMS

Mechanical and nonmechanical systems, when used in pairs, combine the advantages of each and minimize the disadvantages. The complementary operation provides balanced capacity and dependability. The system, designed to meet such requirements as many door openings for long periods and long on-the-road hours, especially under hot summer conditions, combines an electromechanical system with eutectic holdover plates. The location of the components for a plate/blower system is shown in figure 25-15, looking at them from the rear of the truck (A) and in schematic at (B).

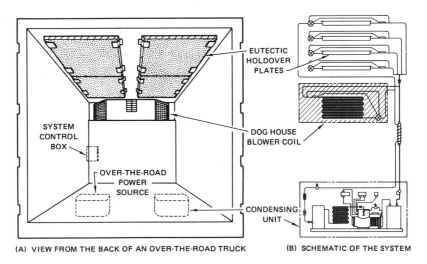

(A) VIEW FROM THE BACK OF AN OVER-THE-ROAD TRUCK   (B) SCHEMATIC OF THE SYSTEM

FIG. 25-15 Plate/blower-coil system

SUMMARY

- Mechanical transport refrigeration systems require unique designing and body construction in order to withstand the damaging effects of weather, severe over-the-road operation stresses, and cargo requirements.
- Mechanical transport refrigeration systems may be classified as:
  - ✓ Self-contained systems.
  - ✓ Systems operated by the vehicle engine.
  - ✓ Systems driven by power takeoff units.
  - ✓ Combination mechanical and nonmechanical systems.
- Mechanical transport refrigeration systems balance the heat load or lower the compartment temperature by removing the Btu's resulting from compartment heat leakage and the heat of respiration of the cargo.
- Mechanical transport refrigeration systems use the complete refrigeration cycle the same as commercial refrigeration. A liquid refrigerant absorbs heat as it evaporates to a gas and gives off heat as it reliquefies. The refrigerant circulates through the evaporator, condenser and receiver tank, involving other components like the compressor, control valves, power sources, and backup units to carry on all functions of the system.
- Mechanical units are also designed to provide cargo heating when required.
- The high-side components of self-contained refrigeration systems are mounted outside of the cargo compartment; the evaporator, inside.
- The ARI provides design specifications against which manufacturers may compare the characteristics of their products for conformance and certification according to industrial standards established by ARI.
- Diesel-powered mechanical transport refrigeration units which transport a pre-chilled load alternate between low-speed cool and low-speed heat.

- Gasoline-powered mechanical transport refrigeration units operate on a stop-run cycle.
- Automatic coil defrosting in large nosemount units is initiated by sensing the pressure drop across the coil with a differential air switch.
- Cylinder unloading is used to reduce the cooling range to provide a more uniform temperature, also resulting in economy of operation.
- Hermetic units are used in conjunction with self-contained systems for standby and dockside cooling.
- Reverse refrigeration is used on large self-contained units for the defrost and heating modes.
- Self-contained units that operate on electricity use an underside motor-generator set for over-the-road operation. At dockside, the unit is plugged into an electrical source to operate.
- Self-contained units are adaptable for transport refrigerated piggyback multiple container application.
- Mechanical and nonmechanical transport refrigeration systems are combined to provide complementary balanced capacity and dependability.

ASSIGNMENT: UNIT 25 REVIEW AND SELF-TEST

A. SELF-CONTAINED REFRIGERATION SYSTEMS

1. State five conditions that must be considered in designing, installing and maintaining mechanical transport refrigeration systems.
2. Explain briefly the function of a self-contained refrigeration system in balancing a heat load.
3. Trace the flow of a heated refrigerant gas in maintaining a controlled temperature in a mechanical transport refrigeration system.
4. Trace the circulation of air in the cargo compartment during the cooling cycle.
5. Refer to a refrigeration technical manual.
   (a) Select a phantom outline, cutaway section, or other illustration of a self-contained refrigeration system.
   (b) Make a simple tracing outlining the major components for both the high and low sides.
   (c) Indicate how the unit is powered.
   (d) Show the major components by identifying and lettering the part names.
   (e) Explain the function and operation of four major components in the system.
6. Examine the ARI symbol and the manufacturer's certified capacity for a particular model as shown on page 481. Then, explain the difference in Btuh capacity between high-temperature operation and low-temperature operation.
7. Explain (for a large mechanical transport refrigeration unit):
   (a) What causes breathing of a prechilled load, and
   (b) The function of cylinder unloading.
8. Locate an illustration of a large nosemount unit. Identify the unit and then tell briefly how the evaporator coil may be automatically and/or manually defrosted.

MODEL	COMPRESSOR	CONTROLLER
400C		Partlow LFC
	(Serviceable Hermetic)	

CAPACITY (ARI Standard Ratings — Cooling Only)
Ambient @ 100F (37.8C) — Speed 1 750 rpm

Cooling

Evaporator Return Air Temp	Btuh (Kcal/hr)
0 F(−17.8C)	17,500 (4428)
35F(1.7C)	32,000 (8096)

9. Study the electrical circuit for a vertical profile self-contained nosemount unit.
   (a) Trace the circuit.
   (b) Letter the major refrigeration components.
   (c) Follow the circuitry of each major component and describe the electrical safety features incorporated in the system.
   (d) Identify whether the safety features in the electrical circuit form a series or parallel circuit.
10. State three reasons why piggyback applications of refrigeration containers are functional.

B. COMPRESSOR DRIVEN BY THE VEHICLE ENGINE

1. Review a manufacturer's descriptive manual of a refrigeration system with the compressor driven by the vehicle engine.
   (a) Make a freehand tracing of the system, labeling the major drive parts between the engine and the compressor.
   (b) Trace and describe the way in which the remaining components are used in the refrigeration cycle.
   (c) Tell what safety features are incorporated to guard against mechanical failure or for overnight storage operation.

C. COMPRESSOR DRIVEN BY POWER TAKEOFF

1. Describe two different types of systems for driving a compressor, other than self-contained units.

D. COMBINATION MECHANICAL AND NONMECHANICAL TRANSPORT REFRIGERATION SYSTEMS

1. Explain briefly the complementary function of the eutectic holdover plates in a combination plate/blower transport refrigeration system.

UNIT 26:
AUTOMOTIVE AIR-CONDITIONING

Constant temperature, humidity and airflow control (within precise limits) are essential to new discoveries of materials, processes and products as well as to advances in technology, economic growth and individual comfort.

Aerospace vehicles; inspection laboratories where ultraprecise measurements, surface texture, and other characteristics are studied for applications in microminiature parts and components; and environmental control of computer room installations are a few everyday applications that depend on air-conditioning systems.

HUMAN CONSIDERATIONS IN DESIGNING AIR-CONDITIONING SYSTEMS

Since air conditioning provides for the comfort of people, consideration must be given to a few simple physiological factors. These affect the design and performance requirements of air-conditioning equipment and materials.

The continuous removal of body heat takes place primarily through three natural processes that usually occur at the same time. These processes are:

- Conduction-convection
- Radiation
- Evaporation

The methods by which the heat travels depend upon the relationship between the individual and the surrounding environment.

Effect of Moisture on Evaporation

When air is hot and saturated with moisture particles, the body is deprived of the cooling effect which accompanies the evaporation of perspiration from the skin surface. The person feels uncomfortably warm. Thus, one function of an air conditioner is to remove excess moisture and produce a dry air. These conditions permit evaporation and produce a cooling effect.

Heat Conduction and Convection Currents and Radiation

An individual who enters a heated car receives heat from the air instead of getting rid of excess heat. If the air within the car is cool, heat will travel from the skin to the air. The moving air (convection currents) continues to bring cool air into contact with the skin and move away air that has picked up heat from the body surface. When the air is dry, the convection currents also facilitate evaporation.

In an air conditioned car that has been cooled, the upholstery, metal, plastics, and other surfaces are cooler than the person who enters. Under these conditions, heat is radiated from the body to the cooler surfaces, providing a cooling sensation.

Comfort Requires Balance

A comfortable environment is maintained when heat is removed at the rate at which it is being formed. There must be a balance among the processes of evaporation, conduction-convection and radiation. For instance, if the air is chilled but is not dried, the removal of part of the heat by evaporation cannot take place. Only the effects of conduction-convection and radiation are felt. Even though heat is removed under these conditions at the correct rate, failure to cool and dry the air simultaneously makes the individual feel uncomfortably damp and cold.

Thus, the air conditioner must remove moisture from the air as well as to cool it. The evaporator must operate at a low enough temperature to condense water vapor out of the air. This temperature is referred to as the *dew point.*

This cold air would be too cold and would upset the balance if it were discharged into the passenger compartment. Instead, this dry but too-cold air is usually passed through the car heater to bring it up to a more comfortable temperature to produce balanced heat removal from the occupants.

The *air-mix* system illustrated in figure 26-1 has the capability of producing the desired changes in the air prior to its discharge into the passenger compartment.

APPLICATIONS OF REFRIGERATION TO AUTOMOTIVE AIR CONDITIONING

Although automotive air conditioning (for practical purposes) was seriously considered in the 1940s, it did not become popular until the 1960s. Today, air conditioning is included in more than 80% of the cars manufactured. Thus, what had been considered a luxury in the early years has turned out to be extremely practical and a necessity.

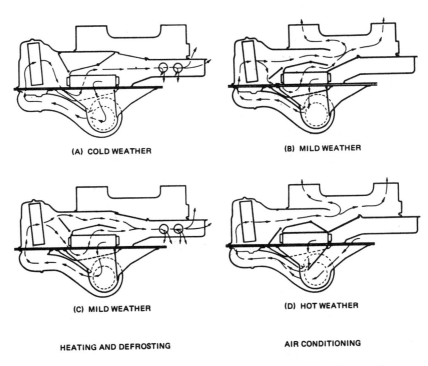

(A) COLD WEATHER

(B) MILD WEATHER

(C) MILD WEATHER

(D) HOT WEATHER

HEATING AND DEFROSTING

AIR CONDITIONING

FIG. 26-1 Air-mix system of heating and cooling

Purposes of Automotive Air Conditioning

Regardless of exterior size (mass), an automobile requires a considerable quantity of refrigeration capacity on a hot summer day to keep the interior temperature at a comfortable level. Conversely, as an automobile is driven at a high speed on a cold winter day, a large quantity of heat is needed to maintain comfort in the car. The refrigeration system used in automotive air conditioning is driven by the car engine for cooling. The heated liquid from the engine cooling system usually provides the heat requirements.

During summertime operation, the automobile air conditioner serves to reduce the humidity of the air inside the car. In addition, moisture that collects on the evaporator picks up a considerable amount of dust and pollen from the car interior. These air pollutants are carried away as the collected moisture drains from the evaporator underneath the car. This process of cleaning the air provides one advantage of an air conditioner in relation to the heating system.

OPERATION OF THE AUTOMOBILE AIR-CONDITIONING SYSTEM

A description of how an automotive air-conditioning system operates follows. A belt driven *compressor* is usually mounted on the engine. The *condenser* is located and mounted in front of the car radiator. During operation, the compressor pumps heat-laden refrigerant vapor from the *evaporator.*

The refrigerant is compressed to a high temperature vapor and is pushed under pressure to the condenser. The high-pressure (high-temperature) vapor in the con-denser gives off heat to the atmosphere, changing the vapor to a high-pressure liquid. This liquid is filtered and stored in the liquid refrigerant receiver until there is a need by the evaporator.

A thermostatic expansion valve is used to meter liquid refrigerant from the liquid receiver to the evaporator, thus controlling the refrigerant flow to the evaporator. With the lowering of the pressure of the refrigerant, it begins to boil and changes to a vapor. The refrigerant picks up heat (during this process) from the warm air that passes through the evaporator. Heat is given off to the atmosphere by the condenser as the heat travels from the evaporator to the compressor to the condenser.

A magnetic clutch mechanism is installed to prevent the air-conditioning system from running continuously. Under continuous operation, the car temperature would drop to an uncomfortable level with possible frosting over and damage to the system. This condition is controlled by the *thermostat* that regulates the electrical power flowing to the *electromagnetic clutch* in the compressor. The clutch makes it possible for the compressor to be freewheeling.

The clutch is operated by a thermostat that opens the circuit of an electromagnetic clutch on the compressor. Under this condition, the compressor pulley rotates while the crankshaft remains stationary. No cooling takes place when the electromagnetic clutch is not engaged.

Cooling Capacities

Automobile air-conditioning units range in size from 12,000 Btu (1 ton) to 48,000 Btu (4 tons). The specifications of the air-conditioning unit must match the requirements of each particular car. Undercapacity reduces the cooling capability. Overcapacity is uneconomical and, in addition, causes too frequent cycling. Generally, car units are selected to maintain an interior car temperature of 15°F to 20°F below the ambient outdoor temperature.

TYPES OF AUTOMOTIVE AIR-CONDITIONING SYSTEMS AND COMPONENTS

Three basic cycle and mechanical air-conditioning systems are in use on automobiles.

- Pressure-operated low-side pressure regulators
- Pressure-operated bypass
- Solenoid-operated bypass

Low-Side Pressure Regulators

An *evaporator pressure-controlled regulator valve* is installed in the suction line of a low-side pressure control system, figure 26-2. The valve is designed to hold a constant pressure (temperature) in the evaporator. When the evaporator pressure/temperature goes below a specific setting, the valve closes automatically, holding the evaporator at a constant temperature/pressure.

Pressure-Operated, Hot-Gas Bypass Valve

A cross-sectional view of a *pressure-operated bypass valve* is shown in figure 26-3. This type of valve is connected between the compressor discharge (high side) and compressor suction line (low side).

As the pressure difference reaches the setting of the valve, the bypass valve opens and bypasses hot vapor from the high side to the low side. The valve opens when the evaporator pressure is lowered and closes when the pressure is increased. A specified pressure is maintained in the evaporator by feeding hot gas vapor into the low-pressure side.

Evaporator temperature is controlled by manual adjustment of the valve for pressure.

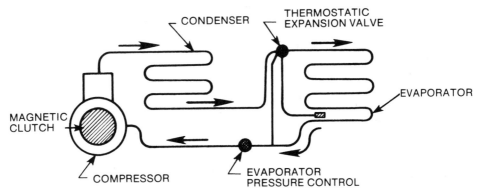

FIG. 26-2 Major components in a low-side system with an evaporator pressure-controlled suction line

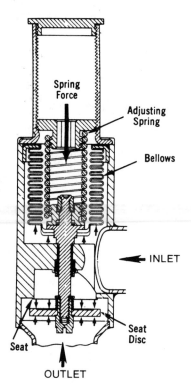

FIG. 26-3 Principles of operation of a suction pressure regulator valve, adaptable to automotive A/C applications *(Courtesy of Sporlan Valve Company)*

Solenoid-Operated, Hot-Gas Bypass Valve

In a *solenoid-operated, hot-gas bypass valve* system, a thermostat is mounted on the evaporator. The thermostat opens a solenoid to bypass hot gas from the high side to the low side when the temperature of the evaporator reaches 33°F.

A *thermostat-sensing bulb* is located in the outlet airflow of the evaporator. The thermostat requirements are met as the air temperature is lowered to 32°F; allowing hot gas from the condenser to bypass back into the suction line. When the thermostat points are open, the solenoid is in the closed position. The points on the thermostat close on temperature drop.

Expansion Valve

A metering device is built into automobile air-conditioning systems to throttle high-pressure liquid refrigerant to low-pressure liquid refrigerant in the evaporator. A *thermostatic expansion valve (TEV)* serves as a metering device. The TEV operates on the temperature at the evaporator outlet and on low-side (suction) pressure. Principles of operation, design and construction features, installation, and servicing of the thermostatic expansion valve were detailed in earlier units.

Suction Pressure Control Valve

Most systems use a *suction pressure control valve* to maintain a specific pressure and an above-freezing temperature in the evaporator to prevent it from freezing. The pressure maintained by a suction pressure control valve is independent of the compressor low-side pressure and of the cooling requirements.

Referring to figure 26-3, a diaphragm or bellows in the valve acts only in response to evaporator pressure. When the pressure is above 29 to 31 psi (R12), the valve opens. When the pressure goes below these settings, the valve closes. Design features of the suction pressure control valve require manual adjustment to control the evaporator temperature.

There are three common types of suction pressure control valves.

- Suction-Throttling Valve (STV)
- Evaporator Pressure Regulator (EPR)
- Pressure-Operated Altitude Valves (POA)

Suction pressure control valves are designed to keep the evaporator above freezing temperature. These valves also prevent the moisture (which accumulates on the evaporator) from freezing.

AIR DISTRIBUTION

There are unique problems of air distribution in an automobile, such as:

- Draft possibilities increase with low roof design
- Confined and limited space within a vehicle
- Airflow restrictions created by the seats and head- and armrests

A popular design for the distribution of air inside an automobile locates the air-conditioning system under the instrument panel on the passenger side of the fire wall. Other installations require the air-conditioning system to be mounted on the engine side of the fire wall, using a series of ducts to control the flow of air.

Heating and cooling ducts are fabricated of metal or plastic materials. While high air velocity is involved, air movement noises are not as critical as they are within a home or business environment. Most duct systems consist of:

- A fresh air inlet
- A return air inlet
- An evaporator housing
- A plenum chamber
- A drain pan and connections
- Conditioned air outlets (defrost and deicer) and grilles
- Manual or power operated dampers to change the airflow

Figure 26-4 illustrates schematically a duct system in which *doors* are used to control the direction of airflow. The evaporator and heater core are designed to use the same duct system for the distribution of air. Vacuum-powered diaphragms are constructed to operate *dampers* (doors) that control the direction of airflow in most automotive air-conditioning systems. The *diaphragm* serves to move the door in one direction; a spring, in another direction.

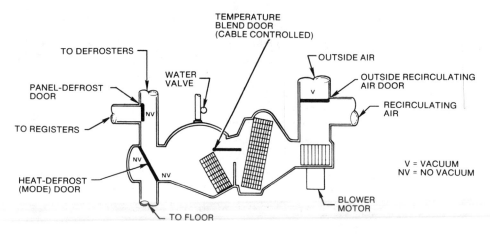

FIG. 26-4 Duct system using "doors" to control the direction of airflow in the ducts
(Courtesy of Ford Customer Service Division, Ford Motor Company)

Some airflow control units are also known as *servomotors*. The pictorial sketch, figure 26-5, shows the main components of a *vacuum-operated power servomotor*. A *damper diverter valve* controls door movement.

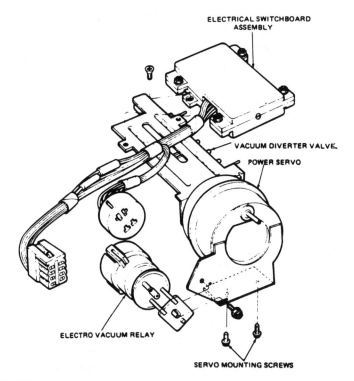

FIG. 26-5 Power servo unit used to control a vacuum-diverter valve

Fans and Fan Motors

Squirrel-cage or *centrifugal radial flow type fans* serve to circulate air in automotive air-conditioning systems. Fan motors are generally designed for flexible mounting to reduce noise. Single-speed, two-speed or three-speed fans are used depending on the application.

Fan motors are DC-12 volt, single- or double-shaft. The use of sealed-in bearings eliminates the need for additional oiling.

MECHANISMS CONTROLLING COMPRESSOR FUNCTIONS

A cutout device on the compressor permits the engine to run without engaging the compressor. The device is a *clutch* on the compressor which permits engaging or disengaging the compressor drive pulley in relation to the compressor crankshaft. The clutch, figure 26-6, operates by forcing an *electromagnetic clutch plate (disk)* against the pulley.

The construction of the compressor pulley and stationary electromagnet and the operation and magnetic field circuit of this clutch are pictured in the cross section drawing, figure 26-6.

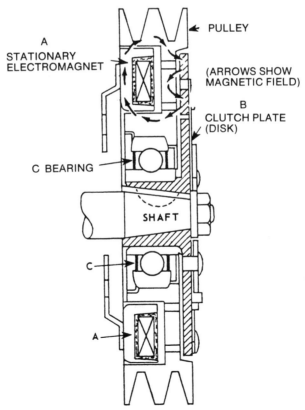

FIG. 26-6 Cross section of a compressor pulley showing major components and internal design features

There are two prime designs of electromagnetic clutches in general use.

- *Revolving magnetic coil* design that requires two carbon brushes to be in contact with two copper rings. These rings are mounted on the coil. The magnetic coil revolves when the compressor revolves.
- *Stationary magnetic coil* design where the coil is mounted on the compressor housing. Two electrical leads are used: one is a ground; the other, the clutch control.

The Warner Plate, Warner Heli-Grip, and Electro Lock Plate types of electromagnetic clutches represent a few of the most commonly manufactured clutches.

BASIC COMPRESSOR DESIGNS AND OPERATING CHARACTERISTICS

Compressors

While there are more than thirty different compressor models for automotive air conditioners, two types are in general use today.

- *Conventional reciprocating type* with crankshaft, connecting rod, piston, and cylinder
- *Swash plate type* that uses a different reciprocating piston and cylinder arrangement. *Reciprocating* refers to the up-and-down, to-and-fro, or back-and forth movement of the piston. The piston may be driven by one of two basic methods: *crankshaft* or by *axial plate*. The term *swash plate* or *wobble plate* is often used with axial plate drives.

Belt-driven compressors are currently being used. These compressors operate at slightly higher RPM than engine speed. At engine idling speeds from 500 to 900 RPM, the compressor revolves at 600 to 1,000 RPM, depending on the engine application. Compressors revolve at 5,000 RPM maximum speed.

Two-Cylinder Compressors

Except for swash plate compressors, there is a similarity of operation among the compressors used in automotive air conditioning. These compressors use 500 viscosity refrigerant oil. A limited amount of refrigerant oil is required (varying from 3 to 7 ounces) depending on the model. A lack of oil can produce bearings, seal or valve problems.

Consideration must be given to the selection of and maintenance of a compressor to ensure that:

- Crankshaft seals are heavy duty.
- All moving parts are balanced for the full range of speeds.
- The efficiency of the compressor remains at a fixed maximum regardless of compressor speed.

Two-cylinder reciprocating type vehicle compressors are used by the Ford Motor Company, American Motors and a few other car manufacturers of "add-on" units. The external features of a rugged, over-the-road air-conditioning system compressor are displayed in figure 26-7. This two-cylinder compressor is designed with a cast-iron

FIG. 26-7 External features of a recip-rocating-type, over-the-road, air-conditioning compressor *(Courtesy of Tecumseh Products Company)*

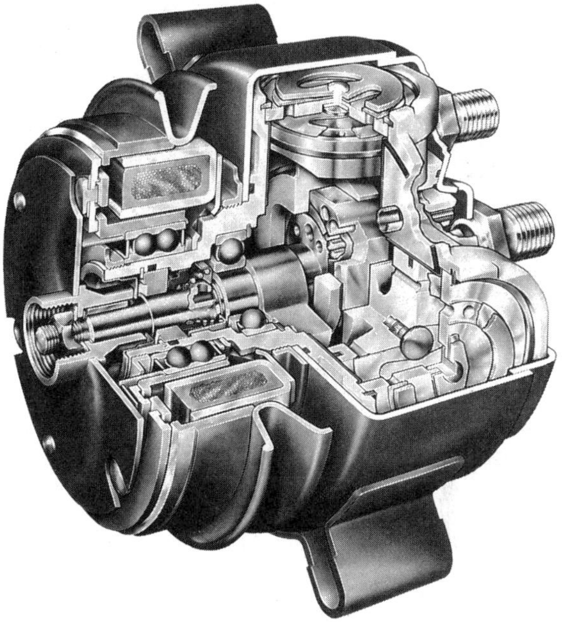

FIG. 26-8 Internal construction of a four-cylinder, reciprocating-type, vehicle, air-conditioning compressor *(Courtesy of Tecumseh Products Company)*

crankcase and cylinder head and is lubricated by differential oil pressure. This heavy-duty compressor is capable of operating under extreme road-pounding, vibration and difficult working conditions.

The internal construction of a large-bore, short stroke, four-cylinder, vehicle air-conditioning system compressor is shown in figure 26-8. The operational speeds range from 1,000 RPM minimum to 6,000 RPM maximum.

The radial arrangement of the cylinders and the short stroke combine to minimize vibration and noise. This model compressor can be mounted in position at any angle through 360° around the crankshaft.

Five-Cylinder Compressor

Another model of compressor is shown in the maintenance drawing, figure 26-9. Internal design features are seen in the two cutaway sections. Important parts are identified by callout numbers, corresponding with the part names.

The placement of the five-cylinder vehicle compressor in an automotive air-conditioning system is shown graphically in figure 26-10. Coded lines are used to represent and to trace high-pressure vapor (___ ___ ___), high-pressure liquid (.___.___.___), low-pressure liquid (_ _ _ _ _ _), and low-pressure vapor (_____) flow.

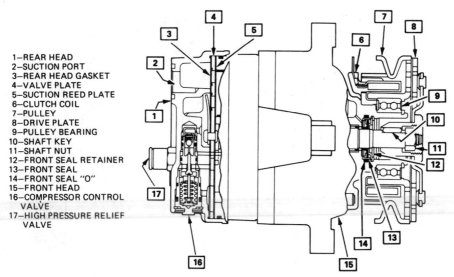

1—REAR HEAD
2—SUCTION PORT
3—REAR HEAD GASKET
4—VALVE PLATE
5—SUCTION REED PLATE
6—CLUTCH COIL
7—PULLEY
8—DRIVE PLATE
9—PULLEY BEARING
10—SHAFT KEY
11—SHAFT NUT
12—FRONT SEAL RETAINER
13—FRONT SEAL
14—FRONT SEAL "O"
15—FRONT HEAD
16—COMPRESSOR CONTROL
    VALVE
17—HIGH PRESSURE RELIEF
    VALVE

FIG. 26-9 **Five-cylinder variable displacement type of vehicle compressor** *(Courtesy of Harrison Radiator Division, General Motors Corporation)*

DESIGN/OPERATING FEATURES OF
SWASH (WOBBLE) PLATE COMPRESSORS

Two basic types of automotive/vehicular/*swash plate compressors* include the five-cylinder and six-cylinder models. There are many design features and operating characteristics of the two models illustrated in figure 26-11, as follows.

- Both bodies are built of die cast aluminum for minimum weight.
- Each compressor requires 8 fluid ounces (236 ml) of 525 viscosity refrigerant oil charge and uses R-12 refrigerant.

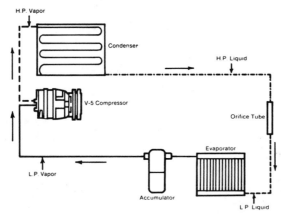

FIG. 26-10 **Location of five-cylinder compressor and liquid/vapor flow in an automotive air-conditioning system** *(Courtesy of Harrison Radiator Division, General Motors Corporation)*

492

- The clutch coil electrical requirements are 12-volt dc, 3.2 amps. However, they may be adapted to 24-volt systems.
- Swash plate compressors are designed for automobiles, trucks, and off-highway equipment and vehicles. The six-pole electromagnetic design permits the use of small diameter pulleys for single-vee or poly-vee pulleys.
- The *low pressure* and *high pressure cutoff switches* and *fan cycling switch* are installed in the high side of the compressor and prevent damage by opening the clutch circuit in case there is total refrigerant charge loss. Also, at low ambient temperatures, the switch opens to avoid compressor operation at temperatures below freezing.
- The *high temperature cutoff switch* opens and turns off the compressor when the discharge pressure exceeds the opening setting.
- The *fan cycling switch* senses the high side pressure and controls the turning on and off of the condenser fan, as required.

Features of Five Cylinder, Variable Displacement Compressors

This compressor, figure 26-11A, is designed to automatically adjust displacement to meet vehicle air-conditioning demands, without cycling. The compressor mechanism consists of a variable angle swash (wobble) plate. There are five axially related cylinders. The compressor senses the evaporator load and changes the displacement to match the load.

A bellows-actuated control valve (positioned in the rear head of the compressor) senses compressor suction pressure and acts as the center of compressor displacement.

The crankcase suction pressure differential controls (in turn) the wobble plate angle and the compressor displacement. The control pressure is lowered with an increase in ambient temperature.

Six Cylinder Swash Plate Compressors

Six cylinder, axial type, (automotive and vehicular) *swash plate compressors* require three sets of opposing cylinders and three double-ended acting pistons, figure 26-11B.

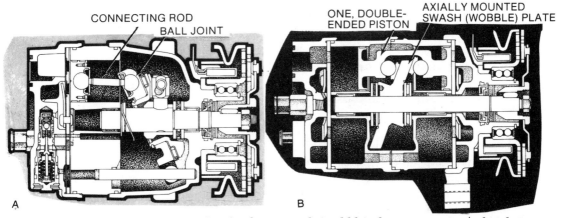

FIG. 26-11   Internal construction details of two swash (wobble) plate automotive/vehicular compressors *(Courtesy of Harrison Radiator Divison, General Motors Corporation)*

The pistons are actuated by a revolving swash plate that is set at an angle on the shaft assembly.

Rotary Van Compressor (Vane Rotary)

Vane rotary compressors provide the greatest cooling capacity per pound of compressor weight in comparison with any other currently available automotive air-conditioning compressor. There are no pistons in a vane rotary compressor and only one valve, a discharge valve. The discharge valve actually serves as a check valve preventing high-pressure refrigerant vapor from entering the compressor during the "off" cycle (or when the compressor is not operating).

Rotary vane compressors and operating characteristics are not new. Two common types have been used for non-automotive applications for years. One rotary vane type has vanes that rotate with the shaft. The second type has stationary vanes.

CONDENSERS

The condenser is mounted in front of the car radiator. The discharge line from the compressor to the condenser may have some kind of mounted vibration absorber. The condenser may be a one-, two- or three-pass finned tube type, usually made of aluminum or copper.

Since all air moving into the engine compartment passes through the condenser first, it is important to keep the condenser clean in order to maintain the efficiency of the air-conditioning system.

RECEIVERS-DRIERS, REFRIGERANT LINES AND EVAPORATORS

A *receiver* is used in most automotive air-conditioning systems. The receiver is positioned between the condenser and the evaporator. The receiver is designed to:

- Store liquid refrigerant during service operations
- Allow for some changes in refrigerant charges and in liquid volume caused by expansion and contraction of the refrigerant as the temperature changes

The receiver usually has a drier chemical inside the outer container shell. The chemical removes moisture from the liquid refrigerant and holds the moisture from passing through the system. Most driers are designed with a *strainer (filter screen),* which serves to remove dirt particles from the refrigerant and oil.

SAFETY PRECAUTIONS

- Liquid receivers usually have a *safety fusible plug.* This plug will open and release refrigerant that is heated to approximately 350°F.
- Some liquid receivers include a sight glass as part of the liquid line outlet; serving as an aid in establishing whether the unit has a low refrigerant charge. When bubbles appear in the sight glass (after the system has been operating a few minutes), this is usually an indication that the system is low on refrigerant or there are other system problems.

Refrigerant Lines

Special *flexible lines* are used to carry refrigerant in automotive air-conditioning systems. These lines are better known as *hoses*. Hoses, covered with a braid to protect them from damage, are designed to be flexible, to withstand high pressures and to be vibration proof. Refrigeration lines are generally made of copper, steel and aluminum.

SAFETY PRECAUTION

- Refrigerant lines (whether flexible or solid) must be carefully routed to prevent rubbing against any part of the car. Wear and corrosion that results from rubbing can cause a leak at the point of contact.

Refrigerant lines are connected to the system in one of three ways: (1) *hose-clamp* fittings, (2) *"O"-ring* fittings, or (3) *flared* fittings. Each of these fitting designs is illustrated by a line drawing, figure 26-12.

Evaporators and Core Assemblies

The evaporator serves to filter and to dehumidify passenger compartment air. Evaporators are generally mounted in a *plenum chamber* and are attached to the engine

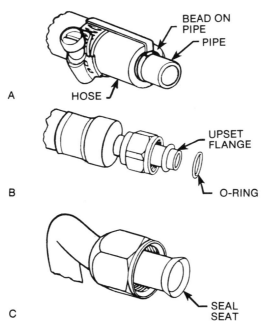

FIG. 26-12 Basic fitting designs for connecting refrigerant lines: (A) hose clamp fitting, (B) O-ring fitting and (C) flare fitting (*Courtesy of Harrison Radiator Divison, General Motors Corporation*)

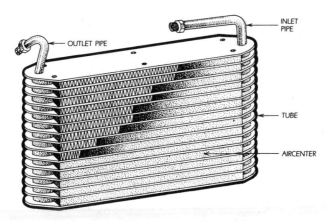

FIG. 26-13   Design features of a staggered rib evaporator core assembly *(Courtesy of Harrison Radiator Division, General Motors Division Corporation)*

compartment, fire wall or dashboard. Automotive air-conditioning evaporators are of the finned, forced-convection type. The metal or plastic housing that encloses the evaporator serves as a plenum and duct for the conditioned air. A moisture pan and/or drain pipe is a part of the unit.

Evaporator cores are generally designed with *parallel* or the newer *staggered rib* assembly feature. *Tubes* and *air centers* are alternately attached as illustrated in figure

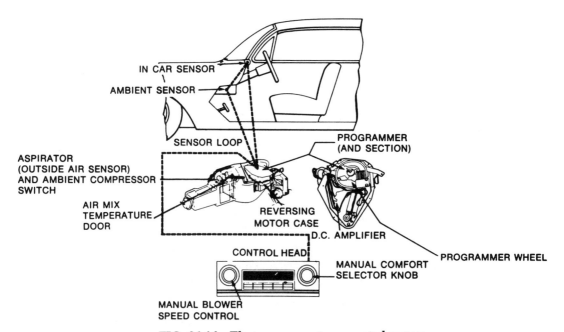

FIG. 26-14   Electroservo motor-operated system

26-13. The inlet and outlet pipes are welded to the core. Although the staggered rib type is lighter in weight, it provides the same heat transfer efficiency as the parallel rib type.

AUTOMATIC CONTROLS OF TEMPERATURE AND CLIMATE

Selected passenger heating and cooling needs may be maintained by a dial-adjustable electronic sensing control unit. Three common systems are used for automatically controlling the temperature and climate within the automobile.

Car producers use adaptations of these and other systems to automatically select the heating or cooling mode and proportion the airflow to maintain a desired temperature regardless of changing conditions outside the automobile compartment.

Electropneumatic System

In the *electropneumatic system*, the outside air temperature, the temperature conditions where the passengers ride, and the car temperature as set on the selector dial, are all monitored by an electrical circuit *(thermistor-resistor loop)*.

The monitored (sensing) information from the loop circuit is processed continuously by a *transistorized computer-amplifier*. The variations in electric current from the amplifier actuate an *electrovacuum transducer*. In turn, the transducer adjusts the vacuum motor unit which controls the heating and cooling levels and the volume of the incoming air under varying driving and weather conditions.

Thermopneumatic System

Four-season air-conditioning systems process hot, humid, dusty, pollen-laden, cold, rainy, snowy and icy air. The automatic climate control system is known as a *thermopneumatic aspirator and vacuum pressure system*. It uses the mechanical principles of the thermostat to monitor vacuum motors which adjust the air valves and switches.

These adjustments regulate the heating and cooling and the volume of air. This sensing device automatically repeats conditions selected on the control head to make the passenger compartment comfortable regardless of driving and weather conditions.

Electroservo System

Another type of automatic temperature control air-conditioning system is motor operated, figure 26-14. A dial-adjustable sensing device permits the selection of comfortable conditions. The unit has two *thermistors* to monitor and to sense both the outside air and the air in the passenger compartment.

The signals from the sensors are amplified in a solid state dc amplifier which operates a reversing motor to maintain the conditions set on the control head. The motor-operated device automatically regulates and repeats the air-conditioning requirements established in the car.

APPLICATION OF AIR CONDITIONING TO VEHICLES OTHER THAN PASSENGER CARS

Extensive needs have developed for installing A/C systems in automotive equipment other than passenger cars. Notable among these are the installations for buses, station wagons, vans, and suburban vehicles.

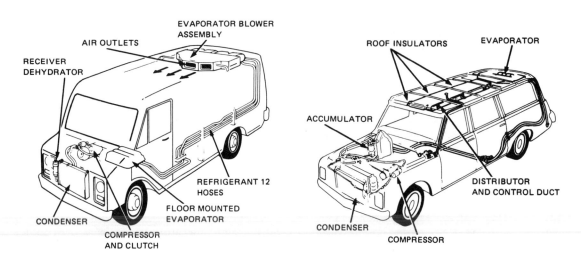

(A) Installation in van   (B) Installation in suburban

FIG. 26-15   Typical A/C automotive installations

Two typical automotive air-conditioning installations are shown in figure 26-15. While identical components are needed, note that the main differences in design and installation are in the locations of the evaporator and the duct work.

SAFETY PRECAUTIONS

- Always exhaust discharge vapor away from the automobile and to the outside. Dangerous chemicals may be produced if large quantities of refrigerant are sucked into the engine or discharged in a room with a live flame.
- Steam cleaning a car or welding should be done at a distance from the refrigeration system. Dangerous and damaging pressures can be built up from the heating action of steam cleaning or welding if either is done near any refrigeration lines or components.

AUTOMOBILE AIR-CONDITIONING TROUBLESHOOTING (SERVICE) CHART

Servicing an automobile air conditioner is not much different than servicing any other air-conditioning unit. Servicing usually begins with a customer-identified problem. The most common problems relate to: (1) a noisy system or compressor, (2) intermittent cooling, (3) insufficient cooling, (4) no cooling, and (5) vibration. Since there may be several causes for each problem, the entire system needs to be checked thoroughly to find the correct cause.

The troubleshooting chart which follows may be used to identify the basic problems and possible corrections. Other manufacturer's troubleshooting charts may be used for specific systems and products.

TROUBLE/PROBLEM	COMMON CAUSE	REMEDY
1. NOISY SYSTEM	Loose belts; squealing	Tighten belts without overtightening
	Worn or frayed belts; noisy	Replace the belts
	Clutch "chatter" (electrical problem)	Check the wiring connections. Repair, as required
	Defective clutch coil	Replace the clutch coil
	Defective clutch or bearings	Replace the clutch or bearings
	Loose compressor mount or braces	Tighten bolts on compressor mount or braces
	Broken compressor mount or braces	Replace compressor mount or braces
	Blower fan hitting	Adjust or reposition blower fan
	Defective idler pulley bearing(s)	Replace idler pulley assembly or bearing(s), as required
2. NOISY COMPRESSOR	Overcharge or undercharge of refrigerant	Purge the system for overcharge until the charge is correct. Locate the leak for undercharge; repair and recharge to correct charge
	Overcharge or undercharge of oil	Correct the oil charge to the required level. Remove excess oil or locate and repair the leak.
	Defective compressor	Check for defect. Repair or replace the compressor
3. INTERMITTENT COOLING	Defective blower motor or control	Replace the blower motor or switch
	Defective clutch coil	Replace the clutch coil
	Loose belt(s)	Tighten belt(s) without overtightening
	Loose ground connection at the clutch coil or clutch brush set	Tighten or replace the ground connection
	Clutch slipping	Check voltage and clutch parts. Correct the voltage; replace the worn parts.
	Thermostat improperly adjusted	Reset the thermostat

TROUBLE/PROBLEM	COMMON CAUSE	REMEDY
	Defective thermostat	Replace the thermostat
	Defective low- or high-pressure control	Replace pressure control(s), as required
	Defective suction pressure regulator	Replace the suction pressure regulator
	Moisture in the system	Purge. Replace drier. Evacuate. Recharge the system
4. IMPROPER COOLING	Defective blower motor	Replace the blower motor
	Clutch slipping; excessive wear	Replace worn clutch parts
	Defective thermostat; short cycle	Replace the thermostat
	Defective low pressure control	Replace the low pressure control
	Defective suction pressure regulator	Replace the suction pressure regulator
	Inadequate air flow from evaporator	Check sticking or binding blend door
	Inadequate air flow over the condensor	Clean the condensor
	Partially clogged screen in the receiver-drier	Replace the receiver-drier
	Partially clogged screen in the expansion valve	Replace the drier
	Loose thermostatic expansion valve bulb	Clean the area. Tighten the bulb. Retape
	Defective thermostatic expansion valve	Replace the valve
	No insulation on TXV remote bulb	Insulate bulb
	Refrigerant overcharge or undercharge	Purge system to correct charge level. Locate leak for undercharge; repair, and recharge to correct level
	Partially clogged accumulator	Replace the accumulator
	Partially clogged receiver-drier	Replace the accumulator

TROUBLE/PROBLEM	COMMON CAUSE	REMEDY
5. NO COOLING	Blown fuse	Check and correct problem. Replace the fuse
	Broken wire	Repair or replace
	Defective clutch coil	Replace the clutch coil
	Defective blower motor	Replace the blower motor
	Defective thermostat	Replace the thermostat
	Defective low pressure control	Replace the low pressure control
	Loose or broken drive belts	Tighten, if loose; replace, if broken
	Defective compressor or suction or discharge plates	Replace plates or compressor
6. IMPROPER COOLING DUE TO LOW REFRIGERANT	No refrigerant or low charge	Check for leak. Repair. Evacuate and recharge
	Leaking compressor shaft seal	Replace shaft seal and gasket set
	Defective hose	Remove and replace hose
	Leaking fusible plug	Replace without trying to repair the fusible plug
	Refrigerant leak in system	Check for leak. Correct the problem.
7. IMPROPER COOLING DUE TO LOW OR NO REFRIGERANT FLOW	Plugged line or hose	Clean the line or hose. Replace, if needed
	Plugged screen (inlet) on the expansion valve	Clean the screen. Replace the drier
	Defective thermostatic expansion valve	Replace the thermostatic expansion valve
	Clogged screen in accumulator	Replace the accumulator
	Clogged expansion tube	Clean or replace the tube. Replace the accumulator
	Clogged screen in the receiver-drier	Replace the receiver-drier
	Excessive moisture in the system	Replace the drier. Evacuate the system. Recharge.
	Defective suction pressure control	Replace the suction pressure control

SUMMARY

- Air conditioning is a term embodying
  - Heating
  - Cooling
  - Dehumidifying
  - Removal of air impurities
  - Elimination of noise
  - Provision for human health, comfort, safety, and security
- Factory-installed automotive A/C systems take in outside air or recirculate inside air as conditions require. Add-on or field-installed systems normally circulate only air from inside the passenger compartment.
- The basic components in automotive A/C systems are identical to any mechanical refrigeration system: system controls, compressor, condenser, metering device, and evaporator.
- The design and operation of automotive A/C systems depend on the application of scientific principles of conduction-convection, evaporation, and radiation to control humidity (moisture), temperature, and air currents.
- The three basic air-conditioning compressors include the double-action piston and the two-cylinder V and in-line reciprocating piston types. The pistons of the double-action compressor are moved by a swash plate instead of a crankshaft.
- An electromagnetic clutch, when energized by the system as refrigeration is called for, couples the drive pulley to the compressor shaft to cause the compressor to start pumping refrigerant.
- Evaporator temperature is controlled by:
  - Cycling the compressor with a thermostatic switch, primarily for add-on units.
  - An evaporator pressure control system in which the compressor runs continuously.
- Three common types of evaporator pressure controls are:
  - The suction throttling valve (STV)
  - Pilot-operated absolute valve (POA)
  - Evaporator-pressure regulator valve (EPR)
- The electropneumatic, thermopneumatic, and electroservo systems sense and control outside air temperature and the heating-cooling control panel requirements in the passenger compartment.
- The A/C system condenser liquefies the very hot, high-pressure refrigerant vapor to a high-pressure warm liquid for the receiver-dehydrator.
- Cooled high pressure liquefied refrigerant is stored, filtered and has moisture removed in the receiver-dehydrator.
- The evaporator cleans, dehumidifies and cools air before it enters the passenger compartment.
- The ambient switch, thermal limiter, superheat switch, and low-pressure cutoff switch are protective devices which either prevent unnecessary compressor operation or other damage resulting from refrigerant or pressure losses.
- The electrical circuitry of the A/C systems provide for current flow for actuating and engaging the clutch coil and to protect the components against damage in case of a malfunction in the system.

- Identical components may be adapted in air-conditioning vehicles other than passenger cars by simply changing the design of the ducts and the location of the evaporator.

ASSIGNMENT: UNIT 26 REVIEW AND SELF-TEST

A. NEWER APPLICATIONS OF REFRIGERATION

1. Identify two new or projected applications or functions of refrigeration to the fields of: (a) medicine, (b) food preservation, and (c) industry.
2. Describe in brief outline form how one projected application of refrigeration in a major industry will revolutionize either the industrial processes or products.

B. HUMAN CONSIDERATIONS IN PLANNING AUTOMOTIVE A/C SYSTEMS

Supply the correct scientific term to complete statements 1 through 3.

1. Hot, moisture-saturated air retards the _____ of skin moisture.
2. A surface or material that is hotter than the body skin _____ heat to the body.
3. Air currents, regardless of whether they are warmer or cooler than the body heat, provide for heating or cooling the body by _____ .

C. AUTOMATIC CONTROL OF TEMPERATURE AND CLIMATE

Select either an electropneumatic, thermopneumatic or electroservo system for the automatic control of an automotive A/C system.

1. Secure and submit a circuit diagram (or similar technical manual drawing or copy) of the control system.
2. Identify the major parts within the controls by using arrows and naming the parts on the diagram.
3. State in outline form what the functions are for (a) the entire control system and (b) each major part.

D. MECHANICAL COMPONENTS OF AUTOMOTIVE A/C EQUIPMENT

Review the illustration on page 504 containing the major components of a typical automotive A/C installation.

1. Name the mechanical components of the system, labeled (A) through (H).
2. Describe the functions of each component identified by letter in the system illustrated.
3. Give two characteristics of R-12 that are important to the lubrication of the moving parts.
4. Select one basic type of compressor from a technical manual.
   a. Make a simple sketch showing how the electromagnetic clutch actuates the compressor.
   b. Describe how the compressor provides the necessary refrigerant to the system.

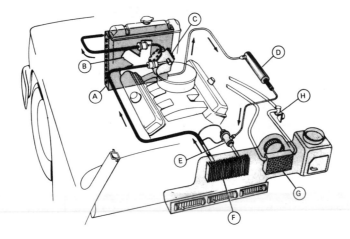

5. Select the functions and/or component location in Column II which describes accurately each component in column I.

Column I
Component

Column II
Functions and/or Component Location

a. Thermostatic expansion valve
b. Suction throttle control
c. Evaporator-pressure regulating valve
d. Thermostatic switch
e. Ambient switch
f. Thermal limiter

(1) Controls pressure clutch engagement to prevent operation when outside temperature is too low.

(2) Installed on the discharge side of the compressor to reduce characteristic compressor noises.

(3) Located in the heater inlet hoses to regulate the flow of coolant to the heater core.

(4) A metering device regulating the supply of liquid refrigerant to the evaporator.

(5) Reacts to temperature changes to intermittently operate the compressor.

(6) A low pressure cutoff switch in the receiver-dehydrator which opens due to a refrigerant pressure drop.

(7) Senses the actual evaporator operating pressure and regulates the cool air output of the system.

(8) Installed in the compressor suction passage, it opens and closes to regulate the flow of refrigerant.

(9) Senses low system pressure and high system gas temperature resulting from refrigerant losses.

6. State the function of the following automotive air-conditioning units.
   a. Low-side pressure control valve.
   b. Pressure-operated, hot-gas, bypass valve
   c. Receiver
7. Name five essential parts of an automotive air-conditioning duct system.
8. Describe briefly the difference between the conventional reciprocating and swash plate types of compressor.
9. Cite four advantages of vane rotary compressors over other compressor designs.
10. Name two refrigeration line fittings.
11. State two design features that are built into automotive air-conditioning units or parts as safety precautions.
12. Circle the number of the function in Column II with the correct system in Column I to which the function relates.

Column I	Column II
System	*Function*

a. Electropneumatic
b. Thermopneumatic
c. Electroservo

(1) Heat-sensing device that monitors and actuates vacuum motors
(2) Actuate and control vacuum motors using a dial-adjustable sensing device.
(3) Metering liquid refrigerant to the evaporator
(4) Motor-operated device for regulating an air-conditioning system; requiring an electrical circuit(thermistor-resistor loop)

E. APPLICATION OF AIR CONDITIONING TO VEHICLES OTHER THAN PASSENGER CARS

Obtain a technical manual for a manufacturer of air conditioning equipment and installations for nonpassenger vehicle systems.

1. Select an illustration of the system and show the principal parts by using arrows and part names on a tracing of the system.
2. Cite the reasons for placing certain of the components in the system in locations in the vehicle which are different than in a passenger car.

F. SERVICING AUTOMOTIVE AIR CONDITIONING SYSTEMS

Troubleshoot the ten automotive air-conditioning problems according to the problem areas of the system and the cause of each problem as given in the table.

Section 6: All-Weather Air-Conditioning Systems

PROBLEM AREA	CAUSE OF PROBLEM	CORRECTIVE ACTION
Noises in System	1. Broken belt	
	2. Loose compression braces	
Noises in Compressor	3. Undercharge of refrigerant	
	4. Moisture in system	
Intermittent Cooling of System	5. Circuit breaker trips on overload	
	6. Loose belts	
Insufficient Cooling	7. Sluggish blower motor	
	8. Insufficient air flow from evaporator	
No Cooling	9. No refrigerant or low charge	
	10. Clogged expansion tube	

UNIT 27:
REFRIGERATED DISTRIBUTION OF ELECTRICITY

Until recently, experiments concerning temperature and its effects have been conducted largely to attain goals at both the high and the low ends of the scale. Particular emphasis has been directed toward attaining absolute zero, the temperature at which all molecular motion is thought to cease. Researchers are successfully reaching lower and lower temperatures as scientists succeed in liquefying more and more gasses.

CRYOGENICS

The temperatures at which various liquids boil at atmospheric pressure are shown in figure 27-1. Temperatures even closer to absolute zero are obtained by reducing the

Fluid	Fahrenheit F	Rankin R	Celsius C	Kelvin K
Water	212	672	100	373
SO_2	14	474	-10	263
R-12	-22	438	-30	243
NH_3	-28	432	-33	240
R-22	-41	419	-43	230
Carbon Dioxide	-109	351	-78	195
R-1150 (Ethylene)	-135	325	-93	180
Beginning of the Cryogenic Range	-250	210	-157	116
Methane	-258	202	-161	112
Oxygen	-297	163	-183	90
Air	-313	147	-192	81
Nitrogen	-320	140	-196	77
Hydrogen	-423	37	-253	20
Helium	-452	8	-270	3
Absolute Zero	-460	0	-273	0

FIG. 27-1 Boiling temperatures of liquids at atmospheric pressure

pressure on a liquid below atmospheric to reach (for example with liquid helium) 0.7°K or –458.4°F. Values still closer to absolute zero (0.001°K or lower) were reached by using liquid helium to cool paramagnetic salts and then using a magnetic field.

The science of producing and applying temperatures below –250°F is called cryogenics. The term *cryogenics* also includes engineering applications as well as the development of instruments, special insulation, and containers for transporting the liquids with a minimal loss by evaporation. Cryogenics is used for low temperature thermometry, LASER frequency tuning, and spectrocopy applications. Other examples in electronics include the cooling of low-noise microwave amplifiers for radar and radio astronomy and satellite communications.

SUPERCONDUCTORS

Of special interest is the startling effect that cryogenic temperatures have upon electrical resistance. Up to this time, regardless of the efficiency of an electrical conductor, it still offered resistance to the flow of current. At ordinary temperatures the resistance causes line drop, reduces the voltage available to operate equipment, and results in energy losses.

Dramatically, a group of metals actually loses all resistance at temperatures near absolute zero. These metals become perfect conductors and are called *superconductors*. A partial list includes zinc, cadmium, tin and lead, to name the familiar ones. Among the newer metals is *niobium*. Strangely, silver and copper (the better conductors at ordinary temperatures) do not become superconductors at extremely low temperatures. As a result, the theories concerning electricity in relation to conductivity are being restudied.

FIG. 27-2 Concept of proposed cryogenic rechill center on an underground transmission line (*Courtesy of Electrical World, McGraw-Hill Publishers*)

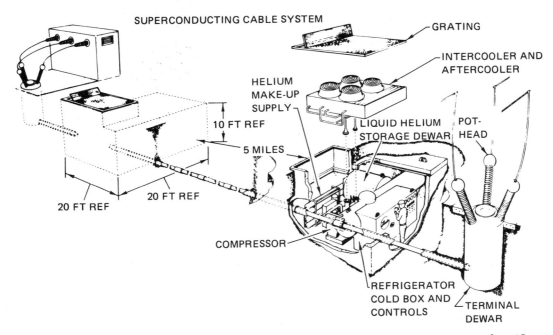

SUPERCONDUCTING CABLE SYSTEM
GRATING
INTERCOOLER AND AFTERCOOLER
HELIUM MAKE-UP SUPPLY
LIQUID HELIUM STORAGE DEWAR
POT-HEAD
10 FT REF
5 MILES
20 FT REF
20 FT REF
COMPRESSOR
REFRIGERATOR COLD BOX AND CONTROLS
TERMINAL DEWAR

FIG. 27-3   Sketch of proposed installation of an underground superconducting ac power line *(Courtesy of Cryogenics, IPC Science and Technology Press)*

REFRIGERATED DISTRIBUTION OF ELECTRICITY

The development of cryogenic conductors is one answer to the problem of how to supply the increasing demand for electrical power and in a more aesthetic way meet ecological standards.

Figure 27-2 represents a concept of a transitional facility. The usual high-voltage transmission lines bring power across the country to the substation where the transition is made to an underground, refrigerated cryogenic conductor distribution circuit.

Basic to such a system is the circulation of a cryogenic fluid to maintain the cable at the desired low temperature for maximum transmission of power with minimum resistance losses. Efforts are being made to design equipment using nitrogen and helium as the refrigerants.

Figure 27-3 shows additional details of the proposed refrigeration stations. The two stations are envisioned as being about five miles apart. The refrigeration system at each station will pump the refrigerant out and back in a 2 1/2-mile *refrigeration loop,* keeping the pumping pressures within reasonable limits.

This particular system uses helium. Although many cryogenic systems are in the engineering/developmental stage, installation and operation of distribution lines were started in the mid-1980s.

The major components of the system include the electrical conductor, cryogenic enclosure, terminal Dewars at both ends of the cable to isolate the cryogenic enviroment and permit the individual conductor phases to be connected into thermally and

509

electrically graded pothead assemblies, and the refrigerator pumping stations. The stations provide a continuous supply of cryogenic refrigerant to intercept and remove heat loads under the cryogenic operating conditions. Although the basic components of the nitrogen and helium systems are similar, there are fundamental differences in the way in which each component functions.

Nitrogen Resistive Cryogenic System

The nitrogen system is referred to as the *resistive cryogenic system*. It usually incorporates aluminum or copper as the conductor in order to take advantage of the decreased electrical resistance of these metals at cryogenic temperatures. The graph in figure 27-4 shows a comparison of resistivity at temperatures from 10 K to 300 K for high-purity copper and aluminum.

Temporary or seasonal overloads are handled easily with this system where the nitrogen is in direct contact with the conductor and the temperature is not critical.

Helium Superconductor Cryogenic System

This system takes advantage of the fact that certain metals have zero resistance when at a temperature that is within a few degrees of absolute zero. Helium is the proposed refrigerant and niobium, or an alloy of niobium, the superconductor. It is very likely that the niobium will be deposited on a copper or aluminum hollow conductor, figure 27-5. This is done in order to handle overloads that might otherwise elevate the temperature enough to cause it to revert to a resistive state, which is only slightly above the temperature at which the helium will be held.

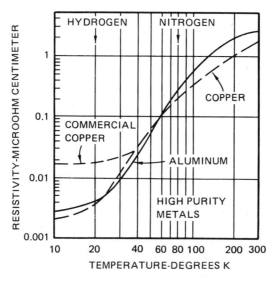

FIG. 27-4 **Decreasing resistivity for high-purity copper and aluminum at cryogenic temperatures** *(Courtesy Cryogenics Branch, General Electric Research and Development)*

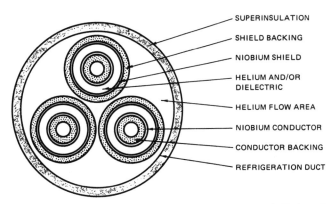

FIG. 27-5 Superconductors in a coaxial pattern *(Courtesy of Union Carbide Corp. Linde Division)*

The deposition on a hollow conductor takes on added significance because the thickness of the niobium or its alloy will be approximately 0.002 of an inch. Even this thin coating will be expected to carry somewhere in the neighborhood of 10,000 amperes of current. Importantly, when niobium is at the temperature of helium, the only factor that limits the amount of current it can carry is the magnetic field that accompanies the current. The magnetic field can be controlled by making the diameter of the conductor of appropriate size, requiring a hollow conductor.

Another problem confronting researchers and designers is thermal expansion and attendant stresses that will occur when the system is first put into operation and in reducing the system temperature from room temperature to one very close to absolute zero. This change will be in excess of 55°F, depending upon station temperature.

Figure 27-6 shows one design that is being investigated. In all probability the copper on which the niobium is deposited will have an invar core because invar has a very low coefficient of expansion.

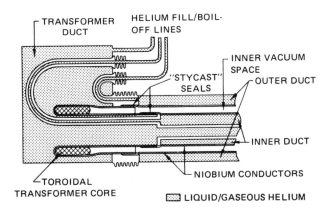

FIG. 27-6 Transformer termination of a superconducting link *(Courtesy Cryogenics, IPC Science and Technology Press)*

FIG. 27-7 Unit for transporting liquid hydrogen *(Courtesy of Union Carbide Corp. Linde Division)*

Cryogenic Envelope. The containment of the cryogenic refrigerant is a critical problem. First, helium (due to its very low molecular weight) intensifies the problem of leakage due to thermal expansion at the nonwelded joints and terminals. Second, the great difference in temperature between that of helium and the station temperature will create a tendency for heat to leak through the envelope toward the helium. This will be in addition to the heat developed by the electricity in the conductor. Helium, as the refrigerant in this case, will need to carry heat away from the conductor. Superinsulation is needed to minimize the job of heat conduction from the outside and to reduce the size of equipment required.

Superinsulation. Liquedfied gases have been produced and transported for many years. Figure 27-7 shows one of the units that has been designed for transporting large quantities of liquid hydrogen. Superinsulation is installed in the walls of this tank car as a special insulation. A range of thermal conductivity values for four selected insulating materials is given in figure 27-8.

Though polyurethane foam insulation as used for insulating refrigerated transport applications was found superior to previously used materials, the large temperature differences between the liquefied gas on the inside of the container and the ambient temperature outside requires insulation with superior qualities.

Superinsulation consists of alternate layers of radiation shields and spacer material, operating in a high vacuum. This design is similar to a thermos bottle which, in turn, is a commercial pattern of the Dewar flask. Superinsulation has a larger number of reflective surfaces (shields) than the original Dewar flask. The spacer material is a paper which holds the reflective foils apart.

The vacuum is maintained by the presence of a *chemical getter* and/or *molecular sieve*. At low temperatures, particularly, the molecular sieve absorbs continually any *outgassing* contaminates as well as inleaking gasses to maintain a vacuum at less than one micron of mercury.

Heat Interceptor. Even with the best insulation there is some heat that leaks in slowly. The amount of heat reaching the superconductor may be minimized with either a liquid nitrogen shield or multishield *heat interceptor*. This improves the economics of the operation.

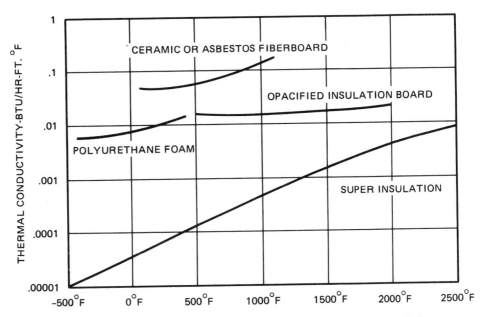

FIG. 27-8 **Examples of thermal conductivity values of insulating materials for ranges between room temperature and stated temperatures** *(Courtesy of Union Carbide Corp. Linde Division)*

Liquid Nitrogen Shield. A layer of liquid nitrogen which completely surrounds the cryogenic cable configuration will be at a temperature where it will intercept any heat leakage from the outside and will reduce the temperature between itself and the cryogenic refrigerant surrounding the cable.

Multishields. Another possibility will be to use thin metal shields that have high thermal conductivity at an intermediate temperature which is produced by a small fraction of helium refrigerant that is allowed to bleed off. An example of this design is illustrated by the comparatively small container in figure 27-9. The multishields are mechanically bonded to the vent tube. As the vaporized helium is vented, it absorbs heat from the multishields and cools them. The multishields, in turn, will absorb heat that leaks in from the outside. The multishields provide an alternate path for the heat so it does not reach the conductor.

Cryogenic Envelope for Superconducting Cable. One concept of superinsulation and thermal shields in the design of the cryogenic containers, that has proven successful when applied to the cryogenic shield for a superconducting cable, incorporates the features shown in figure 27-10.

The vented helium, instead of passing off into the atmosphere, is collected in a manifold and returned to the refrigeration equipment.

Refrigeration Cycle. Refrigeration will be required to remove from the liquid refrigerant any heat that is produced in the transmission and distribution of electrical energy plus any heat that leaks in through the insulation or from other sources.

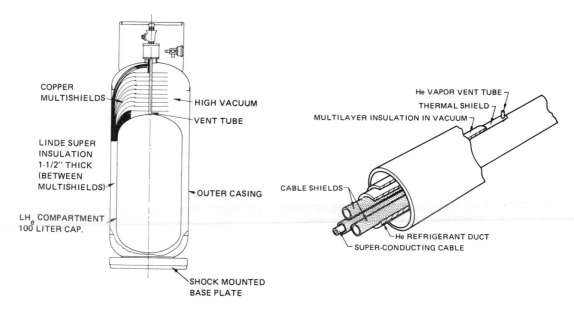

FIG. 27-9 Multishielded storage and shipping container for liquid helium *(Courtesy Union Carbide Corp. Linde Division)*

FIG. 27-10 Conceptual features for a multishielded pipeline for ac superconducting cable

Critical Temperature.   All refrigerants are liquefied by the removal of heat energy. However, the cryogenic refrigerants such as helium, hydrogen, nitrogen, and others, are referred to as *permanent gasses*. There is a temperature above which a gas cannot be liquefied regardless of the amount of pressure that is applied. That temperature is called the *critical temperature*. For ordinary refrigerants the critical temperatures are above 200°F, with the exception of carbon dioxide.

By contrast, the critical temperatures of the common cryogenic refrigerants are: nitrogen, –232.6°F; hydrogen, –394.4°F; and helium, –450.4°F, within less than 10F degrees of absolute zero.

This means that each gas must be cooled below its critical temperature before it can be liquefied. The *critical pressure* is the pressure needed to liquefy the gas at the critical temperature. The refrigeration equipment necessary to keep these refrigerants at cryogenic temperatures is more complex than that needed for the common refrigerants and their applications.

Refrigeration Equipment for Cryogenic Cables.   Restated, the primary purpose of using refrigeration equipment with cryogenic cables is to maintain the cryogenic refrigerant at a temperature below the critical temperature of the cryogenic conductor, by removing heat from the refrigerant. In addition, refrigeration is needed to reliquefy any of the refrigerant gas that has been used in providing a heat interceptor, and/or make up (replace) any refrigerant that is lost through leaks.

LIQUEFICATION OF AIR AND NITROGEN

Air was successfully liquefied before the turn of the century by Karl R. von Linde. The air was compressed at high pressure, cooled by a freezing mixture, and allowed to expand at low pressure. This produced additional cooling. The air thus cooled was circulated back over the high pressure side to further reduce the temperature of the high-pressure gas. The expanded gas was again circulated through the compressor, getting cooler each cycle. Finally, a small part of the supercooled air condensed and collected in the bottom of the Dewar container.

Figure 27-11 shows such a system. Figure 27-12 illustrates a newer system that produces a better *yield* by the addition of another heat exchanger for additional cooling and an *expander* or *expansion engine*. The purpose of the expander is to allow the gas to do external work in expanding in order to remove more heat energy.

When the gas is permitted to expand through an expansion valve (restriction or *porous plug*), it does *internal* work and thus is cooled. This is referred to as the *Joule-Thompson effect* or *Joule-Thompson cooling*.

Nitrogen can be separated from the liquid air by fractional distillation. Also, gaseous nitrogen can be liquefied by the same process as for air.

Mechanical Systems for Liquefying Gases

A multistage refrigeration system in which a single refrigerant is circulated through two or more compressors in series is used for refrigerating some gases. Figure 27-13

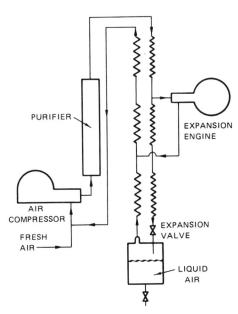

FIG. 27-11 Hampson (simple Linde) system for air liquefication (*Courtesy John Wiley and Sons, Inc.*)

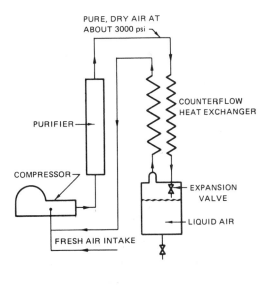

FIG. 27-12 Claude system for air liquefication (*Courtesy of the United States Atomic Energy Commission*)

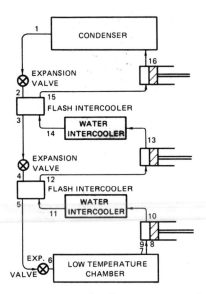

FIG. 27-13   Three-stage system for the production of low temperature

shows a system which is used to produce low temperature. The low temperature chamber is the evaporator of the system. The water intercoolers are used between the compressors.

The *cascade system* is another mechanical one for achieving low temperatures. It has two or more refrigerating systems in which the evaporator coil of one system cools the condenser of the other as shown in figure 27-14. Ordinarily, different refrigerants are

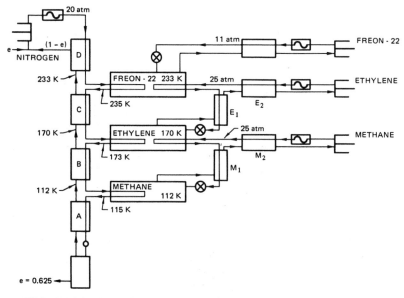

FIG. 27-14   Cascade system using R-22, ethylene and methane

used in each part. This allows the last unit to operate at lower temperatures and pressures than would be possible with a single system.

LIQUEFICATION OF HELIUM FOR SUPERCONDUCTING CRYOGENIC CABLES

Helium was successfully liquefied in 1908 at –268.9°C by the Dutch physicist Kammerlingh Onnes. Present day techniques for the liquefication of helium to produce limited yields of liquid helium combine or employ most of the processes described earlier. Fortunately, only small quantities of helium will need to be reliquefied in the superconducting cryogenic cable, plus the additional cooling that will be needed to maintain the superconductor below its critical temperature.

In a modern helium liquefier-refrigerator unit, the *liquefier-refrigerator* cools helium in a successive series of stages to –273°C and collects a supply of liquid helium at that temperature. A working diagram of a helium liquefier-refrigerator system in cooling two superconducting magnets is shown in figure 27-15.

Helium-Gas, Closed-Cycle Refrigeration Systems

Today, research and production requirements extend to the attainment of extremely low sub-zero temperatures approaching *absolute zero* (–459°F, –273°C, O K).

Operating effectively and consistently down to temperatures of 6 K requires the use of a *working fluid* (helium gas) that remains fluid at temperatures near absolute zero and highly efficient *heat exchangers*.

Closed-cycle helium refrigerators, figure 27-16, that operate at low cryogenic temperatures require an air-cooled, oil-lubricated compressor unit with an oil separation system and room-temperature valves and seals.

Extremely low temperatures are produced by a *regenerator*. This unit extracts heat from the incoming gas, stores it, and then releases the gas to the exhaust stream. A

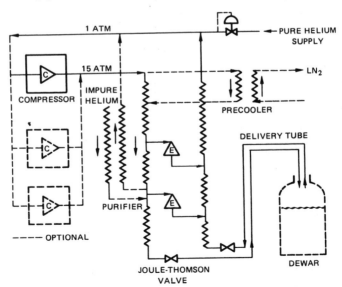

FIG. 27-15 Working diagram of helium liquefier/refrigerator *(Courtesy of CTI-Cryogenics)*

regenerator is a *reversing-flow heat exchanger.* In other words, helium alternately passes through the heat exchanger in either direction. The regenerator readily accepts heat from the higher temperature helium and gives up heat to lower temperature helium.

Single-stage Cryodyne® refrigerator/compressor systems are used to produce temperatures in the 30K to 77K (–320°F, –196°C) range. Refrigeration capacity down to 6K is possible by adding a second stage, figure 27-17 (as diagrammed), or third-stage refrigerator.

Multi-Stage, High-Vacuum Helium Refrigeration Systems

Figure 27-18 shows a state-of-the-technology application of the cooling process (cycle) capability of a high-vacuum helium refrigeration system. This particular system provides a continuous ultra-high (UHV) *environment* in which multi-step processes may be carried on in research, analysis, and production.

Self-contained, ultra-high vacuum *chambers* are installed at *critical junctures* in the system. This arrangement provides separate chambers for the introduction of research samples, storage, processing, and removal under UHV conditions.

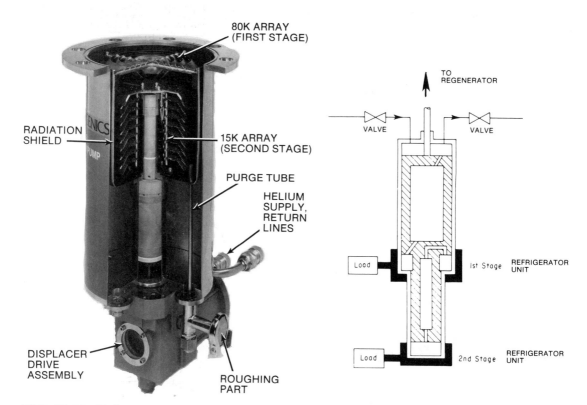

FIG. 27-16 Refrigerator unit of a two-stage 80 K and 15 K), closed loop, cryogenics helium refrigeration system *(Courtesy of CTI-Cryogenics)*

FIG. 27-17 Two-stage refrigeration system with cryogenic temperature capability down to 6 K *(Courtesy of CTI-Cryogenics)*

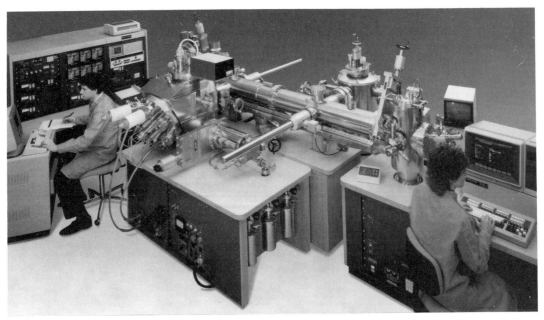

FIG. 27-18 State-of-the-technology research/production application of cooling process capabilities of a multi-stage CRYODINE® helium, high-vacuum refrigeration system *(Courtesy of CTI-Cryogenics)*

Also, the chambers are isolated with *gate valves*. These valves permit each research/production process to be accomplished independently and concurrently. In addition, an *outgassing* technique is used in the load locking of each research sample and holder. Outgassing enhances the cleanliness of the system. The system has capability for "add-on" multiple chambers and multiple-step processes to meet evolving research and production needs.

SUMMARY

- Cryogenics is the science of producing and using temperatures below –250°F.
- Superconductors are those metals whose resistance to the flow of electric current is zero at temperatures close to absolute zero.
- Refrigerated underground systems are being designed for the distribution of large quantities of electricity to centers of high volume need.
- A resistive cryogenic system uses nitrogen to cool the electrical conductors to temperatures where their electrical resistance is very low.
- A cryogenic superconductor system uses helium to cool conductors to within a few degrees of absolute zero where they offer no electrical resistance.
- Good electrical conductors such as pure aluminum or copper may be used with nitrogen in the resistive cryogenic system.
- Good conductors at ordinary temperatures are not necessarily superconductors. New experience and knowledge will necessitate changes in the theory of electrical conductivity.

- Superconductors are lead, zinc, cadmium and tin. Each of these metals has a high resistance to the flow of electricity at ordinary temperatures.
- Niobium, (or one of its alloys) as a superconductor, may be used with helium for the cryogenic superconductor system for the distribution of electrical energy in the future.
- The magnitude of high intensity currents in superconductors is limited only by the magnetic field. A multilayer reflective shield in a vacuum that provides effective insulation for cryogenic temperatures is called superinsulation.
- A multishield made of a high thermal conductivity material and cooled by the cryogenic refrigerant, which is sandwiched between the layers of the superconductor, acts as a heat interceptor and increases the effectiveness of the insulation.
- Liquefication of some gases is achieved with multistage or cascade compressors.
- Liquefication of helium requires additional expansion devices because of its low critical temperature.
- A gas cannot be liquefied at a temperature higher than its critical temperature.
- The complete system for liquefying helium is called a helium liquefier-refrigerator.
- The regenerator of a cryogenic refrigeration system alternately accepts heat from higher temperature helium and gives heat up to lower temperature helium.
- The capability of single-stage, helium-gas, closed-cycle refrigeration systems of 30K to 77K can be increased to 6K and lower temperature levels by adding a second- and third-stage cryodyne refrigerator, as required.

ASSIGNMENT: UNIT 27 REVIEW AND SELF-TEST

A. CRYOGENICS

1. Locate a table of boiling temperatures of liquids at atmospheric pressure in the cryogenic range.
   a. Give the boiling temperature of liquid helium and two other liquids in the cryogenic range in four systems of temperature measurement: Fahrenheit, Rankin, Celsius and Kelvin.
2. Indicate how many Fahrenheit degrees there are in the cryogenic range.

B. NITROGEN AND HELIUM SYSTEMS

1. Identify the major components and their functions in a refrigerated system for the distribution of electricity.
2. Refer to a resistivity chart for high purity copper and aluminum at cryogenic temperatures on the Kelvin scale.
   a. Indicate the microohm centimeter resistivity of copper at
      (1) 200°K,   (2) 100°K,   (3) 20°K,   and (4) 10°K.
   b. Indicate the microohm centimeter resistivity of aluminum at
      (1) 300°K,   (2) 200°K,   (3) 100°K,   and (4) 10°K.

c. State two temperatures at which the resistivity of both copper and aluminum are the same.

d. Compare the resistivity of copper at 30°K to its resistivity at 200°K.

e. State the effect of decreasing the temperature from 200°K to 10°K on the resistivity of copper and aluminum.

C. ELECTRICAL CONDUCTORS AND SUPERCONDUCTORS

1. Select the number(s) of the metal(s) in column II that match(es) each conductive characteristic in column I.

Column I *Conductive characteristic*	Column II *Metal*
a. A superconductor at cryogenic temperatures	(1) Niobium
b. Least resistance of known metals at near absolute temperatures	(2) Aluminum
	(3) Cadmium
c. Best electrical conductor at ordinary temperatures	(4) Zinc
	(5) Silver
d. High resistance to flow of electricity at ordinary temperatures	(6) Tin
	(7) Copper
e. Electrical conductor becomes highly conductive at cryogenic temperatures	(8) Lead

D. REFRIGERATED DISTRIBUTION OF ELECTRICITY

1. Give two advantages of distributing electricity at temperatures close to absolute zero.

2. State two problems related to the containment of cryogenic refrigerant.

3. Select the number of the function or refrigerant in column II that matches the characteristic in column I.

Column I *Characteristic*	Column II *Refrigerant or Function*
a. Resistive cryogenic refrigerant	(1) Heat interceptor
b. Lowest molecular weight	(2) Superinsulation
c. Multishield	(3) Nitrogen
d. Radiation shield in vacuum	(4) Helium
e. Superconductor refrigerant	

4. Describe critical temperature.

5. State the factor that limits current intensity of distributing ac at temperatures close to absolute.

6. Number the following refrigerants in order of the ease with which they can be liquefied: nitrogen, R-12, helium.

7. Select the number of the system in column II which best corresponds with the characteristic in column I.

	Column I *Characteristic*		Column II *System*
a.	Liquefies by pressure and cooling and free expansion of the gas.	(1)	Claude system
b.	Liquefies by pressure, cooling and use of mechanical expander.	(2)	Joule-Thompson
c.	One refrigerant with several compressors.	(3)	Helium Liquefier-Refrigerator
d.	Different refrigerants with several compressors.	(4)	Cascade system
e.	A combination of each system.	(5)	Linde system
		(6)	Multiple-stage system

APPENDIX: HANDBOOK TABLES

Table 1 Physical Characteristics of Eight Common Refrigerants

Selected Physical Data (Performance based on evaporator pressure @ 5°F = 0.02396 psia and condensing pressure @ 86°F = 0.6154 psia)	Refrigerant R-11	Refrigerant R-503	Refrigerant 12**	Refrigerant 22	Ammonia R-717	Water R-718	Carbon Dioxide R-744	-R-502 Mixture of CHClF2 and CClF2CF3
1. Chemical Formula	CCl_3F	$CHF_3/CClF_3$	CCl_2F_2	$CHClF_2$	NH_3	H_2O	CO_2	
2. Molecular Weight	137.4	87.5	120.9	86.5	17.0	18	44.0	1.6
3. Boiling Pt. (°F) at 1 Atm. Pressure	74.9	−126.1	−21.6	−41.4	−28.0	212	−109.3	−49.8
4. Evaporator Pressure at 5°F (psig)	†23.9	16.9	11.8	28.3	19.6	†29.6	316.8	15.3*
5. Condensing Pressure at 86°F (psig)	3.49	132.9	93.3	159.8	154.5	†28.8	1028.3	282.7*
6. Critical Temperature (°F)	388	67	234	205	271.4	706.1	87.8	179.9
7. Critical Pressure (psi absolute)	640	632	597	716	1651.0	3208	1069.9	591.0
8. Compressor Discharge Temp. (°F)	111	14	101	131	210.0	25.68	151.0	99
9. Compression Ratio (86°F/5°F)	6.19	4.65	4.08	4.06	4.94	6.95	3.10	9.9*
10. Specific Volume of Saturated Vapor at 5°F (cu. ft./lb.)	12.2	1.30	1.46	1.25	8.150	1.155×10^{-4}	0.27	0.82
11. Latent Heat of Vaporization at 5°F (Btu/lb.)	83.5	72.7	68.2	93.2	565	1219.9	117	70.75
12. Net Refrig. Effect of Liquid— 86°F/5°F (Btu/lb.)	66.8	53.7	50.0	69.3	474.4	1009.	55.5	45.3
13. Specific Heat of Liquid at 86°F	0.21	∅ 0.28	0.24	0.34	1.10	1.0	.77	
14. Specific Heat of Vapor at Constant Pressure of 1 Atm. & 86°F	0.14	∅ 0.14	0.15	0.15	0.52	0.5	0.2	
15. Coefficient of Performance	5.03	4.10	4.70	4.66	4.76	4.10	2.56	2.01*
16. Horsepower/Ton Refrigeration	0.938	1.16	1.002	1.011	.99	0.62	1.8	2.35*
17. Refrigerant Circulated/Ton Refrig. (lb./min.)	2.99	3.73	4.00	2.89	0.422	0.196	3.6	4.42*
18. Liquid Circulated/Ton Refrig. (cu. in./min.)	56.6	81.6	85.6	68.0	19.6	5.4	167.0	
19. Compressor Displacement/Ton Refrig. (cfm)	36.5	4.83	5.83	3.60	3.44	476.6	0.97	7.17*
20. Toxicity (Underwriter's Laboratories Group No.)	5	6	6	5A	2 Mod.	No	No 5	5A
21. Flammability & Explosivity	None	None	None	None	Yes	No	No	
22. Type of Suitable Compressor	Centrif.	Recip.	Recip./Centrif.	Recip.	Recip.	Centrif.	Recip.	Recip.

∅ @ −30°

† inches of mercury vacuum.
* for evaporating temp. −20°F condensing temp. 120°F return gas temp. 65°F

** R-12 is rapidly being replaced. Physical data is included pending complete changeover.

Table 2   Properties of Saturated Water (R-718)

Temp °F	Pressure		Specific Volume		Enthalpy-Heat Content		
	Gauge in. Hg vac.	Absolute in. Hg abs.	Sat. Liquid	Sat. Vapor	Btu/lb. from 32°F		
			cu. ft./lb.		Saturated Liquid	Latent	Saturated Vapor
-40		0.0038	0.01737	134,430	-177	1043.	1043.
-30		0.0070	0.01738	74,410	-173	1048.	1048.
-20	29.91	0.01	0.01739	42,370	-168	1052.	1052.
-10	29.90	0.02	0.01741	24,750	-164	1057.	1057.
0	29.88	0.04	0.01742	19,090	-159	1061.	1061.
5	29.87	0.05	0.01743	14,810	-157	1063.	1063.
10	29.86	0.06	0.01744	9,060	-154	1066.	1066.
20	29.82	0.10	0.01745	5,662	-149	1070.	1070.
30	29.76	0.16	0.01747	3,608	-144	1074.	1074.
32	29.74	0.18	0.01602	3,304	0.00	1075.	1075.
40	29.71	0.25	0.01602	2,445	8.	1071.	1079.
50	29.56	0.36	0.01602	1,704	18.	1066.	1083.
60	29.40	0.52	0.01603	1,207	28.	1060.	1087.
70	29.18	0.74	0.01605	868	38.	1054.	1092.
80	28.89	1.03	0.01607	633	48.	1049.	1096.
86	28.67	1.25	0.01609	527	54.	1043.	1097.
90	28.50	1.42	0.01610	468.	58.	1043.	1100.
100	27.99	1.93	0.01613	350.	68.	1037.	1105.
110	27.32	2.60	0.01617	265.	78.	1032.	1109.
120	26.47	3.45	0.01620	203.	88.	1025.	1114.
130	25.39	4.53	0.01625	157.	98.	1020.	1118.
140	24.04	5.88	0.01629	123.	108.	1014.	1122.

Table 3   Properties of Saturated Ammonia (R-717)

Temp °F	Pressure		Volume	Density	Enthalpy-Heat Content		
	Gauge lb./sq. in.	Absolute lb./sq. in.	Vapor cu. ft./lb.	Liquid lb./cu. ft.	Btu/lb. from -40°F		
					Liquid	Latent	Vapor
-40	8.7*	10.41	24.86	43.1	0.0	597.6	597.6
-30	1.6*	13.90	18.97	42.65	10.7	590.7	601.4
-28	0.0	14.71	18.00	42.57	12.8	589.3	602.1
-20	3.6	18.30	14.68	42.2	21.4	583.6	605.0
-10	9.0	23.74	11.50	41.8	32.1	576.4	608.5
0	15.7	30.42	9.116	41.3	42.9	568.9	611.8
5	19.6	34.27	8.150	41.15	48.3	565.0	613.3
10	23.8	38.51	7.304	40.85	53.8	561.1	614.9
20	33.5	48.21	5.910	40.4	64.7	553.1	617.8
30	45.0	59.74	4.825	39.9	75.7	544.8	620.5
40	58.6	73.32	3.971	39.45	86.8	536.2	623.0
50	74.5	89.19	3.294	38.96	97.9	527.3	625.2
60	92.9	107.6	2.751	38.5	109.2	518.1	627.3
70	114.1	128.8	2.312	38.0	120.5	508.6	629.1
80	138.3	153.0	1.955	37.45	132.0	498.7	630.7
86	154.5	169.2	1.772	37.16	138.9	492.6	631.5
90	165.9	180.6	1.661	36.9	143.5	488.5	632.0
100	197.2	211.9	1.419	36.35	155.2	477.8	633.0
110	232.3	247.0	1.217	35.85	167.0	466.7	633.7
120	271.7	286.4	1.047	35.25	179.0	455.0	634.0
125	293.1	307.8	.973		185.1	488.9	634.0

*   in. Hg vac. (below atmospheric)

Table 4   Properties of Saturated R-12**

Temp. °F	Pressure Gauge lb./sq.in.	Pressure Absolute lb./sq. in.	Volume Vapor cu.ft./lb.	Density Liquid lb./cu.ft.	Enthalpy - Heat Content Btu/lb. from –40°F Liquid	Latent	Vapor
–40	10.97*	9.31	3.87	94.66	0	72.91	72.91
–30	5.49*	11.99	3.06	93.69	2.11	71.90	74.01
–21.6	0.0	14.71	2.53	92.86	3.93	71.07	74.95
–20	0.57	15.27	2.44	92.69	4.24	70.87	75.11
–10	4.49	19.19	1.97	91.69	6.37	69.82	76.19
0	9.15	23.85	1.61	90.66	8.52	68.75	77.271
5	11.79	26.48	1.46	90.13	9.60	68.20	77.81
10	14.64	29.34	1.32	89.61	10.68	67.65	78.34
20	21.04	35.74	1.10	88.53	12.86	66.52	79.39
30	28.45	43.15	0.92	87.43	15.06	65.36	80.42
40	36.97	51.67	0.77	86.30	17.27	64.16	81.44
50	46.70	61.39	0.66	85.14	19.51	62.93	82.43
60	57.74	72.43	0.56	83.94	21.77	61.64	83.41
70	70.19	84.89	0.48	82.72	24.05	60.31	84.36
80	84.17	98.87	0.41	81.45	26.37	58.92	85.28
86	93.34	108.04	0.38	80.67	27.77	58.05	85.82
90	99.79	114.49	0.36	80.14	28.71	57.46	86.17
100	117.16	131.86	0.31	78.79	31.10	55.93	87.03
110	136.41	151.11	0.27	77.38	33.53	54.31	87.84
120	157.65	172.35	0.23	75.91	36.01	52.60	88.61
130	181.01	195.71	0.20	74.37	38.55	50.77	89.32
140	206.62	221.32	0.18	72.75	41.16	48.81	89.97

* in. Hg vac. (below atmospheric)
** While R-12 is rapidly being replaced, data is included pending complete changeover.

Table 5   Properties of Saturated R-22

Temp. F	Pressure psia	Pressure psig	Volume Vapor cu.ft./lb.	Density Liquid lb./cu.ft.	Heat Content Btu/lb. Liquid	Vapor
–150	0.272	29.37*	141.23	98.34	–25.97	87.52
–125	0.886	28.12*	46.69	96.04	–20.33	90.43
–100	2.398	25.04*	18.43	93.77	–14.56	93.37
– 75	5.610	18.50*	8.36	91.43	– 8.64	96.29
– 50	11.674	6.15*	4.22	89.00	– 2.51	99.14
– 25	22.086	7.39	2.33	86.78	3.83	101.88
– 15	27.865	13.17	1.87	85.43	6.44	102.93
– 10	31.162	16.47	1.68	84.90	7.75	103.46
– 5	34.754	20.06	1.52	84.37	9.08	103.97
0	38.657	23.96	1.37	83.83	10.41	104.47
5	42.888	28.19	1.24	83.28	11.75	104.96
10	47.464	32.77	1.13	82.72	13.10	105.44
25	63.450	48.75	0.86	81.02	17.22	106.83
50	98.727	84.03	0.56	78.03	24.28	108.95
75	146.91	132.22	0.37	74.80	31.61	110.74
86	172.87	158.17	0.32	73.28	34.93	111.40
100	210.60	195.91	0.26	71.24	39.67	112.11
125	292.62	277.92	0.18	67.20	47.37	112.88
150	396.19	381.50	0.12	62.40	56.14	112.73
175	525.39	510.70	0.08	56.14	66.19	110.83
200	686.36	671.66	0.05	44.57	80.86	102.85

* Inches of mercury below one atmosphere.

Table 6 Ventilation Requirements for Compressors

Horsepower	Natural Circulation Room Value Required (cu. ft. of space)		Forced Circulation Air Entering Room (cu. ft./min.)	
	Air-cooled Condenser	Water-cooled Condenser	Air-cooled Condenser	Water-cooled Condenser
1/4	200		110	
1/3	400	190	145	130
1/2	500	260	220	175
3/4	800	360	330	275
1	1,000	475	440	300
1 1/2	1,500	665	660	425
2	2,000	850	880	550
3	3,000	1,100	1,320	780
5		1,300		1,100
7 1/2		1,450		1,400
10		1,600		1,700
15		1,850		2,250
20		2,000		2,750

Table 7 Vapor Pressures of Common Refrigerants

Temp. °F	Ammonia	Refrigerants			Water	Carbon Dioxide	R-502	R-503
		11	12	22				
-50	14.3*	28.9*	15.4*	6.0*		104	0.04	86.1
-45	11.7*		13.3*	2.6*		116	2.1	
-40	8.7*	28.4*	11.0*	0.6		131	4.3	107.9
-35	5.4*	28.1*	8.4*	2.7		147	6.7	120.0
-30	1.6*	27.8*	5.5*	5.0		163	9.4	132.9
-25	1.3	27.4*	2.3*	7.6		181	12.3	146.7
-20	3.6	27.0*	0.6	10.3	29.91*	200	15.5	161.5
-15	6.2	26.5*	2.4	13.3		221	19.0	177.2
-10	9.0	26.0*	4.5	16.6	29.90*	243	22.8	193.9
-5	12.2	25.4*	6.7	20.2		267	26.8	211.7
0	15.7	24.7*	9.2	24.1	29.88*	291	31.2	230.6
5	19.6	23.9*	11.8	28.3	29.87*	318	36.0	250.6
10	23.8	23.1*	14.6	32.9	29.86*	346	41.1	271.8
15	28.4	22.1*	17.7	37.9		376	46.6	294.2
20	33.5	21.1*	21.0	43.3	29.82*	407	52.5	317.9
25	39.0	19.9*	24.6	49.0		441	58.7	342.9
30	45.0	18.6*	28.5	55.2	29.76*	476	65.4	369.4
35	52.6	17.2*	32.6	61.9		513	72.6	397.0
40	58.6	15.6*	37.0	69.0	29.71*	553	80.2	425.9
45	66.3	13.9*	41.7	76.6		594	88.3	456.3
50	74.5	12.0*	46.7	84.7	29.66*	638	96.9	488.6
55	83.4	10.0*	52.0	93.3		684	106.0	523.1
60	92.9	7.8*	57.7	102.5	29.40*	733	115.6	560.1
65	103.0	5.4*	63.8	112.2		784	125.8	600.0
70	114.0	2.8*	70.2	122.5	29.18*	838	136.6	
75	126.0	0.0	77.0	133.4		894	148.0	
80	138.0	1.5	84.2	145.0	28.89*	955	159.9	
85	152.0	3.2	91.8	157.2		1018	172.5	
90	166.0	4.9	99.8	170.1	28.50*		188.8	
95	181.0	6.8	108.3	183.7			199.7	
100	197.0	8.8	117.2	197.9	27.99*		214.4	
105	214.0	10.9	126.6	212.9			229.1	
110	232.0	13.2	136.4	228.7	27.32*		245.8	
115	252.0	15.6	146.8	245.3			262.7	
120	272.0	18.2	157.7	262.6	26.45*		280.3	
125	293.0	21.0	169.1				298.7	
130		24.0	181.0		25.39*		318.0	
135		27.1	193.5				338.1	
140		30.4	206.6		24.04*		359.1	
145		33.9	220.3				381.1	
150		37.7	234.6				403.9	

Vapor pressures are psig except those followed by *. Such pressures are in. Hg vac.
**While R-12 is rapidly being replaced, data is included pending complete changeover.

Table 8   Thermal Transfer Characteristics

Material	Heat Transfer Factors				
	Conductivity (K)	Conductance (C)	Resistance (R)		Surface Coefficient (S)
			(1/K)	(1/C)	
1/4" Asbestos Cement Sheet		4.76		0.21	1.40
1/8" Asphalt Tile		24.80		0.04	
4" Com. Brick	5		0.20		1.40
Cement Plaster	5		0.20		0.93
Gravel agg.	12.6		0.08		1.3
Cinder agg.	4.9		0.22		
8" Conc. Block		0.90		1.11	
12" Sand & Gravel		.78		1.28	
8" Conc. Block		0.58		1.72	
12" Cinder agg.		0.53		1.89	
Copper	2900		0.0004		0.25 (dull) 0.02 (polished)
Corkboard	0.27		3.1		1.25
1/8" Corktile		3.60		0.28	
Glass	5.83		0.17		1.60
Glass Fibre Board	0.25		4.0		
3/8" Gypsum Board		3170		0.27	
Iron	408.0		0.0025		0.25 (polished)
1/8" Linoleum		12.00		0.08	
Mineral or Glass Wool	0.27		3.70		
Multi-foil and Spacers in Vacuum	0.02		50.0		
Perlite Plaster	1.5		0.67		
Soft Woods	0.80		1.25		
Steel	335.0		0.003		0.25 (dull)
Styrofoam	0.22		4.5		
Urethane Foam	0.15		6.67		
Vermiculite	0.68		1.47		1.3
Wood Fibre	0.25		4.0		

Table 9 Wire Sizes and Electrical Characteristics of Copper and Aluminum

| Aluminum | | Wire Size | | Copper | | |
| Capacity in amps RH type | Ohms per 1000 ft. at 20°C or 68°F | American or B & S Gauge (#) | Area in cir. mils (C.M.) | Ohms per 1000 ft. at 20°C or 68°F | Capacity in amps | |
					T type	RH type
195	0.080	0000	212,000	0.049	195	230
165	0.101	000	168,000	0.06	165	200
145	.128	00	133,000	0.078	145	175
125	.161	0	106,000	0.098	125	150
110	.203	1	83,700	0.124	110	130
95	.256	2	66,400	0.156	95	115
80	.323	3	52,600	0.197	80	100
70	.408	4	41,700	0.248	70	85
	.514	5	33,100	0.313		
55	.648	6	26,300	0.395	55	65
	.817	7	20,800	0.50		
40	1.03	8	16,500	0.628	40	45
	1.30	9	13,100	0.792		
30	1.64	10	10,400	1.00	30	30
	2.06	11	8,230	1.26		
20	2.61	12	6,530	1.59	20	20
	3.28	13	5,180	2.0		
15	4.14	14	4,110	2.53	15	15
	5.22	15	3,260	3.18		
6	6.59	16	2,580	4.01		10
	8.31	17	2,050	5.06		
3	10.5	18	1,620	6.38		5

Table 10 Temperature Conversion to Degrees Fahrenheit (°F) or Degrees Celsius (°C)

Note: The center columns of the table include values for degrees Celsius (°C) or degrees Fahrenheit (°F) temperatures to be converted. The right columns indicate °F; left columns, °C.

Examples: To determine the equivalent °F temperature for a given °C temperature, locate the °C in the center column. Read the equivalent °F in the right column. A given temperature of 100° C (center column) = 212.0° F (right column).

To determine the equivalent °C temperature for a given °F temperature, locate the °F in the center column. Read the equivalent °C in the left column. A given temperature of 32° F (center column) is equivalent to °C.

°C	Temperature to be Converted °C or°F	°F	°C	Temperature to be Converted °C or°F	°F	°C	Temperature to be Converted °C or°F	°F
−51.1	−60	−76.0	−26.1	−15	+5.0	−1.1	+30	+86.0
−50.6	−59	−74.2	−25.6	−14	+6.8	−0.6	+31	+87.8
−50.0	−58	−72.4	−25.0	−13	+8.6	0	+32	+89.6
−49.4	−57	−70.6	−24.4	−12	+10.4	+0.6	+33	+91.4
−48.9	−56	−68.8	−23.9	−11	+12.2	+1.1	+34	+93.2
−48.3	−55	−67.0	−23.3	−10	+14.0	+1.7	+35	+95.0
−47.8	−54	−65.2	−22.8	−9	+15.8	+2.2	+36	+96.8
−47.2	−53	−63.4	−22.2	−8	+17.6	+2.8	+37	+98.6
−46.7	−52	−61.6	−21.7	−7	+19.4	+3.3	+38	+100.4
−46.1	−51	−59.8	−21.1	−6	+21.2	+3.9	+39	+102.2
−45.6	−50	−58.0	−20.6	−5	+23.0	+4.4	+40	+104.0
−45.0	−49	−56.2	−20.0	−4	+24.8	+5.0	+41	+105.8
−44.4	−48	−54.4	−19.4	−3	+26.6	+5.5	+42	+107.6
−43.9	−47	−52.6	−18.9	−2	+28.4	+6.1	+43	+109.4
−43.3	−46	−50.8	−18.3	−1	+30.2	+6.7	+44	+111.2
−42.8	−45	−49.0	−17.8	0	+32.0	+7.2	+45	+113.0
−42.2	−44	−47.2	−17.2	+1	+33.8	+7.8	+46	+114.8
−41.7	−43	−45.4	−16.7	+2	+35.6	+8.3	+47	+116.6
−41.1	−42	−43.6	−16.1	+3	+37.4	+8.9	+48	+118.4
−40.6	−41	−41.8	−15.6	+4	+39.2	+9.4	+49	+120.2
−40.0	−40	−40.0	−15.0	+5	+41.0	+10.0	+50	+122.0
−39.4	−39	−38.2	−14.4	+6	+42.8	+10.6	+51	+123.8
−38.9	−38	−36.4	−13.9	+7	+44.6	+11.1	+52	+125.6
−38.3	−37	−34.6	−13.3	+8	+46.4	+11.7	+53	+127.4
−37.8	−36	−32.8	−12.8	+9	+48.2	+12.2	+54	+129.2
−37.2	−35	−31.0	−12.2	+10	+50.0	+12.8	+55	+131.0
−36.7	−34	−29.2	−11.7	+11	+51.8	+13.3	+56	+132.8
−36.1	−33	−27.4	−11.1	+12	+53.6	+13.9	+57	+134.6
−35.6	−32	−25.6	−10.6	+13	+55.4	+14.4	+58	+136.4
−35.0	−31	−23.8	−10.0	+14	+57.2	+15.0	+59	+138.2
−34.4	−30	−22.0	−9.4	+15	+59.0	+15.6	+60	+140.0
−33.9	−29	−20.2	−8.9	+16	+60.8	+16.1	+61	+141.8
−33.3	−28	−18.4	−8.3	+17	+62.6	+16.7	+62	+143.6
−32.8	−27	−16.6	−7.8	+18	+64.4	+17.2	+63	+145.4
−32.2	−26	−14.8	−7.2	+19	+66.2	+17.8	+64	+147.2
−31.7	−25	−13.0	−6.7	+20	+68.0	+18.3	+65	+149.0
−31.1	−24	−11.2	−6.1	+21	+69.8	+18.9	+66	+150.8
−30.6	−23	−9.4	−5.5	+22	+71.6	+19.4	+67	+152.6
−30.0	−22	−7.6	−5.0	+23	+73.4	+20.0	+68	+154.4
−29.4	−21	−5.8	−4.4	+24	+75.2	+20.6	+69	+156.2
−28.9	−20	−4.0	−3.9	+25	+77.0	+21.1	+70	+158.0
−28.3	−19	−2.2	−3.3	+26	+78.8	+21.7	+71	+159.8
−27.8	−18	−0.4	−2.8	+27	+80.6	+22.2	+72	+161.6
−27.2	−17	+1.4	−2.2	+28	+82.4	+22.8	+73	+163.4
−26.7	−16	+3.2	−1.7	+29	+84.2	+23.3	+74	+165.2

Table 10 Temperature Conversion to Degrees Fahrenheit (°F) or Degrees Celsius (°C)

°C	Temperature to be Converted °C or °F	°F	°C	Temperature to be Converted °C or °F	°F	°C	Temperature to be Converted °C or °F	°F
+23.9	+75	+167.0	+36.1	+97	+206.6	+48.3	+119	+246.2
+24.4	+76	+168.8	+36.7	+98	+208.4	+48.9	+120	+248.0
+25.0	+77	+170.6	+37.2	+99	+210.2	+49.4	+121	+249.8
+25.6	+78	+172.4	+37.8	+100	+212.0	+50.0	+122	+251.6
+26.1	+79	+174.2	+38.3	+101	+213.8	+50.6	+123	+253.4
+26.7	+80	+176.0	+38.9	+102	+215.6	+51.1	+124	+255.2
+27.2	+81	+177.8	+39.4	+103	+217.4	+51.7	+125	+257.0
+27.8	+82	+179.6	+40.0	+104	+219.2	+52.2	+126	+258.8
+28.3	+83	+181.4	+40.6	+105	+221.0	+52.8	+127	+260.6
+28.9	+84	+183.2	+41.1	+106	+222.8	+53.3	+128	+262.4
+29.4	+85	+185.0	+41.7	+107	+224.6	+53.9	+129	+264.2
+30.0	+86	+186.8	+42.2	+108	+226.4			
+30.6	+87	+188.6	+42.8	+109	+228.2	+54.4	+130	+266.0
+31.1	+88	+190.4	+43.3	+110	+230.0	+55.0	+131	+267.8
+31.7	+89	+192.2	+43.9	+111	+231.8	+55.6	+132	+269.6
+32.2	+90	+194.0	+44.4	+112	+233.6	+56.1	+133	+271.4
+32.8	+91	+195.8	+45.0	+113	+235.4	+56.7	+134	+273.2
+33.3	+92	+197.6	+45.6	+114	+237.2	+57.2	+135	+275.0
+33.9	+93	+199.4	+46.1	+115	+239.0	+57.8	+136	+276.8
+34.4	+94	+201.2	+46.7	+116	+240.8	+58.3	+137	+278.6
+35.0	+95	+203.0	+47.2	+117	+242.6	+58.9	+138	+280.4
+35.6	+96	+204.8	+47.8	+118	+244.4	+59.4	+139	+282.2

Abridged "Fahrenheit-Celsius Temperature Conversion" Table *(Courtesy of Carrier Transicold Divison, Carrier Corporation)*

Table 11   Copper Conductor Sizes Based on Voltage Drop

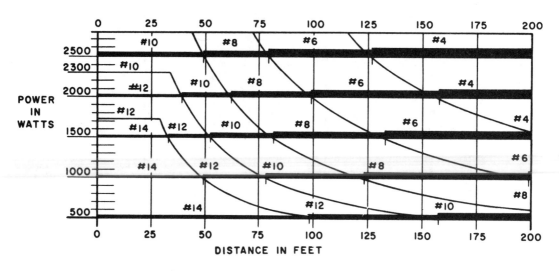

(A)   115-VOLT BRANCH CIRCUIT, 2 % DROP

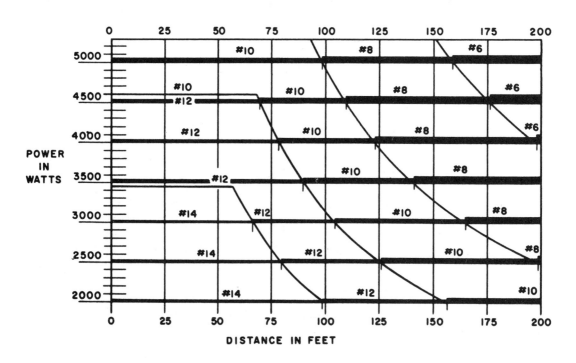

(B) 230-VOLT BRANCH CIRCUIT, 2% DROP

GLOSSARY OF TECHNICAL TERMS

absolute pressure—measured with reference to the point at which the molecules of the refrigerant are thought to have no motion or pressure. Measured in pounds per square inch absolute, inches of mercury, and microns

absolute temperature scales—used for calculating changes in refrigerant vapor pressures

absolute zero—temperature at which all molecular motion ceases, according to the kinetic theory of heat

absorption refrigeration system—one in which the refrigerant, as it is absorbed in another liquid, maintains the pressure difference needed for successful operation of the system

absorption system—uses the ability of a liquid to absorb a vapor to produce the necessary refrigeration conditions

absorptivity—the ability of a material to absorb heat

accumulator—storage unit positioned in the suction line to permit the boiling away of small amounts of liquid refrigerant before entering the compressor

activated alumina—desiccant which operates by adsorption of water molecules

actuators—secondary control mechanisms which function in response to the requirements of the primary group in actually controlling some part of the refrigeration system

adiabatic cooling—method in which paramagnetic salts are precooled, and then demagnetized, thereby producing further cooling

air-cooled condensers—depend on air drawn through tubes and fins for a good distribution of air to cool the refrigerant

air sensor—device used to measure pressure, velocity or moisture content changes in air conditions

alloy—a mixture of metals

alternating-current cycle—the number of times per second the current regularly changes both direction and magnitude

ambient temperature—temperature (usually of the air) surrounding operating equipment

ammeter-voltmeter instrument—used to measure the current intensity in amperes and the potential difference (voltage) of a refrigerating unit

ammonia (NH₃)—one of the earliest compounds used as a refrigerant

annealed—describes the degree of hardness

apparent power—value obtained by multiplying volts by amps; used for calculating power factor

armature—rotating coil of a dc motor

atmospheric pressure—pressure exerted because air has weight. Under normal conditions this is 14.7 lb./sq. in

atom—smallest particle of matter having the properties of the material of which it is composed; not always stable

automatic expansion valve—maintains a constant pressure in an evaporator; acts as a refrigerant control

back-seating valve—one in which when the stem is turned all the way back, the gauge port is closed

barometer—instrument for measuring pressure

basic refrigeration control—device that starts, stops. regulates and/or protects the refrigeration system and its components

boiling point—that temperature at which the vapor pressure of the liquid is equal to the pressure on the liquid

brazing—a process of joining parts by flowing a filler metal at high temperature into the space between two mating parts

Btu—amount of heat energy required to raise the temperature of one pound of water through a change of one F degree. British thermal unit of heat measurement (standard)

burr—fine, feathered edge raised on the inside of tubing by cutting

calorie—amount of heat energy required to raise the temperature of one gram of water through a change of one C degree

capacitive devices—function primarily to store electrical charges

capacitor-start motor—one which has a starting winding in addition to its running winding and has a capacitor in series with the starting winding

capillary—a tube with a very small inside diameter; its diameter and length control the flow of the refrigerant; dividing point between the high side and the low side of the system

carbon—one of the elements used in refrigeration

carbon dioxide (CO_2)—one of the earliest compounds used as a refrigerant

Celsius—the scale of changes of temperature which uses 0° as the freezing point and 100° as the boiling point for water at standard pressure

centrifugal compressor—one in which a rotor (impeller) with many blades rotates in a housing, drawing in vapor and discharging it at a high velocity by centrifugal force

check valve—control device used to regulate fluid flow in one direction

chillers/heaters—a unit that supplies either chilled water for cooling or hot water for heating—used in residential, apartment multiple installations and industrial air-conditioning systems

chill factor—measurement value (number) comparing a varying condition to a known condition. A comparison of temperature, humidity and wind velocity values to a fixed value

circuit breakers—devices used for protection of electrical circuits operated either by magnetic element or thermal and magnetic element

circuits—the flow of a refrigerant through separate rows of tubes rather than through one single tube

circular mils (C.M.)—specification for cross-sectional area of round wires; equal to the square of the diameter (in mils)

closed loop—complete piping circuit that is closed from the atmosphere

closed system—chilled water from the flash tank is pumped through a coil to cool air and is then returned to the flash tank

combination pliers—designed to permit jaws to be opened for gripping larger diameters and also to provide cutting for wires and cotter pins by use of a side-cutting jaw

commutator—the segments of the rotor that are the terminal connections of each coil

compound—the chemical combination of atoms of different elements

compound gauge—device for measuring pressure that might be higher or lower than standard atmospheric pressure

compressor—takes a refrigerant vapor at a low temperature and pressure and raises it to higher temperature and pressure

compressor crankshaft seals—prevent air from entering the compressor, and oil and refrigerant from escaping

compression ratio—the ratio of two pressures: the absolute discharge pressure divided by the absolute suction pressure

concentrators—evaporate excess water from brine which has been diluted by melted ice and frost

condensate—moisture that collects on an evaporator coil

condensation—process by which a vapor is changed into a liquid without changing temperature

condenser (general)—that part of the refrigeration system in which the refrigerant condenses and in so doing gives off heat

condenser (automotive)—the component which liquefies or condenses the hot, high pressure refrigerant vapor from the compressor to a warm, high pressure liquid which flows to the receiver-dehydrator

conductance—the Btu that pass through a material of a specified thickness

conduction—a method by which heat energy is transferred by actual collision of the molecules

conductivity—the ability of metals to conduct heat

conductivity values—indication of the Btu per square foot of area for each Fahrenheit degree of temperature difference per hour for each inch of thickness of the material

congeal—come together and lose viscosity

constant-pressure valve—an automatic expansion valve that holds the pressure at a constant level regardless of the load

contaminant—foreign substances or fine particles in a refrigeration system that may cause damage

convection—a method of transferring heat by the actual movement of heated molecules

convection currents—current movements caused by heat

cooling tower—compartment in a water-cooled system that depends on the evaporation of water to remove heat from the refrigeration system into the atmosphere

crankcase—the housing for the compressor

cryogenics—science of producing and applying temperatures below –250°F

cryogenic superconductor system—uses helium to cool conductors to within a few degrees of absolute zero where they offer no electrical resistance

cycling clutch control system—one in which the compressor is run intermittently to maintain a desired temperature

damper—maintains a required air balance by controlling the air flow in an air distribution system

defrosting evaporator—one in which frost accumulates on the cooling coils when the compressor operates and melts after the compressor shuts off

desiccant—material used in drier to trap moisture from refrigerant

dew point—that temperature at which the air (space) becomes saturated. When the air is cooled to the dew point, water vapor can condense into liquid form (provided its latent heat is removed)

Dewar flask—a container which consists of alternate layers of radiation shields and spacer material in a high vacuum

diamagnetic—those materials that are repelled by a magnetic pole

dichlorodifluoromethane—the chemical compound known as Freon 12 or R-12

dielectric—thin insulating material separating two conductor plates in a capacitor

diesel-powered mechanical transport refrigeration units—alternate between low speed cool and low speed heat

differential—difference of temperature or pressure between the on and off operation of the control

diffuser—directs air in an air distribution system to follow a specific direction

digital electrical test thermometer—an accurate and quick-reading temperature measurement instrument for standard, through-the-wall or other remote readings up to 1,000 feet—used for temperature measurement for heating and air-conditioning systems, freezers, heating and curing ovens, and other applications

direct current(dc)—one which always flows in the same direction; may be steady or fluctuate

direct-expansion evaporator—one that contains only enough liquid to continue boiling as heat is absorbed by it

discharge shutoff valve—a manual valve installed on the compressor

discharge valve—one which controls the flow of refrigerant from the cylinder head of a compressor to the discharge line

double-action piston compressor—one used in automotive A/C systems in which an axial swash plate pressed to the shaft is used to drive the pistons

double thickness flare—indicates that the flare thickness is made up of two thicknesses of tubing

drier—device designed to remove moisture from a refrigerant

drierite—desiccant which operates by chemical action

dry ice—carbon dioxide in the solid state; needs to absorb 246 Btu for complete sublimation

electricity—the flow of electrons through conductors

electromagnetic induction—transforms mechanical energy into electrical energy

electromagnetic clutch mechanism—one which, when engaged, turns the compressor shaft to start piston movement

electromotive force—an electrical pressure which moves through a circuit

electron leak detector—measures the electronic resistance of gas samples

electronic sight glass indicator—an instrument designed with ultrasonic transducers for receiving and transmitting sounds and an audible annunciator—clamps onto the high-side liquid level and indicates when the system being charged is "Full"

electropneumatic air-conditioning system—one in which outside air temperature, temperature conditions in the compartment, and car temperature as set on the selecter dial are all monitored by an electrical circuit

electrons—negative electrical charges

electron vapor pressure—the movement of atoms and their electrons

electroservo air-conditioning system—one in which an adjustable sensing device permits the selection of comfortable conditions; has two thermistors to monitor and sense both the outside air and the air in the passenger compartment

elements—basic building blocks of which everything in the universe and in outer space is composed

eliminator plates—protect refrigerated spaces and air from brine spray

enthalpy—name given to the total heat in the refrigerant at any temperature (with reference to –40°F)

equalizers—connections used with thermostatic expansion valves when the superheat setting of the expansion valve cannot control the amount of refrigerant which flows through the coil

equilibrium—condition existing at saturation; the molecules of the refrigerant in liquid state are changing into the vapor state as rapidly as vapor molecules are changing into the liquid state

evacuation—removing gases from a system or container

evaporation—process by which a liquid changes into a vapor as a result of absorbing heat

evaporative condenser—combines the principles of forced-circulation convection currents with the ability of a vaporizing liquid to absorb heat

evaporator (general)—device in the low-pressure side of a refrigeration system through which the unwanted heat flows; absorbs the heat into the system in order that it may be moved or transferred to the condenser

evaporator (automotive A/C systems)—device that cools, dehumidifies, and takes the pollen and dust from the air before it enters the passenger compartment

evaporator pressure regulator—a limiting control which prevents fluctuations in pressure in the suction line (and hence the evaporator)

evaporator pressure regulator valve—operates by bellows; refrigerant pressure on the diaphragm of the bellows compresses the spring and holds the valve open

evaporator pressure (temperature) control valve system—uses either a suction throttling valve, a pilot-operated absolute valve, or an evaporator pressure regulator valve to control evaporator temperature

evaporator regulating valves—provide independent temperature control for each evaporator

eutectic solution—one which can be made so that it freezes and melts at a specific temperature

expansion valve—metering device which provides a restriction so that there is a steady flow of refrigerant and also maintains the difference of pressure required to change the state of the refrigerant

external analyzer—connection between the evaporator outlet leading to the thermostatic expansion valve

Fahrenheit—the scale of changes of temperature which uses 32° as the freezing point and 212° as the boiling point for water under standard pressure

ferromagnetic—those materials that are strongly attracted by a magnet

field coil—stationary coil of a dc motor

flaring—method of forming or preparing the ends of tubing to connect them directly with or through the use of fittings

flash tank—tank in which the water continuously flashes into steam

flexing disc valve—one type of valve commonly used in compressors

flooded evaporator—one that is full of a liquid refrigerant at all times. Additional liquid is permitted to enter only to replace that which boils away

flux—substance used in soldering or brazing to dissolve any oxides that may have formed on the surfaces to be joined

forced draft cooling tower—cools water by mechanically forcing air through the tower

force of adhesion—the result of attraction between the molecules of solder and the metal part being soldered

four-wire system—electrical distribution circuit for large commercial applications requiring 3-phase (3 Ø), 480/240 and 280/120 volts, or 3-phase for low voltage and lighting installations

free electrons—those farthest from the nucleus which may easily be forced from their orbits

frost—frozen condensation

frosting evaporators—those which always operate at temperature below 32°F

fuses—devices used for protection of electrical circuits; either cartridge or plug type

gasoline-powered mechanical transport refrigeration units—operate on a stop/run cycle

gas volume control—used to regulate the amount of gas needed to produce certain desired temperatures and conditions in domestic absorption automatic control refrigerators

gauge (gage) manifold—one or more gauges with valves; designed to control flow

gauge (gage) pressure— pressure above or below atmospheric pressure

governor—device which maintains a single, steady number of rpm of the compressor

grounding—protection against static charges which sometimes build up on operating equipment

halide leak detector—operates on acetylene to detect vapor leaks of halogen refrigerants

halide refrigerant—R-12, R-22, R-500, R-502 (and other) refrigerants containing halogen chemical substances

heat energy—total energy of all the molecules in a given substance

heat inceptor—used to minimize amount of heat reaching the superconductor

heat of condensation—the heat that is removed per pound of vapor to cause it to condense. It has the same numerical value as the heat of vaporization

heat of respiration—heat given off by cargo

heat of vaporization—the heat required to change one pound of a liquid into its vapor or gaseous form at atmospheric pressure without changing temperature

heat pump—heating/cooling refrigeration system in which valves are used to reverse the refrigerant gas flow

heat transfer—the movement of heat energy from one place to another

helium liquefier-refrigerator—the complete system for liquefying helium

hermetic compressors—those in which the electric motor in encased in a sealed housing containing the compressor

Hertz—electrical measurement representing cycles per second

high-side float—metering system which locates the float and needle valve on the high-pressure side of the refrigeration system

high-vacuum helium refrigeration system—a multi-stage helium system that provides continuous ultra-high vacuum (chambers at critical junctures) environment for conducting independent or concurrent research or industrial production processes

holdover plates—containers that hold the eutectic and provide refrigeration

hollow set-screw (Allen) wrenches—used to remove and install set-screws that are recessed

hot-gas by-pass—piping arrangement designed to move hot refrigerant gas to enter the cooler low-pressure side of a refrigeration system

humidistat—operating control which reacts to variations in humidity

hydrocarbons—made up of different combinations of carbon and hydrogen; among the natural compounds used as refrigerants

hydrostatic pressure—force resulting from the expansion of a heated liquid

ice-making capacity—ability of a refrigerating system to make ice, starting with water at room temperature

impedance—total opposition to the flow of alternating current

induction motor—one in which the rotor requires no outside source of power

inductive devices—designed to convert electrical energy to magnetic and then to mechanical energy

inferred temperature—obtained indirectly from the gauge in the suction service valve of the compressor

insulation—any material that effectively slows down the transfer of heat

insulators—materials that normally deter the flow of electrons

Kelvin scale—measures absolute Celsius temperature

K-factor—term used to describe the conductivity (thermal characteristics) of a material—rate of heat transfer factor

kinetic energy—energy of motion

latent heat—that heat energy which causes a change of state without any change of temperature

latent heat of fusion—the amount of heat which must be added to a pound of material to change its state from a solid to a liquid; or, subtracted from a pound of a liquid to change it to a solid; also, the amount of heat to be added to (or subtracted from) one pound of a substance to cause it to melt (or solidify)

latent heat of vaporization—amount of heat to be added to (or subtracted from) one pound of the refrigerant to cause it to vaporize (or condense); also, the amount of heat energy in a gas which is in addition to that found in the liquid at the same temperature

limiting controls—safety controls

line drop—the voltage needed to push the operating current through just the line

liquid line charging valve—used for high-side charging

liquid line shutoff valve—manual valve installed in liquid line near condenser well to shut off flow of refrigerant between the condenser and the liquid line

liquid nitrogen shield—at room temperature, it can absorb any heat leakage from outside and reduce temperature between itself and cryogenic refrigerant surrounding the cable

liquid nitrogen system—a nonmechanical refrigeration system for transport use

liquid receiver—liquid refrigerant storage container

liquid thermometer—the most common among the temperature indicators

lithium bromide—used in combination with water in absorption cooling systems

low-pressure cutoff switch—senses system pressure only; wired in series with the magnetic clutch

low-side float—metering system which locates a float in the low-pressure side of the refrigeration system

low-temperature transport—refrigerated trucks that maintain temperatures in the range of 0°F and below

magnetic contactor—automatic controller used to control the starting operation in direct current motors

manifolding—in direct-expansion or dry evaporators, the method of circulating the refrigerant through separate rows of tubes

manual control system—an A/C system in which the driver selects heating and cooling by use of a lever which mixes warm and cold air to desired temperature

matter—anything that occupies space and has weight; found in one of three basic states: solid, liquid or gaseous

mercury—liquid commonly used in a liquid thermometer

metallic thermometers—temperature indicator which is operated by either a single metal or a bimetal strip

metering devices—restrict the flow of the refrigerant from the high to the low side; regulate the refrigerant flow according to the needs of the system

methylene chloride (CH₂Cl₂)—a halogenated hydrocarbon which is considered a safe refrigerant

micron—unit of measurement which equals 1/25,000 of one inch (.00004″) or approximately, one millionth of a meter (.0000394″)

microbes—minute living organisms which are a major cause of food spoilage when they multiply

modulating controls—provide for variations by steps as contrasted to the off and on operation of refrigeration systems with ordinary controls

modulating thermostat—used to operate dampers on DX coils and valves for varying the flow of chilled water

modulating thermostatic expansion valve—varies the capacity of the valve in response to variations in load on the system

moisture vapor seal—a tight barrier placed outside the insulation to prevent pushing of moisture through the insulation by vapor pressure

molecule—smallest stable particle of a material having all of its physical properties

Mollier diagram—graphic method of representing the heat quantities contained in and the conditions of a refrigerant at different temperatures

motor starter—switch that is used to connect and disconnect a motor and its power supply

muffler—reduces pumping noises caused as the compressor functions

multiple unit installation—one in which two or more evaporators in different refrigerators are operated from one compressor, or vice versa

national fine threads (NF)—refrigeration fitting threads which confrom in shape and dimension to standards established for the industry for flared fittings

natural convection—natural movement of a fluid or gas produced by differences in temperature

natural draft cooling tower—cools water by moving air at low velocities through the tower

neutrons—combination of a proton and electron having no electrical charge

niobium—a newer metal which is a superconductor

noncondensable gases—those which raise the condensing pressure above that required for the condensing temperature; such as: air, chlorine, oil, water vapors, and combinations of oxygen, hydrogen and nitrogen

nonflexing ring plate type valve—one type of valve commonly used in compressors

nonfrosting evaporators—use only the thermostatic expansion valve type of refrigerant control; operate at a temperature close to freezing

nonmechanical refrigeration systems—those that obtain the required high and low pressures by some method other than a mechanical compressor

normal temperature scales—the Celsius and Fahrenheit scales

oil reservoir—that area in the base of the oil separator where oil is accumulated prior to its return to the compressor

oil separator—device used to separate oil from refrigerant gas, returning the oil to the compressor and allowing the refrigerant to continue on its circuit through the refrigerating system

open-end adjustable wrench—similar in shape to an open-end wrench, but has adjustable jaws

open system—chilled water is sprayed into the air to be cooled and it is then collected in the air washer tank and returned to the flash tank and is again cooled

operating controls—sensitive to changes in the desired conditions such as temperature (or its related pressure) and humidity

operation recorder—makes a graphic record for diagnosis of equipment performance in a given operating time

overload protector—protective device for shutting down a refrigeration system when there is an overload (overcurrent) malfunction

paramagnetic—those materials that are weakly attracted by a magnetic pole

Peltier effect—changes in the direction of an electrical current at a junction produce either heat or cold

permanent gases—the cryogenic refrigerants

pilot-operated absolute valve—throttles the evaporator outlet line as necessary to hold evaporator pressure at a desired level

pilot solenoid—controls the flow of refrigerant by controlling the pressure to the equalizer of the thermostatic expansion valve

pipe loop (heat pump)—a coil or length of pipe buried outdoors in the earth below the frost line—horizontal or vertical length of coil or pipe for ground-source heat pump installations

plenum—sealed chamber at the outlet or inlet of an air handler

pneumatic control systems—employ changes of air pressure between operating and actuating controls

positive-positioning pneumatic-control system—provides positive operation of the damper or valve independently of wear, friction, or other operating conditions

potential difference—required to push electrons through a conductor; measured in volts

potential energy—stored-up energy

power factor—number by which the apparent power is multiplied to calculate true power in watts; numerical value is determined by dividing the watts used in the operation of a device by the volt-amperes taken at the same time

pressure—force on a unit of area exerted by refrigerant vapors; changes as the absolute temperature changes

pressure-operated thermometer—temperature indicator which is controlled by a bellows, a capillary, or a remote sensitive bulb

pressure drop—pressure difference needed to push the refrigerant through a component like the condenser, evaporator or line

pressure drop—pressure difference required to maintain the desired refrigerant flow

pressure relief valve—used to minimize the possibility of explosion when air temperature surrounding a refrigeration system may rise to a point where it causes the pressure of the refrigerant gas to increase to a danger point

pressurestat—operating control which reacts to pressure changes in the evaporator

primary controls—same as operating controls

product heat—heat given off by cargo

protons—positive electrical charges

pumpdown—process for pumping the refrigerant charge into the receiver and/or container, using a compressor

purging—a method of removing air and moisture from a refrigerating system by means of the refrigerant gas pushing some of the air ahead of it and out of the system

purging valves—devices used to remove noncondensable gases from the system

radiation—a method of transferring heat which uses energy waves that move freely through space; these energy waves may be reflected, penetrate the material, or be absorbed

R-factor—a measurment value that describes the thermal characteristics of insulation—expressed as a number per inch for a particular insulator

R-12—commonly used refrigerant for refrigeration systems

R-22—commonly used refrigerant for air-conditioning systems

R-502—commonly used refrigerant consisting of a mixture of R-12 and R-115 for low-temperature refrigeration systems

Rankine scale—measures absolute Fahrenheit temperature

ratchet flare nut box wrench—used where there is limited space in which a wrench handle may be moved

ratchet handle—used with socket wrench; enables user to apply a force in one direction for any fractional part of a revolution and return the handle to its original position without moving the socket

reactance—an additional opposition to the flow of ac beyond its regular resistance

reamer—cutting tool with a series of teeth or sharp cutting edges; used to enlarge a hole

receiver—a container for storing liquid refrigerant

receiver-dehydrator—in automotive A/C systems stores and filters the liquid refrigerant from the condenser and passes it through a desiccant to remove moisture

receiver-drier—component used to store and dry refrigerant in a refrigeration system

reciprocating compressor—one in which a piston travels back and forth in a cylinder

reflectivity—ability of a material to reflect radiant heat

refrigerant—substance which is circulated in a refrigeration system to transfer heat

refrigerant distributors—used to distribute the refrigerant simultaneously to several tubes in parallel

Refrigerant family—safest group of refrigerants; produced by manipulating the atoms in carbon-tetrachloride with those of flourine and hydrogen

refrigerant tables—show the specific heat (c) of many common materials

refrigerating capacity—the ability of a system to remove heat as compared with the cooling effect produced by the melting of ice

refrigerating effect—the amount of heat transferred by one pound of refrigerant as it is circulated in the system

refrigeration—process of removing heat under controlled conditions

relative humidity—a ratio of the amount of water in an air space to the amount of water that the air space can hold at a given temperature

relief valve—control valve designed to open to release liquids at a specified pressure

replusion-induction motor—one in which the repulsion principle is used to bring the rotor up to about 75% of its running speed; it then runs as an induction motor

resistive cryogenic system—uses nitrogen to cool the electrical conductors to temperatures where their electrical resistance is very low

resistors—electrical devices that are designed primarily to produce heat and offer only plain resistance to the flow of the current

restrictor—component that provides resistance to fluid flow

reverse cycle refrigeration—uses rejected heat to produce warmth

rotary compressor—one in which an eccentric rotates within a cylinder

rotor—rotating coil of a dc motor

safety controls—operate to keep the refrigeration system within safe limits of temperature and pressure

safety control valve—provided on absorption refrigerating systems to prevent the waste of gas and to be protected against other dangers which may be present if the gas flame is extinguished

saturation—the condition which exists when the space occupied by the refrigerant is holding as much of the vapor as it can at a particular temperature

secondary refrigerant—chilled liquid-like water which is circulated to distant units where the air is to be cooled in individual rooms

seizing-up—when a movable part or unit becomes secured (temporarily fitted together tightly) and no motion is possible under normal operating conditions

semiconductor—material which conducts better in one direction than the other

semiconductors—materials that combine the properties of a conductor with those of an insulator

sensible heat—heat energy that causes a change of temperature of a solid, liquid or gas; changes the speed with which molecules move

shaded-pole motor—one which has only a single winding on its stator

shell-and-tube—designation of a heat exchanger having straight tubes encased inside a shell

silica gel—desiccant which operates by adsorption of water molecules

silver brazing (soldering)—process of joining parts using hard solders

single-package A/C unit—self-contained A/C unit that includes a heating and cooling coil, condenser, compressor, fan and fan motor—includes the sheet metal duct system (plenums, registers, grilles, etc.)

single thickness flare—the part of the tubing that forms the flare is the thickness of the tubing

sling psychrometer—instrument used for measuring humidity

slugging—condition where large amounts of liquid enter a cylinder of a pumping compressor

socket wrench—used to remove nuts which do not have any obstruction over them

soldering—process of mechanically bonding together of two pieces of material by the force of adhesion

solenoid-operated, hot-gas, by-pass valve (automotive A/C)—a solenoid actuated by a thermostat that is mounted on the evaporator—serves to by-pass hot gas from the high side to the low side when the temperature reaches 33°F

solid-state digital thermometer/pyrometer—permits the measurement of temperature ranging from −40° F to 1999° F—utilizes a microprocessor (digital circuit) and memory storage/arithmetical problem-solving capability to convert thermocouple input voltage—displays temperature differentials on a four-digit °F display

solid-state electronic A/C charging station—portable, programmable charging equipment designed with an electronic flow meter to calculate a preset refrigerant charge; precise measurement capability to establish internal pressure in the system—solenoid controls automatically stop the evacuation process and close the refrigerant valve

solid-state electronic thermistor sensor—a temperature measuring device using electronic solid-state semiconductors (thermistor sensors) to establish changes in current flow that are monitored through electronic circuits—designed to provide temperature readouts and/or to control the starting, stopping, accelerating and decelerating of equipment and processes

solid-state halogen leak detector—an electronic leak detector for all halogen gauges—designed with probes to sense gas leaks—decomposes freon into halogen elements to cause minute electric current flow—amplifies current flow through a solid-state circuit to activate an audible and/or visual signal whenever gas is present

solid-state photoelectric digital tachometer—used to electronically measure the RPM of rotating equipment-decoded RPM impulses appear on a digital readout

solid-state starting relay (SSR)—a self-regulating conducting solid-state device that on start-up serves as a switch to open a motor start circuit

solid-state thermistor vacuum gauge—an instrument that uses an analog display or light emitting diodes (LEDs) to identify ten steps of pressure measurement from normal atmospheric pressure (25,000 microns) to 50 microns at the low end of the range

spanner wrenches—used to make adjustments on packing and other notched nuts

specific heat—the number that results from dividing the amount of heat (Btu or calories) required to change the temperature of a given weight of material by the amount of heat (Btu or calories) to cause the same change of temperature in the same weight of water

split-phase motor—one in which the starting windings contain a much greater number of turns than there are on the running winding

spray header—perforated pipe mounted along the ceiling of the cargo compartment of a transport

spray ponds—method of cooling water for condensers

steam-jet ejector—method used to compress flash vapor by using the energy of a fast moving jet of steam

steam-jet system—uses a device in which the extremely rapid flow of a vapor through a narrow tube reduce the pressure and permits evaporation of a liquid; produces a cooling effect

strong aqua—an ammonia and water solution with a concentration of almost 30% ammonia; used in an ammonia absorption cooling system

sublimation—a change of state from solid to gas without going through the liquid state

subcooling—drop in temperature resulting from the removal of sensible heat from the liquid refrigerant

suction line—runs from evaporator to compressor; returns the heat-laden gases from the evaporator to the compressor

suction pressure control valve (automotive A/C)—prevents the evaporator from freezing by maintaining a specified pressure and an above freezing temperature

suction service valve—manual shutoff valve installed on the compressor

suction throttling valve—the compressor is in continuous operation and the valve is opened and closed by sensing the actual evaporator operating pressure

suction valve—one which controls the flow of refrigerant from the suction line into the cylinder head of a compressor

sulfur dioxide (SO$_2$)—an old refrigerant that was used to recharge units

superconductors—those metals which lose all resistance at temperatures near absolute zero

superheating—the rise in temperature resulting from the addition of heat to the refrigerant· vapor either in the evaporator or the suction line

superheat switch—designed to protect the A/C system compressor against damage when the refrigerant charge is partially or totally lost

superinsulation—alternate layers of radiation shields and spacer material operating in a high vacuum

surge chamber—a drum or container into which liquid enters from the metering device in order to recirculate the refrigerant in a flooded evaporator

swaging—a means of shaping copper tubing so that two pieces may be joined without the use of a fitting

synchronous motor—a motor which has a stationary armature winding connected to the polyphase power line, a moving field winding connected through slip rings to a direct current supply, and a damper winding on the rotor which is short-circuited; similar to the rotor in a squirrel-cage motor

temperature—relative hotness of a material with respect to a fixed reference point

temperature recorder—device which automatically records temperatures over a definite period of time

terminal Dewars—insulated containers used to prevent heat transfer and permit the individual conductor phases to be connected into thermally and electrically graded pothead assemblies

therm—heat quantity measurement value of 100,000 Btu

thermal limiter fuses—designed to protect the A/C system compressor against damage when the refrigerant charge is partially or totally lost

thermal plug-on motor protector—a device used on refrigeration, freezer, water cooler, dehumidifier and other compressor motors—protects the compressor from overheating by sensing abnormal current and motor temperatures

thermobank—a bank for storing heat

thermoelectric indicators—temperature measuring instruments which operate on the principle that minute quantities of electric current may be produced by heating two dissimilar metals which are joined at one end

thermoelectric refrigeration—depends upon passing electrical energy to a couple through two dissimilar semiconductors

thermolator—a device used in the feeding of liquid refrigerant at a low temperature (winter operation) to the expansion valve—protects against liquid return to the compressor and overheating the suction line and discharge refrigerant during summer operation

thermopneumatic air-conditioning system—one which uses the mechanical principles of the thermostat to monitor vacuum motors which adjust the air valves and switches

thermostat—operating control which reacts to temperature

thermostatic expansion valve—control valve which maintains constant superheat in the evaporator; also, used for temperature control; operates on increased pressure resulting from a rise in temperature

thermostatic valve—metering device which responds to pressure change

thermostatic water valve—used to control temperature of gaseous refrigerant in the discharge line

throttling valve—dampens fluctuations of pressure gauge and provides a way to close off the port entirely

torque—twisting force; usually expressed in foot-pounds or inch-pounds; computed by multiplying a force over the distance it is exerted

torque wrench—one designed so that a measured force applied on the wrench handle is transmitted to the socket and, finally to a bolt or nut

transformer—used to either step-up or step-down the potential difference as needed

TRIAC (bidirectional current flow control)—a solid-state two-directional switch that duplicates the functions of a standard open and closed relay, overload protection and start capacitor units—SSRs are available for home, industrial, and commercial applications on 120 volt through 288 volt circuits

true power—electric power that is converted into useful heat or mechanical power; measured in watts by a wattmeter

two-cylinder reciprocating piston (in-line compressor)—used in automotive air-conditioning systems

two-cylinder reciprocating piston (V-type compressor)—used in automotive air-conditioning systems

turbulator—spiral-wound or spiral-shaped piece located in the liquid tube of a heat exchanger

universal element moisture indicators—permits sight inspection of safely dry and dangerously wet conditions—viewing of bubbles to indicate a pressure drop or a low charge of a refrigeration or an air-conditioning system

vacuum pump—used to evacuate systems in preparation for charging them with a refrigerant

valve retainer—device which limits the lift of the valve

vane rotary compressor (automotive A/C)—a non-piston compressor with only a discharge (check) valve which prevents high-pressure refrigerant vapor from entering the compresor when it is not operating

vaporization—change of state from liquid to vapor or gas

vapor lock—condition where liquid flow is impeded; caused by vapor trapped in a liquid line

vars—unit for measuring reactive power

viscosity—resistance of oil to flow

volts-amps—units used for measuring potential difference

volts—unit used for measuring potential difference

water control valve—used in A/C systems to regulate the flow of coolant to the heater core

water-ice refrigeration system—heat is absorbed as ice melts, producing the desired cooling effect

water regulator valve—varies the flow of cooling water to the compressor in response to variations in the need for cooling

INDEX

A

Absolute temperature, 359
Absorption
    Faraday's experiments and, 10–11
    refrigeration and, 9–14, 72
Absorption cycle, 13
Absorption refrigeration systems,
        12–14, 72, 428–36
    commercial, 430–31
    continuous operating, 429–30
    domestic, 431–36
        automatic defroster devices on,
            434–35
        gas volume control, 433
        ice cube makers, 435–36
        safety control for, 434
        unit illustrated, 432
Absorptivity, defined, 112
Actuating controls, 340, 343–46
Adjustable pliers, 34
Adjustable spanner wrenches, 31
Air-acetylene, 64
Air conditioning
    applications of, 5
    automotive. *See* automotive air
        conditioning
    built-up winter-summer, 421–23
    chiller/heater units,
        construction of, 420–21
        cooling cycle of, 419–20
        gas fired, 416, 418
        heating cycle of, 420
    condensers, water cooled, 415
    control terms, 410
    cooling load, estimating, 407, 409
    cooling towers and, 415
    duct, sizing and location of, 410,
        422
    evaporative condenser systems,
        415, 416
    filters, 423
    heat pumps and, 402–403
    heating load, estimating, 407, 409
    introduction of, 3
    system design, factors affecting,
        407
    through-the-wall-mounted,
        412–14
    units,
        single-package, 410
        split-package, 410–12
    window mounted, 412–14
    See also refrigeration

Air Conditioning and Refrigeration
        Institute (ARI), 469
Air-cooled condensers, 200–204
    base-mounted, 200–201
    discharge line pulsations and,
        203–204
    remote, 201–203
        refrigerant piping for, 203
Air, liquefication of, 515–17
Air-to-air heat pumps, 399
Alcohol, cooling effects of, 9
Allen wrenches, 29, 31
Alloys
    brazing, 64
    soldering, 64
Alternating current
    capacitors and, 301–302, 305
    circuits, measuring power in,
        313–15
    distribution circuits, 312–13
    electric motors, distribution
        circuits, 312–13
    induction motors and, 315–22
    inductive devices and, 304–305
    inductor and capacitor together
        on, 305–307
    motor controls, 328–33
    potential differences and, 315
Alternator, 282
Altitude, atmospheric pressure and,
        135
Aluminum, heat conductivity of, 91
Ammonia
    cooling effects of, 9–10
    leak testing for, 266
    as a refrigerant, 237
        precautions with, 237–38
Annealing, defined, 55
Apparent power, 303
ARI. *See* Air Conditioning and
        Refrigeration Institute
Atmospheric pressure
    altitude and, 135
    normal, 134
Atmospheric temperature, baromet-
        ric readings of, 134
Atom, described, 276–77
Automotive air conditioning,
        482–506
    air distribution and, 487–89
    applications of, 483–84,
        497–98
    compressor,

    design, 490–91
    functions, 489–90
    swash plate, 492–94
condensers, 494
electropneumatic system, 497
electroservo system, 497
human considerations in designing,
        482–83
operating characteristics of, 490–91
operation of, 484–85
temperature and climate controls,
        497
thermopneumatic system, 497
troubleshooting chart, 498–501
types of, 485–87

B

Bacon, Francis, food preservation and,
        2
Ball peen hammer, 33, 35
Barometer, 134
Barometric condenser, 438, 439
Barometric readings, atmospheric
        pressure and, 134
Bastard cut files, 39
Bits, 36–38
Black, Joseph, mechanical refrigeration
        and, 15
Boiling. *See* vaporization
Boiling point
    condensing pressures and, 137–38
    defined, 137
    pressure and, 16
Bottoming tap, 42
Box wrenches, 25–26
Brazing
    alloys, 64
    equipment, 64
    joint design for, 66, 67
    preparation for, 65–66
    swaging and, 67
    techniques, 65–66
Brine-spray method of defrosting, 195
British Thermal Unit (Btu), 75, 77,
        118–20
Built-up winter-summer air condition-
        ing systems, 421–23

C

Calorie, describe, 120
Capacitor-start motors, 320
Capacitors, 297, 301–302
    alternating current and, 305

symbols for, 297
Capillary tube control, refrigerants and, 14, 228–29
Carrier, Willia, air conditioning and, 3
Central air conditioning units, 414–15
Centrifugal,
  compressors, 166–68
  refrigeration equipment, 166–68
Change of state
  latent heat and, 129–35
  pressures effect on, 132
Charging refrigeration systems, 250–75
  adapters and, 255
  adding refrigerant to, 262
  compressor shutoff valves and, 251–52
  discharge gauges and, 253
  leak testing, 265–67
  liquid line charging valve and, 253
  liquid line shutoff valves and, 252
  manifolds shutoff valves and, 250–51
  manifolds testing and, 257–58
  oil gauges and, 253
  pressure release valves and, 255–57
  programmable charging station for, 263–64
  service gauge and, 257–58
  special service valves and, 255
  suction gauges and, 253
  throttling valves and, 253–54, 255
  through the high side, 262–63
  through the low side, 262
Charging station, programmable, 263–64
Check valves, 386
Chiller/heater units
  cooling cycle of, 419–20
  gas fired, 416, 418
    construction of, 420–21
  heating cycle of, 420
Chisels, 35–36
Circuit breakers, 287
Circuits
  alternating current and,
    measuring power in, 313–15

conductors for, 289
distribution, alternating current and, 312–13
elements of, 289
single-phase, 313
sizing of, 291–92
testing a, 288–89
wire cross-section area, 290–91
Circular mil, 290
Claw hammer, 33, 35
Coils
  cooling,
    design of, 96–97
    forced convection, 95
  float, flooded evaporators and, 184–85
Combination pliers, 34
Commercial refrigeration units, cycles in, 18–19
Compound gauges, 49, 139
Compression cycle refrigeration machines, 394–95
Compression fittings, 62
Compression ratio, compressors, 170
Compression shutoff valves,
    charging systems and, 251–52
Compression systems
  capillary tube system and, 14
  thermostatically controlled expansion valves (TEV) and, 14–15
Compressor valves, 157–59
  flexing disc type, 158–59
  nonflexing ring plate type, 157–58
  reed type, 158–59
Compressors
  automotive air conditioning and, 489–94
    design and operation of, 490–91
    swash plate, 492–94
  centrifugal, 166–68
  compression ratio, 170
  controlling, 170–72
    cylinder bypass method, 172
    cylinder unloading method, 172
    hot gas bypass method, 171–72
    variable speed motors, 170–71

heads of, 159
hermetic, 161–63
  troubleshooting chart, 168–70
lubrication of, 173–74
overload protectors and, 326
reciprocating, 154–61
  operating cycle of, 155
rotary, 163–65
  lubrication and, 165
  with stationary blade, 163–65
safety springs of, 159
service seals of, 160–61
shaft seals and, 159–60
ventilation of, 95–96
Concrete, heat conductivity of, 91
Condensation
  defined, 185
  heat and, 77–78
  heat of, vaporization and, 16
Condensers
  air-cooled, 200–204
    base-mounted, 200–201
    discharge line pulsations and, 203–204
    refrigerant piping for remote, 203
    remote, 201–203
  automotive air conditioning, 494
  barometric, 438, 439
  cooling water for, 210–14
    cooling towers, 211
    forced draft cooling towers, 213–14
    natural draft cooling towers, 211–13
    spray ponds, 210–11
    water treatment and, 214
  evaporative, 208–209
  forced convection, 92–93
  heat conduction and, 87–88
  multiple unit, 388–90
    control of, 388–90
  noncondensable gases and, 209–210
    purging valves and, 210
  shell and tube type, 88
  steam-jet refrigeration systems and, 437, 438
  surface type, 438
  types of, 89
  water-cooled, 93–94, 204–207
    air conditioning and, 415

cleaning tubes in, 207
shell-and-coil, 204–205
shell-and-tube, 207
tube-in-tube, 205–206
water-cooled unit, 89
Condensing pressure, boiling point
and, 137–38
Conduction, insulation against,
107–110
Conductors
electrical, 280
heat transfer through, 85–87
Control system, direct-coupled,
343–44
Control valves, evaporators and,
382
Controls
actuating, 340, 343–46
humidity, 340
limiting, 340–41
modulating, 377–79
operating, 339, 340–43
pneumatic, 380–82
primary, 339
safety, 340–41
differential, 350
high-pressure cutoff, 347–48
low-pressure, 347
oil pressure failure control,
348–50
suction line regulators,
350–52
secondary, 340, 343–46
temperature, 368
Convection
currents, 91
described, 90
forced, condensers and, 92–93
heat transfer by, 90–97
calculating, 97
compressor ventilation and,
95–96
cooling coil design and,
96–97
evaporator types and, 95
forced convection condensers, 92–93
forced convection cooling
coils and, 95
water chillers, 94
water-cooled condensers,
93–94
insulation against, 110–11
refrigerators and, 91–92

Cooling coils
design of, 96–97
dry type, 86
forced convection, 95
Cooling effect, 77
Cooling load, air conditioning,
estimating, 407, 409
Cooling towers
air conditioning and, 415
condensers and, 211
forced draft, condensers and,
213–14
natural draft, condensers and,
211–13
Copper, heat conductivity of, 91
Corkboard, heat conductivity of, 91
Crankcase, 155
Crimping pliers, 33, 34
Critical temperature, defined, 138
Cryogenic
helium superconductor system,
510–14
critical temperature of, 514
cryogenic envelope of, 512
equipment for, 514
heat interceptor of, 512
liquid nitrogen shield, 513
multishields, 513
refrigeration cycle of, 513
superconducting cable, 513
superinsulation and, 512
nitrogen resistive system, 510
Cryogenics, 507–508
Current relays, 324
Cutoff, high pressure, 347–48
Cutoff switch, low-pressure, 343
Cutting pliers, 33, 34
Cylinder bypass method, compressors and, 172
Cylinder unloading method,
compressors and, 172

D

Defrosting evaporators, 187–88
brine-spray method, 195
electric method, 195–96
hot gas method, 190–92
hot gas thermobank method,
192–94
importance of, 186
manual, 188–89
pressure method, 189
reverse-cycle method, 194
supplementary heat method, 190

temperature method, 189–90
time shut-down method, 190
water method, 195
Desiccants
operation of, 243–46
properties of, 242–43
Dew point temperature, insulation
and, 105–107
Diagonal pliers, 34
Die, 39, 40
sets, 41
stock, 39, 40
Differential, 350
Digital manifold gauge, 141
Direct-coupled control system,
343–44
Direct current
capacitors and, 301
electric motors, 311–12
starting devices, 311–12
Direct expansion evaporators, 178,
179–81
Discharge gauges, charging systems
and, 253
Discharge valve, compressors and,
155
Disconnect switches, motors, 332
Distribution circuits, alternating
current, 312–13
Double thickness flares, 62, 63
Drawings, thread designation on, 41
Drills
cutting edges and points of,
37–38
size systems, 36
twist, 36–38
Dry ice, refrigeration and, 450–52
Dry-type cooling coil, 86
Dryseal
pipe thread, 43
pressure-tight connection, 42, 43
Dual-voltage motors, 322
Duct, air conditioning, sizing and
location of, 410, 422

E

Electric method of defrosting,
195–96
Electric motors, 299
alternating current, distribution
circuits, 312–13
direct current, 311–12
starting devices, 311–12
Electric power, 302–304

Index

Electrical devices
  relays, 298–99
  resistors, 298
  solenoids, 298–99
  symbols for, 297
Electricity
  basics of, 276–79
  charges,
    attraction and repulsion of,
      278
    negative, 278
    neutralized, 279
    positive, 278
  circuit breakers and, 287
  circuits, 280
    conductors for, 289
    elements of, 289
    protection of, 285–87
    sizing of, 291–92
    testing of, 288–89
    wire cross-section area,
      290–91
  conductors, 280
  current, 280
    alternating, 283
    direct, 282, 283
  distributing, 283–85
    three-wire system of, 284–85
  electromotive force, 281–82
    sources of, 282
  fuses and, 285–86
  insulators, 280
  potential difference, 281–82
  refrigerated distribution of,
    509–514
  semiconductors and, 280
  static,
    discharging, 279
    producing, 278–79
  transformers, step-up, 283–84
  transmitting, 283–85
    line losses and, 284
Electromotive force, 281–82
  sources of, 282
Electron theory, 276–78
Electronic sight glass indicator
  described, 261–62
  measuring refrigerants using,
    261
Electrons, 235, 276
Electropneumatic system, automo-
    tive air conditioning, 497
Electroservo system, automotive air
    conditioning, 497

Elements, structure of, 234–35
Energy
  kinetic, 129
  potential, 129
  refrigeration and, 72, 75
Enthalpy, 131
Equipment
  brazing, 64
  categories of, 20–21
  soldering, 64
Ether, cooling effects of, 9
Eutectic solution refrigeration,
    452–56
Evacuation of refrigeration systems,
    258–59
Evaporation
  defined, 9
  Faraday's experiments and,
    10–11
  refrigeration and, 9–14
Evaporative condensers, 200,
    208–209
Evaporators
  control valves for, 382
  defrosting, 187–88
    brine-spray method, 195
    electric method, 195–96
    hot gas method, 190–92
    hot gas thermobank method,
      192–94
    importance of, 186
    manual, 188–89
    pressure method, 189
    reverse-cycle method, 194
    supplementary heat method,
      190
    temperature method, 189–90
    time shut-down method, 190
    water method, 195
  direct expansion, 178, 179–81
    design and operation of, 181
    water chiller and, 181–82
  flooded, 178, 183–84
    float coils and, 184–85
  frosting, 186–87
  humidity and, 185–86
  manifolding and, 181
  multiple, 382
  nonfrosting, 187
  plate-type, 86
  pressure regulators, 351–52
  regulator valves, 385
  solenoid valves and, 386–88
  types of, 95, 178–85

Expansion valves
  automatic, refrigerant control
    and, 222–24
  hand operated, refrigerant
    control and, 219
  thermostatic,
    capacity, 377–79
    capillary tube controls and,
      228–29
    equalizers and, 225–26
    heat pump, 396–98
    refrigerant control and, 224–
      29
    testing, 226–28
Expendable refrigerant refrigera-
    tion, 456–59
External pipe wrenches, 31, 32

F
Faraday, Michael
  absorption experiments, 10–11
  distilling apparatus, 11
  evaporation experiments,
    10–11
Female-pipe fittings (FPT), 60
Fiber glass, heat conductivity of, 91
Files, 39
Filler metals, 64
Filters, air conditioning, 423
Fittings, 60–62
  compression, 62
  flaring, purpose of, 61–62
  threaded, 60–61
Flare nut wrenches, 26, 27
Flares
  double thickness, 62, 63
  single thickness, 61–62
Flaring, fittings, purpose of, 61–62
Flat files, 39
Float coils, flooded evaporators
    and, 184–85
Float systems
  high side, refrigerant control
    and, 220–21
  low side, refrigerant control and,
    221–22
Flooded evaporators, 178, 183–84
Food preservation, early experi-
    ments with, 2
Forced
  convection condensers, 92–93
  convection cooling coils, 95
FPT. *See* female-pipe fittings
Fractional drill sizes, 36

Freezing, 129
Freon, 239
Frosting evaporators, 186–87
Fuses, electricity and, 285–86
Fusion
    heat of, 78
        latent, 77, 78
        latent heat of, 129–30

G

Gauges, 48–51
    charging systems and, 253
    compound, 139
    digital manifold, 141
    high pressure, 139
    pressure,
        principles of, 136–39
        refrigerating systems and, 139–41
    wire, sizes of, 20
Generator, 282
Genetron, 239
Ground heat pumps, 399, 401
Guide plane, die sections and, 39–40

H

Hacksaws, 38
Half-round files, 39
Halogen leak detector, 266–67
Hammers, 33–34, 35
Hand tools, 20–21
    See also specific type of hand tool
Hard solder, 64
    preparation for, 65–66
Heat
    absorbing, 98–99
    British Thermal Unit (Btu) and, 75
    condensation and, 77–78
        condenser operation and, 87–88
    latent,
        change of state and, 129–35
        effects of, 76–77
        vaporization and, 77
    reflecting, 98–99
    refrigeration and, 75
    sensible. *See* sensible heat
    specific, 121
        values, 121–22
    temperature and, comparison of, 75–76
    total, 131
    transmitting, 98–99

Heat conductivity of materials, 91
Heat content, 131
Heat energy
    measuring amount of, 118–25
        British Thermal Unit (Btu), 118–20
        converting units of, 120
        metric system of, 120
    pressure and, 78–80
    volume and, 78–80, 80
Heat exchange, 87
    schematic drawing, 88
Heat flow
    area and thickness influence on, 88–89
    concepts and principles of, 83–85
Heat of compression, 76
Heat of fusion, 78
Heat of vaporization, 77
Heat pumps, 394–406
    air conditioning options, 402–405
    air-to-air, 399
    coils, indoor/outdoor, 396
    compression cycle, 394–95
    design features of, 395
    ground, 399, 401
        energy sources for, 401–402
    heating cycle of, 400
    liquid line filter-driers, 398–99
    metering devices, 396–98
    operating principles, 395–96
    refrigerant flow in, 397, 400
    thermostatic controls for, 403
    water-coiled, 399, 401
    water-to-water, 402
Heat transfer
    by conduction, 85–90
    by convection, 90–97
    calculating rate of, 90
    concepts and principles of, 83–85
    convection and,
        calculating, 97
        compressor ventilation and, 95–96
        cooling coil design and, 96–97
        evaporator types and, 95
        forced convection condensers, 92–93
    forced convection cooling coils and, 95

water-cooled condensers, 93–94
    insulation and, 103–113
    radiation and, 97–99
    through conductors, 85–87
Heating load, air conditioning, estimating, 407, 409
Helium
    liquefaction of, 517–19
    superconductor cryogenic system and, 510–14
Hermetic compressors, 161–63
    troubleshooting chart, 168–70
Hermetic units, 322–23
High-pressure
    cutoff, 347–48
    gauges, 139
High-side float system, refrigerant control and, 221–22
Hollow set-screw (Allen) wrenches, 29, 31
Hot gas
    bypass method, compressors and, 171–72
    method of defrosting, 190–92
    thermobank method of defrosting, 192–94
Humidity
    controllers, 363
    evaporators and, 185–86
    measurement of, 363
Hydrocarbons, as a refrigerant, 238–39

I

Ice cube makers, 435–36
Ice-making
    capacity, 77
    machines, 2–4
    refrigerating capacity and, 146–47
Induction motors
    alternating current and, 315–22
    rotating magnetic field and, development of a, 317
        generating a, 316
    single-phase, 319
    stators and, magnetic field of, 316–17
    three-phase, 316
Inductive devices
    alternating current and, 304–305
    principles affecting, 299–300

Inductors, 297
 symbol for, 297
Industrial refrigeration units, cycles
 in, 18–19
Insulation
 against conduction, 107–110
 against convection, 110–11
 conditions within, 104–106
 described, 103
 dew point temperature and,
  105–107
 heat transfer and, 103–113
 moisture and, 104–105
 reflective,
  defined, 112
  radiant heat energy and,
   111–13
 ultra-low temperatures and, 107
 vapor pressures effect, 106
 vapor seals and, 105
Insulators, 85
 electrical, 280
Internal pipe wrenches, 31–32
IR drop, 289
Iron, heat conductivity of, 91
Isobutane, leak testing for, 265

J

Joints
 design of, brazing and, 66, 67
 soldered, requirements of, 65

K

K factor, 107
Kilocalorie, 120
Kilogram calorie, 120
Kinetic energy, 74, 129

L

Large calorie, 120
Latent heat
 change of state and, 129–35
  barometric readings of atmospheric pressure and,
   134
  computing units of absolute
   pressure, 135
  condensing temperatures and,
   133–34
  normal atmospheric pressure
   and, 134
  pressure above atmospheric
   and, 135–36
  pressure and, 132

 pressure below atmospheric,
  136
 refrigerants boiling point
  and, 133–34
 effects of, 76–77
 of fusion, 77, 78, 129–30
 of vaporization, 77, 129,
  130–31
 defined, 15
Leak detector, halogen, 266–67
Leeuwenhoek, Anton van, food
 preservation and, 2
Left cut shears, 43
Letter drill sizes, 36
Limiting controls, 340–41,
 346–52
Line drop, 289
Linear measuring tools, 45
 precision, 46–48
Liquid feedback, 76
Liquid line,
 charging valve, charging
  systems and, 253
 filter-driers, 398
 shutoff valves, charging systems
  and, 252
Liquid nitrogen refrigeration,
 456–59
 advantages of, 459
 operation of system, 457–58
Liquid slugging, 76
Liquids
 boiling temperatures of, 137,
  507
 sensible heat of, 117–18
Low-level condensers, 438, 439
Low-pressure
 control, 347
 cutoff switch, 343
Low-side float system, refrigerant
 control and, 220–21
Lubrication, compressors, 173–74

M

Magnetic field
 developing a, 317
 generating a, 316
 rotor and, 317–18
Magnetic induction, 283
Magnetic systems, low temperature
 production and, 441–43
Male-pipe fittings (MPT), 60
Manifolds shutoff valves, charging
 systems and, 250–51

Manual defrosting, 188–89
Materials, heat conductivity of,
 91
Matter
 defined, 72
 molecules in, freedom of, 74
 refrigeration and, 72
 structure of, 73
Measuring tools, 20–21
 linear, 45
 precision, 46–48
Mechanical refrigeration
 pressure and, 17
 system illustrated, 17
  commercial, 18
 See also refrigeration
Mechanical transport refrigeration,
 466–81
 compressor,
  power takeoff driven, 478
  vertical engine driven,
   477–78
 electromechanical systems of, 476
 nonmechanical combination, 478
 nosemount,
  cooling cycle of, 475
  large capacity, 470–72
  vertical profile, 472–75
 principles of, 466–67
 self-contained, 467–75
  applications of, 476–77
 units, capacity of, 469–70
Metal cutting shears, 43
Methyl chloride, leak testing for, 265
Metric system of heat measurement,
 120
Micrometer
 measuring with, 46–47
 metric, 47–48
Mill files, 39
Modulating
 motor, 389
 temperature controller, 377
 thermostats, 377
Moisture
 insulation and, 104–105
 refrigerants and, 240–46
  effect of, 240–41
  removal of, 241
Molecules, 72, 74
 arrangement of, 73–75
 described, 73
 movement of, 73–75
Mollier diagram, 147

Motors
    alternating current and, controls
        for, 328–33
    capacitor-start, 320
    disconnect switches, 332
    dual-voltage, 322
    electric, 299
    hermetic units, 322–23
    induction,
        alternating current and,
            315–22
        single-phase, 319
    overload protectors and, 327
    repulsion-induction, 321–22
    shaded-pole, 320–21
    split-phase, 319–20
    starters,
        magnetic across-the-line,
            329–30
        magnetic reduced voltage,
            330–31
        manual, 329
        reversing, 332–33
    synchronous, 319
    variable speed, compressors and,
        170–71
MPT. *See* male-pipe fittings
Mufflers, described, 163

N
Needle nose pliers, 34
Neutrons, 235, 277
Nitrogen
    liquefication of, 515–17
    resistive cryogenic system and,
        510
Noncondensable gases
    condensers and, 209–210
        purging valves and, 210
Noncycling system, 345
Nonfrosting evaporators, 187
Nonmechanical refrigeration
        systems. *See* refrigeration
        systems, nonmechanical
Nonmechanical transport refrigera-
        tion, 447–65
    combination systems, 460–62
    dry ice, 450–52
    eutectic solution, 452–56
    liquid nitrogen system, 456–59
        advantages of, 459
        operation of, 457–58
    mechanical combination system,
        478

water-ice, 447–50
See also mechanical transport
    refrigeration
Number drill sizes, 36

O
Offset screwdrivers, 45
Oil
    charging systems and, 253
    pressure failure control, 348–50
    separators, 173–74
    viscosity of, 173
Open-end adjustable wrenches, 28,
    29
Open-end wrenches, 26, 27
Operating controls, 339, 340–43
Overload protectors, 326–28
    inherent type of, 326
    terminal type of compressor, 326
    thermal plug-in motor, 327
    TRIAC, 327–28

P
Peltier effect, 439
Perkins, Jacob, ice-making machines
    and, 2
Pipe wrenches, 31–32
Plate-type evaporator, 86
Pliers, 33, 34
Plug tap, 42
Pneumatic control systems, 380–82
Potential
    difference, 281–82
    energy, 129
Power
    apparent, 303
    electric, 302–304
    factor, 303
    reactive, 303
    tools, 20–21, 45
    true, 302–303
Precision linear measurements,
    46–48
Pressure
    above atmospheric, 135–36
    absolute, computing units of, 135
    below atmospheric, 136
    boiling point and, 16
    change of state and, 132
    condensing, boiling point and,
        137–38
    heat energy and, 78–80
    mechanical refrigeration and, 17
    residual absolute, 137

Pressure gauges
    principles of, 136–39
    refrigerating systems and,
        139–41
    location of, 139–41
Pressure method of defrosting, 189
Pressure release valves, charging
        systems and, 255–57
Primary controls, 339
Primary winding, 283
Protons, 235, 277
Pump-down
    described, 250–51
    system, 344
    restarting, 343
Punches, 35–36
Purging a system, 259
Pyrometer, 367

R
R refrigerant, leak testing for, 266
Radiant heat energy, reflective
        insulation and, 111–13
Radiation, heat transfer by, 97–99
Rankine temperature scale, 359–60,
        360–61
Ratchet flare nut wrench, 26, 27
Ratchet handles, 29, 30
Ratcheting screwdrivers, 45
Reactive power, 303
Reciprocating compressors, 154–61
    operating cycle of, 155
Recorders, operation, 380
Recycling relay, 389
Reflective insulation, defined, 112
Reflectivity, defined, 112
Refrigerants, 12, 239
    absorption systems and, 13
    adding to a system, 262
    ammonia as a, 237
        precautions with, 237–38
    capillary tube control, compres-
        sion system using, 14
    chemistry of, 234–36
    circulation rate of, 145
    control devices,
        expansion valves, 220,
            222–29
        high-side float system,
            221–22
        low-side float system,
            220–21
        types and functions of,
            219

controls, pressure used for, 80
defined, 9
desiccants,
    operation of, 243–46
    properties of, 242–43
Faraday's experiments and, 10–11
hydrocarbons as, 238–39
identification of, 239–40
liquid, subcooling, 117–18
measuring, electronic sight glass
    indicator and, 261–62
moisture and, 240–46
    effect of, 240–41
    removal of, 241
properties of,
    chemical, 237
    physical, 237
qualities of, 236–37
    density and, 236
    heat and, 236
    pressure and, 236
    temperature and, 236
    volume and, 236
saturated, 142
selecting, vapor pressure and,
    138–39
selection table, 139
tables of, 123–25
    using in calculations, 124–25
transferring, 259–60
Refrigerating
capacity, ice-making and, 146–47
effect, 142–44
machines, compression cycle,
    394–95
Refrigeration
absorption and, 9–14
chemical symbols in, 235–36
cycles, commercial and industrial
    units, 18–19
elementary system illustrated, 10
energy and, 72
    application of, 75
evaporation and, 9–14
heat and, application of, 75
historical sketch of, 1
ice-making machines and, 2–4
matter and, 72
mechanical,
    pressure and, 15–18, 17
    system illustrated, 17, 18
systems classification, 19–20
temperature and, relating, 76
uses of, 4–7

air-conditioning, 5
    industrial applications, 5–7
    See also air-conditioning; me-
        chanical refrigeration
Refrigeration service valve
    wrenches, 32
Refrigeration systems
absorption, 12–14, 72, 428–36
    commerical, 430–31
    continuous operating, 429–30
    domestic, 431–36
charging, 250–75
    adapters and, 255
    adding refrigerant to, 262
    discharge gauges and, 253
    halogen leak detector, 266
    leak testing of, 265–67
    liquid line charging valve
        and, 253
    liquid line shutoff valves and,
        252
    manifolds testing and,
        257–58
    oil gauges and, 253
    pressure release valves and,
        255–57
    programmable charging
        station for, 263–64
    service gauge and, 257–58
    special service valves and,
        255
    suction gauges and, 253
    throttling valves and,
        253–54, 255
    through the high side, 262–63
    through the low side, 262
    valves and, 250–57
circuit requirements for, 287–92
components of,
    capacitors, 301–302
    electric motors, 299
    relays, 298–99
    resistors, 297, 298
    solenoids, 298–99
evacuation principles, 258–59
magnetic, 441–43
mechanical, nonmechanical
    versus, 436
nonmechanical, 428–46
    absorption systems, 428–36
    mechanical versus, 436
    transport. See nonmechani-
        cal transport refrigeration
steam-jet, 436–38

    condenser, 437
    ejector, 437
    flash tank, 437–38
    operating principles, 436–37
    thermoelectric, 439–41
        applications of, 441
    troubleshooting, 267–70
Refrigerators
absorption cycle, 13
absorption system, 12–14
    illustrated, 12
absorption-type, control for, 371
convection and, 91–92
development of, 3
heat reaching, 108
Regulators
evaporator pressure, 351–52
suction line, 350–52
Relays, 298-299
current, 324
recycling, 389
solid-state starting, 324–25
thermal, 324
voltage, 325–26
Repulsion-induction motors, 321–22
Residual absolute pressure, 137
Resistors, 297, 298
symbol for, 297
Reverse-cycle
defrosting, 194
refrigeration, 20
Reversing starters, 332–33
Right cut shears, 43
Rock wool, heat conductivity of, 91
Rotary compressors, 163–65
lubrication and, 165
with stationary blade, 163–65
Rotor, magnetic field and, 317–18
Round files, 39
Rubbing alcohol, cooling effects of, 9

S
Safety controls, 340–41, 346–52
differential, 350
high-pressure cutoff, 347–48
low-pressure, 347
oil pressure failure control,
    348–50
suction line regulators, 350–52
Safety springs, compressors and, 159
Screwdrivers, 43–45
Seals, shaft, compressors and,
    159–60
Secondary controls, 340, 343–46

Secondary winding, 283
Seizing-up, defined, 137
Semiconductors, 280
Sensible heat, 75
    determining amount for a given
        charge, 122
    of liquids, 117–18
    of solids, 117
    of vapor, 118
Service seals, compressors and,
    160–61
Shaded-pole motors, 320–21
Shaft seals, compressors and,
    159–60
Shears, metal cutting, 43
Sheet metal, cutting, 43
Short cycling, 345
Shutoff valves
    compressor, charging systems
        and, 251–52
    liquid line, charging systems
        and, 252
    manifolds, charging systems
        and, 250–51
Side cutting pliers, 34
Sight glass indicator, electronic,
    261–62
Silver, heat conductivity of, 91
Silver solder, 64
Single thickness flares, 61–62
Single-phase circuits, 313
Slip-joint pliers, 34
Socket wrenches, 29
Soft face hammer, 33, 35
Soft solder, preparation for, 65–66
Soft soldering, 64
Soldered joints, requirements of,
    65
Soldering
    alloys, 64
    defined, 63
    equipment, 64
    preparation for, 65–66
    principles of, 63–68
    swaging and, 67
    techniques, 64–65
Solenoid valve
    evaporators and, 386–88
    opening of, 342
    pilot, 345–46
Solidification, 129
Solids, sensible heat of, 117
Specific heat, 121
    values, 121–22

Split-phase motors, 319–20
Spray ponds, condensers and,
    210–11
Square files, 39
Starters
    enclosures for, 331
    magnetic,
        across-the-line, 329–30
        reduced voltage, 330–31
    manual, 329
    reversing, 332–33
Starting relays, solid state, 324–25
Starting tap, 41
Steam-jet refrigeration systems,
    436–38
    condenser, 437, 438
    cooling air, systems for, 438
    ejector, 437
    flash tank, 437–38
    operating principles, 436–37
Steel, heat conductivity of, 91
Steel rules, 45
Step controller, 389
Step-down transformers, 284
Step-up transformers, 283–84
Straight cut shears, 43
Straight pipe threads, 43
Strap wrenches, 32
Strippers, 33
Subcooling, 76, 87
Suction gauges, charging systems
    and, 253
Suction line regulators, 350–52
Suction pressure
    hold-back valves, 352
    regulating valves, 382–85
    valves, thermostatic, 385–86
Suction valve, compressors and,
    155
Superconductors, 508
    liquefication of helium and,
        517–19
Superheat
    defined, 363
    measuring, 365
Superheating, 76, 87
Superinsulation, 512
Supplementary heat method of
    defrosting, 190
Surface type condensers, 438
Swaging, 67–68
Swash plate compressors, 492–94
Swiss files, 39
Synchronous motors, 319

T
Tachometers, photoelectric digital,
    333–34
Tap, 39
    drill hole, 39
    drill size table, 40, 41
    sets, 41, 42
        features of, 41–42
    wrench, 39
Taper pipe threads, 43
Taper tap, 41
Temperature
    absolute, 359
    ambient, 356
    controls, 368
    critical, defined, 138
    heat and, comparison of,
        75–76
    indicators, 361–63
        digital test thermometer,
            366–67
        liquid-operated, 361
        metallic thermometers,
            363–64
        pressure-operated thermome-
            ters, 364–65
        stem-type thermometers,
            361–62
        thermoelectric indicator,
            365–66
    measuring systems, 355–58
    method of defrosting, 189–90
    operating, 356
    programmable, 371–73
    recorders, 368, 371
    refrigeration and, relating, 76
    scales, 356
        absolute, 359–61
        fixed reference points on,
            355–56
        Kelvin, 359–60
        Rankine, 360–61
    values, comparing, 356–57
Test instruments, 48–51
    categories of, 21
TEV. *See* thermostatically
        controlled expansion
        valves (TEV)
Thermal relays, 324
Thermistor
    defined, 49
    sensors, 366
    vacuum gauge, 49–50
        operation of, 50–51

Thermoelectric
    refrigeration systems, 439–41
    temperature indicator, 365–66
Thermometers, 48
    digital test, 366–67
    metallic, 363–64
    pressure-operated, 364–65
    stem-type, 361–62
Thermopneumatic system, automotive
    air conditioning, 497
Thermostatic expansion valves
    capillary tube controls and,
        228–29
    equalizers and, 225–26
    refrigerant control and, 224–29
    testing, 226–28
Thermostatic suction-pressure valves,
    385–86
Thermostatic water valves, 370–71
Thermostatically controlled expan-
    sion valves (TEV), com-
    pression systems and, 14–15
Thermostats
    heat pumps and, 403
    heating-cooling, 371
    modulating, 377
    motor control, 368–73
Thread, designation of on drawings,
    41
Thread form, 60
Thread-cutting tools, 39–41
Threaded fittings, 60–61
Threading die, 39
Three-square files, 39
Throttling valves, charging systems
    and, 253–54, 255
Time shut-down method of defrosting,
    190
Timer controls, 380
Timers, 380
Tinner's snips, 43
Tools
    categories of, 20–21
    hand, 20–21
    linear measuring, 45
    measuring, 20–21
    power, 20–21, 45
Torque wrenches, 27–28
Total heat, 131
    leakage per day, 108
        described, 108, 110
Transformers
    parts of, 283
    step-down, 284

step-up, 283–84
TRIAC, overload protectors, 327–28
Triangular files, 39
True power, 302–303
Tubing, 55–59
    brushes, 65, 66
    cutting, 56–57
    forming, 57–59
        inside bending tool, 58
        minimum safe bending radii,
        58
        outside bend, 59
        tube bender, 59
    properties of, 55–56
    sizes and weights, soft copper, 56
Turbulator, described, 88
Twist drills, 36-38

V
Vacuum, drawing a, 259
Vacuum gauges, 49
    solid-state electronic thermistor,
        49–50
    operation of, 50–51
Valves
    check, 386
    compressor, 157–59
    control, evaporators and, 382
    discharge, compressors and, 155
    evaporator regulator, 385
    expansion,
        refrigerant control and, 219,
        222–29
        thermostatic, 377–79, 396–98
    manifolds shutoff, charging
        systems and, 250–51
    pressure release, charging systems
        and, 255–57
    service, compressors and, 160–61
    solenoid,
        evaporators and, 386–88
        pilot, 345–46
    special service, charging systems
        and, 255
    suction, compressors and, 155
    suction pressure,
        hold-back, 352
        regulating, 382–85
    thermostatic water, 370–71
    thermostatically controlled
        expansion, compression
        systems and, 14–15
    water regulating, 379–80
Vapor seals, insulation and, 105

Vaporization
    heat of, condensation and, 16
    latent heat of, 77, 129, 130–31
        defined, 15
Vapors, sensible heat of, 118
Vise grip, 34
Volt-Ohm-Ammeter, advanced clip-
    on, 314
Voltage relays, 325–26

W
Water, heat conductivity of, 91
Water chillers
    convection heat transfer and, 94
    direct expansion evaporators and,
        181–82
Water, method of defrosting, 195
Water-regulating valves, 379–80
Water-coiled heat pumps, 399, 401
Water-cooled condensers, 93–94, 200,
    204–207
    air conditioning and, 415
    cleaning tubes in, 207
    shell-and-coil, 204–205
    shell-and-tube, 207
    tube-in-tube, 205–206
Water-to-water heat pumps, 402
Watt, James, mechanical refrigeration
    and, 15
Wattmeter, clamp-on, application of,
    314–15
Winding, primary, 283
Window glass, heat conductivity of, 91
Window-mounted air condition units,
    412–14
    central units, 414–15
    operating principles, 413–14
Wire gauges sizes, 290
Wobble plate compressors, 492–94
Wood, heat conductivity of, 91
Wrenches, 25–32
    adjustable spanner, 31
    Allen, 29, 31
    box, 25-26
    flare nut, 26, 27
    open-end, 26, 27
        adjustable, 28, 29
    pipe, 31–32
    ratchet flare nut, 26, 27
    ratchet handles and, 29, 30
    refrigeration service valve, 32
    socket, 29
    strap, 32
    torque, 27–28